LONDON MATHEMATICAL SOCIETY
MONOGRAPHS NEW SERIES

Series Editors
Peter M. Neumann   H. G. Dales

LONDON MATHEMATICAL SOCIETY MONOGRAPHS NEW SERIES

Previous volumes of the LMS Monographs were published by Academic Press, to whom all enquiries should be addressed. Volumes in the New Series will be published by Oxford University Press throughout the world.

NEW SERIES

1. *Diophantine inequalities* R. C. Baker
2. *The Schur multiplier* Gregory Karpilovsky
3. *Existentially closed groups* Graham Higman and Elizabeth Scott
4. *The asymptotic solution of linear differential systems* M. S. P. Eastham
5. *The restricted Burnside problem* Michael Vaughan-Lee
6. *Pluripotential theory* Maciej Klimek
7. *Free Lie algebras* Christophe Reutenaur
8. *The restricted Burnside problem (2nd edition)* Michael Vaughan-Lee
9. *The geometry of topological stability* Andrew du Plessis and Terry Wall
10. *Spectral decompositions and analytic sheaves* J. Eschmeier and M. Putinar
11. *An atlas of Brauer characters* C. Jansen, K. Lux, R. Parker, and R. Wilson
12. *Fundamentals of semigroup theory* John M. Howie
13. *Area, lattice points, and exponential sums* M. H. Huxley
14. *Super-real fields* H. G. Dales and W. H. Woodin

Super-Real Fields

Totally Ordered Fields with Additional Structure

H. Garth Dales

Department of Pure Mathematics
University of Leeds

and

W. Hugh Woodin

Department of Mathematics
University of California at Berkeley

CLARENDON PRESS • OXFORD
1996

This book has been printed digitally and produced in a standard specification in order to ensure its continuing availability

OXFORD
UNIVERSITY PRESS

Great Clarendon Street, Oxford OX2 6DP

Oxford University Press is a department of the University of Oxford.
It furthers the University's objective of excellence in research, scholarship, and education by publishing worldwide in

Oxford New York

Auckland Cape Town Dar es Salaam Hong Kong Karachi
Kuala Lumpur Madrid Melbourne Mexico City Nairobi
New Delhi Shanghai Taipei Toronto

With offices in

Argentina Austria Brazil Chile Czech Republic France Greece
Guatemala Hungary Italy Japan South Korea Poland Portugal
Singapore Switzerland Thailand Turkey Ukraine Vietnam

Oxford is a registered trade mark of Oxford University Press in the UK and in certain other countries

Published in the United States
by Oxford University Press Inc., New York

Reprinted 2006

ISBN 0-19-853991-6

Preface

Our purpose in this book is to study a class of totally ordered (real) fields with additional structure, the *super-real fields*. The ordered fields that interest us are 'larger than the real line $\mathbb{R}$', in that they are η_1-fields properly containing $\mathbb{R}$: an ordered field K is an *η_1-field* if, given countable subsets A and B of K such that $a < b$ for each $a \in A$ and $b \in B$, there exists $c \in K$ with $a < c < b$ for each $a \in A$ and $b \in B$. An η_1-field contains infinitesimal and infinitely large elements; certainly $\mathbb{R}$ itself is not an η_1-field. We shall concentrate throughout on the class of real-closed η_1-fields.

Very well-known members of the class of real-closed η_1-fields are the *ultrapowers*; these are ordered fields of the form $\mathbb{R}^\kappa/\mathcal{U}$, where κ is a cardinal and $\mathcal{U}$ is a free ultrafilter on κ. Ultrapowers have proved to be of enormous importance since they were first studied near the beginning of this century; they are the foundation on which non-standard analysis is built, and are of great significance in model theory and many other areas of mathematics. A natural generalization of ultrapowers, the hyper-real fields, has also been studied, for example in the classic text of Gillman and Jerison (1960). Let X be a completely regular topological space, and let $C(X)$ denote the algebra of all continuous, real-valued functions on X. *Hyper-real fields* have the form $C(X)/M$ for a maximal ideal M in $C(X)$ such that $C(X)/M$ is not isomorphic to $\mathbb{R}$. A hyper-real field is a real-closed η_1-field, and hitherto no example of a real-closed η_1-field which is not isomorphic to a hyper-real field was known.

Now let P be a (non-maximal) prime ideal in $C(X)$, and denote by A_P the integral domain $C(X)/P$, so that A_P is an ordered algebra with respect to the quotient order structure; a preliminary study of A_P is given in Chapter 14 of Gillman and Jerison (1960). Our *super-real fields* are the quotient fields K_P of the algebras A_P with a natural order; it is easily proved that super-real fields are real-closed, and they are usually (but not always) η_1-fields. We shall seek to classify super-real fields, to determine how the properties of the prime ideal P are reflected in the structure of the field K_P, and to give examples of super-real η_1-fields that distinguish between various classes.

We were led to our study of super-real fields by consideration of important problems that arise in Banach algebra theory on the normability of certain integral domains. Thus our results have significance within analysis; for this reason we have striven to make our work accessible to analysts. These normability problems will be described in our summary, below, of Chapter 5. We regret to say that we have left open what is in our opinion the most significant question in this area: whether or not the algebra $\widetilde{\mathbf{R}}^{\#}$ of finite elements in the real-closed field $\widetilde{\mathbf{R}}$ (see below) is normable.

The class of super-real η_1-fields is contained in the class of all real-closed η_1-fields and contains the class of hyper-real fields, and hence the class of ultrapowers. It was known that there are hyper-real fields which are not ultrapowers, and we shall prove that there are real-closed η_1-fields which are not super-real. To which of the two known classes consisting of all real-closed η_1-fields and of all ultrapowers is our class of super-real fields the 'closer'? We shall show that the class of super-real fields is 'close' to the latter class in that much of the extra structure that is well-known for ultrapowers is also a feature of super-real fields. On the other hand, the class of super-real fields is 'genuinely bigger' than that of ultrapowers because it contains the field $\widetilde{\mathbf{R}}$, which is of particular interest to us because it is an η_1-field which is universal in the class of real-closed fields of weight $\aleph_1$, and it is a Cauchy complete field. Indeed, $\widetilde{\mathbf{R}}$ is a hyper-real field, but it is not an ultrapower. The question whether or not the class of super-real η_1-fields is strictly larger than the class of hyper-real fields is subtle; we shall discuss it below.

In the second half of this book, we shall associate to each super-real η_1-field K_P an *additional structure*: this is a model $\mathfrak{N}_{(P)}$ analogous to a model of analysis. In the case where K_P is an ultrapower, $\mathfrak{N}_{(P)}$ is a well-known and standard 'non-standard model of analysis', but in general we shall obtain a wider class of structures; our constructions can be construed as a theory of *continuous ultrapowers.* The class of models that we shall obtain contains the non-standard models of analysis which are given by the generic extensions of the models specified by ultrapowers, and also the 'cut models', which are well known in model theory. In the general case, the model $\mathfrak{N}_{(P)}$ is not necessarily a model of any of the customary fragments of analysis.

There are many set-theoretic entanglements that surround our path; we shall seek to be careful about stating the underlying hypotheses of our theorems. However, we wish to stress that our work is written in such a way that, for most of the book, the only background in set theory and logic that is required is a knowledge of the basic theory of ordinal and cardinal numbers and some facility with transfinite induction. Our policy is to give full and explicit proofs of all our theorems in a language and notation accessible to non-logicians, and to avoid an assumption of a background in set theory. We shall sustain this policy almost throughout, but it does crumble near the end of Chapter 11. Further, set theorists and model theorists should be reassured that the modest background in analysis that is required is fully explained.

To minimize the set-theoretic complications, we shall focus in the last three chapters on η_1-fields that have weight $\aleph_1$, and we shall assume the generalized continuum hypothesis, GCH. We shall show that, even in this restricted setting, the identification of super-real and hyper-real fields is far from straightforward. Our main concept used in the construction of examples is that of 'rigidity': a consequence of the rigidity of a super-real field K_P is that the additional structure $\mathfrak{N}_{(P)}$ is uniquely determined up to isomorphism just by the structure of K_P as a totally ordered set (and, in particular, the structure of K_P as a ring is determined). It is through the construction of rigid super-real η_1-fields K_P that we can transfer properties of the prime ideal P to the field.

Our work begins with an introduction to order structures in Chapter 1; we shall quickly define the fundamental notions of α_1- and η_1- totally ordered sets. Our intuition is that α_1-sets are 'rather small' and that η_1-sets are 'rather large'; there are no sets which are simultaneously α_1-sets and η_1-sets. We shall define *β_1-sets* to be ordered sets which are the union of an increasing family of α_1-sets, and we shall prove that there is an (essentially unique) β_1-η_1-set; this set is Sierpiński's set $\mathbf{Q}$, and it has an explicit representation as a specific set of dyadic sequences of length ω_1.

We shall often refer to the *weight* $w(S)$ of a totally ordered set S: this is the minimum cardinality of an order-dense subset of S. A feature of our approach is the use of *gaps* in totally ordered sets: a (κ, λ)-gap $\langle A, B\rangle$ in a totally ordered set S (where κ and λ are cardinals) is a pair $\{A, B\}$ of subsets of S such that $a < b$ for a in A and b in B, such that A has cofinality κ and B has coinitiality λ, and such that there is no element c in S with $a < c < b$ for all a in A and b in B.

We shall also discuss ordered groups in Chapter 1. Our prototypical examples are the *Hahn groups* $\mathfrak{F}(\mathbb{R}, S)$ and $\mathfrak{F}_{(1)}(\mathbb{R}, S)$, built on a totally ordered set S. (The former group is 'much larger' than the latter.) The main tool for understanding totally ordered groups is Hahn's embedding theorem, from 1907: each totally ordered group G is isomorphic to a subgroup of $\mathfrak{F}(\mathbb{R}, \Gamma_G)$, where Γ_G is the value set of G. We have the notions of *α_1-groups*, *β_1-groups*, and *η_1-groups*, and we shall prove Esterle's theorem of 1984 that every divisible β_1-η_1-group is isomorphic to $\mathbf{G}$, defined to be $\mathfrak{F}_{(1)}(\mathbb{R}, \mathbf{Q})$.

In Chapter 2 we shall turn to our main object of study throughout this book, the class of (totally) ordered fields. (We shall adopt the special convention that a 'field' is an algebra over $\mathbb{R}$; this gives a distinct flavour to our study.) For example, in the case where G is a totally ordered group, $\mathfrak{F}(\mathbb{R}, G)$ and $\mathfrak{F}_{(1)}(\mathbb{R}, G)$ are ordered fields. The fields which we shall be considering will be 'very large', containing a wealth of infinitesimal and infinitely large elements; for an ordered field K, the subalgebras of finite and of infinitesimal elements are denoted by $K^{\#}$ and K^0, respectively. In this chapter, we shall touch on Artin–Schreier theory and valuation theory for fields. The analogue of Hahn's embedding theorem for ordered fields is Kaplansky's embedding theorem, from 1943: an ordered field K is isomorphic to a subfield of $\mathfrak{F}(\mathbb{R}, \Gamma_K)$, where Γ_K is the value group of K. Our proof of Kaplansky's theorem is built on the notion of gaps, and may be more concise than earlier proofs.

Naturally, we have the notions of *α_1-fields*, *β_1-fields*, and *η_1-fields*. The two specific fields that we shall introduce in this chapter are $\mathbf{R}$ and $\hat{\mathbf{R}}$, defined to be $\mathfrak{F}_{(1)}(\mathbb{R}, \mathbf{G})$ and $\mathfrak{F}(\mathbb{R}, \mathbf{G})$, respectively; both these fields are real-closed η_1-fields, and $\mathbf{R}$ is a β_1-field. Indeed, we shall prove Esterle's theorem that every real-closed β_1-η_1-field is isomorphic to $\mathbf{R}$. In ZFC + CH, the theory of β_1-fields becomes the theory of ordered fields of cardinality $\aleph_1$.

Chapter 3 discusses the *Cauchy completion* $\widetilde{K}$ of an ordered field K, and, in particular, introduces the fundamentally important field $\widetilde{\mathbf{R}}$, giving an easy, explicit definition of this field; with GCH, $\widetilde{\mathbf{R}}$ is a real-closed, Cauchy complete, η_1-field of cardinality $\aleph_2$ and weight $\aleph_1$, and each field with these properties is isomorphic to $\widetilde{\mathbf{R}}$. Thus we see $\widetilde{\mathbf{R}}$ as a natural generalization of the real line $\mathbb{R}$.

Chapter 4 studies the algebra $C(X)$ for a completely regular topological space X, and, in particular, the special case $C(\Omega)$, where Ω is a compact space; we regard $C(\Omega)$ as a Banach algebra. The theory of the first part of this chapter is mostly contained in (Gillman and Jerison 1960), but the second part develops new material. In particular, we shall define ultrafilters, hyper-real fields, and super-real fields K_P, where we may suppose that P is a non-maximal, prime ideal in an algebra $C(\Omega)$. At one point in our work, we hoped for a characterization of the super-real η_1-fields in the class of real-closed η_1-fields, but, as we shall see eventually, such a characterization seems unattainable.

There are various additional properties that a prime ideal P in an algebra $C(\Omega)$ may enjoy. For example, P may be a *z-ideal*; indeed every hyper-real field is isomorphic to a super-real field K_P, where P is a non-maximal, prime z-ideal in some $C(\Omega)$. The property of being a z-ideal has a surprising significance in our work. How can we determine by looking at a super-real field whether or not it arises from a prime z-ideal? In fact, we shall also introduce various other special properties that a prime ideal in an algebra $C(\Omega)$ may have; we shall define *valuation primes*, *strongly convex primes*, *cut primes*, and *exponential primes*, and hence we shall establish a variety of corresponding classes of super-real fields. Naturally we are led to a classification scheme, and the particular question of deciding when two apparently distinct classes are in fact equal.

The above questions have several natural variants. First, we must specify whether 'isomorphism' means isomorphism as ordered sets, as ordered groups, as ordered rings, or as ordered fields. Second, we must specify the theory in which we are working; the answers to our question may depend on the set-theoretic assumptions that we make. Finally, whilst we shall strive to prove theorems (in the theory ZFC), we may have to content ourselves with establishing the 'relative consistency' (see below) of some assertions.

A first version of our attempt at a classification scheme will be given at the end of Chapter 4 on page 104; successive refinements will be given in later chapters, on pages 248 and 338. It will be seen that even our final classification scheme is not fully complete. It will also be seen that, whilst we have immediately established several obvious containments between our classes in Chapter 4, we do not succeed in adding any further containments (even up to 'relative consistency') in the later chapters. Of course, this may be either because of our weakness or because there are no such containments. Our major results establish that certain of our classes are distinct (perhaps only up to relative consistency); these results do not seem to be obvious, and some of our constructions are quite challenging, with a considerable technical apparatus. Our major open question (see Question 12 of Chapter 12) is whether or not every super-real η_1-field K_P arising from a prime z-ideal P is already isomorphic to a hyper-real field; we feel that this is plausible, at least if we restrict attention to fields of weight $\aleph_1$.

In Chapter 5 we come to the questions of normability that were the seed of our study. Earlier, in Dales and Woodin (1987), we discussed the question whether or not there is a discontinuous homomorphism from the Banach algebra $C(\Omega)$ into any other Banach algebra for each infinite compact space Ω. It was noted that there is such a homomorphism if we assume the continuum hypothesis (and so work in the theory ZFC + CH), but we proved that the existence of such a homomorphism cannot be established in the theory ZFC itself; this question is thus independent of ZFC. The construction of a discontinuous homomorphism from $C(\Omega)$ proceeds by first showing that the algebra $\mathbf{R}^{\#}$ of finite elements in $\mathbf{R}$ is normable (as a theorem of ZFC), and then by showing that (with CH) there is a non-maximal, prime ideal P in $C(\Omega)$ such that $C(\Omega)/P$ can be embedded in $\mathbf{R}^{\#}$, and hence is normable. (The full details of this proof will be given in (Dales 1997).) It is a natural question to ask (in the theory ZFC + CH) for which prime ideals P an algebra $A_P = C(\Omega)/P$ is normable; more generally, we ask when $K^{\#}$ is normable for an arbitrary ordered field K. Known results reduce this latter question to the case where $|K| = \aleph_2$ and $|\Gamma_K| = \aleph_1$. We shall show in Chapter 5 that $K^{\#}$ is normable for each such K if and only if $\widehat{\mathbf{R}}^{\#}$ is normable, and so $\widehat{\mathbf{R}}$ is a 'test case' for our question. However, we cannot decide whether or not $\widehat{\mathbf{R}}^{\#}$ is normable. Further, we cannot resolve the particularly interesting question whether or not the smaller algebra $\widetilde{\mathbf{R}}^{\#}$ is normable. These explicit questions have relevance to the question of the normability of algebras $C(\Omega)/P$ only if there are super-real fields K with $|K| = \aleph_2$ and $|\Gamma_K| = \aleph_1$. In particular, we ask if $\widehat{\mathbf{R}}$ or $\widetilde{\mathbf{R}}$ is such a super-real field.

A related question from Banach algebra theory is to ask which complex, non-unital integral domains are normable. We shall show, extending a theorem of Esterle, that each complex, non-unital integral domain of cardinality $\mathfrak{c}$ can be embedded in an algebra of the form $c_0(\mathbb{C})/\mathcal{U}$ for some ultrafilter $\mathcal{U}$ on $\mathbb{N}$, and hence, with CH, can be normed. Our result establishes an important 'universality' property of the family of integral domains of the form $c_0(\mathbb{C})/\mathcal{U}$.

A key property of super-real fields K_P, to be developed in Chapter 6, is that there is an *operational calculus* for K_P; in the case where P is a z-ideal, all functions $F \in C(\mathbb{I})$ 'operate' on K_P to give an element $F_{(P)}(a) \in K_P$ for each $a \in K_P$ with $0 \leq a \leq 1$, but, in the case of a general prime ideal P, we can only be sure that Lipschitz functions F operate. The property of having an operational calculus distinguishes super-real fields from more general real-closed fields. The operational calculus leads, in particular, to the definition of an *exponential* on a convex subgroup of $(K_P, +)$ for each super-real field K_P. The possession of such an exponential is an invariant of ordered fields; since there is no exponential on $\widehat{\mathbf{R}}$, we shall deduce that the real-closed η_1-field $\widehat{\mathbf{R}}$ is not isomorphic to any super-real field. There is an exponential on $\widetilde{\mathbf{R}}$ and so, at this stage, we do not know whether or not $\widetilde{\mathbf{R}}$ is a super-real field.

In Chapter 7 we shall show by a fairly straightforward example that there are indeed super-real η_1-fields K_P such that $|K_P| = \aleph_2$ and $|\Gamma_{K_P}| = \aleph_1$ (with GCH); with the assistance of an axiom called $\diamondsuit$, we shall show that there is a hyper-real field with these properties, and we shall claim that it is relatively consistent with ZFC that there is also such an ultrapower. Thus our question on the normability of quotients $C(\Omega)/P$ is not vacuously solved: there are explicit algebras for which the question is open.

Chapter 8 has two aims. The first is to prove one of our (rare) positive results about the order-structure of super-real fields. The result is the *gap theorem*: for each super-real η_1-field K, there is exactly one cardinal λ, namely the cofinality of K^0, such that there is an $(\aleph_0, \lambda)$-gap in K. Using this theorem, we shall exhibit a real-closed η_1-field which is not even order-isomorphic to any super-real field.

The second aim of Chapter 8 is to begin the construction of the 'additional structure' in super-real fields that is referred to in our subtitle. Our claim is that there can be no serious understanding of super-real η_1-fields and the possible classification of these fields other than through a full study of this additional structure.

We shall start by introducing some extra structure in the special case where Q is a (non-maximal, prime) z-ideal in $C(\Omega)$; the method, involving Q-partitions, is close to that which applies in the case of ultrapowers. As a first step, we form an 'infinite version' S_Q of each subset S of $\mathbb{R}$ inside the super-real field K_Q, and show, for example, that $\mathbb{Q}_Q$ is order-dense in K_Q. We shall also define subsets $S_{(Q)}$ of K_Q corresponding to certain sets S: these sets are conceptually distinct from the sets S_Q (although they coincide in some cases), being constructed inside K_Q in an analogous way to that in which S is constructed as a subset of $\mathbb{R}$. For example, $\mathbb{N}_{(Q)}$ is the set of natural numbers in K_Q, and $\mathbb{N}_{(Q)}$ is a 'non-standard model of arithmetic'; $\mathbb{N}_{(Q)}$ as a subset of K_Q 'acts like' $\mathbb{N}$ as a subset of $\mathbb{R}$. It is technically convenient for us to introduce $\mathbb{D}_{(Q)}$, corresponding to the dyadic rationals in the closed unit interval $\mathbb{I}$, and $\varepsilon_r^{(Q)}$ for $r \in \mathbb{I}_{(Q)} \setminus \mathbb{D}_{(Q)}$, corresponding to the function ε_r which, for $r \in \mathbb{I} \setminus \mathbb{D}$, gives the coefficients of the binary expansion of r. The functions $\varepsilon_r^{(Q)}$ define a very special subset of the family of all functions from $\mathbb{N}_{(Q)}$ into $\{0,1\}$; it is the properties of this family that are the key to all our later constructions. We continue in Chapter 8 to build $\{0,1\}^{<\mathbb{N}}_{(Q)}$, the analogue of the set $\{0,1\}^{<\mathbb{N}}$ of finite sequences taking values in $\{0,1\}$.

As we have stated, these additional structures arise in a very straightforward way in the case where Q is a z-ideal; in the second half of Chapter 8, we shall carry forward similar constructions for an arbitrary prime ideal P, and this is a significantly more subtle business. For example, we now have two analogues of $\mathbb{N}$ in K_P, called $\mathbb{M}_{(P)}$ and $\mathbb{N}_{(P)}$; the former is closed under multiplication and the latter may not be, but nevertheless we shall argue that $\mathbb{N}_{(P)}$ is the correct analogue. The additional structures introduced in Chapter 8 are required for the proof of our gap theorem, and are also necessary as the first stage of the analysis that will occupy the remainder of the book.

At the end of Chapter 6 we left open the question whether or not $\widetilde{\mathbf{R}}$ is a super-real field; we shall return to this topic in Chapter 9. The presentation of $\widetilde{\mathbf{R}}$ as a super-real field cannot be trivial: for example, we shall show that $\widetilde{\mathbf{R}}$ is not isomorphic to any ultrapower. However, the main theorem of Chapter 9 will establish that (with GCH) there is a topological space X (of the form $\mathbb{N} \times \{0,1\}^{\mathbb{N}\times I}$ for an index set I of cardinality $\aleph_2$) and a maximal ideal M in $C(X)$ such that the hyper-real field $C(X)/M$ *is* isomorphic to $\widetilde{\mathbf{R}}$. Thus the question of the normability of algebras $C(\Omega)/P$ cannot be resolved without a decision whether or not the specific algebra $\widetilde{\mathbf{R}}^{\#}$ is normable.

In the case where P is a prime z-ideal, necessarily $w(K_P) = |\Gamma_P|$, where Γ_P is the value group of K_P. In the second half of Chapter 9, we shall seek a super-real η_1-field K_P such that $w(K_P) > |\Gamma_{K_P}|$; we shall prove that it is relatively consistent with ZFC + GCH that such a field exists, and so it cannot be proved in ZFC + GCH that every super-real η_1-field is isomorphic to a hyper-real field, thus distinguishing between two classes.

In Chapter 10 we shall extend the structures obtained in Chapter 8 by defining an analogue $\{0,1\}^{\mathbb{N}}_{(P)}$ related to a prime ideal P of the set $\{0,1\}^{\mathbb{N}}$ (equivalently, the power set $\mathcal{P}(\mathbb{N})$ of $\mathbb{N}$). Here certain pathologies occur, even in the case where P is a z-ideal. For example, it is not always the case that $f \vee g \in \{0,1\}^{\mathbb{N}}_{(P)}$ whenever $f, g \in \{0,1\}^{\mathbb{N}}_{(P)}$. This leads us to the concept of *weak Cauchy completeness*, which will be studied in this chapter. Essentially, these pathologies trace back to the fact that we are working with continuous ultrapowers, and there may be 'rather few' continuous functions on a compact space Ω. For the work of this chapter, it is necessary to introduce basic notions of model theory: we shall define and briefly discuss *models of number theory* and *models of analysis*.

In Chapter 11 we shall introduce the key notions of *rigidity* and *solidity*; the latter is the device that we shall use to achieve rigidity. Using the machinery developed in Chapter 9, we shall prove that (with GCH) there is a super-real η_1-field of weight $\aleph_1$ which is rigid, and we shall prove that it is relatively consistent with the theory ZFC that there is a super-real η_1-field K_P of weight $\aleph_1$ which is not even isomorphic as a totally ordered set to any super-real field of the form K_Q for which Q is a prime z-ideal in an algebra $C(\Omega)$. We must confess that, in the second half of Chapter 11, we shall abandon our erstwhile policy of assuming that the reader has little knowledge of set theory and will be forced to rely on the technique of 'forcing' to prove our relative-consistency results. Perhaps surprisingly, we shall also use the technique of forcing to prove as Theorem 11.25 that it is true in ZFC + GCH that there is a rigid hyper-real field.

We shall conclude in Chapter 12 with a final discussion and a list of some questions that we have not resolved.

Further details of, and references to, the sources and history of our results are given in the notes at the end of each chapter. We believe that all the unattributed results, including essentially all the results from Chapter 5 onwards (as well as some earlier results), are original.

Let us explain further the theory in which we are working in this book. Without specific mention, we shall throughout work in the theory ZFC. This is the basic language of set theory, implicitly assumed by most mathematicians in their work, and consisting of the Zermelo-Fraenkel axioms, together with the Axiom of Choice; in this theory, Zorn's lemma and the well-ordering principle hold. Frequently, we shall go beyond ZFC by assuming the generalized continuum hypothesis, GCH, and occasionally we shall go beyond ZFC + GCH, but the wilder shores of large cardinals and forcing will make only a fleeting appearance, at least until Chapter 11 is reached.

Let T be a theory in a certain language $\mathcal{L}$, and let φ be a sentence of $\mathcal{L}$. Then there is a clear notion of T *proving* φ, written $T \vdash \varphi$. A theory T is *inconsistent* if there is a sentence φ such that $T \vdash (\varphi \wedge \neg\varphi)$; otherwise T is *consistent.* A sentence φ is *relatively consistent* with T either if $T + \varphi$ is consistent or if T is inconsistent, and φ is *independent* of T if $T \nvdash \varphi$ and $T \nvdash \neg\varphi$, so that, in the case where T is consistent, φ is independent of T if and only if both φ and $\neg\varphi$ are relatively consistent with T. (For a gentle expansion of these remarks, see Dales and Woodin (1987, Chapter 4).) It cannot be proved that ZFC is consistent, but for a number of sentences φ one can prove that φ is relatively consistent with ZFC. In the case where a sentence φ is relatively consistent with ZFC, we can assert that the negation $\neg\varphi$ cannot be proved as a theorem of ZFC (unless ZFC be inconsistent).

It will often be the case that we cannot resolve the questions that interest us in the theory ZFC, and so we shall appeal to sentences φ that are relatively consistent with, but independent of, ZFC to assist us. For example, the continuum hypothesis (CH) and the generalized continuum hypothesis (GCH) are well known to be independent of the theory ZFC; these independence results are proved by using the technique of forcing in Chapter 7 of Dales and Woodin (1987) and many other texts. Many of our results will be proved in the theory ZFC + CH or ZFC + GCH; in this case the statement will be marked with CH or GCH, respectively. In Chapter 7, we shall also appeal to a combinatorial principle called diamond ($\diamondsuit$); this sentence is known to be independent of ZFC and to imply CH, and so a result proved in ZFC + $\diamondsuit$ is weaker than the same result proved in ZFC + CH.

Finally, and especially in the second half of Chapter 11, we shall establish that certain theorems are relatively consistent with ZFC, not by deducing them from a well-known and standard extra axiom, but by proving directly their relative consistency using the techniques of forcing.

An index of terminology is given at the end of this book; we also include a full index of notation on pages 352 – 356. However, we do mention here some specific points of notation, which might otherwise cause confusion: first, following the tradition in analysis, we set

$$\mathbb{N} = \{1, 2, 3, \ldots\},$$

whereas the first ordinal is 0; second, we denote set inclusion by $\subset$, so that '$A \subset B$' allows the possibility that $A = B$ (and we must write $A \subsetneq B$ to exclude the case of equality); third, T^S denotes the set of functions from a set S into a set T; fourth, for $f \in T^S$, $f \mid S_0$ is the restriction of f to a subset S_0 of S, whereas, for $f \in T^{\mathbb{N}}$ and $k \in \mathbb{N}$, $f \upharpoonright k$ denotes $f \mid \{1, \ldots, k\}$.

We shall write $\operatorname{dom} F$ and $\operatorname{ran} F$ for the domain and range, respectively, of a function F, and we shall frequently write $F(S)$ for the set $\operatorname{ran}(F \mid S)$, in the tradition of analysis, trusting to context to arbitrate the potential ambiguity of this notation.

Let S and T be sets. A *partial function* from S to T is a function F with $\operatorname{dom} F \subset S$ and $\operatorname{ran} F \subset T$, and the *graph* of such a function is

$$\operatorname{Gr} F = \{(s, F(s)) : s \in \operatorname{dom} F\}.$$

A subset R of $S \times T$ is the graph of a partial function from S to T if and only if $t_1 = t_2$ whenever $(s, t_1) \in R$ and $(s, t_2) \in R$. We shall slightly abuse language by saying that 'R is a function (with domain in S)' in this case. We shall distinguish throughout between a sequence $(x_n : n \in \mathbb{N})$ in a set X and a subset $\{x_n : n \in \mathbb{N}\}$; the former is an element of the set $X^{\mathbb{N}}$, and it is often denoted just by (x_n).

This book was written in intermittent intensive bursts over a period of five years. In fact the presentation reflects the evolution of our knowledge as the writing proceeded; for example the gap theorem of Chapter 8 is the residue of a struggle to complete an anticipated proof that $\widetilde{\mathbf{R}}$ is not a super-real field.

Some of the first half of this book was presented in a series of seminars at Berkeley in 1992; it was a privilege for us that these lectures were attended by Professor Irving Kaplansky, at the time of the 50th anniversary of his proof of his embedding theorem. We are grateful to Professor T.Y. Lam for sponsoring these lectures.

We are very grateful to Joanna Dales and to Ann Woodin for tolerating considerable domestic disruption whilst this book was being written; we also thank Joanna for several judicious comments on our grammar. We thank Dr Jan Stegeman of Utrecht for a careful reading of this manuscript and for pointing out a number of errors, and we thank Dr Franz-Viktor Kuhlmann, Dr Salma Kuhlmann, and Professor Peter Vamos for some valuable comments. We owe a special debt of thanks to Mrs Margaret Williams of the University of Leeds for typing substantial parts of this manuscript; we are grateful to Miss Elizabeth Johnston and Mr Keith Mansfield of Oxford University Press for their assistance, especially on the production of the Latex files.

We acknowledge support of grants GR/H88664, GR/J56240, and GR/K39592 from the United Kingdom EPSRC (formerly SERC) to enable the second author to visit Leeds, and from the NSF in the USA, supporting the second author on grant DMS–9322442 and funding the first author to visit Berkeley.

H.G. Dales, Leeds, England W.H. Woodin, Berkeley, California, USA

November 1995

Contents

1
Ordered sets and ordered groups

The fundamental notion that will inform our work is that of *order*: we shall be concerned with ordered sets, ordered groups, and ordered fields, and so, in due course, we shall specify how various algebraic operations interact with order structures. We shall begin, however, by recalling the precise definitions of various orders themselves and by giving some elementary properties of ordered sets. The most important early definition is Definition 1.7, where we introduce α_1-, β_1-, and η_1-sets. In Definition 1.8, we shall introduce an ordered set $\mathbf{Q}$ which is both a β_1-set and an η_1-set, and we shall show in Theorem 1.15 that $\mathbf{Q}$ is the only such set: every β_1-η_1-set is order-isomorphic to $\mathbf{Q}$.

Next we shall turn to ordered groups. A particular way of constructing ordered groups is to build groups $\mathfrak{F}(\mathbb{R}, S)$ and $\mathfrak{F}_{(\sigma)}(\mathbb{R}, S)$ of formal series over an ordered set S; an example is the group $\mathbf{G} = \mathfrak{F}_{(1)}(\mathbb{R}, \mathbf{Q})$. Hahn's embedding theorem (Theorem 1.35) shows that each ordered group embeds in a group of the form $\mathfrak{F}(\mathbb{R}, S)$. We have notions of α_1-, β_1-, and η_1-groups analogous to the corresponding notions for sets (see Definition 1.37), and we shall see in Corollary 1.49 that every divisible β_1-η_1-group is isomorphic to $\mathbf{G}$ in the category of totally ordered groups.

Definition 1.1 *Let S be a set. A* partial order *on S is a binary relation $\leq$ on S such that:*

(i) *if $r \leq s$ and $s \leq t$ in S, then $r \leq t$;*

(ii) *$s \leq s$ for each $s \in S$;*

(iii) *if $r \leq s$ and $s \leq r$ in S, then $r = s$.*

The partially ordered set $(S, \leq)$ is totally ordered *if, further:*

(iv) *for each $r, s \in S$, either $r \leq s$ or $s \leq r$.*

A totally ordered set is also called a chain.

A binary relation on S is identified with a subset of the Cartesian product $S \times S$. Formally, we should always write $(S, \leq)$ for a partially ordered set, thus indicating the binary relation $\leq$; however, we shall usually just write S in this case, and leave the relation to be understood. We sometimes write $s \geq r$ for $r \leq s$.

For example, the inclusion relation $\subset$ is a partial order on the power set $\mathcal{P}(S)$ of each non-empty set S. (Recall that, for us, the notation '$A \subset B$' allows the possibility that $A = B$.) The sets $\mathbb{Z}$ and $\mathbb{R}$ with the natural order are both totally ordered sets.

As a further example, for each non-empty set S the linear space $\mathbb{R}^S$ is a partially ordered set with respect to the relation given by

$$f \leq g \quad \text{if} \quad f(s) \leq g(s) \quad (s \in S)\,,$$

where $f, g \in \mathbb{R}^S$. This is the *natural order* on $\mathbb{R}^S$. It is only a total order in the case where S is a singleton.

There is an equivalent approach to partially ordered sets. A *strict partial order* on a set S is a binary relation $<$ such that:

(i) if $r < s$ and $s < t$ in S, then $r < t$;

(ii) $s \not< s$ for each s in S.

Let $\leq$ be a partial order on S, and set $r < s$ if $r \leq s$ and $r \neq s$. Then $<$ is a strict partial order on S. On the other hand, let $<$ be a strict partial order on S, and set $r \leq s$ in S if $r < s$ or $r = s$. Then $\leq$ is a partial order on S. We shall move freely between these two approaches.

A subset of a partially ordered set is also a partially ordered set with respect to the restriction of the given order.

Let $\{S_\tau : \tau < \sigma\}$ be a family of totally ordered sets indexed by an ordinal σ, and let $s = (s_\tau)$ and $t = (t_\tau)$ be distinct elements of the product set

$$S = \prod\{S_\tau : \tau < \sigma\}\,.$$

Then there is a minimum ordinal τ such that $s_\tau \neq t_\tau$; this ordinal is denoted by $\delta(s,t)$. If $s \neq t$ in S, then we define $s \prec t$ if $s_\tau < t_\tau$ for $\tau = \delta(s,t)$, and we set

$$s \preceq t$$

if $s \prec t$ or $s = t$. The binary relation $\preceq$ is a total order on S, called the *lexicographic order*. In particular, the product $S \times T$ of two totally ordered sets S and T is throughout supposed to have the lexicographic order unless stated otherwise.

Let $(S, \leq)$ be a totally ordered set with subsets S_1 and S_2. Then

$$S_1 << S_2 \quad (\text{respectively, } S_1 \leq S_2)$$

if $s_1 < s_2$ (respectively, $s_1 \leq s_2$) whenever $s_1 \in S_1$ and $s_2 \in S_2$; we write

$$S_1 << s \quad (\text{respectively, } S_1 \leq s)$$

for $S_1 << \{s\}$ (respectively, $S_1 \leq \{s\}$), etc.

Let T be a subset of a partially ordered set S. An element $s \in S$ is an *upper bound* of T if $T \leq s$, and s is the *supremum* of T, written

$$s = \sup T\,,$$

if s is an upper bound of T and if $s \leq u$ for each upper bound u of T. Similarly, we define a *lower bound* and the *infimum* of T, written $\inf T$. The subset T is *bounded above* (respectively, *bounded below*) if T has an upper bound (respectively, a lower bound).

An element s of a partially ordered set S is the *maximum* of S if S is an upper bound of S, and s is a *maximal* element if there is no element t of S with $t > s$. Similarly, we define the *minimum* of S and a *minimal* element of S.

We shall frequently use *Zorn's lemma* in the form: each non-empty, partially ordered set in which each chain has an upper bound has a maximal element.

Definition 1.2 *Let $(S, \leq)$ be a partially ordered set. Then S is* well-ordered *if each non-empty subset of S has a minimum element.*

Thus a well-ordered set is totally ordered. Technically, the empty set $\emptyset$ is a well-ordered set, or, even more pedantically, $(\emptyset, \emptyset)$ is a well-ordered set.

For example, a subset of $(\mathbb{Z}, \leq)$ is well-ordered if and only if it is bounded below.

On many occasions throughout this book, having introduced a structure, we shall quickly proceed to define the 'natural' maps between two examples of such a structure. The following is the first occurrence of this.

Definition 1.3 *Let $(S, \leq)$ and $(T, \leq)$ be two partially ordered sets. A map $\psi : S \to T$ is* isotonic *(respectively,* anti-isotonic*) if*

$$\psi(s_1) < \psi(s_2) \quad (\textit{respectively, } \psi(s_1) > \psi(s_2))$$

in T whenever $s_1 < s_2$ in S, and S and T are order-isomorphic *if there is a bijection $\psi : S \to T$ such that ψ and ψ^{-1} are isotonic; in this latter case ψ is an* order-isomorphism. *A map $\psi : S \to T$ is* order-preserving *if*

$$\psi(s_1) \leq \psi(s_2)$$

in T whenever $s_1 \leq s_2$ in S.

Let S and T be two totally ordered sets. Then

$$S \odot T$$

denotes the totally ordered set which is 'S followed by T'. The set $\{1\}$ is a totally ordered set with just one element, and so $\{1\} \odot \mathbb{R}$ is order-isomorphic to the interval $[0, 1)$ in $\mathbb{R}$, for example.

We shall suppose that the reader is familiar with the elementary theory of ordinal and cardinal numbers. Nevertheless, we briefly review this topic, establishing the notation that we shall use.

Let S be a set. Then S is *countable* if there is an injection from S into $\mathbb{N}$; otherwise S is *uncountable*. (In particular, finite sets are countable.) The set S is *transitive* if each element of S is a subset of S, and S is an *ordinal* if S is transitive and $(S, \in)$ is a totally ordered set, where we regard $\in$ as a strict partial order on S. We use the terminology of classes, and transfer the notions of orderings to classes in the usual way. Let α and β be ordinals. Then $\alpha < \beta$ in the class **Ord** of all ordinals if $\alpha \in \beta$; $(\mathbf{Ord}, \leq)$ is a well-ordered class, and each well-ordered set is order-isomorphic to a unique ordinal. Let S be a set of ordinals. Then $\bigcup\{\alpha : \alpha \in S\}$ is an ordinal, and it is the supremum of S in **Ord**. For each ordinal α, the set $\alpha \cup \{\alpha\}$ is an ordinal; it is the *successor* of α, and it is denoted by $\alpha + 1$. A non-zero ordinal which is not a successor is a *limit* ordinal. The minimum element in the class **Ord** is $\emptyset$, identified with the number 0; the finite ordinals are identified with the elements of $\mathbb{Z}^+$. The ordered set $(\mathbb{Z}^+, \leq)$ is order-isomorphic to the first infinite ordinal ω; the first uncountable ordinal is denoted by ω_1.

Two sets S and T are *equipotent* if there is a bijection from S onto T. An ordinal α is a *cardinal* if, for each ordinal β with $\beta < \alpha$, α and β are not equipotent. For each set S, there is a minimum ordinal which is equipotent to S; this ordinal is the *cardinal* of S, denoted by $|S|$. The ordinals ω and ω_1 are cardinals, and as such they are denoted by

$$\aleph_0 \quad \text{and} \quad \aleph_1 \, ,$$

respectively. The smallest cardinal strictly larger than a given cardinal κ is the *successor* cardinal of κ, denoted by κ^+. We inductively define cardinals $\aleph_\alpha$ for each ordinal α as follows: if $\alpha = \beta + 1$, then $\aleph_\alpha = \aleph_\beta^+$, and, if α is a limit ordinal, then $\aleph_\alpha = \sup\{\aleph_\beta : \beta < \alpha\}$. For each cardinal κ, we define $2^\kappa = |\mathcal{P}(\kappa)|$.

The real line $\mathbb{R}$ has cardinality $2^{\aleph_0}$, and this cardinal will usually be denoted by $\mathfrak{c}$; it is called the *cardinal of the continuum*.

A famous theorem of Cantor implies that $\kappa^+ \leq 2^\kappa$ for each cardinal κ; in particular,

$$\aleph_1 \leq 2^{\aleph_0} \quad \text{and} \quad \aleph_{\alpha+1} \leq 2^{\aleph_\alpha}$$

for each ordinal α. The *continuum hypothesis* (CH) is the assertion that

$$\aleph_1 = 2^{\aleph_0} \, ,$$

and the *generalized continuum hypothesis* (GCH) is the assertion that

$$\aleph_{\alpha+1} = 2^{\aleph_\alpha} \quad (\alpha \in \mathbf{Ord}) \, .$$

We shall frequently use CH, and occasionally we shall use GCH, but usually only in the form that $\aleph_1 = 2^{\aleph_0}$ and $\aleph_2 = 2^{\aleph_1}$; results whose proofs depend on these axioms will be so denoted. For example, with CH, we have

$$2^{\aleph_1} > 2^{\aleph_0} = \mathfrak{c} \, ,$$

but this is not a theorem of ZFC.

We shall use the following elementary facts of cardinal arithmetic:

(i) if κ and λ are non-zero cardinals with $\lambda \leq \kappa$ and κ infinite, then

$$\kappa + \lambda = \kappa\lambda = \kappa \, ;$$

(ii) if κ, λ, and μ are non-zero cardinals, then

$$(\kappa^\lambda)^\mu = \kappa^{\lambda\mu} \, .$$

In particular,

$$\mathfrak{c}^{\aleph_0} = \mathfrak{c} \, .$$

Definition 1.4 *Let $(S, \leq)$ be a totally ordered set. A subset T of S is:*

(i) *an* interval *if, for each $t_1, t_2 \in T$ with $t_1 < t_2$ and each $s \in S$ with $t_1 \leq s \leq t_2$, it follows that $s \in T$;*

(ii) order-dense *in S if, for each $s_1, s_2 \in S$ with $s_1 < s_2$, there exists $t \in T$ with $s_1 \leq t \leq s_2$;*

(iii) cofinal *in S if, for each $s \in S$, there exists $t \in T$ with $t \geq s$;*

(iv) coinitial *in S if, for each $s \in S$, there exists $t \in T$ with $t \leq s$.*

The real line $\mathbb{R}$ with the usual order is, of course, a standard example of a totally ordered set, and it will be the prototype of much of our theory. It contains the set $\mathbb{Q}$ of rational numbers as an order-dense subset, and $\mathbb{N}$ is a cofinal subset of $\mathbb{R}$.

We shall use the usual notations for intervals in a totally ordered set S. For example, if $s_1, s_2 \in S$, then

$$[s_1, s_2] = \{t \in S : s_1 \leq t \leq s_2\} \quad \text{and} \quad (-\infty, s_1] = \{t \in S : t \leq s_1\}.$$

The *order topology* on S is the topology defined by taking as a subbasis the collection of intervals of the form $(-\infty, s)$ and (s, ∞) for $s \in S$. For example, the standard topology on $\mathbb{R}$ is just the order topology derived from the usual order on $\mathbb{R}$. An ordinal α, identified with the interval $[0, \alpha)$ in **Ord**, is always a locally compact space with respect to the order topology; it is compact if and only if α is not a limit ordinal.

Let T be a subset of a totally ordered set S. A *maximal interval of S in T* is an interval I of S which is contained in T and such that I is not properly contained in any other such interval of S. Clearly T is the disjoint union of the maximal intervals of S in T. For example, the maximal intervals of $\mathbb{R}$ in an open subset U are just the connected components of U; however, in general, intervals of S are not necessarily connected in the order topology.

Definition 1.5 *Let $(S, \leq)$ be a totally ordered set. Then the* weight *of S is the minimum cardinality of an order-dense subset in S.*

The weight of S is denoted by $w(S)$. It is clear that, if $w(S) = \kappa$, then $|S| \leq 2^\kappa$. For let T be an order-dense subset in S: the map

$$s \mapsto \{t \in T : t < s\}, \quad S \to \mathcal{P}(T),$$

is then an injection. For example, note that $w(\mathbb{R}) = \aleph_0$, whereas $|\mathbb{R}| = \mathfrak{c}$.

The weight, as well as the cardinality, is an invariant of totally ordered sets under order-isomorphism.

Let σ be an ordinal. A sequence $s = (s_\tau : \tau < \sigma)$ in a set S is a map from σ into S; the *length* of s is σ, written

$$l(s) = \sigma.$$

Suppose that $(S, \leq)$ is a partially ordered set. The sequence s is, respectively, *increasing* if $s_\tau \leq s_\rho$, *strictly increasing* if $s_\tau < s_\rho$, *decreasing* if $s_\rho \leq s_\tau$ and *strictly decreasing* if $s_\rho < s_\tau$ whenever $\tau < \rho < \sigma$. Clearly S is well-ordered if and only if each decreasing sequence of length ω is eventually constant. We note that each infinite, totally ordered set contains either a strictly increasing sequence or a strictly decreasing sequence of length ω.

Let $s = (s_\tau : \tau < \sigma)$ be a sequence of length σ in a set S, and take $\rho < \sigma$. Then

$$s \restriction \rho = (s_\tau : \tau < \rho);$$

clearly, $s \restriction \rho$ is a sequence of length ρ in S which is the *restriction* of the original sequence s to ρ. A sequence t of length $\rho \geq \sigma$ in s is an *extension* of s if $t \restriction \sigma = s$. Next take $s_\sigma \in S$. Then

$$s^\frown s_\sigma = (s_\tau : \tau \leq \sigma)$$

is a sequence of length $\sigma + 1$ in S which extends s. More generally,

$$s^\frown t$$

is the sequence s 'followed by' the sequence t.

Now let $f : \mathbb{N} \to S$ be a function, and take $k \in \mathbb{N}$. Then

$$f \restriction k : \{1, \ldots, k\} \to S$$

is a *restriction* of f. (Recall that, for us, the first element of $\mathbb{N}$ is 1, whereas the first ordinal is 0.)

The *cofinality* (respectively, *coinitiality*) of a non-empty, totally ordered set S is the minimum cardinal κ such that there is a strictly increasing (respectively, strictly decreasing) sequence of length κ whose range is cofinal (respectively, coinitial) in S. We write

$$\operatorname{cof} S = \kappa \quad (\text{respectively}, \operatorname{coi} S = \kappa).$$

Abusing language a little, we say that $(s_\sigma : \sigma < \kappa)$ is cofinal in S, etc. We also set $\operatorname{cof} \emptyset = \operatorname{coi} \emptyset = 0$. Thus $\operatorname{cof} S$ and $\operatorname{coi} S$ are the minimum cardinalities of cofinal and coinitial subsets of S, respectively. For example, $\operatorname{cof} \mathbb{R} = \aleph_0$. A cardinal κ is *regular* if $\operatorname{cof} \kappa = \kappa$; for each cardinal κ, the successor cardinal κ^+ is regular.

We start with a trivial lemma about cofinalities.

Lemma 1.6 *Let S and T be totally ordered sets, and let $\psi : S \to T$ be an order-preserving map. Suppose that $\psi(S)$ is cofinal in T and that T has no maximum element. Then* $\operatorname{cof} S = \operatorname{cof} T$.

Proof Set $\kappa = \operatorname{cof} S$ and $\lambda = \operatorname{cof} T$.

Let $(t_\sigma : \sigma < \lambda)$ be cofinal in T. Then there exists $(s_\sigma : \sigma < \lambda)$ in S with $\psi(s_\sigma) \geq t_\sigma \quad (\sigma < \lambda)$. Take $s \in S$. Since T has no maximum element, there exists $\sigma < \lambda$ with $t_\sigma > \psi(s)$, and then $s < s_\sigma$. Thus $(s_\sigma : \sigma < \lambda)$ is cofinal in S, and so $\kappa \leq \lambda$.

Let $(s_\sigma : \sigma < \kappa)$ be cofinal in S. For each $t \in T$, there exists $s \in S$ with $\psi(s) \geq t$ since $\psi(S)$ is cofinal in T, and there exists $\sigma < \kappa$ with $s_\sigma \geq s$, and then $\psi(s_\sigma) \geq t$. Thus $(\psi(s_\sigma) : \sigma < \lambda)$ is cofinal in T, and so $\lambda \leq \kappa$. □

We now come to a key definition about orders. The terminology that we are now to introduce will be of central importance for us, and will percolate through the remainder of this book.

Definition 1.7 *Let $(S, \leq)$ be a non-empty, totally ordered set. Then S is:*

(i) *an* α_1-set *if each non-empty subset of S has a countable cofinal and coinitial subset;*

(ii) *a* β_1-set *if $S = \bigcup\{S_\nu : \nu < \rho\}$, where $\{S_\nu : \nu < \rho\}$ is a chain of α_1-sets in $(\mathcal{P}(S), \subset)$;*

(iii) *an* η_1-set *if, for each pair $\{S_1, S_2\}$ of subsets of S with $S_1 << S_2$ and $|S_1 \cup S_2| < \aleph_1$, there exists $s \in S$ with $S_1 << s << S_2$;*

(iv) *a* semi-η_1-set *if, for each strictly increasing sequence (s_n) and each strictly decreasing sequence (t_n) such that $\{s_n : n \in \mathbb{N}\} << \{t_n : n \in \mathbb{N}\}$, there exists $s \in S$ with $\{s_n : n \in \mathbb{N}\} << s << \{t_n : n \in \mathbb{N}\}$.*

The intuition is that an α_1-set is 'rather small', the position of an element in the order being 'countably determined', but that an η_1-set is 'rather large', no element being 'countably determined'.

For example, it is easily seen that $\mathbb{R}$ is an α_1-set and a semi-η_1-set, but that it is not an η_1-set; the order-dense subset $\mathbb{Q}$ of $\mathbb{R}$ is not a semi-η_1-set.

Note that a totally ordered set is an α_1-set if and only if it contains no strictly increasing or strictly decreasing sequence of length ω_1. Thus each well-ordered subset of an α_1-set is countable. A set which is a countable union of α_1-sets is an α_1-set; the possibility of β_1-sets which are not α_1-sets arises because the union in (ii), above, need not be countable. (We shall see in Corollary 1.14, below, that each β_1-set can be written in the form $\bigcup\{S_\nu : \nu < \omega_1\}$.)

The notions of α_1-sets and of η_1-sets have obvious generalizations. For example, we say that a totally ordered set is an *α_2-set* if it contains no strictly increasing or strictly decreasing sequence of length ω_2. We shall see shortly that every β_1-set is an α_2-set. Generalizations of η_1-sets will be described in the notes.

Clearly, subsets of α_1-sets and β_1-sets are also α_1-sets and β_1-sets, respectively.

Let T be an order-dense subset of S; we ask if the η_1- and semi-η_1-properties transfer between S and T. Suppose first that S is an η_1-set. Then T is also an η_1-set. For take T_1 and T_2 to be countable subsets of T with $T_1 << T_2$ and $T_1 \cup T_2 \neq \emptyset$. Then there exists $s_1 \in S$ with $T_1 << s_1 << T_2$ and $s_2 \in S$ with $s_1 < s_2 << T_2$. Since T is order-dense in S, there exists $t \in T$ with $s_1 \leq t \leq s_2$, and we have $T_1 << t << T_2$, so that T is an η_1-set. On the other hand, T is not necessarily a semi-η_1-set when S is a semi-η_1-set. If T is a semi-η_1-set, then S is also a semi-η_1-set, but S is not necessarily an η_1-set when T is an η_1-set. Thus, for T order-dense in S:

$$\begin{aligned} S \text{ is an } \eta_1\text{-set} &\Rightarrow T \text{ is an } \eta_1\text{-set}\,; \\ T \text{ is a semi-}\eta_1\text{-set} &\Rightarrow S \text{ is a semi-}\eta_1\text{-set}\,. \end{aligned}$$

If S is an η_1-set, then $\operatorname{cof} S \geq \aleph_1$, and so no ordered set can be both an α_1-set and an η_1-set. We shall, however, now build a set which is both a β_1-set and an η_1-set.

Let $\mathbf{S}$ be the set of all sequences $\alpha = (\alpha_\tau : \tau < \omega_1)$ of length ω_1 such that $\alpha_\tau \in \{0,1\}$ $(\tau < \omega_1)$. Thus $\mathbf{S}$ is the collection of *dyadic sequences of length* ω_1, and

$$\mathbf{S} = \left(\{0,1\}^{\omega_1}, \preceq\right)$$

is a totally ordered set with respect to the lexicographic order, $\preceq$; again, we set $\delta(\alpha, \beta) = \min\{\tau : \alpha_\tau \neq \beta_\tau\}$ for $\alpha, \beta \in \mathbf{S}$ with $\alpha \neq \beta$. We clearly have

$$|\mathbf{S}| = 2^{\aleph_1}.$$

Definition 1.8 *The set* $\mathbf{Q}$ *is the set of elements* $\alpha = (\alpha_\tau)$ *of* $\mathbf{S}$ *such that*

$$\{\tau < \omega_1 : \alpha_\tau = 1\}$$

is non-empty and has a largest member. For each $\sigma < \omega_1$, $\mathbf{Q}_\sigma$ *is the set of elements of* $\mathbf{Q}$ *for which this largest member is strictly less than* σ. *The set* $\mathbf{Q}$ *is* Sierpiński's *set.*

For example, the sequence $\alpha = (1\,1\,1\,\ldots\,0\,0\,\ldots)$, which has $\alpha_\tau = 1$ for $\tau < \omega$ and $\alpha_\tau = 0$ for $\omega \leq \tau < \omega_1$, belongs to $\mathbf{S}$, but not to $\mathbf{Q}$. Note that, for $\alpha = (\alpha_\tau) \in \mathbf{Q}$, we have $\alpha_\tau = 0$ for all but countably many values of τ.

Proposition 1.9 (i) *For each* $\sigma < \omega_1$, $\mathbf{Q}_\sigma$ *is an* α_1*-set.*

(ii) $\mathbf{Q} = \bigcup\{\mathbf{Q}_\sigma : \sigma < \omega_1\}$, *and* $\mathbf{Q}$ *is a* β_1*-set and an* α_2*-set.*

(iii) $\mathbf{Q}$ *is an* η_1*-set.*

(iv) $|\mathbf{Q}| = w(\mathbf{Q}) = \mathfrak{c}$.

(v) $\operatorname{coi} \mathbf{Q} = \operatorname{cof} \mathbf{Q} = \aleph_1$.

Proof (i) Fix $\sigma < \omega_1$, and let A be a non-empty subset of $\mathbf{Q}_\sigma$. We shall prove that A has a countable cofinal subset; the result for a coinitial subset is similar.

We first define a sequence $\beta = (\beta_\zeta : \zeta < \omega_1)$ inductively as follows. Define $\beta_0 = 1$ if there exists $\alpha \in A$ with $\alpha_0 = 1$ and define $\beta_0 = 0$ otherwise. Now suppose that $0 < \tau < \omega_1$ and that β_ζ has been defined for each $\zeta < \tau$. Define $\beta_\tau = 1$ if there exists $\alpha = (\alpha_\zeta : \zeta < \omega_1)$ in A with $\alpha_\zeta = \beta_\zeta$ $(\zeta < \tau)$ and with $\alpha_\tau = 1$, and define $\beta_\tau = 0$ if no such α exists. Clearly $\alpha \preceq \beta$ for each $\alpha \in A$. Now suppose that $\gamma \in \mathbf{S}$ with $\gamma \prec \beta$ and set $\tau = \delta(\gamma, \beta)$. Then $\gamma_\tau = 0$ and $\beta_\tau = 1$ and there exists $\alpha \in A$ with $\alpha \restriction (\tau + 1) = \beta \restriction (\tau + 1)$, so that $\alpha \succ \gamma$. We have shown that $\beta = \sup A$ in $\mathbf{S}$.

Since $\alpha_\zeta = 0$ $(\zeta \geq \sigma)$ for each $\alpha \in A$, necessarily $\beta_\zeta = 0$ $(\zeta \geq \sigma)$. Take τ to be the minimum ordinal such that $\beta_\zeta = 0$ $(\tau \leq \zeta < \omega_1)$; we have shown that $\tau \leq \sigma$. First suppose that τ is a successor ordinal, say $\tau = \rho + 1$, so that $\beta_\rho = 1$. Then there exists $\alpha \in A$ with $\alpha \restriction \tau = \beta \restriction \tau$; since $\beta_\zeta = 0$ $(\zeta \geq \tau)$, we have $\beta = \alpha$, and so the singleton $\{\beta\}$ is cofinal in A. Second suppose that τ is a limit ordinal with $\tau \leq \sigma$, so that $\beta \notin A$. Then the set $X = \{\rho < \tau : \beta_\rho = 1\}$ is countable and cofinal in τ. For each $\rho \in X$, there exists $\alpha^{(\rho)} \in A$ with $\alpha_\zeta^{(\rho)} = \beta_\zeta \quad (\zeta \leq \rho)$. Clearly $(\alpha^{(\rho)} : \rho \in X)$ is a countable cofinal subset in A.

(ii) This is immediate from the definitions and from (i).

(iii) Let A and B be countable subsets of $\mathbf{Q}$ with $A << B$ and $A \cup B \neq \emptyset$. Since $A \cup B$ is countable, there exists $\sigma < \omega_1$ such that $\alpha_\tau = 0 \quad (\tau \geqslant \sigma)$ for all $\alpha = (\alpha_\zeta : \zeta < \omega_1)$ in $A \cup B$. Set $\gamma_\sigma = 1$. For each $\tau < \omega_1$ with $\tau \neq \sigma$, set $\gamma_\tau = 1$ if there exists $\alpha \in A$ with $\alpha_\tau = 1$ and such that $\delta(\alpha, \beta) \geqslant \tau$ for each $\beta \in A$ with $\alpha \prec \beta$, and set $\gamma_\tau = 0$ otherwise. Since $\gamma_\sigma = 1$ and $\gamma_\tau = 0$ for $\tau \in (\sigma, \omega_1)$, the sequence $\gamma = (\gamma_\zeta : \zeta < \omega_1)$ belongs to $\mathbf{Q}$; since $\alpha_\sigma = 0 \quad (\alpha \in A \cup B)$, we have $\gamma \notin A \cup B$. We shall prove that $A << \gamma << B$.

First suppose that $A \neq \emptyset$, and take $\alpha \in A$. Set

$$\tau = \inf\{\delta(\gamma, \beta) : \beta \in A,\ \alpha \preceq \beta\}\,,$$

and take $\beta \in A$ with $\alpha \preceq \beta$ and $\delta(\gamma, \beta) = \tau$. Necessarily $\tau < \sigma$. For each $\beta' \in A$ with $\beta \prec \beta'$, we have $\delta(\gamma, \beta') \geqslant \tau$, and so $\delta(\beta, \beta') \geqslant \tau$. Assume towards a contradiction that $\beta_\tau = 1$. Then $\gamma_\tau = 1$ by the definition of γ_τ, and this contradicts the fact that $\delta(\gamma, \beta) = \tau$. So necessarily $\beta_\tau = 0$, and hence $\gamma_\tau = 1$ and $\beta \prec \gamma$. But now $\alpha \prec \gamma$, and so $A << \gamma$.

Before proving that $\gamma << B$, we make a remark. Suppose that $\tau < \sigma$ and $\gamma_\tau = 1$, and take $\alpha \in A$ with $\alpha_\tau = 1$ and such that $\delta(\alpha, \beta) \geqslant \tau$ for each $\beta \in A$ with $\alpha \prec \beta$. Set $\rho = \delta(\alpha, \gamma)$. We *claim* that $\rho > \tau$. Indeed, since $\alpha \prec \gamma$, we have $\alpha_\rho = 0$ and $\gamma_\rho = 1$, and so $\rho \neq \tau$. Assume that $\rho < \tau$, and take $\alpha' \in A$ with $\alpha'_\rho = 1$ and such that $\delta(\alpha', \beta) \geqslant \rho$ for each $\beta \in A$ with $\alpha' \prec \beta$. If $\alpha' \preceq \alpha$, then $\delta(\alpha', \alpha) \geqslant \rho$, a contradiction because $\alpha_\rho = 0$ and $\alpha'_\rho = 1$. If $\alpha \prec \alpha'$, then $\delta(\alpha, \alpha') \geqslant \tau > \rho$, again a contradiction. So the claim holds.

Suppose that $B \neq \emptyset$, and take $\beta \in B$. Set $\rho = \delta(\beta, \gamma)$, so that $\rho \leq \sigma$. Assume first that $\rho = \sigma$, and let τ be the maximum ordinal with $\beta_\tau = 1$, so that $\tau < \sigma$ and $\gamma_\tau = 1$. By the claim, there exists $\alpha \in A$ with $\alpha_\tau = 1$ and $\delta(\alpha, \gamma) > \tau$. But this is not possible because $\alpha \prec \beta$ and $\beta_\zeta = 0$ for $\zeta > \tau$ and $\beta_\zeta = \gamma_\zeta$ for $\zeta < \sigma$. Thus $\rho < \sigma$. Now assume that $\gamma_\rho = 1$. By the claim again, there exists $\alpha \in A$ with $\alpha_\rho = 1$ and $\delta(\alpha, \gamma) > \rho$. But this implies that $\beta \prec \alpha$, again a contradiction. So $\gamma_\rho = 0$, and hence $\gamma < \beta$. This proves that $\gamma << B$.

(iv) For each $\sigma < \omega_1$, we have $|\mathbf{Q}_\sigma| \leq \mathfrak{c}$, and so $|\mathbf{Q}| \leq \mathfrak{c} \cdot \aleph_1 = \mathfrak{c}$. Let S be an order-dense subset in $\mathbf{Q}$. For each $\alpha \in \{0,1\}^{\mathbb{N}}$, there exist $\beta, \gamma \in \mathbf{Q}$ with $\beta \prec \gamma$ and $\beta \restriction \omega = \gamma \restriction \omega = \alpha$. Thus $|S| \geq 2^{\aleph_0} = \mathfrak{c}$, and we have shown that $\mathfrak{c} \leq w(\mathbf{Q}) \leq |\mathbf{Q}| \leq \mathfrak{c}$.

(v) For $\sigma < \omega_1$, set:

$$\delta_\tau^{(\sigma)} = 1 \quad (\tau = \sigma)\,, \qquad \delta_\tau^{(\sigma)} = 0 \quad (\tau \neq \sigma)\,;$$
$$\alpha_\tau^{(\sigma)} = 1 \quad (\tau \leq \sigma)\,, \qquad \alpha_\tau^{(\sigma)} = 0 \quad (\tau > \sigma)\,.$$

Then $(\delta^{(\sigma)} : \sigma < \omega_1)$ is coinitial in $\mathbf{Q}$ and $(\alpha^{(\sigma)} : \sigma < \omega_1)$ is cofinal in $\mathbf{Q}$. Thus $\operatorname{coi}\mathbf{Q} \leq \aleph_1$ and $\operatorname{cof}\mathbf{Q} \leq \aleph_1$. Since $\mathbf{Q}$ is an η_1-set, $\operatorname{coi}\mathbf{Q} \geq \aleph_1$ and $\operatorname{cof}\mathbf{Q} \geq \aleph_1$. We have proved the result. □

The second key notion that will inform our work is that of a *gap* in an ordered set.

Definition 1.10 *Let* $(S, \leq)$ *be a non-empty, totally ordered set. A pair* $\{A, B\}$ *of subsets of* S *is a* pregap *if*

$$A \cup B \neq \emptyset \quad \text{and} \quad A << B\,;$$

the pregap is denoted by $\langle A, B\rangle$. *A pregap* $\langle A, B\rangle$ *is a* (κ, λ)-pregap *(where* κ *and* λ *are cardinals) if* $\operatorname{cof} A = \kappa$ *and* $\operatorname{coi} B = \lambda$, *and a pregap is* countable *if it is a* (κ, λ)*-pregap with* $\kappa \leq \aleph_0$ *and* $\lambda \leq \aleph_0$. *An element* $s \in S$ interpolates *the pregap* $\langle A, B\rangle$ *if*

$$A << s << B\,;$$

a pregap $\langle A, B\rangle$ *is a* gap *if there is no element* s *of* S *which interpolates* $\langle A, B\rangle$.

(In the definition of a pregap $\langle A, B\rangle$, it may be that either $A = \emptyset$ or $B = \emptyset$.)

Thus a non-empty set S is an η_1-set if and only if there are no countable gaps in $(S, \leq)$, and S is a semi-η_1-set if and only if there are no $(\aleph_0, \aleph_0)$-gaps. In the case where S is an α_1-set, every pregap is countable and, for an α_2-set, every pregap is a (κ, λ)-pregap with $\kappa, \lambda \leq \aleph_1$.

For example, let $A = \{0\}$ and $B = \{1/n : n \in \mathbb{N}\}$ in $\mathbb{R}$. Then $\langle A, B\rangle$ is a $(1, \aleph_0)$-gap in $\mathbb{R}$.

Let S be an order-dense subset in a totally ordered set T. It is clear that, if there is a (κ, λ)-gap in T with κ, λ infinite, then there is a (κ, λ)-gap in S.

Let $(T, \leq)$ be a totally ordered set, and let S be a non-empty subset of T. Choose $t \in T$, and set

$$A_t = \{s \in S : s < t\} = (-\infty, t) \cap S\,,$$
$$B_t = \{s \in S : s \geq t\} = [t, \infty) \cap S\,.$$

Then $\langle A_t, B_t\rangle$ is a gap in $(S, \leq)$: it is the *gap defined by* t.

Proposition 1.11 (i) *Let* S *be an* α_2-η_1*-set. Then* S *contains the following six types of gap:*

$$(0, \aleph_1),\ (\aleph_1, 0),\ (1, \aleph_1),\ (\aleph_1, 1),\ (\aleph_0, \aleph_1),\ (\aleph_1, \aleph_0)\,.$$

The only other possible gap in S *is an* $(\aleph_1, \aleph_1)$*-gap.*

(ii) *Each gap in* $(\mathbf{Q}, \preceq)$ *has one of the above types, and all seven types occur.*

Proof (i) Let $\langle A, B\rangle$ be a (κ, λ)-gap in S, so that $\kappa, \lambda \leq \aleph_1$ because S is an α_2-set. Since S is an η_1-set, the maximum of κ and λ is at least $\aleph_1$. Thus the only possibilities are the seven types of gap listed. It is clear that the first six types of gap do exist in S. For example, S contains a strictly increasing sequence (s_n). Set

$$A = \bigcup\{(-\infty, s_n] : n \in \mathbb{N}\}$$

and $B = S \setminus A$. Then $\langle A, B\rangle$ is an $(\aleph_0, \aleph_1)$-gap.

(ii) By Proposition 1.9, $\mathbf{Q}$ is an α_2-η_1-set. Thus, by (i), each gap in $\mathbf{Q}$ has one of the specified seven types, and the first six types occur. Consider the sequences formed by inserting 0 and 1 at the front of the sequences $\delta^{(\sigma)}$ and $\alpha^{(\sigma)}$ respectively, of Proposition 1.9(v). Then these sequences define an $(\aleph_1, \aleph_1)$-gap in $\mathbf{Q}$. □

There are α_2-η_1-sets S which do not have any $(\aleph_1, \aleph_1)$-gaps.

We now make a very simple remark which, nevertheless, will be the key to several important theorems in that it suggests a method of extending the domain of an isotonic map from a subset of a totally ordered set into another totally ordered set. Let $\psi : S \to T$ be an isotonic map between totally ordered sets S and T, and suppose that $\langle A, B\rangle$ is a pregap in S. Then certainly $\langle \psi(A), \psi(B)\rangle$ is a pregap in T. If $\langle A, B\rangle$ is a (κ, λ)-pregap, then $\langle \psi(A), \psi(B)\rangle$ is a (κ, λ)-pregap in T. Of course, even if $\langle A, B\rangle$ is a gap in S, it may well be that there exists $t \in T$ which interpolates the pregap $\langle \psi(A), \psi(B)\rangle$, and so $\langle \psi(A), \psi(B)\rangle$ is not necessarily a gap in T.

Our first main result is to show that, up to order-isomorphism, $\mathbf{Q}$ is the only β_1-η_1-set. The structure of the proof will be a model for two later proofs, involving groups and fields, respectively. The first lemma uses the above remark to extend a given isotonic map.

Let S and T be non-empty sets, and let S_0 be a subset of S. For $f : S \to T$, define

$$(f \mid S_0)(s) = f(s) \quad (s \in S_0).$$

Then $f \mid S_0 : S_0 \to T$ is the *restriction* of f to S_0. Given a map $g : S_0 \to T$, a map $f : S \to T$ such that $f \mid S_0 = g$ is an *extension* of g to S.

Lemma 1.12 *Let S be an α_1-set, and let T be an η_1-set. Then each isotonic map from a subset of S to T has an extension which is an isotonic map from S into T.*

Proof Let S_0 be a subset of S, and let $\psi_0 : S_0 \to T$ be an isotonic map. Define $\mathcal{F}$ to be the family of pairs (U, φ) such that $S_0 \subset U \subset S$ and $\varphi : U \to T$ is an isotonic map with $\varphi \mid S_0 = \psi_0$. Set

$$(U_1, \varphi_1) \preccurlyeq (U_2, \varphi_2)$$

in $\mathcal{F}$ if $U_1 \subset U_2$ and $\varphi_2 \mid U_1 = \varphi_1$. Clearly $(S_0, \psi_0) \in \mathcal{F}$. Each chain in $(\mathcal{F}, \preccurlyeq)$ has an obvious upper bound, and so, by Zorn's lemma, there is a maximal element, say (U, φ), in $\mathcal{F}$.

We *claim* that $U = S$. To obtain a contradiction, assume that $U \neq S$, and take $s \in S \setminus U$. Let $\langle A_s, B_s\rangle$ be the gap in U defined by s. Since S is an α_1-set, $\langle A_s, B_s\rangle$ is a countable pregap, and so $\langle \varphi(A), \varphi(B)\rangle$ is a countable pregap in T. Since T is an η_1-set, there exists $t \in T$ with

$$\varphi(A) << t << \varphi(B).$$

Set $\overline{\varphi}(u) = \varphi(u) \quad (u \in U)$ and $\overline{\varphi}(s) = t$. Then $\overline{\varphi} : U \cup \{s\} \to T$ is an isotonic map extending φ, and $(U \cup \{s\}, \overline{\varphi}) \succcurlyeq (U, \varphi)$ in $\mathcal{F}$, a contradiction of the assumed maximality of (U, φ).

Thus $U = S$, and φ is the required extension of ψ_0 to S. □

The above lemma shows, in particular, that every α_1-set can be isotonically embedded into an arbitrary η_1-set. We now extend this by showing that every β_1-set can be isotonically embedded into an η_1-set.

Theorem 1.13 *Let S be a β_1-set, and let T be an η_1-set. Then there is an isotonic map from S into T. In particular, there are isotonic maps from S into $\mathbf{Q}$ and from $\mathbf{Q}$ into T.*

Proof Set $S = \bigcup\{S_\nu : \nu < \rho\}$, where $\{S_\nu : \nu < \rho\}$ is a chain of α_1-sets. We shall inductively define a family of maps $\psi_\nu : S_\nu \to T$ for each $\nu < \rho$. We start with an isotonic map $\psi_0 : S_0 \to T$. Suppose that there is an isotonic map $\psi_\nu : S_\nu \to T$. Then, by Lemma 1.12, there is an isotonic map $\psi_{\nu+1} : S_{\nu+1} \to T$ with $\psi_{\nu+1} \mid S_\nu = \psi_\nu$. Next, suppose that ν is a limit ordinal with $\nu < \rho$ and that, for each $\zeta < \nu$, $\psi_\zeta : S_\zeta \to T$ is an isotonic map such that $\psi_\zeta \mid S_\tau = \psi_\tau$ whenever $\tau < \zeta < \nu$. Then the formulae $\varphi \mid S_\zeta = \psi_\zeta \quad (\zeta < \nu)$ define an isotonic map $\varphi : \bigcup\{S_\zeta : \zeta < \nu\} \to T$. Again by Lemma 1.12, φ extends to an isotonic map $\psi_\nu : S_\nu \to T$ such that $\psi_\nu \mid S_\zeta = \psi_\zeta \quad (\zeta < \nu)$.

Eventually we obtain a family of isotonic maps $\psi_\nu : S_\nu \to T$ for each $\nu < \rho$ such that $\psi_\nu \mid S_\zeta = \psi_\zeta$ whenever $\zeta < \nu$, and these partial maps fit together to give the required isotonic embedding of $S = \bigcup S_\nu$ into T. □

Let T be an η_1-set. Then there is an isotonic map from $\mathbb{R}$ into T, and so, necessarily, $|T| \geqslant \mathfrak{c}$.

Corollary 1.14 *Let S be a β_1-set. Then there is a chain $\{S_\sigma : \sigma < \omega_1\}$ of α_1-subsets of S such that*

$$S = \bigcup\{S_\sigma : \sigma < \omega_1\}.$$

Proof By the theorem, there is an isotonic map $\psi : S \to \mathbf{Q}$. Set

$$S_\sigma = \psi^{-1}(\mathbf{Q}_\sigma) \quad (\sigma < \omega_1).$$

By Proposition 1.9(i), each $\mathbf{Q}_\sigma$ is an α_1-set, and so S_σ is an α_1-set. □

It follows immediately from Corollary 1.14 that the union of a family of size $\aleph_1$ of β_1-subsets of a given totally ordered set is itself a β_1-set and that every β_1-set is an α_2-set.

Let S be a β_1-set. Since S embeds into $\mathbf{Q}$ and $|\mathbf{Q}| = \mathfrak{c}$, necessarily $|S| \leq \mathfrak{c}$. On the other hand, it is clear that each totally ordered set of cardinality $\aleph_1$ is a β_1-set, and so, with CH, a totally ordered set S is a β_1-set if and only if $|S| \leq \aleph_1$. It may be said that the β_1-condition replaces a cardinality condition on a totally ordered set in the case where CH fails. We shall obtain theorems in the theory ZFC involving β_1-sets which become known results in the theory ZFC + CH about sets of cardinality $\aleph_1$.

The ordinal ω_2, regarded as an ordered set, is not a β_1-set. Thus, if CH be false, then ω_2 is an example of an ordered set of cardinality at most $\mathfrak{c}$ which is not a β_1-set.

Note that Theorem 1.13 asserts that $\mathbf{Q}$ is *universal* in the class of β_1-sets: $\mathbf{Q}$ is a β_1-set, and every β_1-set is order-isomorphic to a subset of $\mathbf{Q}$.

Our isomorphism theorem, below, is deduced from the embedding result, Theorem 1.13, by a rather standard 'back-and-forth' argument. It relies on the simple fact that the countable union of α_1-subsets is also an α_1-subset.

Theorem 1.15 *Let S and T be two β_1-η_1-sets. Then S and T are order-isomorphic. In particular, each β_1-η_1-set is order-isomorphic to $\mathbf{Q}$.*

Proof Set $S = \bigcup\{S_\sigma : \sigma < \omega_1\}$ and $T = \bigcup\{T_\sigma : \sigma < \omega_1\}$, where $\{S_\sigma : \sigma < \omega_1\}$ and $\{T_\sigma : \sigma < \omega_1\}$ are chains of α_1-subsets of S and T, respectively.

We shall construct by transfinite induction a family

$$\{(U_\sigma, V_\sigma, \varphi_\sigma, \psi_\sigma) : \sigma < \omega_1\}$$

such that, for each $\sigma < \omega_1$:

$(\mathrm{i})_\sigma$ U_σ is an α_1-subset of S with $S_\sigma \subset U_\sigma$;

$(\mathrm{ii})_\sigma$ V_σ is an α_1-subset of T with $T_\sigma \subset V_\sigma$;

$(\mathrm{iii})_\sigma$ $\varphi_\sigma : U_\sigma \to V_\sigma$ and $\psi_\sigma : V_\sigma \to S$ are isotonic maps such that $\psi_\sigma \circ \varphi_\sigma$ is the identity on U_σ;

$(\mathrm{iv})_\sigma$ if $\tau < \sigma$, then $U_\tau \subset U_\sigma$, $V_\tau \subset V_\sigma$, $\varphi_\sigma \mid U_\tau = \varphi_\tau$, $\psi_\sigma \mid V_\tau = \psi_\tau$, $\psi_\tau(V_\tau) \subset U_\sigma$, and $\varphi_\sigma \circ \psi_\tau$ is the identity on V_τ.

To begin the inductive construction, take $U_0 = S_0$, and let $\varphi_0 : U_0 \to T$ be an isotonic map. Set $V_0 = \varphi_0(U_0) \cup T_0$. By Lemma 1.12, there is an isotonic map $\psi_0 : V_0 \to S$ such that $\psi_0 \mid \varphi_0(U_0) = \varphi_0^{-1}$. Clearly $(U_0, V_0, \varphi_0, \psi_0)$ satisfies $(\mathrm{i})_0$ – $(\mathrm{iv})_0$.

Now suppose that $\sigma < \omega_1$, and that a family

$$\{(U_\tau, V_\tau, \varphi_\tau, \psi_\tau) : \tau < \sigma\}$$

has been constructed so that $(\mathrm{i})_\tau$–$(\mathrm{iv})_\tau$ hold for each $\tau < \sigma$. Set

$$U'_\sigma = \bigcup\{\psi_\tau(V_\tau) : \tau < \sigma\}$$

and $U_\sigma = U'_\sigma \cup S_\sigma$. Then U_σ is an α_1-set in S (we are using the fact that U'_σ is defined by a countable union) with $S_\sigma \subset U_\sigma$. Define $\varphi'_\sigma : U'_\sigma \to T$ by the formulae

$$\varphi'_\sigma \mid \psi_\tau(V_\tau) = \psi_\tau^{-1} \quad (\tau < \sigma).$$

Then φ'_σ is well defined. Take $\tau < \sigma$. Since $\varphi_\tau(U_\tau) \subset V_\tau$ and

$$\psi_\tau \mid \varphi_\tau(U_\tau) = \varphi_\tau^{-1},$$

we have $U_\tau \subset \psi_\tau(V_\tau)$ and $\psi_\tau^{-1} \mid U_\tau = \varphi_\tau$, and this shows that $U_\tau \subset U'_\sigma$ and that $\varphi'_\sigma \mid U_\tau = \varphi_\tau$. By Lemma 1.12, there is an isotonic map $\varphi_\sigma : U_\sigma \to T$ such that $\varphi_\sigma \mid U'_\sigma = \varphi'_\sigma$. Set

$$V_\sigma = \varphi_\sigma(U_\sigma) \cup T_\sigma,$$

so that V_σ is an α_1-set in T with $T_\sigma \subset V_\sigma$.

For each $\tau < \sigma$, necessarily $\psi_\tau(V_\tau) \subset U_\sigma$ and $\varphi_\sigma \mid \psi_\tau(V_\tau) = \psi_\tau^{-1}$, and so $V_\tau \subset \varphi_\sigma(U_\sigma)$ and $\varphi_\sigma^{-1} \mid V_\tau = \psi_\tau$. By Lemma 1.12 again, there is an isotonic map $\psi_\sigma : V_\sigma \to S$ such that $\psi_\sigma \mid \varphi_\sigma(U_\sigma) = \varphi_\sigma^{-1}$. Clearly the new family

$\{(U_\tau, V_\tau, \varphi_\tau, \psi_\tau) : \tau < \sigma + 1\}$ satisfies (i)$_\tau$–(iv)$_\tau$ for $\tau < \sigma + 1$, and so the induction continues.

Define $\varphi : S \to T$ by the formulae

$$\varphi \mid U_\sigma = \varphi_\sigma \quad (\sigma < \omega_1).$$

Then φ is a well-defined, isotonic map. For each $\sigma < \omega_1$, we have $T_\sigma \subset V_\sigma$ and $\psi_\sigma(V_\sigma) \subset U_{\sigma+1}$, and so $T_\sigma \subset \varphi_{\sigma+1}(U_{\sigma+1}) \subset \varphi(S)$. Hence $T = \varphi(S)$, and φ is the required order-isomorphism. □

It follows that two η_1-sets of cardinality $\aleph_1$ are order-isomorphic.

We shall require the following two results about totally ordered sets. The first, which shows that the property of having an order-dense subset which is a β_1-set is hereditary, will be used at just one point, in Chapter 3.

Proposition 1.16 *Let T be a subset of a totally ordered set S. Suppose that S contains a β_1-set which is order-dense in S. Then T also contains a β_1-set which is order-dense in T.*

Proof Take U to be a β_1-set which is order-dense in S. To show that T contains the required subset, it suffices to suppose that $U \cap T = \emptyset$, noting that the union of two β_1-sets is a β_1-set.

Define

$$V = \{t \in T : (t, \infty) \cap T \quad \text{has a minimum}\};$$

for each $t \in V$, the minimum element of $(t, \infty) \cap T$ is t^+. We first *claim* that either $V = \emptyset$ or V is a β_1-set. For suppose that $V \neq \emptyset$. For each $t \in V$, there exists $u_t \in U$ with $t \le u_t \le t^+$ because U is order-dense in S, and in fact we have $t < u_t < t^+$ because $U \cap T = \emptyset$. It follows that the map

$$t \mapsto u_t, \quad V \to U,$$

is isotonic, and so V is a β_1-set. This establishes the claim.

We write $U = \bigcup\{U_\sigma : \sigma < \omega_1\}$, where $\{U_\sigma : \sigma < \omega_1\}$ is a chain of α_1-subsets of S. For each $\sigma < \omega_1$, $S \setminus U_\sigma$ is the disjoint union of maximal intervals of S in $S \setminus U_\sigma$; form a set T_σ by choosing one point of $T \cap I$ whenever I is a maximal interval of S in $S \setminus U_\sigma$ and $T \cap I \neq \emptyset$. Clearly T_σ is an α_1-subset of T.

Define

$$W = V \cup \bigcup\{T_\sigma : \sigma < \omega_1\}.$$

Then W is a β_1-set of T, and we now *claim* that W is order-dense in T. Indeed, take $t_1, t_2 \in T$ with $t_1 < t_2$. If $t_1 \in V$, then $[t_1, t_2] \cap W \neq \emptyset$. Now suppose that $t_1 \notin V$. Then there exists $t_3 \in T$ with $t_1 < t_3 < t_2$. Since U is order-dense in S and $U \cap T = \emptyset$, there exist $u_1, u_2 \in U$ with

$$t_1 < u_1 < t_3 < u_2 < t_2.$$

Choose $\sigma < \omega_1$ with $u_1, u_2 \in U_\sigma$, and let I be the maximal interval of S in $S \setminus U_\sigma$ with $t_3 \in I$. Then $u_1 << I << u_2$ and $T_\sigma \cap I \neq \emptyset$, and so $(u_1, u_2) \cap T_\sigma \neq \emptyset$, whence $[t_1, t_2] \cap W \neq \emptyset$.

Thus W is the required subset of T. □

The second result is a special case of a theorem of Erdős and Rado.

A *tree* is a partially ordered set $(T, \leq)$ such that, for each $t \in T$, the subset $\{s \in T : s < t\}$ is well-ordered. For example, take σ to be an ordinal and set

$$\mathcal{T}_\sigma = \{0,1\}^{<\sigma} = \bigcup\{\{0,1\}^\tau : \tau < \sigma\},$$

the family of binary sequences of length at most σ. For $\alpha, \beta \in \mathcal{T}_\sigma$, set $\alpha \leq \beta$ if α is a restriction of β. Then $(\mathcal{T}_\sigma, \leq)$ is a tree; it is the *complete* $\{0,1\}$-*tree of height* σ.

Proposition 1.17 *(Erdős–Rado) Let κ be an infinite cardinal, and let $(S, \leq)$ be a totally ordered set of cardinality $(2^\kappa)^+$. Then S contains either a strictly increasing or a strictly decreasing sequence of length κ^+.*

Proof Set $\lambda = \kappa^+$. We shall construct a subtree T of the above tree $\mathcal{T}_\lambda$, and, to each $a \in T$, we shall associate a non-empty interval I_a of S by the following procedure.

Since $|S| \geq 2$, we may divide S into two non-empty intervals S_0 and S_1 such that $S_0 \cap S_1 = \emptyset$ and $S_0 \cup S_1 = S$; the sequences (0) and (1) of length 1 are placed in T. Now suppose that $a \in T$. If $|I_a| = 1$, then no proper extension of a is added to T. If $|I_a| \geq 2$, then I_a is divided into two non-empty intervals $I_{a^\frown 0}$ and $I_{a^\frown 1}$ of S such that $I_{a^\frown 0} \cap I_{a^\frown 1} = \emptyset$ and $I_{a^\frown 0} \cup I_{a^\frown 1} = I_a$, and the two extensions $a^\frown 0$ and $a^\frown 1$ of a are placed in T. If σ is a limit ordinal and $a \in \{0,1\}^\sigma$ is such that $a \upharpoonright \tau \in T$ for each $\tau < \sigma$, then we add a to T if

$$\bigcap\{I_{a\upharpoonright\tau} : \tau < \sigma\} \neq \emptyset$$

(but not otherwise), and set

$$I_a = \bigcap\{I_{a\upharpoonright\tau} : \tau < \sigma\}.$$

We see that, for each $a \in T$ with $l(a) = \sigma$ and each $\tau < \sigma$, necessarily $a \upharpoonright \tau \in T$ (so that T is a tree), and $I_{a\upharpoonright\tau} \subsetneq I_a$.

Assume towards a contradiction that each sequence $a \in T$ has $l(a) \leq \kappa$. Then, for each $s \in S$, there is a unique $a \in \mathcal{T}_\lambda$ such that $\{s\} = I_a$, and hence

$$|S| \leq |\mathcal{T}_\lambda| \leq 2^\kappa \cdot \lambda = 2^\kappa,$$

a contradiction of the fact that $|S| = (2^\kappa)^+$. Thus there exists $a \in T$ with $l(a) = \lambda$. Set $J_\tau = I_{a\upharpoonright\tau}$ $(\tau < \lambda)$.

The family $\{J_\tau : \tau < \lambda\}$ is a strictly decreasing family of non-empty intervals in S. For each $\tau < \lambda$, choose $s_\tau \in J_\tau \setminus J_{\tau+1}$, and set

$$A = \{\tau < \lambda : s_\tau << J_{\tau+1}\}, \qquad B = \{\tau < \lambda : J_{\tau+1} << s_\tau\}.$$

Then $\{s_\tau : \tau \in A\}$ is strictly increasing and $\{s_\tau : \tau \in B\}$ is strictly decreasing in S. For each $\tau < \lambda$, either $\tau \in A$ or $\tau \in B$ because $J_{\tau+1}$ is an interval, and so at least one of these two sequences has length $\lambda = \kappa^+$. □

We now consider ordered groups. Throughout, all groups are abelian and will be written additively; the identity element of a group will be denoted by 0. A group G is *divisible* if, for each $x \in G$ and each $n \in \mathbb{N}$, there exists $y \in G$ with $ny = x$. A map $\psi : G \to H$ between two groups G and H is a *group morphism* if

$$\psi(x + y) = \psi(x) + \psi(y) \quad (x, y \in G).$$

Definition 1.18 *Let $(G, +)$ be a group, and let $\leq$ be a binary relation on G. Then $(G, +, \leq)$ is an* ordered group *if $\leq$ is a partial order on G and if*

$$x + z < y + z$$

whenever $x, y, z \in G$ and $x < y$.

For example, for each non-empty set S, $\mathbb{R}^S$ is an ordered group with respect to addition of functions and the natural order.

Let $(G, \leq)$ be an ordered group. Then we define

$$G^+ = \{x \in G : 0 \leq x\}, \quad G^- = \{x \in G : x \leq 0\}.$$

Now suppose that $(G, \leq)$ is a lattice, so that, for each $x, y \in G$, the pair $\{x, y\}$ has a supremum $x \vee y$ and an infimum $x \wedge y$. For $x \in G$, set

$$x^+ = x \vee 0, \quad x^- = x \wedge 0, \quad |x| = x \vee (-x).$$

Then $x^+ \in G^+$, $x^- \in G^-$, $x = x^+ + x^-$, and $|x| = x^+ - x^-$ for all $x \in G$. Also it is easily verified that

$$|x + y| \leq |x| + |y| \quad (x, y \in G).$$

For example, the ordered group $\mathbb{R}^S$ described above is a lattice with respect to the operations

$$(f \vee g)(s) = f(s) \vee g(s), \quad (f \wedge g)(s) = f(s) \wedge g(s) \quad (s \in S),$$

and f^+, f^-, and $|f|$ coincide with the pointwise definitions of these functions.

Clearly each totally ordered group is a lattice.

Let G be an ordered group which is a lattice. Then a subset S of G is *absolutely convex* if $x \in S$ whenever $x \in G$ and $|x| \leq |y|$ for some $y \in S$. For example, let G be a totally ordered group, and let $a \in G^+$. Then the interval $[-a, a]$ is an absolutely convex subset of G.

Let G be a totally ordered, divisible group. Then, for each $x \in G$ and $n \in \mathbb{N}$, there is a unique element $y \in G$ with $ny = x$, and so there is a map

$$(\alpha, x) \mapsto \alpha x, \quad \mathbb{Q} \times G \to G,$$

with respect to which G is a linear space over $\mathbb{Q}$.

Two totally ordered groups G and H are *isomorphic*, written $G \simeq H$, if there is an order-isomorphism from G onto H which is also a group morphism.

Definition 1.19 *Let $(G, +, \leq)$ be a totally ordered group. The* base number *of G, written $\delta(G)$, is given by*

$$\delta(G) = \operatorname{coi}(G^+ \setminus \{0\}).$$

Proposition 1.20 *Let G be a totally ordered group such that $\delta(G) = \kappa$. Then*

$$|G| \leq w(G)^{\kappa}.$$

Proof Let S be order-dense in G with $|S| = w(G)$, and let $(a_\sigma : \sigma < \kappa)$ be a strictly decreasing, coinitial sequence in $G^+ \setminus \{0\}$. For each $x \in G$ and $\sigma < \kappa$, choose $s_\sigma \in S$ with

$$x - a_\sigma \leq s_\sigma \leq x + a_\sigma.$$

The map $x \mapsto (s_\sigma : \sigma < \kappa)$, $G \to S^\kappa$, is an injection, and so $|G| \leq |S|^\kappa$. □

Corollary 1.21 (CH) *Let G be a totally ordered group with $|G| = \aleph_2$ and $w(G) = \aleph_1$. Then $\delta(G) = \aleph_1$.*

Proof It follows from Proposition 1.20 that $\aleph_1 \leq \delta(G)$ because $\aleph_1^{\aleph_0} = \aleph_1$ with CH. Certainly $\delta(G) \leq w(G) = \aleph_1$. □

Our main tool for understanding totally ordered groups will be *Hahn's embedding theorem*: we shall associate to a totally ordered group G a 'value set' Γ, define a collection $\mathfrak{F}(\mathbb{R}, \Gamma)$ of 'formal power series' over Γ, and embed G in $\mathfrak{F}(\mathbb{R}, \Gamma)$. First we define the value set of G.

Definition 1.22 *Let $(G, +, \leq)$ be a totally ordered group, and let $x, y \in G$. Set:*

(i) $x = \mathbf{o}(y)$ *if* $n\,|x| \leq |y|$ *for all* $n \in \mathbb{N}$;

(ii) $x = \mathbf{O}(y)$ *if* $|x| \leq n\,|y|$ *for some* $n \in \mathbb{N}$;

(iii) $x \sim y$ *if both* $x = \mathbf{O}(y)$ *and* $y = \mathbf{O}(x)$.

Clearly, for each $y \in G$, $\{x \in G : x = \mathbf{o}(y)\}$ and $\{x \in G : x = \mathbf{O}(y)\}$ are absolutely convex subsets of G. It is also clear that $\sim$ is an equivalence relation on G; each equivalence class (other than $\{0\}$) is the union of an interval contained in $G^+ \setminus \{0\}$ and an interval contained in $G^- \setminus \{0\}$.

Definition 1.23 *Let $(G, +, \leq)$ be a totally ordered group. The set*

$$\Gamma = \Gamma_G = (G \setminus \{0\})/\sim$$

of equivalence classes is the value set *of G, and the elements of Γ are the* archimedean classes *of G. The quotient map from $G \setminus \{0\}$ onto Γ is denoted by v: it is the* archimedean valuation *on G. Set $v(x) \leq v(y)$ for $x, y \in G \setminus \{0\}$ if $y = \mathbf{O}(x)$.*

It is easily checked that $\leq$ is well defined on Γ, that $(\Gamma, \leq)$ is a totally ordered set, that

$$v(x + y) \geq \min\{v(x), v(y)\} \quad (x, y \in G), \tag{1.1}$$

and that

$$v(x + y) = \min\{v(x), v(y)\} \text{ if } v(x) \neq v(y) \text{ or if } x, y \in G^+. \tag{1.2}$$

We adjoin an element ∞ to Γ, and adopt the convention that $s < \infty$ for each $s \in \Gamma$, so that $(\Gamma \cup \{\infty\}, \leq)$ is a totally ordered set. We then extend v to a map $v : G \to \Gamma \cup \{\infty\}$ by setting $v(0) = \infty$. The extended map is also referred to as the archimedean valuation. A totally ordered group G is *archimedean* if $|\Gamma_G| \leq 1$.

Let us note the following obvious remark: for an ordered group G with value set Γ,

$$w(G) \geq |\Gamma| \, .$$

Let $(G, +, \leq)$ be a totally ordered group, and consider elements x and y of $G \setminus \{0\}$ with $v(x) < v(y)$. Then $x > 0$ if and only if $x - y > 0$. For, if $x > 0$, then $x = |x| > |y| \geq y$, and so $x - y > 0$, and, if $x - y > 0$, then $x - y = |x - y| \geq |x| - |y| > 2\,|y| - |y| = |y| \geq -y$, and so $x > 0$. In particular,

$$0 < x \leq y \text{ in } G \;\Rightarrow\; v(x) \geq v(y) \text{ in } \Gamma_G \, .$$

Note also that, if $x, y \in G \setminus \{0\}$ with $v(x) = v(y)$ and $x < y$, then v is constant on the interval $[x, y]$ if and only if $0 \notin [x, y]$.

Let G and H be totally ordered groups with archimedean valuations v_G and v_H, respectively, and let $\psi : G \to H$ be an isotonic group morphism. For each $x \in G$, $\psi(|x|) = |\psi(x)|$. Suppose that $x, y \in G$ with $v_G(x) = v_G(y)$. Then $v_H(\psi(x)) = v_H(\psi(y))$.

Let H be a totally ordered group with a subgroup G, and take $x, y \in G$. Then $x = \mathbf{O}(y)$ in G if and only if $x = \mathbf{O}(y)$ in H. Thus the archimedean classes of G are exactly the intersections with G of the archimedean classes of H; we regard Γ_G as an ordered subset of Γ_H, and then $v_H \mid G = v_G$. We shall often write v for either v_G or v_H in this circumstance. If G and H are totally ordered groups with $G \simeq H$, then Γ_G and Γ_H are order-isomorphic, and, in particular, $|\Gamma_G| = |\Gamma_H|$.

One of the themes that we shall explore throughout this work is the relation between gaps in G and in Γ_G. The next proposition is our first result in this area.

Proposition 1.24 *Let G be a totally ordered group with value set Γ. Suppose that Γ has a (κ, λ)-gap with κ and λ infinite. Then G has a (κ, λ)-gap.*

Proof By hypothesis, there is a strictly increasing sequence $(s_\sigma : \sigma < \kappa)$ and a strictly decreasing sequence $(t_\tau : \tau < \lambda)$ in Γ such that

$$s_\sigma < t_\tau \quad (\sigma < \kappa, \ \tau < \lambda)$$

and such that there is no element $s \in \Gamma$ with

$$s_\sigma < s < t_\tau \quad (\sigma < \kappa, \ \tau < \lambda) \, . \tag{1.3}$$

Choose $(x_\sigma : \sigma < \kappa)$ and $(y_\tau : \tau < \lambda)$ in $G^- \setminus \{0\}$ such that

$$v(x_\sigma) = s_\sigma \quad (\sigma < \kappa) \quad \text{and} \quad v(y_\tau) = t_\tau \quad (\tau < \lambda) \, .$$

Clearly $(x_\sigma : \sigma < \kappa)$ is strictly increasing and $(y_\tau : \tau < \lambda)$ is strictly decreasing in G, and these two sequences define a (κ, λ)-pregap, say $\langle A, B \rangle$, in G. We shall show that $\langle A, B\rangle$ is, in fact, a gap. Indeed, assume towards a contradiction that the element z of G interpolates $\langle A, B\rangle$. Then

$$|x_\sigma| > |z| > |y_\tau| \quad (\sigma < \kappa,\ \tau < \lambda)\,,$$

and so

$$v(x_\sigma) \leq v(z) \leq v(y_\tau) \quad (\sigma < \kappa,\ \tau < \lambda)\,.$$

Since κ and λ are infinite cardinals, we have $v(x_\sigma) < v(x_{\sigma+1}) \quad (\sigma < \kappa)$ and $v(y_{\tau+1}) < v(y_\tau) \quad (\tau < \lambda)$, and so

$$s_\sigma < v(z) < t_\tau \quad (\sigma < \kappa,\ \tau < \lambda)\,,$$

a contradiction of (1.3). Thus $\langle A, B\rangle$ is a gap in G. □

We now give a construction of some groups $\mathfrak{F}(\mathbb{F}, S)$ and $\mathfrak{F}_{(\sigma)}(\mathbb{F}, S)$ that originates with Hahn in a monumental paper of 1907.

Let S be a non-empty set, and let $f \in \mathbb{R}^S$. The *support* of f is the set

$$\mathrm{supp} f = \{s \in S : f(s) \neq 0\}\,.$$

Definition 1.25 *Let S be a totally ordered set, let σ be an ordinal, and let $\mathbb{F} = \mathbb{Q}$, $\mathbb{R}$, or $\mathbb{C}$. Then:*

$$\mathfrak{F}(\mathbb{F}, S) = \{f \in \mathbb{F}^S : \ \mathrm{supp} f \ \text{ is well-ordered}\}\,;$$
$$\mathfrak{F}_{(\sigma)}(\mathbb{F}, S) = \{f \in \mathfrak{F}(\mathbb{F}, S) : |\,\mathrm{supp} f| < \aleph_\sigma\}\,.$$

In particular, $\mathfrak{F}_{(0)}(\mathbb{F}, S)$ is the set of functions in $\mathbb{F}^S$ of finite support, and $\mathfrak{F}_{(1)}(\mathbb{F}, S)$ is the set of elements of $\mathbb{F}^S$ with countable, well-ordered support. Since

$$\mathrm{supp}(f - g) \subset \mathrm{supp}\, f \cup \mathrm{supp}\, g \quad (f, g \in \mathbb{F}^S)\,,$$

$\mathfrak{F}(\mathbb{F}, S)$ and each $\mathfrak{F}_{(\sigma)}(\mathbb{F}, S)$ are subgroups of $\mathbb{F}^S$. Indeed, each of these sets is an $\mathbb{F}$-linear space, and, in particular, is a divisible group. Since each well-ordered subset of an α_1-set is countable, $\mathfrak{F}_{(1)}(\mathbb{F}, S) = \mathfrak{F}(\mathbb{F}, S)$ for each α_1-set S.

To give an example, let us consider the case where S is the real line $\mathbb{R}$ (with the usual order) and $\mathbb{F} = \mathbb{R}$. A countable, well-ordered subset of $\mathbb{R}$ has the form $\{s_n : n \in \mathbb{N}\}$ for an increasing sequence (s_n), and $\mathfrak{F}_{(1)}(\mathbb{R}, \mathbb{R}) \setminus \{0\}$ is the collection of functions $f : \mathbb{R} \to \mathbb{R}$ such that $f(s_n) \in \mathbb{R} \setminus \{0\} \quad (n \in \mathbb{N})$ for such a sequence (s_n) and $f(s) = 0$ otherwise.

Definition 1.26 *For $f \in \mathfrak{F}(\mathbb{F}, S)$ with $f \neq 0$, set*

$$v(f) = \inf \mathrm{supp}\, f\,.$$

The map v is the Hahn valuation *on $\mathfrak{F}(\mathbb{F}, S)$. In the case where $\mathbb{F} = \mathbb{R}$ or $\mathbb{Q}$, set*

$$f > 0 \quad \text{if} \quad f(v(f)) > 0 \quad \text{in} \quad \mathbb{F}\,.$$

That is, we specify that f be positive if it is positive at the first, or starting, point of its support. Then

$$(\mathfrak{F}(\mathbb{F}, S),\, +\,,\, \leq)$$

is a totally ordered group, and $f \sim g$ in $\mathfrak{F}(\mathbb{F}, S)$ if and only if $v(f) = v(g)$. Thus we can identify S as the value set of $\mathfrak{F}(\mathbb{F}, S)$ and of each $\mathfrak{F}_{(\sigma)}(\mathbb{F}, S)$, and we can identify the Hahn valuation v with the archimedean valuation.

Definition 1.27 *Let S be a totally ordered set. Then $(\mathfrak{F}(\mathbb{R}, S), +, \leq)$ is the* Hahn group *of S.*

We introduce the notation δ_s for the characteristic function of the set $\{s\}$ for each $s \in S$. If $s < t$ in S, then $\delta_s > \delta_t$ in $\mathfrak{F}(\mathbb{R}, S)$, and so the map $s \mapsto \delta_s$, $S \to \mathfrak{F}(\mathbb{R}, S)$, is anti-isotonic; the map $s \mapsto -\delta_s$, $S \to \mathfrak{F}(\mathbb{R}, S)$, is isotonic.

A stronger form of part (ii) of the following proposition will be given in Theorem 1.40(ii).

We denote by $[S]^\omega$ the family of countable subsets of a set S.

Proposition 1.28 *Let S be a totally ordered set.*

(i) *Suppose that $|S| = \kappa \geqslant \mathfrak{c}$. Then $\left|\mathfrak{F}_{(1)}(\mathbb{R}, S)\right| = \kappa$.*

(ii) *Suppose that $|S| = \kappa \geq \aleph_0$. Then $|\mathfrak{F}(\mathbb{R}, S)| \leq 2^\kappa$.*

(iii) *Suppose that* $\operatorname{coi} S = \kappa$. *Then* $\operatorname{cof} \mathfrak{F}_{(1)}(\mathbb{R}, S) = \operatorname{cof} \mathfrak{F}(\mathbb{R}, S) = \kappa$.

(iv) *Suppose that* $\operatorname{cof} S = \kappa$. *Then* $\delta(\mathfrak{F}_{(1)}(\mathbb{R}, S)) = \delta(\mathfrak{F}(\mathbb{R}, S)) = \kappa$.

Proof (i) For each $T \in [S]^\omega$, we have $\left|\mathbb{R}^T\right| \leq \mathfrak{c}^{\aleph_0} = \mathfrak{c}$. There is an obvious injection from $[S]^\omega$ into $S^{\mathbb{N}}$, and so $|[S]^\omega| \leq \kappa^{\aleph_0} = \kappa$. Thus

$$\left|\mathfrak{F}_{(1)}(\mathbb{R}, S)\right| \leq \mathfrak{c} \cdot \kappa = \kappa\,.$$

Since there is an injective map from S into $\mathfrak{F}_{(1)}(\mathbb{R}, S)$, we have

$$\left|\mathfrak{F}_{(1)}(\mathbb{R}, S)\right| \geq \kappa\,.$$

(ii) For each $T \in \mathcal{P}(S)$, we have $\left|\mathbb{R}^T\right| \leq 2^\kappa$, and $|\mathcal{P}(S)| = 2^\kappa$, and so $|\mathfrak{F}(\mathbb{R}, S)| \leq 2^\kappa \cdot 2^\kappa = 2^\kappa$.

(iii) Let $(s_\sigma : \sigma < \kappa)$ be a coinitial sequence in S. Then $(\delta_{s_\sigma} : \sigma < \kappa)$ is cofinal in both $\mathfrak{F}_{(1)}(\mathbb{R}, S)$ and $\mathfrak{F}(\mathbb{R}, S)$.

(iv) Let $(s_\sigma : \sigma < \kappa)$ be a cofinal sequence in S. Then $(\delta_{s_\sigma} : \sigma < \kappa)$ is coinitial in both $\mathfrak{F}_{(1)}(\mathbb{R}, S)^+ \setminus \{0\}$ and $\mathfrak{F}(\mathbb{R}, S)^+ \setminus \{0\}$. □

As a preliminary to Hahn's embedding theorem, we shall show how to 'fill a gap' in an ordered group.

Let $(G, +, \leq)$ be a totally ordered, divisible group. We consider the set

$$G \times \mathbb{Q}$$

with the group operation

$$(x, \alpha) + (y, \beta) = (x + y, \alpha + \beta) \qquad (x, y \in G,\ \alpha, \beta \in \mathbb{Q})\,.$$

Then $G \times \mathbb{Q}$ is clearly a divisible group. We identify G with the subgroup $G \times \{0\}$ of $G \times \mathbb{Q}$. Let A be a subset of G, and take $x \in G$ and $\alpha \in \mathbb{Q}$. Then we define

$$x + \alpha A = \{x + \alpha a : a \in A\}\,. \tag{1.4}$$

If $\langle A, B\rangle$ is a gap in G and if $x \in G$ and $\alpha \in \mathbb{Q}^+$, then $\langle x + \alpha A,\ x + \alpha B\rangle$ is a gap in G; it is a 'translated and rescaled' version of the original gap. We describe an order on $G \times \mathbb{Q}$ as follows. Intuitively, we arrange that $(0, 1)$ interpolates a gap $\langle A, B\rangle$ by saying that $(0, 1) > (a, 0)$ for each $a \in A$ and $(0, 1) < (b, 0)$ for each $b \in B$. More generally $(x, \alpha) > (0, 0)$ in $G \times \mathbb{Q}$ (in the case where $\alpha \geq 0$) if 0 is 'below' the gap $\langle x + \alpha A, x + \alpha B\rangle$. The precise definition of the order on $G \times \mathbb{Q}$ is as follows.

Definition 1.29 *Let $(G, +, \leq)$ be a totally ordered, divisible group, and let $\langle A, B\rangle$ be a gap in G. Then*

$$(x, \alpha) > (0, 0) \quad \text{in} \quad G \times \mathbb{Q}$$

if $\alpha > 0$ and there exists $a \in A$ such that $x + \alpha a \geq 0$, or if $\alpha < 0$ and there exists $b \in B$ such that $x + \alpha b \geq 0$, or if $\alpha = 0$ and $x > 0$.

We have $(x, \alpha) > (y, \beta)$ in $G \times \mathbb{Q}$ if $(x - y, \alpha - \beta) > (0, 0)$ in $G \times \mathbb{Q}$. It is easily checked that $(G \times \mathbb{Q}, +, \leq)$ is a totally ordered, divisible group; it is denoted by

$$G_{\langle A,B\rangle}\,.$$

The natural embedding of G in $G_{\langle A,B\rangle}$ is isotonic, and $A << (0, 1) << B$ in $G_{\langle A,B\rangle}$.

Now suppose that G is a proper, divisible subgroup of a totally ordered, divisible group $\overline{G}$. Take $x \in \overline{G} \setminus G$, and suppose that x defines the gap $\langle A, B\rangle$ in G. Then we may identify the subgroup $G \oplus \mathbb{Q}x$ of $\overline{G}$ with $G \times \mathbb{Q}$, and the relative ordering on $G \oplus \mathbb{Q}x$ from $\overline{G}$ coincides with the ordering on $G_{\langle A,B\rangle}$ defined above.

Definition 1.30 *Let G be a totally ordered group, and let $\langle A, B\rangle$ be a pregap in G. Then $\langle A, B\rangle$ is a* pseudo-gap *if, for each $x \in G$, there exist $a \in A$ and $b \in B$ such that v is constant on the interval $[a - x,\ b - x]$ in G. The group G is* pseudo-complete *if there are no pseudo-gaps.*

For example, set

$$A = \{r \in \mathbb{Q}^+ : r^2 < 2\}\,, \quad B = \{r \in \mathbb{Q}^+ : r^2 > 2\}\,.$$

Then $\langle A, B\rangle$ is a pseudo-gap in $(\mathbb{Q}, +)$, and so $(\mathbb{Q}, +)$ is not pseudo-complete.

In fact, every pseudo-gap $\langle A, B\rangle$ in a totally ordered group G is a gap. For assume that $x \in G$ with $A << x << B$. Then, for each $a \in A$ and $b \in B$, we have $a - x < 0 < b - x$, but $v(a - x) \neq v(0)$, and so v is not constant on $[a - x,\ b - x]$, a contradiction.

Suppose that $\langle A, B\rangle$ is a pseudo-gap, and that $\langle A, B\rangle$ is a (κ, λ)-gap. Then necessarily κ and λ are infinite. For assume that A has a maximum element a_0. By applying the definition with $x = a_0$, we see that there exist $a \in A$ and $b \in B$ with v constant on $[a - a_0,\ b - a_0]$. But $0 \in [a - a_0,\ b - a_0]$, a contradiction.

A pseudo-gap is a special type of gap. The terminology in the above definition is a little unfortunate because it suggests, incorrectly, that there are pseudo-gaps which are not gaps. The reason for our choice of the word is that our pseudo-gaps correspond to the 'pseudo-convergent sequences which are not convergent' of the seminal paper of Kaplansky (1942), and so our 'pseudo-complete ordered fields' will coincide with the same class in the existing literature.

Intuitively, a pseudo-gap is a gap $\langle A, B\rangle$ such that there is a cofinal subset A_0 of A and a coinitial subset B_0 of B with $\langle A_0, B_0\rangle$ contained in an interval within one archimedean class, and such that the same is true of any translate of $\langle A, B\rangle$.

The first result about pseudo-gaps is that they have a certain 'upwards persistence'.

Proposition 1.31 *Let G be a subgroup of a totally ordered group H, and suppose that $\langle A, B\rangle$ is a pseudo-gap in G which is a gap in H. Then $\langle A, B\rangle$ is a pseudo-gap in H.*

Proof Take $x \in H$, say $x > 0$. Since $\langle A, B\rangle$ is a gap in H, either there exists $a_0 \in A$ with $a_0 \geq x$ or there exists $b_0 \in B$ with $b_0 \leq x$; suppose, in fact, that the former case occurs.

Since $\langle A, B\rangle$ is a pseudo-gap in G, there exist $a \in A$ and $b \in B$ such that v is constant on the interval $[a - a_0,\, b - a_0] \cap G$ of G. It follows immediately that $v(a - a_0) = v(b - a_0)$ and that $0 \notin [a - a_0,\, b - a_0]$. Thus there exists $n \in \mathbb{N}$ with

$$0 < a - a_0 < b - a_0 \leq n(a - a_0)\,.$$

If $a_0 - x \leq a - a_0$, then

$$b - x = (b - a_0) + (a_0 - x) \leq (n+1)(a - a_0) \leq (n+1)(a - x)\,,$$

and, on the other hand, if $a - a_0 \leq a_0 - x$, then

$$b - x \leq (n+1)(a_0 - x) < (n+1)(a - x)\,,$$

and so, in either case,

$$0 < a - x < b - x \leq (n+1)(a - x)\,.$$

It follows that $v(b - x) = v(a - x)$ and that $0 \notin [a - x,\, b - x]$, and hence v is constant on the interval $[a - x,\, b - x]$ in H.

A similar argument applies in the case where $b_0 \leq x$ for some $b_0 \in B$.

Thus $\langle A, B\rangle$ is a pseudo-gap in H. □

The next proposition gives a key property of pseudo-gaps. It shows that pseudo-gaps are exactly those gaps which are 'value-preserving' when the gap is interpolated by a new element.

Proposition 1.32 *Let $\langle A, B\rangle$ be a gap in a totally ordered, divisible group G, and set $H = G_{\langle A,B\rangle}$. Then $\langle A, B\rangle$ is a pseudo-gap in G if and only if $\Gamma_H = \Gamma_G$.*

Proof Suppose that $\langle A, B\rangle$ is a pseudo-gap, and take $(x, 1) \in H$. Then there exist $a \in A$ and $b \in B$ such that v is constant on $[a + x,\, b + x]$ in G. Since $(a, -1) < (0, 0)$ in H, we have $(a + x, 0) < (x, 1)$ in H. Similarly we have

$(x, 1) < (b + x, 0)$ in H. Since $v(a + x) = v(b + x)$ and $0 \notin [a + x,\, b + x]$, it follows that $v((x, 1)) = v(a + x) \in \Gamma_G$. Since $v((x, \alpha)) = v((x/\alpha, 1))$ for each $\alpha \in \mathbb{Q} \setminus \{0\}$, we see that $v((x, \alpha)) \in \Gamma_G$ in each case, and so $\Gamma_H = \Gamma_G$.

Conversely, suppose that $\Gamma_H = \Gamma_G$. Take $x \in G$. Then there exists $c \in G$ with $v((x, 1)) = v(c)$, say $c \in G^+ \setminus \{0\}$. Take $\alpha, \beta \in \mathbb{Q}^+ \setminus \{0\}$ with

$$\alpha c < (x, 1) < \beta c$$

in H. Since $(x - \alpha c, 1) > (0, 0)$, there exists $a \in A$ with $a + x \geq \alpha c$, and, similarly, there exists $b \in B$ with $b + x \leq \beta c$. This implies that v is constant on the interval $[a + x,\, b + x]$, and so $\langle A, B\rangle$ is a pseudo-gap. □

Theorem 1.33 *Let S be a totally ordered set. Then the group $\mathfrak{F}(\mathbb{R}, S)$ is pseudo-complete.*

Proof Set $G = \mathfrak{F}(\mathbb{R}, S)$. To obtain a contradiction, assume that there is a pseudo-gap $\langle A, B\rangle$ in G. For convenience, we may suppose that

$$A \cup B \subset G^+ \setminus \{0\}\,.$$

We shall inductively define elements $s_\sigma \in S$ and $\alpha_\sigma \in \mathbb{R}$ for ordinals σ such that $s_\rho > s_\sigma$ whenever $\rho > \sigma$, and such that there exist $f_\sigma \in A$, $g_\sigma \in B$, and $h_\sigma \in G$ with $v(f) = s_\sigma$ for $f \in [f_\sigma - h_\sigma,\, g_\sigma - h_\sigma]$ and $v(h_\rho - h_\sigma) \geq s_\sigma$ whenever $\rho > \sigma$.

Initially take $f_0 \in A$ and $g_0 \in B$ such that v is constant on $[f_0, g_0]$, say $v(f) = s_0$ for $f \in [f_0, g_0]$, and take $h_0 = 0$.

Suppose now that $f_\sigma, g_\sigma, h_\sigma$, and s_σ have been defined. Clearly there exists $\alpha_\sigma \in \mathbb{R}$ such that

$$\begin{cases} f(s_\sigma) \leq \alpha_\sigma & (f \in [f_\sigma - h_\sigma,\, g_\sigma - h_\sigma] \cap A)\,, \\ g(s_\sigma) \geq \alpha_\sigma & (g \in [f_\sigma - h_\sigma,\, g_\sigma - h_\sigma] \cap B)\,. \end{cases}$$

Set $h_{\sigma+1} = h_\sigma + \alpha_\sigma \delta_{s_\sigma}$, and choose $f_{\sigma+1} \in A$ and $g_{\sigma+1} \in B$ such that v is constant on

$$[f_{\sigma+1} - h_{\sigma+1},\, g_{\sigma+1} - h_{\sigma+1}]\,;$$

we may suppose that $f_{\sigma+1} \geq f_\sigma$ and that $g_{\sigma+1} \leq g_\sigma$, so that

$$v(f_{\sigma+1} - h_\sigma) = v(g_{\sigma+1} - h_\sigma) = s_\sigma\,.$$

Assume that $f_{\sigma+1}(s_\sigma) < \alpha_\sigma$. It cannot be that $g_{\sigma+1}(s_\sigma) = \alpha_\sigma$, for, if this were true, we would have

$$v(g_{\sigma+1} - h_{\sigma+1}) > s_\sigma = v(f_{\sigma+1} - h_{\sigma+1})\,,$$

and this is not the case. But now $g_{\sigma+1}(s_\sigma) > \alpha_\sigma$ and

$$0 \in [f_{\sigma+1} - h_{\sigma+1},\, g_{\sigma+1} - h_{\sigma+1}]\,,$$

a contradiction. We conclude that $f_{\sigma+1}(s_\sigma) = \alpha_\sigma$. Similarly we also have $g_{\sigma+1}(s_\sigma) = \alpha_\sigma$. Set

$$s_{\sigma+1} = v(f_{\sigma+1} - h_{\sigma+1})\,.$$

Then $s_{\sigma+1} > s_\sigma$ and $v(f) = s_{\sigma+1}$ for $f \in [f_{\sigma+1} - h_{\sigma+1},\, g_{\sigma+1} - h_{\sigma+1}]$, and clearly $v(h_\rho - h_\tau) \geq s_\tau$ whenever $\tau < \rho \leq \sigma + 1$.

Suppose next that σ is a limit ordinal and that $f_\zeta, g_\zeta, h_\zeta$, and s_ζ have been defined for $\zeta < \sigma$. Define h_σ by setting $h_\sigma(s) = h_\zeta(s)$ if $s < s_\zeta$ for some $\zeta < \sigma$ and $h_\sigma(s) = 0$ if $s > s_\zeta$ for each $\zeta < \sigma$. Then h_σ is well defined, $h_\sigma \in G$, and $v(h_\sigma - h_\zeta) \geq s_\zeta \quad (\zeta < \sigma)$. By hypothesis, there exist $f_\sigma \in A$ and $g_\sigma \in B$ such that v is constant on the interval $[f_\sigma - h_\sigma,\, g_\sigma - h_\sigma]$, say $v(f_\sigma - h_\sigma) = s_\sigma$. Assume towards a contradiction that there exists $\zeta < \sigma$ such that $s_\sigma < s_{\zeta+1}$. If $f_\sigma(s_\sigma) > h_{\zeta+1}(s_\sigma) = h_\sigma(s_\sigma)$, then $f_\sigma > g_{\zeta+1}$, which is not possible. So $f_\sigma(s_\sigma) < h_{\zeta+1}(s_\sigma)$, and hence $f_\sigma < f_{\zeta+1}$. Thus $f_{\zeta+1} - h_\sigma \in [f_\sigma - h_\sigma,\, g_\sigma - h_\sigma]$, a contradiction of the fact that

$$v(f_{\zeta+1} - h_\sigma) = v(f_{\zeta+1} - h_{\zeta+1}) = s_{\zeta+1} > s_\sigma = v(f_\sigma - h_\sigma)\,.$$

Thus $s_\sigma \geq s_{\zeta+1} > s_\zeta$ for all $\zeta < \sigma$.

This completes the inductive definition. But clearly there is no strictly increasing sequence in S of length greater than $|S|$, and so we have the required contradiction. □

The converse of Theorem 1.33 holds in the following sense: if G is a proper, divisible subgroup of $\mathfrak{F}(\mathbb{R}, S)$ such that $\delta_s \in G$ for each $s \in G$, then G is not pseudo-complete. For take $f \in \mathfrak{F}(\mathbb{R}, S)\backslash G$, and let $\langle A, B\rangle$ be the gap in G defined by f. Set $H = G_{\langle A,B\rangle}$, identified with $G \oplus \mathbb{Q}f$ in $\mathfrak{F}(\mathbb{R}, S)$. Then $\Gamma_H = \Gamma_G = S$, and so, by Proposition 1.32, $\langle A, B\rangle$ is a pseudo-gap.

Definition 1.34 *Let $(G, +, \leq)$ and $(H, +, \leq)$ be totally ordered groups such that Γ_G is subset of Γ_H and the restriction of the order on Γ_H to Γ_G is the given order on Γ_G. A map $\psi : G \to H$ is* completely isotonic *if it is isotonic and if $v(\psi(x)) = v(x) \quad (x \in G)$.*

Theorem 1.35 *(Hahn's embedding theorem) Let G be a totally ordered, divisible group with value set Γ. Suppose that G_0 is a divisible subgroup of G with value set Γ_0, and that*

$$\psi_0 : G_0 \to \mathfrak{F}(\mathbb{R}, \Gamma_0) \subset \mathfrak{F}(\mathbb{R}, \Gamma)$$

is a completely isotonic group morphism. Then there is a completely isotonic group morphism $\psi : G \to \mathfrak{F}(\mathbb{R}, \Gamma)$ such that $\psi \mid G_0 = \psi_0$ and such that $\delta_s \in \psi(G)$ for each $s \in \Gamma \setminus \Gamma_0$.

Proof Consider the family $\mathcal{F}$ of pairs (H, φ) such that H is a divisible subgroup of G with $H \supset G_0$ and $\varphi : H \to \mathfrak{F}(\mathbb{R}, \Gamma_H)$ is a completely isotonic group morphism such that $\varphi \mid G_0 = \psi_0$ and such that

$$\delta_s \in \psi(H) \quad (s \in \Gamma_H \setminus \Gamma_0)\,.$$

Clearly $\mathcal{F}$ has a maximal element for the obvious ordering $\preccurlyeq$ in $\mathcal{F}$ (similar to the one in Lemma 1.12), say the maximal element is (H, φ).

We *claim* that $H = G$. To obtain a contradiction, assume that $H \neq G$, and take $x \in G \setminus H$. Then the gap defined by x in H is $\langle A, B\rangle$, say.

First, suppose that $\langle A, B\rangle$ is a pseudo-gap in H. Then $\langle\varphi(A), \varphi(B)\rangle$ is a pseudo-gap in $\varphi(H) \subset \mathfrak{F}(\mathbb{R}, \Gamma_H)$. It cannot be that $\langle\varphi(A), \varphi(B)\rangle$ is a gap in $\mathfrak{F}(\mathbb{R}, \Gamma_H)$, for, if it were, it would be a pseudo-gap by Proposition 1.31, and this is impossible because, by Theorem 1.33, $\mathfrak{F}(\mathbb{R}, \Gamma_H)$ is pseudo-complete. Thus there exists $f \in \mathfrak{F}(\mathbb{R}, \Gamma_H)$ with

$$\varphi(A) << f << \varphi(B)\,.$$

Define

$$\overline{\varphi} : y + \alpha x \mapsto \varphi(y) + \alpha f\,, \quad H \oplus \mathbb{Q}x \to \mathfrak{F}(\mathbb{R}, \Gamma_H)\,.$$

Then $\overline{\varphi}$ is a completely isotonic group morphism, and $(H \oplus \mathbb{Q}x, \overline{\varphi}) \succcurlyeq (H, \varphi)$ in $\mathcal{F}$, a contradiction of the supposed maximality of (H, φ).

Second, suppose that $\langle A, B\rangle$ is not a pseudo-gap in H. By Proposition 1.32, there exist $s \in \Gamma \setminus \Gamma_H$ and $y_0 + \alpha_0 x \in H \oplus \mathbb{Q}x$ with $v(y_0 + \alpha_0 x) = s$; necessarily $\alpha_0 \neq 0$. Set $\overline{\varphi}(y_0 + \alpha_0 x) = \delta_s$, so that

$$\overline{\varphi}(x) = (\delta_s - \varphi(y_0))/\alpha_0\,,$$

and extend $\overline{\varphi}$ to $H \oplus \mathbb{Q}x$ as before. The value set of $H \oplus \mathbb{Q}x$ is $\Gamma_H \cup \{s\}$, and so $(H \oplus \mathbb{Q}x, \overline{\varphi}) \in \mathcal{F}$, again a contradiction.

Thus $H = G$, and $\varphi : H \to \mathfrak{F}(\mathbb{R}, \Gamma)$ is the required map. □

Corollary 1.36 *Let G be a totally ordered, divisible group with value set Γ. Then there is a completely isotonic group morphism $\varphi : G \to \mathfrak{F}(\mathbb{R}, \Gamma)$ such that*

$$\mathfrak{F}_{(0)}(\mathbb{R}, \Gamma) \subset \varphi(G) \subset \mathfrak{F}(\mathbb{R}, \Gamma)\,. \qquad \square$$

In fact, it is straightforward to see that every totally ordered group can be embedded in a totally ordered, divisible group, and hence can be embedded in a group of the form $\mathfrak{F}(\mathbb{R}, \Gamma)$.

We next develop a theory of α_1-, β_1-, and η_1-groups analogous to the theory of α_1-, β_1-, and η_1-sets, which was previously explained.

Definition 1.37 *Let $(G, +, \leq)$ be a totally ordered group. Then G is:*

(i) *an* α_1-group *(respectively,* α_2-group*) if $(G, \leq)$ is an α_1-set (respectively, α_2-set);*

(ii) *a* β_1-group *if $G = \bigcup\{G_\nu : \nu < \rho\}$, where $\{G_\nu : \nu < \rho\}$ is a chain of α_1-subgroups;*

(iii) *an* η_1-group *if $(G, \leq)$ is an η_1-set;*

(iv) *a* semi-η_1-group *if $(G, \leq)$ is a semi-η_1-set.*

Let G be a divisible β_1-group. Then we can write $G = \bigcup\{G_\nu : \nu < \rho\}$, where $\{G_\nu : \nu < \rho\}$ is a chain of divisible α_1-subgroups.

Let G be a totally ordered group with $|G| \leq \aleph_1$. Since the subgroup of G generated by a countable subset is countable, G is a β_1-group. Thus the theory of β_1-groups becomes, in ZFC + CH, the theory of totally ordered groups of cardinality at most $\aleph_1$.

When working in the theory ZFC (without CH), some care is required with the notion of a β_1-group: it is not obvious that a totally ordered group which is a β_1-set is always a β_1-group, and we think that this is unlikely to be true.

Let G be a semi-η_1-group with value set Γ, and suppose that Γ is infinite. Then, for each $s, t \in \Gamma$ with $s < t$, the interval (s,t) in Γ is an η_1-set. To see this, assume towards a contradiction that $\langle A, B\rangle$ is a countable gap, say it is a (κ, λ)-gap. By Proposition 1.24, at least one of κ and λ is finite, say $\kappa = 0$ or $\kappa = 1$. Suppose in addition that λ is infinite. If $\kappa = 0$, take $s_0 = s$; if $\kappa = 1$, take $s_0 = \max A$. Then there exists $x_0 \in G^+$ and (y_n) contained in G^+ with $v(x_0) = s_0$ and $(v(y_n))$ coinitial in B. Since G is a semi-η_1-group, there exists $z \in G^+$ with $y_m < z < x_0/n \quad (m, n \in \mathbb{N})$, and so $A << v(z) << B$ in Γ. The other cases are covered similarly.

In particular, Γ_G is a semi-η_1-set whenever G is a semi-η_1-group; also Γ_G is an α_1-set or β_1-set whenever G is an α_1-group or β_1-group, respectively.

Let $G = \mathfrak{F}_{(1)}(\mathbb{R}, \omega_2)$. Then G is not a β_1-group because ω_2 is not a β_1-set. However, in the case where $\mathfrak{c} \geq \aleph_2$, we see that $|G| = \mathfrak{c}$. Thus, if CH fails, then G is an example of a totally ordered group of cardinality at most $\mathfrak{c}$ which is not a β_1-group.

The following result classifies the value sets of a certain class of groups; recall that sets such as $\mathbb{N} \times \mathbf{Q}$ are supposed to have the lexicographic order.

Theorem 1.38 *Let G be an infinite, divisible semi-η_1-group which is a β_1-group. Then the value set Γ_G of G is order-isomorphic to exactly one of the following list of totally ordered sets:*

(1) $\{1\}$; (2) $\mathbf{Q}$; (3) $\{1\} \odot \mathbf{Q}$; (4) $\mathbf{Q} \odot \{1\}$;

(5) $\{1\} \odot \mathbf{Q} \odot \{1\}$; (6) $\mathbb{N} \times \mathbf{Q}$; (7) $\mathbb{Z}^- \times \mathbf{Q}$;

(8) $\mathbb{Z} \times \mathbf{Q}$; (9) $\{1\} \odot (\mathbb{N} \times \mathbf{Q})$; (10) $(\mathbb{Z}^- \times \mathbf{Q}) \odot \{1\}$.

Proof Since G is a semi-η_1-group, the above remarks show that there are no $(\aleph_0, \aleph_0)$-gaps, no $(1, \aleph_0)$-gaps, no $(\aleph_0, 1)$-gaps, and no $(1,1)$-gaps in Γ. Thus the only possible countable gaps in Γ are $(1,0)$-gaps, $(0,1)$-gaps, $(\aleph_0, 0)$-gaps, and $(0, \aleph_0)$-gaps.

Suppose that Γ_G is finite. Then $\Gamma_G = \{1\}$, giving possibility (1).

Now suppose that Γ_G is infinite. If $\operatorname{cof} \Gamma_G = \operatorname{coi} \Gamma_G \geq \aleph_1$, then Γ_G is an η_1-set, and so $\Gamma_G \simeq \mathbf{Q}$ by Theorem 1.15, giving possibility (2). If $\operatorname{cof} \Gamma_G \geq \aleph_1$ and $\operatorname{coi} \Gamma_G = 1$, then Γ_G has a minimum element, say γ, and $\Gamma_G \setminus \{\gamma\}$ is a β_1-η_1-set, whence $\Gamma_G \simeq \{1\} \odot \mathbf{Q}$, giving possibility (3). If $\operatorname{cof} \Gamma_G \geq \aleph_1$ and $\operatorname{coi} \Gamma_G = \aleph_0$, then $\Gamma_G \simeq \mathbb{Z}^- \times \mathbf{Q}$, giving possibility (7). The other possibilities arise from the various other permitted values of $\operatorname{cof} \Gamma_G$ and $\operatorname{coi} \Gamma_G$. □

We shall see later that the above ten possibilities do arise.

The next theorem shows that various properties of totally ordered sets can be 'transmitted upwards' to $\mathfrak{F}(\mathbb{R}, S)$ or $\mathfrak{F}_{(1)}(\mathbb{R}, S)$; it will be a key step when we construct more complicated algebraic structures with desirable order properties.

Theorem 1.39 *Let S be a non-empty, totally ordered set.*

(i) *Suppose that S is an η_1-set. Then $\mathfrak{F}(\mathbb{R}, S)$ is an η_1-group, and $\mathfrak{F}_{(\sigma)}(\mathbb{R}, S)$ is an η_1-group for each $\sigma \geq 1$.*

(ii) *Suppose that κ is an infinite cardinal and that S contains no strictly increasing or strictly decreasing sequence of length κ^+. Then $\mathfrak{F}(\mathbb{R}, S)$ contains no strictly increasing or strictly decreasing sequence of length κ^+, and*

$$|\mathfrak{F}(\mathbb{R}, S)| \leq 2^{\kappa}.$$

(iii) *Suppose that S is an α_1-set (respectively, an α_2-set). Then $\mathfrak{F}(\mathbb{R}, S)$ is an α_1-group (respectively, an α_2-group).*

(iv) *Suppose that S is a β_1-set. Then $\mathfrak{F}_{(1)}(\mathbb{R}, S)$ is a β_1-group.*

Proof The Hahn valuation on $\mathfrak{F}(\mathbb{R}, S)$ is denoted by v.

(i) We shall prove that $\mathfrak{F}_{(1)}(\mathbb{R}, S)$ is an η_1-group; the other cases are proved by essentially the same argument. Thus, set $G = \mathfrak{F}_{(1)}(\mathbb{R}, S)$, and let A and B be countable subsets of G with $A << B$.

Suppose first that A and B are non-empty, that A has no maximum, and that B has no minimum, so that $\langle A, B \rangle$ is an $(\aleph_0, \aleph_0)$-pregap. Then there exist a cofinal sequence (f_m) in A and a coinitial sequence (g_n) in B with

$$f_m < f_{m+1} < g_{n+1} < g_n \quad (m, n \in \mathbb{N}).$$

Set $s_n = v(f_n - g_n) \quad (n \in \mathbb{N})$, so that (s_n) is increasing in S, and set

$$T = \bigcup \{(-\infty, s_n) : n \in \mathbb{N}\}.$$

Define $h \in \mathbb{R}^S$ by setting $h(s) = f_m(s) = g_m(s)$ if $s < s_m$ for some $m \in \mathbb{N}$, and $h(s) = 0$ if $s \geq s_m$ for all $m \in \mathbb{N}$. Then h is well defined on T, and $\operatorname{supp} h$ is a countable, well-ordered subset of T, and so $h \in G$.

Fix $m \in \mathbb{N}$. Suppose that there exists $k > m$ such that $v(f_k - f_m) \in T$, say $v(f_k - f_m) = s'_k < s_n$ for some $n \geq k$. Set $t = v(f_n - f_m)$. Clearly, $t < s_n$ (for, if $t \geq s_n$, then f_m and f_n agree on $(-\infty, s'_k]$ and

$$f_k(s'_k) > f_m(s'_k) = f_n(s'_k),$$

whence $f_k > f_n$, a contradiction of the fact that $n \geq k$ and $f_n \geq f_k$) and $h(s) = f_n(s) = f_m(s)$ for $s < t$. However $h(t) = f_n(t) > f_m(t)$. Thus $h > f_m$. Similarly, if $v(g_k - g_m) \in T$ for some $k > m$, then $h < g_m$.

We have so far found $h \in G$ with $A << h << B$ except in either of the two cases:

(a) there exists $m \in \mathbb{N}$ with $f_k \mid T = f_m \mid T \quad (k \geq m)$;

(b) there exists $m \in \mathbb{N}$ with $g_k \mid T = g_m \mid T \quad (k \geq m)$.

Suppose that case (a) occurs, and set

$$t_k = v(f_k - f_m) \quad (k > m),$$

so that $s_n \le t_k \quad (n \in \mathbb{N},\ k > m)$. If $s_n < t_k \quad (n \in \mathbb{N},\ k > m)$, then there exists $t_0 \in S$ with $s_n < t_0 < t_k \quad (n \in \mathbb{N},\ k > m)$ because S is an η_1-set. In this case, define h by setting

$$h(s) = f_m(s) \quad (s < t_0), \quad h(t_0) = f_m(t_0) + 1, \quad h(s) = 0 \quad (s > t_0);$$

we have $h \in G$ and $A << h << B$. If there exist $n_0, k_0 \in \mathbb{N}$ with $k_0 > m$ such that

$$s_n = t_k \quad (n \ge n_0,\ k \ge k_0),$$

take the common value to be s_0. Then $(f_k(s_0) : k \ge k_0)$ is a strictly increasing sequence in $\mathbb{R}$, and $(g_k(s_0) : k \ge m)$ is a decreasing sequence in $\mathbb{R}$ such that $f_k(s_0) < g_k(s_0) \quad (k \ge k_0)$, and so there exists $\alpha \in \mathbb{R}$ with

$$f_j(s_0) < \alpha \le g_k(s_0) \quad (j, k \ge k_0);$$

set $h(s) = f_m(s) \quad (s < s_0)$ and $h(s_0) = \alpha$. If $\alpha < g_k(s_0) \quad (k \ge k_0)$, take $h(s) = 0 \quad (s > s_0)$, so that $h \in G$ and $A << h << B$. If $(g_k(s_0))$ is eventually equal to α, say for $k \ge k_1$, then there exists $u_0 \in S$ with

$$s_0 < u_0 < v(g_k - g_{k_1}) \quad (k > k_1);$$

again this uses the fact that S is an η_1-set. Define

$$h(s) = g_{k_1}(s) \quad (s < u_0), \quad h(u_0) = g_{k_1}(u) - 1, \quad h(s) = 0 \quad (s > u_0);$$

once again, we see that $h \in G$ and that $A << h << B$.

Case (b), above, is dealt with similarly.

Second, suppose that A has a maximum, say f_0, and that B is non-empty and has no minimum, say (g_n) is a strictly decreasing sequence in B. Set

$$s_n = v(g_n - f_0) \quad (n \in \mathbb{N}),$$

so that (s_n) is an increasing sequence in S. Since S is an η_1-set, there exists $s_0 \in S$ with $s_n < s_0 \quad (n \in \mathbb{N})$. Define

$$h(s) = f_0(s) \quad (s < s_0), \quad h(s_0) = f_0(s_0) + 1, \quad h(s) = 0 \quad (s > s_0).$$

Then $h \in G$ and $f_0 < h < g_n \quad (n \in \mathbb{N})$, so that $A << h << B$.

Third, suppose that $A = \emptyset$. Since S is an η_1-set and B is countable, there exists $s_0 \in S$ with $s_0 < v(g) \quad (g \in B)$. Define

$$h = -\delta_{s_0}.$$

Then $h << B$ in G.

All other cases are dealt with by trivial variations of the above, and so we have proved that G is an η_1-set.

(ii) To obtain a contradiction, assume that $(f_\sigma : \sigma < \kappa^+)$ is a strictly increasing sequence in $\mathfrak{F}(\mathbb{R}, S)$. We may suppose that $f_0 > 0$.

Fix $\sigma < \kappa^+$, and consider the sequence $(v(f_\tau - f_\sigma) : \sigma < \tau < \kappa^+)$ in S. The sequence is decreasing in S, and so, by hypothesis, it is eventually constant, say this constant value is s_σ.

It is further clear that the sequence $(s_\sigma : \sigma < \kappa^+)$ is increasing in S, and so this sequence is eventually constant, say

$$s_\sigma = s \quad (\sigma_0 < \sigma < \kappa^+);$$

we take $\tau_0 < \kappa^+$ such that $v(f_\tau - f_{\sigma_0}) = s \quad (\tau_0 < \tau < \kappa^+)$. The sequence $(f_\tau(s) : \tau_0 < \tau < \kappa^+)$ is an increasing sequence of length $\kappa^+ \geq \aleph_1$ in $\mathbb{R}$, and so this sequence is also eventually constant, say

$$f_\tau(s) = \alpha \quad (\tau_1 \leq \tau < \kappa^+),$$

where $\tau_1 > \tau_0$. Since the sequence $(v(f_\tau - f_{\tau_1}) : \tau_1 < \tau < \kappa^+)$ eventually takes the value s, there exists $\tau_2 > \tau_1$ such that $v(f_{\tau_2} - f_{\tau_1}) = s$. But this is a contradiction because $f_{\tau_2}(s) = f_{\tau_1}(s) = \alpha$.

Thus $\mathfrak{F}(\mathbb{R}, S)$ contains no strictly increasing or strictly decreasing sequence of length κ^+. It follows from the Erdős–Rado theorem, Proposition 1.17, that $|\mathfrak{F}(\mathbb{R}, S)| \leq 2^\kappa$.

(iii) This follows from the special case of (ii) in which $\kappa = \aleph_0$ (respectively, $\kappa = \aleph_1$).

(iv) By Corollary 1.14, there is a chain $\{S_\sigma : \sigma < \omega_1\}$ of α_1-subsets of S with

$$S = \bigcup\{S_\sigma : \sigma < \omega_1\}.$$

By (ii), each $\mathfrak{F}(\mathbb{R}, S_\sigma)$ is an α_1-group. We identify each $\mathfrak{F}(\mathbb{R}, S_\sigma)$ as a subgroup of $\mathfrak{F}(\mathbb{R}, S)$; since each well-ordered subset of S_σ is countable, in fact we have $\mathfrak{F}(\mathbb{R}, S_\sigma) \subset \mathfrak{F}_{(1)}(\mathbb{R}, S)$. Further, each countable subset of S is contained in S_σ for some $\sigma < \omega_1$, and so

$$\mathfrak{F}_{(1)}(\mathbb{R}, S) = \bigcup\{\mathfrak{F}(\mathbb{R}, S_\sigma) : \sigma < \omega_1\}.$$

Hence $\mathfrak{F}_{(1)}(\mathbb{R}, S)$ is a β_1-group. □

By combining the above result with Hahn's embedding theorem, we obtain the following theorem.

Theorem 1.40 *Let G be a totally ordered group with value set Γ_G.*

(i) *Suppose that Γ_G is an α_1-set. Then G is an α_1-group.*

(ii) *Suppose that κ is an infinite cardinal and Γ_G contains no strictly increasing or strictly decreasing sequence of length κ^+. Then G contains no strictly increasing or strictly decreasing sequence of length κ^+, and $|G| \leq 2^\kappa$.* □

Proposition 1.41 *Let S be a totally ordered set with $\operatorname{cof} S \geq \aleph_1$, and let T be a β_1-set. Then there is no isotonic map from $\mathfrak{F}(\mathbb{R}, S)^+$ into $\mathfrak{F}_{(1)}(\mathbb{R}, T)$.*

Proof Set $F = \mathfrak{F}(\mathbb{R}, S)$ and $G = \mathfrak{F}_{(1)}(\mathbb{R}, T)$. By Theorem 1.39(iv), we can write

$$G = \bigcup\{G_\nu : \nu < \omega_1\},$$

where $\{G_\nu : \nu < \omega_1\}$ is a chain of α_1-subgroups of G.

To obtain a contradiction, assume that $\psi : F^+ \to G$ is an isotonic map. Then ψ induces an isotonic map $\overline{\psi} : S \to T$ such that

$$v(\psi(f)) = \overline{\psi}(v(f)) \quad (f \in F^+),$$

and we may suppose that $\overline{\psi}(S) = T$.

We inductively construct elements $f_\sigma \in F^+$, intervals $I_\sigma \subset F^+$, and $s_\sigma \in S$ for $\sigma < \omega_1$ such that the following hold whenever $\tau < \sigma < \omega_1$:

$(\mathrm{i})_\sigma$ $s_\tau < s_\sigma$;

$(\mathrm{ii})_\sigma$ $f_\sigma(s) = f_\tau(s)$ for each $s \in S$ with $s < s_\tau$;

$(\mathrm{iii})_\sigma$ $I_\sigma = \{f_\sigma + h : h \in F^+,\ v(h) \geq s_\sigma\}$;

$(\mathrm{iv})_\sigma$ $\psi(I_\sigma) \cap G_\sigma = \emptyset$.

To do this, we give the construction of f_σ, I_σ, and s_σ under the hypothesis that f_τ, I_τ, and s_τ have been defined for each $\tau < \sigma$. Set $I'_\sigma = \bigcap\{I_\tau : \tau < \sigma\}$ (where $I'_0 = F^+$). Since $\operatorname{cof} S \geq \aleph_1$ and $\sigma < \omega_1$, we see that $\operatorname{cof} I'_\sigma \geq \aleph_1$, and so it is not the case that $\psi(I'_\sigma) \subset G_\sigma$. Take $f_\sigma \in I'_\sigma$ with $\psi(f_\sigma) \in G \setminus G_\sigma$, so that $f_\sigma(s) = f_\tau(s)$ for $s \in (-\infty, s_\tau)$, and let $\langle A, B\rangle$ be the gap in G_σ defined by f_σ. Since G_σ is an α_1-set, $\langle A, B\rangle$ is a countable gap, and so there exists $s_\sigma \in S$ with $s_\sigma > s_\tau$ for $\tau < \sigma$ (vacuous if $\sigma = 0$) and such that

$$\{v(f) : f \in A \cup B\} << \overline{\psi}(s_\sigma),$$

again using the fact that $\operatorname{cof} S \geq \aleph_1$. Set

$$I_\sigma = \{f_\sigma + h : h \in F^+,\ v(h) \geq s_\sigma\}.$$

Then $A << \psi(I_\sigma) << B$ in G, and so $\psi(I_\sigma) \cap G_\sigma = \emptyset$. This shows that f_σ, I_σ, and s_σ have the required properties, and so the inductive construction continues.

Now define $f : S \to \mathbb{R}$ so that $f(s) = f_\sigma(s)$ if $s < s_\sigma$ for some $\sigma < \omega_1$, and $f(s) = 0$ if $s > s_\sigma$ for all $\sigma < \omega_1$. Then f is well defined, and clearly $f \in F^+$. Indeed, $f \in I_\sigma$ for each $\sigma < \omega_1$, and so, by (iv), $\psi(f) \notin G_\sigma$ for each $\sigma < \omega_1$, a contradiction of the fact that $G = \bigcup\{G_\sigma : \sigma < \omega_1\}$.

Thus there is no isotonic map $\psi : F^+ \to G$. □

We now define a group **G** which plays an analogous role in the class of totally ordered groups to that which the set **Q** does in the class of totally ordered sets.

Definition 1.42 *For each $\sigma < \omega_1$, set $\mathbf{G}_\sigma = \mathfrak{F}(\mathbb{R}, \mathbf{Q}_\sigma)$. Also, set*

$$\mathbf{G} = \mathfrak{F}_{(1)}(\mathbb{R}, \mathbf{Q}).$$

The following result is a consequence of Propositions 1.9 and Theorem 1.39.

Proposition 1.43 (i) **G** *is a divisible η_1-group.*

(ii) *For each non-zero ordinal σ with $\sigma < \omega_1$, $\mathbf{G}_\sigma$ is an α_1-group.*

(iii) $\mathbf{G} = \bigcup\{\mathbf{G}_\sigma : 0 < \sigma < \omega_1\}$, *and* **G** *is a β_1-group.* □

In particular, **G** is an α_2-group.

Proposition 1.44 (i) $|\mathbf{G}| = w(\mathbf{G}) = \mathfrak{c}$.

(ii) $\operatorname{cof} \mathbf{G} = \delta(\mathbf{G}) = \aleph_1$.

(iii) $\mathbf{G}$ *contains the following types of gap, and no other types:*

$$(0, \aleph_1),\ (\aleph_1, 0),\ (1, \aleph_1),\ (\aleph_1, 1),\ (\aleph_0, \aleph_1),\ (\aleph_1, \aleph_0),\ (\aleph_1, \aleph_1).$$

Proof By Proposition 1.9(v) and Proposition 1.28(iv), $\delta(\mathbf{G}) = \aleph_1$. By Proposition 1.43, $\mathbf{G}$ is a β_1-η_1-set, and so, by Theorem 1.15, $\mathbf{G}$ is order-isomorphic to $\mathbf{Q}$. By Proposition 1.9 and Proposition 1.11, all the remaining results stated hold for $\mathbf{Q}$, and they are all preserved by an order-isomorphism. □

We now give a development for divisible, totally ordered groups which parallels that of Lemma 1.12 to Theorem 1.15 for totally ordered sets.

Lemma 1.45 *Let G be a divisible α_1-group, let H be a divisible η_1-group, and let G_0 be a divisible subgroup of G. Then each isotonic group morphism from G_0 to H has an extension which is an isotonic group morphism from G into H.*

Proof Suppose that G_1 is a proper, divisible subgroup of G, that

$$\psi_1 : G_1 \to H$$

is an isotonic group morphism, and that $x \in G \setminus G_1$ defines the gap $\langle A, B\rangle$ in G_1. Since G is an α_1-group and H is an η_1-group, there exists $y \in H$ which interpolates the pregap $\langle \psi_1(A), \psi_1(B)\rangle$. Clearly ψ_1 extends to a unique group isomorphism

$$\overline{\psi}_1 : G_1 \oplus \mathbb{Q}x \to \psi_1(G_1) \oplus \mathbb{Q}y$$

with $\overline{\psi}_1(x) = y$, and $\overline{\psi}_1$ is isotonic when $G_1 \oplus \mathbb{Q}x$ and $\psi_1(G_1) \oplus \mathbb{Q}y$ have the relative orders from G and H, respectively.

The remainder of the proof is the same as that in Lemma 1.12. □

Theorem 1.46 *Let G be a divisible β_1-group, and let H be a divisible η_1-group. Then there is an isotonic group morphism from G into H. In particular, there are isotonic group morphisms from G into $\mathbf{G}$ and from $\mathbf{G}$ into H.*

Proof This follows from Lemma 1.45 by the argument of the proof of Theorem 1.13. □

Corollary 1.47 *Let G be a divisible β_1-group. Then there is a chain $\{G_\sigma : \sigma < \omega_1\}$ of divisible α_1-subgroups of G such that*

$$G = \bigcup \{G_\sigma : \sigma < \omega_1\}.$$

Proof By the theorem, there is an isotonic map which is a group morphism ψ from G into $\mathbf{G}$. Set $G_\sigma = \psi^{-1}(\mathbf{G}_\sigma)$ for $\sigma < \omega_1$. □

Note that Theorem 1.46 asserts that the group $\mathbf{G}$ is *universal* in the class of divisible β_1-groups. Corollary 1.49, below, shows that, just as $\mathbf{Q}$ is 'really' the only β_1-η_1-set, $\mathbf{G}$ is the only β_1-η_1-group.

Theorem 1.48 *Let G be a semi-η_1-group which is also a β_1-group. Then there is a completely isotonic group morphism from G onto $\mathfrak{F}_{(1)}(\mathbb{R}, \Gamma_G)$.*

Proof Set $G = \bigcup\{G_\sigma : \sigma < \omega_1\}$, where $\{G_\sigma : \sigma < \omega_1\}$ is a chain of divisible α_1-subgroups of G. Denote the value set of G_σ by Γ_σ, and regard $\{\Gamma_\sigma : \sigma < \omega_1\}$ as a chain of ordered subsets of Γ_G.

We say that G is expressed as a *valid union* if

$$G = \bigcup\{H_\sigma : \sigma < \omega_1\},$$

where $\{H_\sigma : \sigma < \omega_1\}$ is a chain of divisible subgroups of G, and if, for each $\sigma < \omega_1$, we have $G_\sigma \subset H_\sigma$ and $\Gamma_{H_\sigma} = \Gamma_\sigma$. By Theorem 1.40(i), each H_σ is an α_1-group, and, by Corollary 1.36, we may suppose that

$$\mathfrak{F}_{(0)}(\mathbb{R}, \Gamma_\sigma) \subset G_\sigma \subset H_\sigma \subset \mathfrak{F}(\mathbb{R}, \Gamma_\sigma)$$

for each $\sigma < \omega_1$ in this situation.

Let $\mathcal{H} = \{H_\sigma : \sigma < \omega_1\}$ and $\mathcal{H}' = \{H'_\sigma : \sigma < \omega_1\}$ be two families defining valid unions. Then we say that $\mathcal{H} \preccurlyeq \mathcal{H}'$ if $H_\sigma \subset H'_\sigma \quad (\sigma < \omega_1)$. The family of valid unions is a partially ordered set with respect to the order $\preccurlyeq$. Consider a chain of valid unions in the collection of all valid unions, and, for each $\sigma < \omega_1$, let H_σ be the union of the corresponding divisible subgroups in the members of the chain. Then H_σ is a divisible subgroup of G with $G_\sigma \subset H_\sigma \subset \mathfrak{F}(\mathbb{R}, \Gamma_\sigma)$, and so $\Gamma_{H_\sigma} = \Gamma_\sigma$. Thus $\{H_\sigma : \sigma < \omega_1\}$ is a valid union which is an upper bound for the given chain. It follows from Zorn's lemma that there is a maximal element in this partially ordered collection, say it is $\{H_\sigma : \sigma < \omega_1\}$.

We *claim* that each H_σ is pseudo-complete. To obtain a contradiction, assume that $\langle A, B\rangle$ is a pseudo-gap in H_σ for some $\sigma < \omega_1$, say $\langle A, B\rangle$ is a (κ, λ)-gap. Then κ and λ are infinite. Since H_σ is an α_1-group, we have $\kappa = \lambda = \aleph_0$. Since G is a semi-η_1-group, there exists $x \in G \setminus H_\sigma$ with

$$A << x << B.$$

Let τ be the least ordinal such that there exists an element $x \in H_\tau$ such that $A << x << B$. By Proposition 1.31, $\langle A, B\rangle$ is a pseudo-gap in H_ζ for each ζ with $\sigma \le \zeta < \tau$. Set

$$\overline{H}_\zeta = \begin{cases} H_\zeta \oplus \mathbb{Q}x & (\zeta \ge \sigma), \\ H_\zeta & (\zeta < \sigma), \end{cases}$$

(so that $\overline{H}_\zeta = H_\zeta$ for $\zeta \ge \tau$). By Proposition 1.32, the value set of $\overline{H}_\zeta$ is Γ_ζ for each ζ with $\sigma \le \zeta < \tau$, and this is immediate for other values of ζ. Thus $G = \bigcup\{\overline{H}_\zeta : \zeta < \omega_1\}$ expresses G as a valid union. By the maximality of the family $\{H_\zeta : \zeta < \omega_1\}$, we have $\overline{H}_\zeta = H_\zeta \quad (\zeta < \omega_1)$, a contradiction, and so each H_σ is pseudo-complete, as claimed.

For each $\sigma < \omega_1$, H_σ is a pseudo-complete, divisible subgroup of $\mathfrak{F}(\mathbb{R}, \Gamma_\sigma)$ containing $\mathfrak{F}_{(0)}(\mathbb{R}, \Gamma_\sigma)$, and so $H_\sigma = \mathfrak{F}(\mathbb{R}, \Gamma_\sigma)$.

We have now identified G by an isotonic group morphism with the subset

$$\bigcup\{\mathfrak{F}(\mathbb{R}, \Gamma_\sigma) : \sigma < \omega_1\}$$

of $\mathfrak{F}(\mathbb{R},\Gamma)$. Since each Γ_σ is an α_1-set, $\mathfrak{F}_{(1)}(\mathbb{R},\Gamma_\sigma) = \mathfrak{F}(\mathbb{R},\Gamma_\sigma)$, and so this subset is exactly $\mathfrak{F}_{(1)}(\mathbb{R},\Gamma)$. □

Corollary 1.49 *(Esterle) Let G_1 and G_2 be divisible β_1-η_1-groups. Then there is a completely isotonic group morphism from G_1 onto G_2.*

Proof The ordered set Γ_G is a β_1-η_1-set, and so, by Theorem 1.15, Γ_G is order-isomorphic to $\mathbf{Q}$. The result now follows from the theorem. □

It follows that each divisible β_1-η_1-group is isomorphic to $\mathbf{G}$, and that any two η_1-groups of cardinality $\aleph_1$ are isomorphic.

The above result has an extension which classifies divisible groups which are both β_1-groups and semi-η_1-groups. We require some preliminary notation.

Let $\{G_\lambda : \lambda \in \Lambda\}$ be a family of groups. Then

$$G = \prod\{G_\lambda : \lambda \in \Lambda\}$$

is a group with respect to the coordinatewise operations. In the case where each G_λ is a totally ordered, divisible group and Λ is well-ordered, G is a totally ordered, divisible group with respect to the lexicographic order.

Let G be a totally ordered group. Then

$$c_{00}(G) = \{(x_n) \in G^{\mathbb{N}} : x_n = 0 \ \text{ eventually}\}\,;$$

the set $c_{00}(G)$ is a subgroup of $G^{\mathbb{N}}$, and it is a divisible group if G is divisible. We say that

$$0 \prec^* (x_n)$$

in $c_{00}(G)$ if there exists $n_0 \in \mathbb{N}$ such that $x_{n_0} > 0$ and $x_n = 0 \quad (n > n_0)$, so that we are ordering $c_{00}(G)$ by looking at the last non-zero term of each of its elements; the group $(c_{00}(G), \preceq^*)$ is a totally ordered group.

There is an extension of the definition of $\delta(G)$ in the case where G is divisible and $\delta(G) = \aleph_0$: set

$$\delta(G) = (\aleph_0, 1) \quad \text{or} \quad \delta(G) = (\aleph_0, \aleph_0)$$

according as there is or is not an element $x \in G^+ \setminus \{0\}$ such that $\{x/n : n \in \mathbb{N}\}$ is coinitial in $G^+ \setminus \{0\}$. Thus $G = (\aleph_0, 1)$ if and only if Γ_G has a maximum element. Similarly, there is an extension of the definition of $\operatorname{cof} G$ in the case where $\operatorname{cof} G = \aleph_0$: set

$$\operatorname{cof} G = (\aleph_0, 1) \quad \text{or} \quad \operatorname{cof} G = (\aleph_0, \aleph_0)$$

according as there is or is not an element $x \in G^+$ such that $\{nx : n \in \mathbb{N}\}$ is cofinal in G.

Clearly, the values of $\delta(G)$ and $\operatorname{cof} G$ in this extended sense are invariants of the group G under isomorphism, as is the weight $w(G)$.

Table 1.1 is a list of certain totally ordered groups G, the corresponding values of these invariants, and the value group Γ_G.

Theorem 1.50 *Let G be a divisible semi-η_1-group which is a β_1-group. Then G is isomorphic to exactly one of the groups in Table 1.1.*

Proof It is easily checked that each of the groups on the list is a divisible semi-η_1-group which is a β_1-group. For any two such groups, at least one of the three invariants distinguishes the two groups, and so no two groups on the list are isomorphic.

Now let G be an arbitrary divisible semi-η_1-group which is a β_1-group. By Theorem 1.38, Γ_G is order-isomorphic to exactly one of the ordered sets in the right-hand column of Table 1.1, and so, by Theorem 1.48, G is isomorphic to the corresponding group in the left-hand column. $\square$

Table 1.1

	G	$w(G)$	$\delta(G)$	cof G	Γ_G
1	$(\mathbb{R}, \leq)$	$\aleph_0$	$(\aleph_0, 1)$	$(\aleph_0, 1)$	$\{1\}$
2	$(\mathbf{G}, \leq)$	$\mathfrak{c}$	$\aleph_1$	$\aleph_1$	$\mathbf{Q}$
3	$(\mathbb{R} \times \mathbf{G}, \preceq)$	$\mathfrak{c}$	$\aleph_1$	$(\aleph_0, 1)$	$\{1\} \odot \mathbf{Q}$
4	$(\mathbf{G} \times \mathbb{R}, \preceq)$	$\mathfrak{c}$	$(\aleph_0, 1)$	$\aleph_1$	$\mathbf{Q} \odot \{1\}$
5	$(\mathbb{R} \times \mathbf{G} \times \mathbb{R}, \preceq)$	$\mathfrak{c}$	$(\aleph_0, 1)$	$(\aleph_0, 1)$	$\{1\} \odot \mathbf{Q} \odot \{1\}$
6	$(\mathbf{G}^{\mathbb{N}}, \preceq)$	$\mathfrak{c}$	$(\aleph_0, \aleph_0)$	$\aleph_1$	$\mathbb{N} \times \mathbf{Q}$
7	$(c_{00}(\mathbf{G}), \preceq^*)$	$\mathfrak{c}$	$\aleph_1$	$(\aleph_0, \aleph_0)$	$\mathbb{Z}^- \times \mathbf{Q}$
8	$(c_{00}(\mathbf{G}^{\mathbb{N}}), \preceq^*)$	$\mathfrak{c}$	$(\aleph_0, \aleph_0)$	$(\aleph_0, \aleph_0)$	$\mathbb{Z} \times \mathbf{Q}$
9	$(\mathbb{R} \times \mathbf{G}^{\mathbb{N}}, \preceq)$	$\mathfrak{c}$	$(\aleph_0, \aleph_0)$	$(\aleph_0, 1)$	$\{1\} \odot (\mathbb{N} \times \mathbf{Q})$
10	$(c_{00}(\mathbf{G} \times \mathbb{R}), \preceq^*)$	$\mathfrak{c}$	$(\aleph_0, 1)$	$(\aleph_0, \aleph_0)$	$(\mathbb{Z}^- \times \mathbf{Q}) \odot \{1\}$

We conclude with a technical result about $\mathbf{G}$ to be used in Chapter 5.

Proposition 1.51 *Let $x \in \mathbf{G}^+$. Then there is a group morphism*

$$\pi : \mathbf{G} \to \mathbf{G}$$

such that $\pi(\mathbf{G}) \cap [-x, x] = \{0\}$.

Proof Set $s = v(x) \in \mathbf{Q}$, and set $I = (-\infty, s)$. Then I is a β_1-η_1-set in $\mathbf{Q}$, and so, by Theorem 1.15, I is order-isomorphic to $\mathbf{Q}$. Thus there is a group morphism π from $\mathfrak{F}_{(1)}(\mathbb{R}, \mathbf{Q})$ onto $\mathfrak{F}_{(1)}(\mathbb{R}, I)$. For each $y \in \mathbf{G}$ with $-x \leq y \leq x$, we have $v(y) \geq s$, and so $\mathfrak{F}_{(1)}(\mathbb{R}, I) \cap [-x, x] = \{0\}$. Thus π is as required. $\square$

Notes 1.52 The initial definitions in this chapter are mostly quite standard. A totally ordered set is sometimes said to be *linearly ordered* or *simply ordered.* For the theory of ordinal and cardinal numbers, see Jech (1978) or Kunen (1980).

The notions of α_1-sets and β_1-sets are given by Gillman and Jerison (1960, 13L), but our terminology for β_1-sets is not used there. The definition of η_1-sets is an old one with a distinguished history. We believe it was first given by Hausdorff (1908); see also Hausdorff (1914, pp.180-1). Hausdorff proved that two η_1-sets of cardinality $\aleph_1$ are order-isomorphic. These sets are discussed by Gillman and Jerison (1960) and Rosenstein (1982). The terminology for semi-η_1-sets is new.

The set $\mathbf{Q}$ was introduced by Sierpiński. Proposition 1.9 is proved in Gillman and Jerison (1960, Chapter 13).

There is a natural generalization (already given by Hausdorff) of the notion of an η_1-set. Let S be a totally ordered set, and let α be an ordinal. Then S is an *η_α-set* if, for each pregap $\langle A, B\rangle$ in S with $|A \cup B| < \aleph_\alpha$, there exists $s \in S$ with $A << s << B$. Let $\mathbf{S}_{(\alpha)}$ be the set of all dyadic sequences of length ω_α, and let $\mathbf{Q}_{(\alpha)}$ be the set of elements x of $\mathbf{S}_{(\alpha)}$ such that $\{\tau < \omega_\alpha : x_\tau = 1\}$ is non-empty and has a largest member (so that $\mathbf{Q}_{(\alpha)}$ is an analogue of $\mathbf{Q} = \mathbf{Q}_{(1)}$). It was proved by Hausdorff that, if $\aleph_\alpha$ is a regular cardinal, then $\mathbf{Q}_{(\alpha)}$ is an η_α-set. Every $\eta_{\alpha+1}$-set has cardinality at least $2^{\aleph_\alpha}$; there is an $\eta_{\alpha+1}$-set of cardinality $\aleph_{\alpha+1}$ if and only if $2^{\aleph_\alpha} = \aleph_{\alpha+1}$. Further, any two η_α-sets of cardinality $\aleph_\alpha$ are order-isomorphic, and the ordered set $\mathbf{Q}_{(\alpha)}$ contains every totally ordered set of cardinality at most $\aleph_\alpha$. For the above results, see Rosenstein (1982, 9, §4).

Specific consideration of β_1-sets and β_1-groups originates in the lectures of J. R. Esterle (1984), where the precise definitions were first formulated and Theorem 1.15 was proved. The 'back-and-forth' argument of Theorem 1.15 is used by Gillman and Jerison (1960, Chapter 13), for example.

Hahn's paper in which $\mathfrak{F}(\mathbb{F}, S)$ and $\mathfrak{F}_{(\sigma)}(\mathbb{F}, S)$ are defined is Hahn (1907).

General forms of the well-known Erdős–Rado theorem, Proposition 1.17, are given in Drake (1974, 7, §2), Jech (1978, §29), and Kunen (1980, VIII, (B1)).

Our approach to Hahn's embedding theorem through pseudo-gaps may be original, although it is clearly related to earlier work; Proposition 1.32, which is central to our approach, is new. Kaplansky proved Proposition 1.33 in his seminal paper of 1942 (using his different, but equivalent, definition of 'pseudo-complete'). For other work on the theory of gaps in ordered sets, see Frankiewicz and Zbierski (1994).

Hahn's embedding theorem, Theorem 1.35, was first proved in Hahn (1907) by a very long method. A proof is given in Fuchs (1963, IV.5), and some other references and history of the theorem are contained in this book.

Part (i) of Theorem 1.39 is essentially a special case of Alling (1962a, Corollary 1.2), Esterle (1977, Lemma 1.1), and Ribenboim (1965). Parts (iii) and (iv) of the theorem are due to Esterle (1977, Lemma 1.2). Corollary 1.49 is also due to Esterle (1984); our approach is somewhat different from Esterle's because of our use of the theory of gaps, but the general method of proving Theorem 1.48 is an extension of that of Esterle. The classification of divisible semi-η_1-groups which are β_1-groups, which is given in Theorem 1.50, is new.

2
Ordered fields

In the previous chapter, we gave a certain development for ordered sets and ordered groups. In this chapter, we shall add another layer of structure, and we shall discuss ordered fields. In many ways, our development will follow the earlier story, but naturally at a few points we shall have to work a little harder to take into account all the structures that must be preserved in our embedding and isomorphism theorems.

Throughout, we shall only consider fields which are associative, linear algebras over a ground field $\mathbb{F}$, where $\mathbb{F} = \mathbb{R}$, $\mathbb{Q}$, or $\mathbb{C}$. The identity of a field is denoted by 1; we shall always suppose that $1 \neq 0$. Let K be a field over the ground field $\mathbb{F}$. Then the elements $\alpha 1$ for $\alpha \in \mathbb{F}$ constitute a copy of $\mathbb{F}$ in K, and we identify this subfield with $\mathbb{F}$.

In this book we shall be almost exclusively concerned with algebras and fields over the ground field $\mathbb{R}$, and so we make the important convention that *the use of the word 'algebra' or 'field' by itself implies that the object in question is an algebra over* $\mathbb{R}$. Similarly, if we refer to, say, isomorphisms between fields, we imply that the map in question is a real algebra isomorphism, or, equivalently, that the map identifies the two copies of $\mathbb{R}$ in the fields. Also, by our convention, a subfield of a field necessarily contains $\mathbb{R}$. Occasionally, we shall consider fields over $\mathbb{Q}$ or $\mathbb{C}$; these fields will be explicitly referred to as *rational fields* and *complex fields*, respectively. Of course, a ring isomorphism between two rational fields is necessarily a $\mathbb{Q}$-algebra isomorphism, but it is possible for two fields to be isomorphic as rings, and hence as $\mathbb{Q}$-algebras, but not isomorphic as fields. A small number of our results which are stated for fields actually apply to the more general class of rational fields; this will not usually be pointed out.

A complex field may be regarded as a field or as a rational field, and a field may be regarded as a rational field.

An algebra A is *unital* if it has an identity; the identity of such an algebra is sometimes denoted by e or e_A.

Definition 2.1 *Let* $(K, +, \cdot)$ *be a rational field, and let* $\leq$ *be a binary relation on* K. *Then* $(K, +, \cdot, \leq)$ *is an* ordered rational field *if* $(K, \leq)$ *is a totally ordered set, if* $(K, +, \leq)$ *is an ordered group, and if*

$$ab \in K^+ \quad (a, b \in K^+)\,.$$

A field which is an ordered rational field is an ordered field.

Note the (slightly awkward) convention that the order on an ordered field is necessarily a total order, but that the order on an ordered group may be a partial order which is not total.

Let K be an ordered field. It is easily checked that

$$|ab| = |a|\,|b| \quad (a, b \in K)\,.$$

Let K and L be ordered fields. An *embedding* of K in L is an algebra monomorphism from K into L which is isotonic. A surjective embedding is an *isomorphism.* In the case where there is an isomorphism from K onto L, K and L are *isomorphic*, and we write

$$K \cong L\,.$$

We review here some basic facts of the Artin–Schreier theory of ordered fields. Let A be an algebra. Then $A[X]$ denotes the algebra of polynomials with coefficients in A. In the case where A is a subalgebra of an algebra B and $b \in B$,

$$A[b] = \{p(b) : p \in A[X]\}\,;$$

b is *algebraic* over A if there exists $p \in A[X] \setminus \{0\}$ with $p(b) = 0$, and b is *transcendental* if it is not algebraic. The algebra B is an *algebraic* extension of A if each $b \in B$ is algebraic over A; otherwise, B is a *transcendental* extension of A. A field K is *algebraically closed* if there is no field L which is a proper algebraic extension of K; equivalently, K is algebraically closed if and only if each non-constant $p \in K[X]$ has a root in K. An ordered field K is *real-closed* if it has no proper algebraic extension to an ordered field, or, equivalently, if the complexification $K_{\mathbb{C}}$ of K is algebraically closed, or, again equivalently, if every positive element in K is a square and every polynomial over K of odd degree has a root in K. The *Artin–Schreier theorem* asserts that every ordered field K has an algebraic extension to a real-closed field $\mathcal{R}K$ whose order is an extension of the order on K, and that $\mathcal{R}K$ is unique up to an isomorphism that leaves the elements of K fixed.

Let K be a real-closed ordered field, and take $a \in K^+ \setminus \{0\}$ and $n \in \mathbb{N}$. Clearly there is a unique element $b \in K^+ \setminus \{0\}$ such that $b^n = a$, and so there is a map

$$\psi : \alpha \mapsto a^\alpha, \ \mathbb{Q}^+ \to K^+ \setminus \{0\}\,,$$

such that $\psi(0) = 1$, $\psi(1) = a$, and

$$\psi(\alpha + \beta) = \psi(\alpha)\psi(\beta) \quad (\alpha, \beta \in \mathbb{Q}^+)\,.$$

Suppose that K and L are ordered fields with K real-closed, and that θ is a monomorphism from K into L. Then θ is automatically isotonic, and hence an embedding, for, if $a > 0$ in K, then there exists $b \in K \setminus \{0\}$ with $b^2 = a$, and then $\theta(a) = \theta(b)^2 > 0$ in L. It follows that the order on a real-closed field K is uniquely specified, and that the restriction of the order on an ordered field to $\mathbb{R}$ coincides with the usual order on $\mathbb{R}$.

Proposition 2.2 *Let L be an ordered field which, as a totally ordered set, has weight κ, where $\kappa \geq \mathfrak{c}$. Then L contains an order-dense subfield K of cardinality κ. If L is a real-closed field, then we may suppose that K is real-closed.*

Proof Let S be an order-dense subset of L with $|S| = \kappa$, and let K be the smallest subfield of L containing S. Then $|K| = \kappa$ (because $\kappa \geq \mathfrak{c}$), and K is order-dense in L. If L is real-closed, replace K by its real-closure, identified with a subfield of L; it is still true that $|K| = \kappa$. □

Definition 2.3 *Let K be an ordered rational field. An element $a \in K$ is:*

(i) *an* infinitesimal *if $a = \mathbf{o}(1)$;*

(ii) finite *if $a = \mathbf{O}(1)$;*

(iii) infinitely large *if it is not finite.*

Thus a is infinitely large if and only if $|a| > n$ for each $n \in \mathbb{N}$.

The sets of infinitesimals and of finite elements in K are denoted by

$$K^0 \quad \text{and} \quad K^{\#},$$

respectively. Clearly, K^0 and $K^{\#}$ are subrings of K, and both K^0 and $K^{\#}$ are absolutely convex subsets of K; further K^0 is an ideal in $K^{\#}$.

Let K be an ordered field, and let Γ_K be the value set of the ordered group $(K, +, \leq)$, with the archimedean valuation v. Define $+$ on Γ_K by the formula

$$v(a) + v(b) = v(ab) \quad (a, b \in K \setminus \{0\}) .$$

It is easy to check that $+$ is well defined, and that $(\Gamma_K, +, \leq)$ is a totally ordered group; it is the *value group* of K. We have $v(1) = 0$, the identity of the group Γ_K. Also we see immediately that

$$K^0 = \{a \in K : v(a) > 0\} \quad \text{and} \quad K^{\#} = \{a \in K : v(a) \geq 0\} .$$

Let $\mathrm{Inv}K^{\#}$ denote the set of elements of $K^{\#}$ whose inverses are also in $K^{\#}$. Then

$$\mathrm{Inv}K^{\#} = \{a \in K^{\#} : v(a) = 0\} = K^{\#} \setminus K^0 ,$$

and so K^0 is the unique maximal ideal of $K^{\#}$. The quotient $K^{\#}/K^0$ is the *residue field* of K. Clearly we have $v(a) = v(b)$ if and only if $a/b \in \mathrm{Inv}K^{\#}$. The field K is the quotient field of the integral domain $K^{\#}$.

Let K be an ordered field. For each $a \in \mathrm{Inv}K$, there exists $\alpha \in \mathbb{R} \setminus \{0\}$ such that $a \in \alpha 1 + K^0$. Indeed, also suppose that $a \in K^+$, and set

$$\alpha = \sup \{\beta > 0 : a > \beta 1\} .$$

Since $v(a) = 0$, the set $\{\beta > 0 : a > \beta 1\}$ is non-empty and bounded above, and so α is well defined. For each $n \in \mathbb{N}$, we have

$$-1/n \leq a - \alpha 1 \leq 1/n ,$$

and so $n\,|a - \alpha 1| \leq 1$, whence $a - \alpha 1 \in K^0$. It follows that $K^{\#} = \mathbb{R}1 + K^0$, and so the residue field of the ordered fields that we are considering is necessarily $\mathbb{R}$; this is a difference from most established theory in this area.

Definition 2.4 *Let A be an integral domain with identity. Then A is a* valuation ring *if, for each $a, b \in A$, either $a \in bA$ or $b \in aA$.*

Thus, in a valuation ring, the set of principal ideals, and hence the set of all ideals, is totally ordered by inclusion. A valuation ring necessarily has a unique maximal ideal, and so it is a *local algebra.* The unique maximal ideal of a valuation ring A is denoted by M_A. The *residue field* of A is the field A/M_A; we shall be concerned with rings for which this field is $\mathbb{R}$ itself.

Definition 2.5 *Let A be a (real or complex) algebra which is a valuation ring. Then A is a* (real *or* complex) valuation algebra *if $A/M_A \cong \mathbb{R}$ or $A/M_A \cong \mathbb{C}$, respectively.*

Let A be a real or complex algebra. A *character* on A is a homomorphism from A onto $\mathbb{R}$ or $\mathbb{C}$, respectively. The kernel of a character on a unital algebra is clearly a maximal ideal in the algebra, and so, in the case where A is a real valuation algebra, there is a unique character on A, and its kernel is M_A; this character is denoted by φ_A.

Proposition 2.6 *Let K be an ordered field. Then $K^\#$ is a real valuation algebra, and its unique maximal ideal is K^0. Let $L = \mathcal{R}K$. Then*

$$\Gamma_L = \{x/n : n \in \mathbb{N},\ x \in \Gamma_K\}. \tag{2.1}$$

Suppose that Γ_K is an α_1-group (respectively, a β_1-group). Then Γ_L is an α_1-group (respectively, a β_1-group).

Proof Take $a, b \in K^\# \setminus \{0\}$, and suppose that $v(a) \geq v(b)$. Set

$$c = ab^{-1} \in K.$$

Then $v(a) = v(b) + v(c)$, and so $v(c) \geq 0$, i.e. $c \in K^\#$. Thus $a \in bK^\#$, and so $K^\#$ is a real valuation ring. The residue field of $K^\#$ is $\mathbb{R}$, and so $K^\#$ is a valuation algebra.

Take $x \in \Gamma_K$ and $n \in \mathbb{N}$. Then there exists $a \in K^+$ with $v(a) = x$, and there exists $b \in L^+$ with $b^n = a$. We have $nv(b) = x$, and so $x/n \in \Gamma_L$. Now take $a \in L \setminus \{0\}$. Then there is a non-zero polynomial $p \in K[X]$ such that $p(a) = 0$, say

$$p = \sum_{j=0}^{m} a_j X^j,$$

where $a_0, \ldots, a_m \in K$ and $a_m \neq 0$. By (1.2) there exist $j, k \in \{0, \ldots, m\}$ with $j < k$ such that $v(a_j a^j) = v(a_k a^k)$. We have

$$(k - j)v(a) = v(a_j) - v(a_k) \in \Gamma_K,$$

and so (2.1) is established.

Since the countable union of α_1-sets is an α_1-set, the final clause of the result follows. □

It follows from Proposition 2.6 that Γ_K is a divisible group whenever K is a real-closed field.

Let G be a totally ordered group, let σ be an ordinal, and let $\mathbb{F} = \mathbb{Q}$, $\mathbb{R}$, or $\mathbb{C}$. Then the $\mathbb{F}$-linear spaces $\mathfrak{F}(\mathbb{F}, G)$ and $\mathfrak{F}_{(\sigma)}(\mathbb{F}, G)$ were defined in Definition 1.25. We now define a product on these spaces.

Definition 2.7 *Let G be a totally ordered group. For $f, g \in \mathfrak{F}(\mathbb{F}, G)$, set*

$$(f * g)(s) = \sum\{f(t_1)g(t_2) : t_1, t_2 \in G,\ \ t_1 + t_2 = s\} \quad (s \in G)\,. \tag{2.2}$$

Let $f, g \in \mathfrak{F}(\mathbb{F}, G)$. For the sake of completeness, we give the standard argument that $(f * g)(s)$ is given by a finite sum and that $f * g$ belongs to $\mathfrak{F}(\mathbb{F}, G)$.

Indeed, set $S_f = \operatorname{supp} f$ and $S_g = \operatorname{supp} g$, and take $s \in S_f + S_g$. Assume towards a contradiction that $\{(t_{1,n}, t_{2,n}) : n \in \mathbb{N}\}$ is a set of distinct elements of $S_f \times S_g$ such that $t_{1,n} + t_{2,n} = s \quad (n \in \mathbb{N})$. Then $\{t_{1,n} : n \in \mathbb{N}\}$ and $\{t_{2,n} : n \in \mathbb{N}\}$ are infinite sets in S_f and S_g, respectively, and so each contains either a strictly increasing or a strictly decreasing subsequence. But this is not possible, for neither contains a strictly decreasing subsequence because S_f and S_g are well-ordered, and, if (t_{1,n_j}) were strictly increasing in S_f, then (t_{2,n_j}) would be strictly decreasing in S_g. Thus $(f * g)(s)$ is defined by a finite sum.

Second, assume that (s_n) is a strictly decreasing sequence in $S_f + S_g$, say $s_n = t_{1,n} + t_{2,n}$, with $(t_{1,n}) \subset S_f$ and $(t_{2,n}) \subset S_g$. Then either $\{t_{1,n} : n \in \mathbb{N}\}$ or $\{t_{2,n} : n \in \mathbb{N}\}$ is infinite, say $\{t_{1,n} : n \in \mathbb{N}\}$ is infinite. Since S_f is well-ordered, $(t_{1,n})$ contains a strictly increasing subsequence, and so $(t_{2,n})$ contains a strictly decreasing subsequence, which again is not possible. Hence $S_f + S_g$ is well-ordered, and so

$$\operatorname{supp}(f * g) \subset S_f + S_g$$

is also well-ordered.

It is now easy to verify that $*$ is a commutative, associative product, that $\mathfrak{F}(\mathbb{F}, G)$ is an algebra over $\mathbb{F}$, and that $\mathfrak{F}_{(\sigma)}(\mathbb{F}, G)$ is a subalgebra of $\mathfrak{F}(\mathbb{F}, G)$ for each ordinal σ.

Definition 2.8 *The algebra $\mathfrak{F}(\mathbb{F}, G)$ is the* formal power series algebra *or* Hahn algebra *over G.*

Let v be the Hahn valuation on $\mathfrak{F}(\mathbb{F}, G)$, and take $f, g \in \mathfrak{F}(\mathbb{F}, G) \setminus \{0\}$, say $v(f) = s$ and $v(g) = t$. Then we see immediately from the definition of $*$ that

$$(f * g)(s + t) = f(s)g(t)\,, \quad v(f * g) = v(f) + v(g)\,.$$

It follows that $\mathfrak{F}(\mathbb{F}, G)$ is an integral domain, and that, if $f, g > 0$ in $\mathfrak{F}(\mathbb{F}, G)$, then $f * g > 0$.

Temporarily, write X^s for the characteristic function of $\{s\}$ for $s \in G$. Then $X^s \in \mathfrak{F}(\mathbb{F}, G)$ and $X^s * X^t = X^{s+t} \quad (s, t \in G)$. One can think of a typical element of $\mathfrak{F}(\mathbb{F}, G)$ as having the form

$$\sum\{\alpha_s X^s : s \in G\}\,,$$

where $\alpha_s \in \mathbb{F} \quad (s \in G)$ and $\alpha_s = 0$ save for s in a well-ordered subset of G. The formula for the product is consistent with this symbolism.

For example, $\mathfrak{F}(\mathbb{F}, \mathbb{Z})$ is the set of Laurent series of the form

$$\sum_{k=n_0}^{\infty} \alpha_k X^k ,$$

where $n_0 \in \mathbb{Z}$ and $\alpha_k \in \mathbb{F}$ for $k \geq n_0$. This algebra is also denoted by $\mathbb{F}((X))$ in the literature. Set $K = \mathfrak{F}(\mathbb{R}, \mathbb{Z})$. Then

$$v\left(\sum_{k=n_0}^{\infty} \alpha_k X^k\right) = \inf\{k \in \mathbb{Z} : \alpha_k \neq 0\} , \tag{2.3}$$

and $K^{\#} = \mathbb{R}[[X]]$, the algebra of formal power series in one variable over $\mathbb{R}$. This field K is certainly not real-closed: there is no element $a \in K$ with $a * a = X$. We shall obtain real-closed fields by replacing the 'small' group $\mathbb{Z}$ by much 'larger', divisible groups G. First we require some preliminaries on valuations.

Let G be a totally ordered group. It is again convenient to adjoin an element ∞ to G: extend the algebraic operations on G to $G \cup \{\infty\}$ by setting $x \leq \infty$ and $x + \infty = \infty$ for $x \in G \cup \{\infty\}$.

Definition 2.9 *Let A be a unital algebra over a field $\mathbb{F}$, and let G be a totally ordered group.* A valuation on A with values in G *is a map $v : A \to G \cup \{\infty\}$ such that:*

(i) $v(a + b) \geq \min\{v(a), v(b)\} \quad (a, b \in A)$;

(ii) $v(ab) = v(a) + v(b) \quad (a, b \in A)$;

(iii) $v(\alpha a) = v(a) \quad (a \in A, \alpha \in \mathbb{F} \setminus \{0\})$;

(iv) *if $a, b \in A \setminus \{0\}$ with $v(a) = v(b)$, then there exists $\alpha \in \mathbb{F}$ such that $v(a - \alpha b) > v(a)$.*

It follows that $v(a) = \infty$ if and only if $a = 0$.

For example, let G be a totally ordered group, and let v be the Hahn valuation on $\mathfrak{F}(\mathbb{R}, G)$. Then v is a valuation on $\mathfrak{F}(\mathbb{R}, G)$ with values in G. To see this, note that v trivially satisfies clauses (i)–(iii) of Definition 2.9. To verify that v satisfies clause (iv), take $f, g \in \mathfrak{F}(\mathbb{R}, G) \setminus \{0\}$ with $v(f) = v(g)$, say $v(f) = s$, and set $\alpha = f(s)/g(s)$. Then $v(f - \alpha g) > s$, as required.

Let v be a valuation on A with values in G, and suppose that $a, b \in A$ with $v(a) \neq v(b)$. Then it follows that

$$v(a + b) = \min\{v(a), v(b)\} .$$

Proposition 2.10 *Let A be a unital integral domain over $\mathbb{F}$ with quotient field K, and let $v : A \to G \cup \{\infty\}$ be a valuation on A. Then the formula*

$$\overline{v} : a/b \mapsto v(a) - v(b) , \quad K \setminus \{0\} \to G ,$$

extends v to a valuation on K.

Proof We first verify that $\overline{v}$ is well defined. For suppose that $a_1/b_1 = a_2/b_2$ in $K \setminus \{0\}$. Then $a_1 b_2 = a_2 b_1$ in A, and so $v(a_1) + v(b_2) = v(a_2) + v(b_1)$, whence $\overline{v}(a_1/b_1) = \overline{v}(a_2/b_2)$, as required.

Next take $x = a_1/b_1$ and $y = a_2/b_2$, and suppose that $v(a_1b_2) \geq v(a_2b_1)$, say, so that $\overline{v}(x) \geq \overline{v}(y)$. Then

$$\begin{aligned}\overline{v}(x+y) &= v(a_1b_2 + a_2b_1) - v(b_1b_2)\\ &\geq v(a_2b_1) - v(b_1b_2)\\ &= v(a_2) - v(b_2)\\ &= \overline{v}(y) = \min\{\overline{v}(x), \overline{v}(y)\}\,,\end{aligned}$$

and so $\overline{v}$ satisfies (i) of Definition 2.9.

That $\overline{v}$ satisfies clauses (ii) and (iii) of Definition 2.9 is immediately verified.

Finally take $x = a_1/b_1$ and $y = a_2/b_2$ in $K \setminus \{0\}$ with $\overline{v}(x) = \overline{v}(y)$. Then $v(a_1b_2) = v(a_2b_1)$, and so there exists $\alpha \in \mathbb{K}$ with

$$v(a_1b_2 - \alpha a_1b_1) > v(a_1b_2)\,.$$

Hence

$$\begin{aligned}\overline{v}\left(\frac{a_1}{b_1} - \frac{\alpha a_2}{b_2}\right) &= v(a_1b_2 - \alpha a_2b_1) - v(b_1b_2)\\ &> v(a_1b_2) - v(b_1b_2)\\ &= v(a_1) - v(b_1) = \overline{v}\left(\frac{a_1}{b_1}\right),\end{aligned}$$

and so $\overline{v}$ satisfies clause (iv) of Definition 2.9. □

The connection between valuation algebras and valuations is as follows. We only sketch the details because this is not very important for us.

Let A be a valuation algebra over $\mathbb{F}$ with quotient field K and maximal ideal M_A. Then the set of invertible elements in A is $\operatorname{Inv} A = A \setminus M_A$. We set

$$\Gamma_A = (K \setminus \{0\}, \cdot)/\operatorname{Inv} A\,,$$

and take v to be the quotient map from $K \setminus \{0\}$ to Γ_A. Set $v(a) \leq v(b)$ for $a, b \in K \setminus \{0\}$ if $b \in aA$, and set

$$v(a) + v(b) = v(ab) \quad \text{in} \quad \Gamma_A\,.$$

Then the operation $+$ is well defined, Γ_A is a totally ordered group, and v is a valuation (setting $v(0) = \infty$).

Conversely, let A be an integral domain with a valuation v, and let K be the quotient field of A. By Proposition 2.10, there is a valuation $\overline{v}$ on K extending v, and

$$B = \{a \in K : \overline{v}(a) \geq 0\}$$

is a valuation ring; the unique maximal ideal of this algebra is

$$M_B = \{a \in K : \overline{v}(a) > 0\}\,,$$

and it is an immediate consequence of (iv) of Definition 2.9 that the residue field of B is $\mathbb{F}$.

Definition 2.11 *Let B be a unital algebra with subalgebra A, and suppose that v and w are valuations on A and B, respectively, with values in totally ordered groups $G = v(A \setminus \{0\})$ and $H = w(B \setminus \{0\})$, respectively. Then (B, w) is:*

(i) *an* extension *of (A, v) if G is an ordered subgroup of H and if $w \mid A = v$;*

(ii) *an* immediate extension *if, further, for each $b \in B \setminus \{0\}$, there exists $a \in A$ with $w(b - a) > w(b)$.*
A unital algebra A with valuation v is maximal *if $A = B$ whenever (B, w) is an algebra with a valuation which is an immediate extension of (A, v).*

Let (A, v) be a valuation algebra which is a subalgebra of a valuation algebra (B, w). Then (B, w) is an extension of (A, v), and (B, w) is an immediate extension if and only if the totally ordered groups Γ_A and Γ_B coincide. To see this, set $\varphi = \varphi_B \mid A$. Then φ is a character on A, and so $M_A = \ker \varphi = M_B \cap A$ and $\operatorname{Inv} A = A \cap \operatorname{Inv} B$. Thus B is an extension of A. If B is an immediate extension, then $\Gamma_A = \Gamma_B$, for, if $b \in B \setminus \{0\}$ and $a \in A$ is such that $w(b - a) > w(b)$, then

$$w(a) = \min\{w(b - a), w(b)\} = w(b),$$

and $w(b) \in \Gamma_A$. Conversely, suppose that $\Gamma_A = \Gamma_B$, and take $b \in B \setminus \{0\}$. There exist $a \in A \setminus \{0\}$ and $c \in \operatorname{Inv} B$ with $b = ac$, and we have

$$b - \varphi_B(c)a = a(c - \varphi_B(c)e_B) \in aM_B = bM_B .$$

Since $\varphi_B(c)a \in A$, B is an immediate extension.

For example, let $A = \mathbb{R}[X]$ and $B = \mathbb{R}[[X]]$, and let v be the valuation on B given in (2.3). Then B is an immediate extension of A, and so A is not maximal.

Proposition 2.12 *Let G be a totally ordered group. Then $(\mathfrak{F}(\mathbb{R}, G), v)$ is a maximal algebra with a valuation.*

Proof Set $A = \mathfrak{F}(\mathbb{R}, G)$, and let (B, w) be an immediate extension of (A, v).

To obtain a contradiction, assume that there exists $b \in B \setminus A$, and set

$$S = \{w(b - a) : a \in A\},$$

a subset of $\Gamma_B = G$. Since (B, w) is an immediate extension of (A, v), S has no maximum element. We now define a certain element f of A. If $x \in G$ with $S << x$, set $f(x) = 0$. If $x \in G$ and it is not true that $S << x$, then there exists $a \in A$ with $w(b - a) > x$: set $f(x) = a(x)$. If also $a' \in A$ with $w(b - a') > x$, then $w(a - a') > x$, and so $a(x) = a'(x)$. Thus $f(x)$ is well defined. Let (x_n) be a decreasing sequence in $\operatorname{supp} f$, and take $a \in A$ with $x_1 < w(b - a)$. Then (x_n) is contained in $\operatorname{supp} a$, and so (x_n) is eventually constant. Thus $\operatorname{supp} f$ is well-ordered, and hence $f \in A$.

Since $f \in A$, we have $w(b - f) \in S$. But, for each $x \in S$, necessarily $f(x) = b(x)$, and so $S << w(b - f)$. This is a contradiction, and thus $B = A$, as required. □

To give a full proof of our main theorem, Theorem 2.15, we include as a preliminary a proof of a classical result called the *fundamental theorem for the extension of places.*

Theorem 2.13 *Let A be a complex algebra with a character φ_A such that A is a unital subalgebra of a complex field K. Then there is a complex valuation algebra B with quotient field K such that $A \subset B \subset K$ and $\varphi_B \mid A = \varphi_A$.*

Proof Let $\mathcal{F}$ be the family of all pairs (V, ψ), where V is a subalgebra of K containing A and ψ is a character on V such that $\psi \mid A = \varphi_A$. Set

$$(V_1, \psi_1) \preccurlyeq (V_2, \psi_2)$$

in $\mathcal{F}$ if $V_1 \subset V_2$ and $\psi_2 \mid V_1 = \psi_1$. It is clear that each chain in $(\mathcal{F}, \preccurlyeq)$ has an upper bound, and so, by Zorn's lemma, $(\mathcal{F}, \preccurlyeq)$ has a maximal element, say (V, ψ). Set $L = \ker \psi$, a maximal ideal in V.

Take $a \in V \setminus L$, and set $b = a^{-1}$. Define

$$\psi' : \sum_{k=0}^{n} a_k b^k \mapsto \sum_{k=0}^{n} \psi(a_k)\psi(a)^{-k}, \quad V[b] \to \mathbb{C}.$$

Then ψ' is well defined, for, if $\sum_{k=0}^{n} a_k b^k = 0$ in $V[b]$, then $\sum_{k=0}^{n} a_k a^{n-k} = 0$ in V, and so

$$\sum_{k=0}^{n} \psi(a_k)\psi(a)^{-k} = 0.$$

It is now immediate that ψ' is a character on $V[b]$ and that $(V, \psi) \preccurlyeq (V[b], \psi')$ in $\mathcal{F}$. Thus $V[b] = V$ by the maximality of (V, ψ), and so $b \in V$ and $a \in \operatorname{Inv} V$. It follows that $L = M_V$ and that ψ is the unique character on V.

Take $x \in K \setminus \{0\}$. We *claim* that either $e \notin L[x]$ or $e \notin L[x^{-1}]$, where e is the identity of A. For assume towards a contradiction that both $e \in L[x]$ and $e \in L\left[x^{-1}\right]$. Then there exist $a_0, \ldots, a_m, b_0, \ldots, b_n \in L$ with $a_m b_n \neq 0$ such that

$$a_0 + a_1 x + \cdots + a_m x^m = e, \tag{2.4}$$

$$b_0 + b_1 x^{-1} + \cdots + b_n x^{-n} = e. \tag{2.5}$$

We may suppose that m and n are minimal for equations (2.4) and (2.5), respectively, to hold. Suppose that $m \geq n$, say. Now $e - b_0 \in \operatorname{Inv} V$, and so, from (2.5),

$$x^n = (e - b_0)^{-1}(b_1 x^{n-1} + \cdots + b_n) = p(x),$$

say, where $p \in L\left[X\right]$ and the degree of p is at most $n - 1$. Substituting in (2.4), we obtain an equation of the same type with the degree of the polynomial on the left-hand side at most $m - 1$. This contradicts the minimality of m. Thus either $e \notin L\left[x\right]$ or $e \notin L\left[x^{-1}\right]$.

Suppose, in fact, that $e \notin L\left[x\right]$. Since $L\left[x\right]$ is a proper ideal in $V\left[x\right]$, there is a maximal ideal, say M, in $V\left[x\right]$ with $L\left[x\right] \subset M$. Consider the maps

$$\sigma : p \longmapsto \psi(p(0)), \quad V\left[X\right] \to \mathbb{C},$$

and

$$\tau : p \longmapsto p(x), \;\; V[X] \to K\,.$$

First suppose that $\ker \tau \subset \ker \sigma$. Then there is a character χ on $V[x]$ such that $\chi(p(x)) = \psi(p(0))$. Since χ extends ψ, $(V, \psi) \preccurlyeq (V[x], \chi)$ in $\mathcal{F}$, and so, by the maximality of (V, ψ), we have $x \in V$. Second, suppose that $\ker \tau \not\subset \ker \sigma$. Then there exist $\alpha_0, \ldots, \alpha_n \in \mathbb{C}$ with $\alpha_0 \alpha_n \neq 0$ and $q \in L[X]$ such that

$$\alpha_0 + \alpha_1 x + \cdots + \alpha_n x^n + q(x) = 0\,.$$

Since $e \notin L[x]$, necessarily $n \geq 1$. Take $\beta_1, \ldots, \beta_n \in \mathbb{C}$ such that

$$\alpha_0 + \alpha_1 X + \cdots + \alpha_n X^n = \alpha_n (X - \beta_1) \cdots (X - \beta_n)$$

in $\mathbb{C}[X]$. Then

$$(x - \beta_1 e) \;\cdots\; (x - \beta_n e) \in M\,.$$

Since M is a maximal ideal in $V[x]$, there exists $j \in \{1, \ldots, n\}$ with $x - \beta_j e \in M$. Thus $x \in \mathbb{C}e + M$, and $V[x] \subset (\mathbb{C}e + L)(\mathbb{C}e + M) = \mathbb{C}e + M$. So M has codimension 1 in $V[x]$, and there is a character χ on $V[x]$ with $\ker \chi = M$ and $\chi \mid V = \psi$. Again, it follows that $x \in V$.

Similarly, if $e \notin L[x^{-1}]$, then $x^{-1} \in V$.

We have shown that, for each $x \in K \setminus \{0\}$, either $x \in V$ or $x^{-1} \in V$. Thus, for each $a, b \in V \setminus \{0\}$, either $a \in bV$ or $b \in aV$, and so V is a valuation ring with quotient field K. Since V has a character, V is a valuation algebra (with $M_V = L$ and $\varphi_V = \psi$). We set $B = V$ for the result. □

We say that a complex integral domain is *algebraically closed* if its quotient field is an algebraically closed field.

Corollary 2.14 *Let A be a complex valuation algebra. Then there is an algebraically closed valuation algebra B which is an algebraic extension of A such that $\varphi_B \mid A = \varphi_A$.*

Proof Let L be the quotient field of A, and let K be the algebraic closure of L. Then A is a unital subalgebra of K, and A has a character φ_A. By the theorem, there is a valuation algebra B with quotient field K such that $A \subset B \subset K$ and $\varphi_B \mid A = \varphi_A$. The algebra B is algebraically closed, and each element of B is algebraic over A. □

Theorem 2.15 *Let G be a totally ordered group, and let σ be a non-zero ordinal. Then:*

(i) *$\mathfrak{F}(\mathbb{R}, G)$ is a field, and $\mathfrak{F}_{(\sigma)}(\mathbb{R}, G)$ is a subfield of $\mathfrak{F}(\mathbb{R}, G)$;*

(ii) *$(\mathfrak{F}(\mathbb{R}, G), \leq)$ is an ordered field;*

(iii) *the ordered group G is the value group of each of the ordered fields $\mathfrak{F}(\mathbb{R}, G)$ and $\mathfrak{F}_{(\sigma)}(\mathbb{R}, G)$;*

(iv) *(Mac Lane) if G is a divisible group, then $\mathfrak{F}(\mathbb{R}, G)$ and $\mathfrak{F}_{(\sigma)}(\mathbb{R}, G)$ are real-closed fields.*

Proof Set $\mathfrak{F} = \mathfrak{F}(\mathbb{R}, G)$. So far, we know that $\mathfrak{F}$ is a unital integral domain, and that the Hahn valuation v is a valuation on $\mathfrak{F}$ with values in G.

Let K be the quotient field of $\mathfrak{F}$, and let $\overline{v}$ be the valuation on K extending v which is given by Proposition 2.10. Take $f, g \in \mathfrak{F} \setminus \{0\}$, and set $s = v(f)$, $t = v(g)$, and $\alpha = f(s)/g(t)$. Then $\alpha\delta_{s-t} \in \mathfrak{F}$, and

$$\overline{v}\left(\frac{f}{g} - \alpha\delta_{s-t}\right) = v(f - \alpha\delta_{s-t} * g) - v(g) > v(f) - v(g) = \overline{v}\left(\frac{f}{g}\right).$$

Thus $(K, \overline{v})$ is an immediate extension of $(\mathfrak{F}, v)$. By Proposition 2.12, we have $\mathfrak{F} = K$, and so $\mathfrak{F}$ is a field.

It is now clear that $\mathfrak{F}$ is an ordered field with value group G.

We next seek to show that $\mathfrak{F}(\mathbb{R}, G)$ is real-closed in the case where G is a divisible group. In fact, we shall show the equivalent result that $\mathfrak{F}(\mathbb{C}, G)$, the complexification of $\mathfrak{F}(\mathbb{R}, G)$, is algebraically closed in this case.

Set $A = \mathfrak{F}(\mathbb{C}, G)$, a complex valuation algebra. By Corollary 2.14, there is an algebraically closed valuation algebra B which is an algebraic extension of A. Now B is an extension of A in the sense of Definition 2.11 because $G \subset \Gamma_B$ and $A \cap \operatorname{Inv}B = \operatorname{Inv}A$. Since B is algebraic over A and G is a divisible group, it follows from Proposition 2.6 that $G = \Gamma_B$, and so B is an immediate extension of A. By Proposition 2.12, we have $A = B$, and so A is algebraically closed.

We have proved Theorem 2.15 for $\mathfrak{F}(\mathbb{R}, G)$ itself.

Now let σ be a non-zero ordinal. First take $f \in \mathfrak{F}_{(\sigma)}(\mathbb{R}, G) \setminus \{0\}$. Since $|\operatorname{supp} f| < \aleph_\sigma$, the subgroup G_f of G generated by $\operatorname{supp} f$ has $|G_f| < \aleph_\sigma$. Since $\mathfrak{F}(\mathbb{R}, G_f)$ is a field, f has an inverse, say g, in $\mathfrak{F}(\mathbb{R}, G_f)$, and we may regard g as an element of $\mathfrak{F}_{(\sigma)}(\mathbb{R}, G)$. Clearly $f * g = 1$ in $\mathfrak{F}_{(\sigma)}(\mathbb{R}, G)$, and so $\mathfrak{F}_{(\sigma)}(\mathbb{R}, G)$ is a field. Similarly it is shown that $\mathfrak{F}_{(\sigma)}(\mathbb{R}, G)$ is real-closed in the case where G is a divisible group.

This completes the proof of Theorem 2.15. □

Let K be the field $\mathfrak{F}(\mathbb{R}, G)$. Then we know that $K^{\#} = \{f \in K : v(f) \geq 0\}$; by analogy, we define

$$\mathfrak{F}(\mathbb{C}, G)^{\#} = \{f \in \mathfrak{F}(\mathbb{C}, G) : v(f) \geq 0\},$$

the set of *finite elements* of $\mathfrak{F}(\mathbb{C}, G)$, but we note that, in the case of this complex field, there is no intrinsic way of defining this subalgebra.

In Chapter 1, we explained how to 'fill a gap' in an ordered group. We must now carry through the same process in an ordered field. First we briefly recall some standard definitions and facts concerning transcendence degrees with respect to ordered fields. Let K and L be fields. Then L is an *extension field* of K if K is a subfield of L. In this case, let S be a subset of L. Then the subfield of L generated by $K \cup S$ (i.e. the intersection of all the subfields of L containing $K \cup S$) is the *subfield generated by S over K*; this subfield is denoted by $K(S)$. In the case where $S = \{x_1, \ldots, x_n\}$, $K(S)$ is denoted by $K(x_1, \ldots, x_n)$; it is easily checked that $K(x_1, \ldots, x_n)$ consists of all elements of L of the form

$$p(x_1, \ldots, x_n)/q(x_1, \ldots, x_n),$$

where $p, q \in K[X_1, \ldots, X_n]$, the polynomial algebra in n indeterminates over K, and $q(x_1, \ldots, x_n) \neq 0$.

Let L be an extension field of K, and take $x \in L$. A subset S of L is *algebraically dependent over* K if there exist $n \in \mathbb{N}$ and $p \in K[X_1, \ldots, X_n]$ with $p \neq 0$ such that $p(s_1, \ldots, s_n) = 0$ for some distinct elements $s_1, \ldots, s_n \in S$; otherwise, S is *algebraically independent over* K. A *transcendence base for* L *over* K is a subset S of L which is algebraically independent over K and such that every element of L is algebraic over $K(S)$. An algebraically independent set which is maximal in the family of all such subsets of K is a transcendence base, and the cardinality of a transcendence base is an invariant, called the *transcendence degree of* L *over* K. Let K and L be algebraically closed complex fields. Then K and L are isomorphic as complex fields if and only if they have the same transcendence degree over $\mathbb{C}$.

For example, L has transcendence degree 1 over K if L is an algebraic extension of $K(x)$ for some transcendental element x of L. A transcendental extension of K of transcendence degree 1 is obtained by forming the polynomial ring $K[X]$ with indeterminate X and taking its quotient field $K(X)$. The field $K(X)$ is an extension field of K called the *field of rational functions over* K. Clearly X is transcendental over K, and $K(X)$ has transcendence degree 1 over K.

An algebra $(A, +, \cdot, \leq)$ is an *ordered algebra* if $(A, +, \leq)$ is an ordered group and if $ab \in A^+$ whenever $a, b \in A^+$.

Proposition 2.16 *Let* $(K, \leq)$ *be a real-closed field, and let* $\langle A, B\rangle$ *be a gap in* K. *Then there is an ordering* $\leq$ *on the field* $K(X)$ *such that:*

(i) $(K(X), \leq)$ *is an ordered field;*

(ii) *the natural embedding of* $(K, \leq)$ *in* $(K(X), \leq)$ *is isotonic;*

(iii) $A << X << B$ *in* $(K(X), \leq)$.

The ordering on $K(X)$ *is uniquely specified by these conditions.*

Proof Since K is real-closed, each monic polynomial in $K[X]$ is a product of linear factors of the form $X - x_1$ and of irreducible quadratics of the form $(X - x_2)^2 + x_3^2$, where $x_1, x_2, x_3 \in K$. Set $X - x_1 > 0$ in $K[X]$ if there exists $a \in A$ with $a \geq x_1$, and set $(X - x_2)^2 + x_3^2 > 0$. For an arbitrary polynomial p in $K[X]$, set $p > 0$ if the number of negative factors in the expansion of p (where the coefficient of the highest power of X is regarded as a factor) is even. Clearly we have $A << X << B$ in $K[X]$.

Let $p \in K[X]$. By considering the factors of p, we see that $p > 0$ if and only if there exist $a \in A$ and $b \in B$ with $p(c) > 0$ for each $c \in [a, b]$. It follows easily that $(K[X], +, \cdot, \leq)$ is a totally ordered algebra.

The order on $K[X]$ is extended to $K(X)$ by setting $p/q > 0$ if $pq > 0$; clearly $(K(X), +, \cdot, \leq)$ is an ordered field, and the order is uniquely specified by the given conditions. □

The field $\mathcal{R}(K(X))$ with the extension of the ordering described in the above proposition is denoted by

$$K^{\langle A,B\rangle}\,.$$

Let K be a real-closed subfield of a field $(L, \leq)$, where $L = \mathcal{R}(K(a))$ for some $a \in L \setminus K$. Then there is an isomorphism

$$\iota : K(a) \to K(X)$$

of fields which is the identity on K with $\iota(a) = X$. Let $\langle A, B \rangle$ be the gap in K defined by a. Then ι extends to an isomorphism of the ordered fields L and $K^{\langle A,B \rangle}$. Thus the ordering on L is determined by the ordering on $K \cup \{a\}$.

The analogue for ordered fields of Hahn's embedding theorem (Theorem 1.35) for totally ordered groups is called Kaplansky's embedding theorem.

We shall require the following remark in the proof of the theorem. Let K be a real-closed field, and let Γ be a totally ordered group containing Γ_K as an ordered subgroup. Suppose that $\theta : K \to \mathfrak{F}(\mathbb{R}, \Gamma)$ is a monomorphism and that there is a subset S of K such that $\mathcal{R}(\mathbb{R}(S)) = K$ and $v(\theta(x)) = v(x) \quad (x \in S)$. Then θ is completely isotonic. Indeed, θ is certainly isotonic. To see that θ is completely isotonic, it suffices to show that, for each $a \in K$, there exists $b \in K$ such that $v(b) = v(a)$ and $v(\theta(b)) = v(b)$. This is clear for $a \in \mathbb{R}(S)$, and it follows for $a \in K$ by Proposition 2.6.

A gap in an ordered field K is a *pseudo-gap* if it is a pseudo-gap in the underlying ordered group $(K, +)$. By Proposition 1.32, gaps in totally ordered, divisible groups are pseudo-gaps if and only if they are 'value-preserving'; we shall see in Corollary 2.18 that the analogous characterization applies for field extensions, but this does not seem to be obvious, and we can only establish it *after* proving Kaplansky's embedding theorem.

Theorem 2.17 *(Kaplansky's embedding theorem) Let K be a real-closed field with value group Γ. Suppose that K_0 is a real-closed subfield of K with value group Γ_0, and that*

$$\theta_0 : K_0 \to \mathfrak{F}(\mathbb{R}, \Gamma_0) \subset \mathfrak{F}(\mathbb{R}, \Gamma)$$

is a completely isotonic homomorphism. Then there is a completely isotonic homomorphism $\theta : K \to \mathfrak{F}(\mathbb{R}, \Gamma)$ such that $\theta \mid K_0 = \theta_0$ and such that $\delta_s \in \theta(K)$ for each $s \in \Gamma \setminus \Gamma_0$.

Proof Consider the family $\mathcal{F}$ of pairs (L, θ), where L is a real-closed subfield of K with $L \supset K_0$, and $\theta : L \to \mathfrak{F}(\mathbb{R}, \Gamma_L)$ is a completely isotonic homomorphism such that $\theta \mid K_0 = \theta_0$ and such that $\delta_s \in \theta(L) \quad (s \in \Gamma_L \setminus \Gamma_0)$. Again, we may suppose that (L, θ) is a maximal element of $\mathcal{F}$ for the obvious ordering, $\preccurlyeq$. To obtain a contradiction, assume that $L \neq K$. There are two cases to consider.

Case 1 Suppose that $\Gamma_L = \Gamma$.

We choose $a \in K^+ \setminus L$, and we let $\langle A, B \rangle$ be the gap in L defined by a. Since L is real-closed, a is transcendental over L. By Proposition 1.32, $\langle A, B \rangle$ is a pseudo-gap, and so, by the argument in the proof of Theorem 1.35, there exists $f \in \mathfrak{F}(\mathbb{R}, \Gamma_L)$ with $\theta(A) << f << \theta(B)$. Define a monomorphism

$$\overline{\theta} : L(a) \to \mathfrak{F}(\mathbb{R}, \Gamma_L) = \mathfrak{F}(\mathbb{R}, \Gamma)$$

by requiring that $\overline{\theta} \mid L = \theta$ and $\overline{\theta}(a) = f$. Then $\overline{\theta}$ extends to a monomorphism $\overline{\theta} : \mathcal{R}(L(a)) \to \mathfrak{F}(\mathbb{R}, \Gamma_L)$. Certainly $v(f) = v(a)$, and so, by the remark before the theorem, $\overline{\theta}$ is completely isotonic.

Case 2 Suppose that $\Gamma_L \neq \Gamma$.

There exists $s_0 \in \Gamma \setminus \Gamma_L$, and we now choose $a \in K^+ \setminus L$ with $v(a) = s_0$. Define

$$\overline{\theta} : L(a) \to \mathfrak{F}(\mathbb{R}, \Gamma_L \oplus \mathbb{Q}s_0) \subset \mathfrak{F}(\mathbb{R}, \Gamma)$$

by requiring that $\overline{\theta} \mid L = \theta$ and $\overline{\theta}(a) = \delta_{s_0}$. Set $M = \mathcal{R}(L(a))$. The group $\Gamma_L \oplus \mathbb{Q}s_0$ is divisible, and so, by Mac Lane's theorem, Theorem 2.15(iv), the field $\mathfrak{F}(\mathbb{R}, \Gamma_L \oplus \mathbb{Q}s_0)$ is real-closed. Thus, $\overline{\theta}$ extends to a monomorphism from M into $\mathfrak{F}(\mathbb{R}, \Gamma_L \oplus \mathbb{Q}s_0)$. Since $a > 0$, we have $a^\alpha \in M$ for each $\alpha \in \mathbb{Q} \setminus \{0\}$, and clearly $\overline{\theta}(a^\alpha) = \delta_{\alpha s_0}$ and $v(\overline{\theta}(a^\alpha)) = \alpha s_0 \quad (\alpha \in \mathbb{Q} \setminus \{0\})$. Thus $\overline{\theta}$ is completely isotonic, and $\delta_t \in \theta(M)$ for each $t \in \Gamma_M \setminus \Gamma_0$.

In each of the two cases, we have constructed an element $(L(a), \overline{\theta})$ of $\mathcal{F}$ with $(L(a), \overline{\theta}) \succcurlyeq (L, \theta)$ and $a \notin L$. But this contradicts the assumed maximality of (L, θ). Thus $L = K$, and $\theta : L \to \mathfrak{F}(\mathbb{R}, \Gamma)$ is the required map. □

Corollary 2.18 *Let $\langle A, B \rangle$ be a gap in a real-closed field K, and set*

$$L = K^{\langle A, B \rangle}.$$

Then $\langle A, B \rangle$ is a pseudo-gap if and only if $\Gamma_L = \Gamma_K$.

Proof Set $\mathfrak{F} = \mathfrak{F}(\mathbb{R}, \Gamma_K)$. By the theorem, we may regard K as a subfield of $\mathfrak{F}$ such that $\delta_s \in K \quad (s \in \Gamma_K)$.

Suppose that $\langle A, B \rangle$ is a pseudo-gap in K. Then $\langle A, B \rangle$ is not a gap in $\mathfrak{F}$, say $A << f << B$ in $\mathfrak{F}$. There is an embedding $\theta : L \to \mathfrak{F}$ such that θ is the identity on K and $\theta(X) = f$. Thus $\Gamma_L = \Gamma_K$.

Conversely, suppose that $\Gamma_L = \Gamma_K$. Then the value set of the totally ordered group $K_{\langle A, B \rangle}$ is Γ_K, and so $\langle A, B \rangle$ is a pseudo-gap by Proposition 1.32. □

We now come to definitions about the order structure on a field analogous to those for a group given in Definition 1.37.

Definition 2.19 *Let K be an ordered field. Then K is:*

(i) *an* α_1-field *(respectively, an* α_2-field*) if $(K, \leq)$ is an α_1-set (respectively, an α_2-set);*

(ii) *a* β_1-field *if $K = \bigcup\{K_\nu : \nu < \rho\}$, where $\{K_\nu : \nu < \rho\}$ is a chain of α_1-subfields of K;*

(iii) *an* η_1-field *if $(K, \leq)$ is an η_1-set;*

(iv) *a* semi-η_1-field *if $(K, \leq)$ is a semi-η_1-set.*

For example, $\mathbb{R}$ itself is a semi-η_1-field.

Again some care is needed in the definition of a β_1-field. Because of our convention about the meaning of the word 'field', the subfields K_ν which appear in (ii), above, must contain $\mathbb{R}$. We do not know whether a field which is a union of a chain of rational subfields each of which is an α_1-set is necessarily a β_1-field. Also, we think it unlikely that an ordered field which is a β_1-set or a β_1-group is necessarily a β_1-field. Note that, if K is a β_1-field, then $|K| = \mathfrak{c}$.

We shall occasionally consider rational ordered fields K such that $(K, \leq)$ is an η_1-set; we shall term these η_1-*rational fields*. This is the concept to which the term 'η_1-field' is usually applied in the earlier literature, but we stress again that, for us, the terms 'η_1-field' and 'semi-η_1-field' imply that the ground field of the field is $\mathbb{R}$.

It is clear that Γ_K is a semi-η_1-group (respectively, a β_1-group) whenever K is a semi-η_1-field (respectively, a β_1-field). Let G be a totally ordered group of cardinality $\mathfrak{c}$ which is not a β_1-group. Then $\mathfrak{F}_{(1)}(\mathbb{R}, G)$ is a field of cardinality $\mathfrak{c}$ which is not a β_1-field.

We start our discussion of η_1-fields with a minor result that gives an easy way of showing that a semi-η_1-field is actually an η_1-field. The base number $\delta(K)$ of an ordered field K is defined to be the base number of the totally ordered group $(K, +, \leq)$, as in Definition 1.19. Clearly $\delta(K) = \operatorname{cof} K$.

Proposition 2.20 *Let K be a semi-η_1-field with $\delta(K) \geq \aleph_1$. Then K is an η_1-field.*

Proof Let A and B be countable subsets of K with $A << B$. We must find $c \in K$ with $A << c << B$; we may suppose that $0 << A$ and $0 << B$.

If $A = \emptyset$ and B has a minimum b, set $c = b/2$. If $B = \emptyset$ and A has a maximum a, set $c = 2a$. If A has a maximum a and B has a minimum b, set $c = (a + b)/2$.

If B has no minimum, then B contains a strictly decreasing, coinitial sequence, say (b_n). If $A = \emptyset$, there exists $c \in K$ with $0 < c << \{b_n : n \in \mathbb{N}\}$ because $\delta(K) \geq \aleph_1$. If A has a maximum a, take $x \in K$ with

$$0 < x << \{b_n - a : n \in \mathbb{N}\},$$

and set $c = x + a$.

If A has no maximum, then A contains a strictly increasing, cofinal sequence (a_n). If $B = \emptyset$, take $x \in K$ with $0 < x < a_n^{-1}$, and set $c = x^{-1}$. If B has no minimum, there exists $c \in K$ with $\{a_n : n \in \mathbb{N}\} << c << \{b_n : n \in \mathbb{N}\}$ because K is a semi-η_1-set.

In each case, $A << c << B$, and so K is an η_1-field. □

Proposition 2.21 *Let K be a real-closed semi-η_1-field which is also a β_1-field. Then the value group of K is isomorphic to exactly one of the following list of totally ordered groups:*

$$(1)\ \{0\};\quad (2)\ \mathbf{G};\quad (3)\ (\mathbf{R} \times \mathbf{G} \times \mathbb{R}, \preceq);\quad (4)\ \left(c_{00}\left(\mathbf{G}^{\mathbb{N}}\right), \preceq^*\right).$$

Proof Suppose that $K \cong \mathbb{R}$. Then $\Gamma_K = \{0\}$. If $K \ncong \mathbb{R}$, then Γ_K is an infinite, divisible semi-η_1-group which is a β_1-group, and so, by Theorem 1.50, Γ_K is isomorphic to exactly one of the groups G in Table 1.1 (page 34). However, Γ_K has the additional properties that $\delta(\Gamma_K) = \operatorname{cof} \Gamma_K$ (in the extended sense of Table 1.1), and that each open interval in Γ_K is an η_1-set. Thus we have $\Gamma_K \simeq \mathbf{G}$ or $\Gamma_K \simeq (\mathbb{R} \times \mathbf{G} \times \mathbb{R}, \preceq)$ or $\Gamma_K \simeq (c_{00}(\mathbf{G}^{\mathbb{N}}), \preceq^*)$. □

We shall see in Chapter 4 that each of the four possibilities in the above result does arise, even when K is required to belong to the special class of 'super-real' fields.

We now introduce some specific notation for three important fields.

Definition 2.22 *For $\sigma < \omega_1$, set $\mathbf{R}_\sigma = \mathfrak{F}(\mathbb{R}, \mathbf{G}_\sigma)$. Also, set*

$$\mathbf{R} = \mathfrak{F}_{(1)}(\mathbb{R}, \mathbf{G}), \quad \mathbf{C} = \mathfrak{F}_{(1)}(\mathbb{C}, \mathbf{G}), \quad \widehat{\mathbf{R}} = \mathfrak{F}(\mathbb{R}, \mathbf{G}).$$

Theorem 2.23 (i) *Both $\mathbf{R}$ and $\widehat{\mathbf{R}}$ are real-closed η_1-fields with value group $\mathbf{G}$.*

(ii) *Each $\mathbf{R}_\sigma$ is an α_1-field, and $\mathbf{R} = \bigcup\{\mathbf{R}_\sigma : \sigma < \omega_1\}$ is a β_1-field.*

(iii) $|\mathbf{R}| = w(\mathbf{R}) = \mathfrak{c}$ *and* $\operatorname{cof} \mathbf{R} = \delta(\mathbf{R}) = \aleph_1$.

(iv) $\left|\widehat{\mathbf{R}}\right| = w(\widehat{\mathbf{R}}) = 2^{\aleph_1}$ *and* $\operatorname{cof} \widehat{\mathbf{R}} = \delta(\widehat{\mathbf{R}}) = \aleph_1$.

(v) *$\widehat{\mathbf{R}}$ is an α_2-field, and both $\mathbf{R}$ and $\widehat{\mathbf{R}}$ contain the following types of gap, and no other types:*

$$(0, \aleph_1),\ (\aleph_1, 0),\ (1, \aleph_1),\ (\aleph_1, 1),\ (\aleph_0, \aleph_1),\ (\aleph_1, \aleph_0),\ (\aleph_1, \aleph_1).$$

(vi) *$\widehat{\mathbf{R}}$ is not order-isomorphic to $\mathbf{R}$.*

(vii) *There is an isotonic map from $\widehat{\mathbf{R}}$ into each order-dense subset of itself.*

(viii) *$\widehat{\mathbf{R}}$ does not contain an order-dense subset which is a β_1-set.*

(ix) *The field $\mathbf{C}$ is algebraically closed, and $|\mathbf{C}| = \mathfrak{c}$.*

Proof (i) By Proposition 1.43(i), $\mathbf{G}$ is a divisible η_1-group, and so, by Theorems 2.15(iv) and 1.39(i), $\mathbf{R}$ and $\widehat{\mathbf{R}}$ are real-closed η_1-fields with value group $\mathbf{G}$.

(ii) By Proposition 1.43(ii), each $\mathbf{G}_\sigma$ is an α_1-group, and so, by Theorem 1.39(iii), each $\mathfrak{F}(\mathbb{R},\mathbf{G}_\sigma)$ is an α_1-field. Hence $\mathbf{R}$ is a β_1-field.

(iii) By (i) and (ii), $\mathbf{R}$ is a β_1-η_1-set, and hence, by Theorem 1.15, $\mathbf{R}$ is order-isomorphic to $\mathbf{Q}$. Thus the result follows from Proposition 1.9(iv) and (v).

(iv) By Proposition 1.43(iii), $\mathbf{G}$ is a β_1-set, and hence an α_2-set. It follows from Theorem 1.39(ii) that $\left|\widehat{\mathbf{R}}\right| \leq 2^{\aleph_1}$.

By Proposition 1.44(iii), the group $\mathbf{G}$ contains a strictly increasing sequence, say $(x_\sigma : \sigma < \omega_1)$, which is bounded above, say by y. The set

$$S = \{x_\sigma : \sigma < \omega_1\} \cup \{y\}$$

is a well-ordered subset of $\mathbf{G}$. Let T be an order-dense subset of $\widehat{\mathbf{R}}$, and, for each $f \in \mathbb{R}^{\omega_1}$, define

$$f_1(x_\sigma) = f_2(x_\sigma) = f(\sigma) \quad (\sigma < \omega_1), \quad f_1(y) = 0, \quad f_2(y) = 1,$$
$$f_1 \mid (\mathbf{G} \setminus S) = f_2 \mid (\mathbf{G} \setminus S) = 0.$$

Then $f_1 < f_2$ in $\widehat{\mathbf{R}}$, and so there exists $h \in T$ with $f_1 \le h \le f_2$. Clearly $h(x_\sigma) = f(\sigma) \quad (\sigma < \omega_1)$, and so

$$|T| \ge \left|\mathbb{R}^{\omega_1}\right| = 2^{\aleph_1}.$$

Thus $w(\widehat{\mathbf{R}}) = 2^{\aleph_1}$.

By Propositions 1.28(iii) and 1.44(ii), $\operatorname{cof} \widehat{\mathbf{R}} = \aleph_1$, and so $\delta(\widehat{\mathbf{R}}) = \aleph_1$.

(v) The fields $\mathbf{R}$ and $\widehat{\mathbf{R}}$ are η_1-fields and, by Proposition 1.39(iii), both are α_2-sets, where we use the fact that $\mathbf{G}$ is an α_2-set. By Proposition 1.44(iii), $\mathbf{G}$ contains an $(\aleph_1, \aleph_1)$-gap, and so, by Proposition 1.24, $\mathbf{R}$ and $\widehat{\mathbf{R}}$ both contain $(\aleph_1, \aleph_1)$-gaps. The result follows from Proposition 1.11(i).

(vi) By Proposition 1.41, there is no isotonic map from $\widehat{\mathbf{R}}$ into $\mathbf{R}$.

(vii) The totally ordered sets $\mathbf{G}$ and $\mathbf{G}^- \setminus \{0\}$ are both β_1-η_1-sets, and hence are order-isomorphic. Thus $\widehat{\mathbf{R}}$ and $\mathfrak{F}(\mathbb{R}, \mathbf{G}^- \setminus \{0\})$ are order-isomorphic.

Now let S be an order-dense subset of $\widehat{\mathbf{R}}$. For each $f \in \mathfrak{F}(\mathbb{R}, \mathbf{G}^- \setminus \{0\})$, there exists $\overline{f} \in S$ with $\overline{f} \mid (\mathbf{G}^- \setminus \{0\}) = f$, and the map $f \mapsto \overline{f}$ is isotonic. Thus there is an isotonic map from $\widehat{\mathbf{R}}$ into S.

(viii) Assume towards a contradiction that S is a β_1-set which is order-dense in $\widehat{\mathbf{R}}$. By (vii), there is an isotonic map from $\widehat{\mathbf{R}}$ into S, and so $\widehat{\mathbf{R}}$ is a β_1-set. By the isomorphism theorem, Theorem 1.15, $\mathbf{R}$ and $\widehat{\mathbf{R}}$ are order-isomorphic. But this is a contradiction of (vi).

(ix) This is clear from (i) and (iii). □

Theorem 2.24 *Let K be a field such that Γ_K is a β_1-group. Then there is an embedding of K in $\widehat{\mathbf{R}}$.*

Proof Set $L = \mathcal{R}K$. By Kaplansky's embedding theorem, Theorem 2.17, there is an embedding of L in $\mathfrak{F}(\mathbb{R}, \Gamma_L)$. By Proposition 2.6, Γ_L is a divisible β_1-group, and so, by Theorem 1.44, there is an isotonic group morphism from Γ_L into $\mathbf{G}$. Thus there is an embedding of L (and hence of K) in $\mathfrak{F}(\mathbb{R}, \mathbf{G}) = \widehat{\mathbf{R}}$. □

The following easy example will be referred to in Theorem 8.63.

Proposition 2.25 *There is a real-closed η_1-field K with $|K| = \aleph_2$ such that K contains both an $(\aleph_0, \aleph_1)$-gap and an $(\aleph_0, \aleph_2)$-gap.*

Proof Let S be the totally ordered set $\omega \odot \omega_2^*$, where ω_2^* is the ordinal ω_2 with the reverse of the usual order, and set $T = S \times \mathbf{G}$ with the lexicographic order. Clearly T contains an $(\aleph_0, \aleph_2)$-gap. By Proposition 1.44(iii), $\mathbf{G}$ contains an $(\aleph_0, \aleph_1)$-gap, and so T contains an $(\aleph_0, \aleph_1)$-gap. Also T is an η_1-set because $\mathbf{G}$ is an η_1-set, and $|T| = \aleph_2$.

Set $G = \mathfrak{F}_{(1)}(\mathbb{R}, T)$ and $K = \mathfrak{F}_{(1)}(\mathbb{R}, G)$. It follows from Theorem 1.39(i) that G is an η_1-group and K is an η_1-field. By Proposition 1.24, G and K both have an $(\aleph_0, \aleph_1)$-gap and an $(\aleph_0, \aleph_2)$-gap. By Proposition 1.28(i),

$$|K| = |G| = |T| = \aleph_2 .$$

□

The following two results do not seem to be immediately obvious, but they do follow from Kaplansky's embedding theorem.

Proposition 2.26 *Let K be an α_1-field. Then $\mathcal{R}K$ is also an α_1-field.*

Proof Since K is an α_1-field, Γ_K is an α_1-group. Set $L = \mathcal{R}K$. By Proposition 2.6, Γ_L is an α_1-group, and so, by Theorem 1.39(iii), $\mathfrak{F}(\mathbb{R}, \Gamma_L)$ is an α_1-field. By Theorem 2.27, L is a subfield of $\mathfrak{F}(\mathbb{R}, \Gamma_L)$, and so L is an α_1-field. □

Proposition 2.27 *Let K be a real-closed subfield of a field L, and let $a \in L \backslash K$. Suppose that K is an α_1-field. Then $K(a)$ is an α_1-field.*

Proof Again, Γ_K is an α_1-group. Set $s = v(a) \in \Gamma_L$. In the case where $v(b_0 - a) \notin \Gamma_K$ for some $b_0 \in K$, we may suppose that $s \notin \Gamma_K$, and so, in each case

$$v(b - a) \in \Gamma_K \cup \{s\} \quad (b \in K).$$

Let $p \in K[X]$, say $p = b_0 rs$, where $b_0 \in K$, r is a product of linear terms $b - X$, and s is a product of irreducible quadratics of the form $(X - c)^2 + d^2$. By (1.2),

$$v((a - c)^2 + d^2) = 2\min\{v(a - c), v(d)\},$$

and so $v(p(a)) \in \Gamma_K + \mathbb{Z}^+ s$. It follows that $\Gamma_{K(a)} \subset \Gamma_K + \mathbb{Z}s$, an α_1-group, and hence, by Theorem 1.39(iii), $K(a)$ is an α_1-field. □

Corollary 2.28 *Let K be a real-closed field with transcendence degree at most $\aleph_1$ over $\mathbb{R}$. Then K is a β_1-field.*

Proof Let $\{a_\sigma : \sigma < \omega_1\}$ be a transcendence base for K over $\mathbb{R}$. Set $K_0 = \mathbb{R}$, and inductively define the fields K_σ for $\alpha < \omega_1$ by setting $K_{\sigma+1} = \mathcal{R}(K_\sigma(a_\sigma))$ and $K_\sigma = \bigcup\{K_\zeta : \zeta < \sigma\}$ in the case where σ is a limit ordinal. Suppose that K_σ is an α_1-field. Then $K_{\sigma+1}$ is an α_1-field by Propositions 2.26 and 2.27. If σ is a limit ordinal (with $\sigma < \omega_1$) and each K_ζ for $\zeta < \sigma$ is an α_1-field, then K_σ is an α_1-field. Thus $\{K_\sigma : \sigma < \omega_1\}$ is a chain of α_1-fields, and $K = \bigcup\{K_\sigma : \sigma < \omega_1\}$ is a β_1-field. □

We have noted that a β_1-field K has $|K| \leq \mathfrak{c}$; the above result shows that K is a β_1-field in the case where K is a real-closed field with $|K| = \aleph_1$. Thus the theory of β_1-fields becomes, in ZFC + CH, the theory of ordered fields of cardinality $\aleph_1$.

The final stage of our theory in the present chapter is the development of results analogous to those given for totally ordered sets in Lemma 1.12 to Theorem 1.15 and for totally ordered groups in Lemma 1.45 to Corollary 1.49.

Lemma 2.29 *Let K be a real-closed α_1-field, and let L be real-closed η_1-field. Suppose that K_0 is a real-closed subfield of K, and that there is an embedding $\theta_0 : K_0 \to L$. Then there is an embedding $\theta : K \to L$ such that $\theta \mid K_0 = \theta_0$.*

Proof Suppose that K_1 is a proper, real-closed subfield of the field K, that $\theta_1 : K_1 \to L$ is an embedding, and that $a \in K \setminus K_1$ defines the gap $\langle A, B\rangle$ in K_1. Set $L_1 = \theta_1(K_1)$. Since K is an α_1-field and L is an η_1-field, there exists $b \in L$ which interpolates the pregap $\langle \theta_1(A), \theta_1(B)\rangle$. Since K_1 and L_1 are real-closed, a and b are transcendental with respect to K_1 and L_1, respectively, and so θ_1 extends to a unique isomorphism $\overline{\theta}_1$ from $K_1(a)$ onto $L_1(b)$. Since the orderings on $K_1(a)$ and on $L_1(b)$ are determined by the orderings on $K_1 \cup \{a\}$ and $L_1 \cup \{b\}$, respectively, the map $\overline{\theta}_1$ is isotonic. Next $\overline{\theta}_1$ extends to an embedding from $\mathcal{R}(K_1(a))$ onto $\mathcal{R}(L_1(b))$.

The remainder of the proof is the same as that in Lemma 1.45. □

Theorem 2.30 *Let K be a β_1-field, and let L be a real-closed η_1-field. Then there is an embedding of K in L. In particular, there is an embedding of K in* **R**.

Proof By hypothesis, there is a chain $\{K_\nu\}$ of α_1-subfields of K such that $K = \bigcup K_\nu$, and clearly $\mathcal{R}K = \bigcup \mathcal{R}K_\nu$. By Proposition 2.26, each $\mathcal{R}K_\nu$ is an α_1-field. The result then follows from the lemma, as before. □

Note that Theorem 2.30 asserts that **R** is *universal* in the class of real-closed β_1-fields.

Corollary 2.31 *Let K be a real-closed β_1-field. Then there is a chain $\{K_\sigma : \sigma < \omega_1\}$ of real-closed α_1-subfields of K such that*

$$K = \bigcup \{K_\sigma : \sigma < \omega_1\}.$$

Proof By the theorem, there is an embedding $\theta : K \to \mathbb{R}$. Set $K_\sigma = \theta^{-1}(\mathbf{R}_\sigma)$ for $\sigma < \omega_1$. □

It follows immediately from Corollary 2.31 that the union of a family of size $\aleph_1$ of β_1-subfields of a given field is itself a β_1-subfield.

Theorem 2.32 *Let K be a real-closed semi-η_1-field which is a β_1-field. Then there is a completely isotonic isomorphism from K onto $\mathfrak{F}_{(1)}(\mathbb{R}, \Gamma_K)$.*

Proof Set $K = \bigcup\{K_\sigma : \sigma < \omega_1\}$, where $\{K_\sigma : \sigma < \omega_1\}$ is a chain of real-closed α_1-subfields of K. We formulate the notion of a *valid union* in an analogous way to that in Theorem 1.48, and take a maximal such union, $K = \bigcup\{L_\sigma : \sigma < \omega_1\}$, as before.

We *claim* that each L_σ is pseudo-complete. Assume towards a contradiction that $\langle A, B\rangle$ is a pseudo-gap in L_σ for some $\sigma < \omega_1$, and let τ be the least ordinal such that there exists $a \in L_\tau$ with $A << a << B$. Set $\overline{L}_\zeta = \mathcal{R}(L_\zeta(a))$ $(\zeta \geq \sigma)$ and set $\overline{L}_\zeta = L_\zeta$ $(\zeta < \sigma)$. We now use Corollary 2.18 to see that the value group of $\overline{L}_\zeta$ is the same as the value group of L_ζ for each ζ with $\sigma \leq \zeta < \omega_1$.

The remainder of the proof follows as in Theorem 1.48. □

Corollary 2.33 *(Esterle) Let K be a real-closed β_1-η_1-field. Then there is an isomorphism from K onto $\mathbf{R}$.*

Proof The value group Γ_K is a divisible β_1-η_1-group, and so, by Corollary 1.49, Γ_K is isomorphic to $\mathbf{G}$. The result now follows from the theorem. □

Corollary 2.34 (CH) (i) *There is an embedding of each real-closed field of cardinality $\mathfrak{c}$ in $\mathbf{R}$.*

(ii) *Each real-closed η_1-field of cardinality $\mathfrak{c}$ is isomorphic to $\mathbf{R}$.*

Proof By Corollary 2.28, a real-closed field of cardinality $\aleph_1$ is a β_1-field, and so, by Theorem 2.30, there is an embedding of it in $\mathbf{R}$; by Corollary 2.33, the field is isomorphic to $\mathbf{R}$ in the case where it is an η_1-field. □

Corollary 2.35 *Let K be a real-closed semi-η_1-field which is a β_1-field. Then exactly one of the following occurs:*

(1) $w(K) = \aleph_0$, *and* $K \cong \mathbb{R}$;

(2) $\operatorname{cof} K = \aleph_1$, *and* $K \cong \mathbf{R}$;

(3) $\operatorname{cof} \Gamma_K = (\aleph_0, 1)$, *and then* $K \cong \mathfrak{F}_{(1)}(\mathbb{R}, (\mathbb{R} \times \mathbf{G} \times \mathbb{R}, \preceq))$;

(4) $\operatorname{cof} \Gamma_K = (\aleph_0, \aleph_0)$ *and then* $K \cong \mathfrak{F}_{(1)}(\mathbb{R}, (c_{00}(\mathbf{G}^{\mathbb{N}}), \preceq^*))$.

Proof By the theorem, $K \cong \mathfrak{F}_{(1)}(\mathbb{R}, \Gamma_K)$. The four possibilities for Γ_K are listed in Proposition 2.21. If $w(K) = \aleph_0$, then $|\Gamma_K| \leq \aleph_0$, and so $K \cong \mathbb{R}$. If $\operatorname{cof} K = \aleph_1$, then K is an η_1-field by Proposition 2.20, and so $K \cong \mathbf{R}$. □

Of course, the fact that a semi-η_1-field of weight $\aleph_0$ is isomorphic to $\mathbb{R}$ is immediately obvious.

In Theorem 2.23(ii), we expressed $\mathbf{R}$ as the union $\mathbf{R} = \bigcup\{\mathbf{R}_\sigma : \sigma < \omega_1\}$. There is another way of expressing $\mathbf{R}$ as a union; this will be useful later.

Theorem 2.36 *There is an ordinal ρ_0 such that $\mathbf{R}$ can be expressed in the form*

$$\mathbf{R} = \bigcup\{\mathbf{R}^{(\sigma)} : \sigma < \rho_0\},$$

where:

(i) $\mathbf{R}^{(0)} = \mathbb{R}$;

(ii) *for each* $\sigma < \rho_0$, $\mathbf{R}^{(\sigma+1)} = \mathcal{R}(\mathbf{R}^{(\sigma)}(a_\sigma))$ *for some* $a_\sigma \in \mathbf{R} \setminus \mathbf{R}^{(\sigma)}$ *such that a_σ defines a countable gap in* $\mathbf{R}^{(\sigma)}$;

(iii) *for each limit ordinal* $\sigma < \rho_0$, $\mathbf{R}^{(\sigma)} = \bigcup\{\mathbf{R}^{(\tau)} : \tau < \sigma\}$.

Proof Consider the family $\mathcal{F}$ of pairs $(\rho, \{\mathbf{R}^{(\sigma)} : \sigma < \rho\})$, where ρ is an ordinal (necessarily with $\rho \leq \mathfrak{c}$) and $\{\mathbf{R}^{(\sigma)} : \sigma < \rho\}$ is a family satisfying (i), (ii), and (iii) (with ρ for ρ_0). There is an obvious ordering on $\mathcal{F}$, and each chain in this ordering has an upper bound, and so $\mathcal{F}$ has a maximal element, corresponding to ρ_0, say. Set $S = \bigcup\{\mathbf{R}^{(\sigma)} : \sigma < \rho_0\}$.

We *claim* that S is an η_1-set. For assume towards a contradiction that $\langle A, B\rangle$ is a countable gap in S. Then there exists $x \in \mathbf{R}$ with $A << x << B$. Set $\mathbf{R}^{(\sigma)} = \mathcal{R}(S(x))$. Then

$$(\rho_0 + 1, \{\mathbf{R}^{(\sigma)} : \sigma < \rho_0 + 1\})$$

is an element of $\mathcal{F}$ which exceeds the specified maximum element of $\mathcal{F}$, a contradiction. Thus S is an η_1-set.

Since S is a subfield of $\mathbf{R}$, S is a β_1-field.

By Corollary 2.33, $S \cong \mathbf{R}$, and so $\mathbf{R}$ itself can be expressed in the required form. □

Notes 2.37 The basic theory of the extensions of fields is given in Hungerford (1980, Chapter VII). For the Artin–Schreier theory of ordered fields, see Jacobson (1980), for example. The first main result of this theory is that a rational field K can be ordered if and only if it is formally real, in the sense that, if $a_1, \ldots, a_n \in K$ with $a_1^2 + \cdots + a_n^2 = 0$, then necessarily $a_1 = \cdots = a_n = 0$. For a survey of ordered fields (from a different point of view to ours), see Lam (1980).

The formal power series algebra $\mathfrak{F}(\mathbb{F}, G)$ was introduced by Hahn (1907); see Fuchs (1963) and Priess-Crampe (1983), for example, for further details.

Definition 2.9 differs slightly from the usual definition of a valuation by the insertion of clause (iv). This condition ensures that a valuation defines a valuation algebra, and not just a valuation ring. A classic text on valuations is Schilling (1950). Forms of the fundamental theorem for the extension of places, Theorem 2.13, are given in Bourbaki (1985, VI, §1, Théorème 2), Jacobson (1980, II, §9.8), Lang (1958, p.8), and Zariski and Samuel (1960, Chapter VI).

Part (i) of Theorem 2.15 is a slight extension of a classical result of Hahn (Fuchs 1963; Priess-Crampe 1983, II . 5 . Satz 3). Part (iv) is *Mac Lane's theorem* (Mac Lane 1939); see Alling (1962*a*, I, 2.3), Alling (1987, 2.5.23), and Priess-Crampe (1983, II . 5 . Satz 6).

The argument in Proposition 2.16 is very close to that given in Lemma 13.12 of Gillman and Jerison (1960).

Kaplansky's embedding theorem, Theorem 2.17, was given in the seminal paper on totally ordered fields (Kaplansky 1942) of half a century ago. In view of the importance of this result, we feel that it deserves more than the modest number of expositions that have been given in texts. Our approach, based on the notion of gaps, is different from that used before. It should be noted however that we have considered only the special case of Kaplansky's theorem in which the ground field is $\mathbb{R}$. Kaplansky himself considered more general fields, and obtained his results under a certain 'Hypothesis A' on the ground field; the meaning of Hypothesis A was clarified in Whaples (1957) and Kuhlmann *et al.* (1986). The theory of valued fields is presented, in a more general setting than ours, in Kuhlmann (1996).

The fields $\mathbf{R}$ and $\widehat{\mathbf{R}}$ of Definition 2.22 were introduced by Esterle (1984), and Theorem 2.30 and Corollary 2.33 were first proved in those lectures. Again, our present approach is different.

The fact that any two real-closed η_1-fields of cardinality $\aleph_1$ are isomorphic as rings is (Gillman and Jerison 1960, Theorem 13.13). The extra fact that they are isomorphic as fields was proved in Johnson (1976). In the presence of CH, this latter result is exactly Corollary 2.33.

Let α be a non-zero ordinal. An ordered field K is an *η_α-field* if $(K, \leq)$ is an η_α-set. We have remarked in the notes to Chapter 1 that there is a theory of η_α-sets. The route that we used to build η_1-fields from η_1-sets can be used to build η_α-fields from η_α-sets (Alling 1962*a*): if S is an η_α-set with $|S| = \aleph_\alpha$, then $\mathfrak{F}_{(\alpha)}(\mathbb{R}, S)$ is an η_α-set with $\left|\mathfrak{F}_{(\alpha)}(\mathbb{R}, S)\right| = \aleph_\alpha$. Thus, let $\aleph_\alpha$ be a regular cardinal. Then $\mathbf{Q}_{(\alpha)}$ (defined in Notes 1.52) is an η_α-set, $\mathbf{G}_{(\alpha)} = \mathfrak{F}_{(\alpha)}(\mathbb{R}, \mathbf{Q}_{(\alpha)})$ is an η_α-group, and $\mathbf{R}_{(\alpha)} = \mathfrak{F}_{(\alpha)}(\mathbb{R}, \mathbf{G}_{(\alpha)})$ is an η_α-field with $\left|\mathbf{R}_{(\alpha)}\right| = \aleph_\alpha$ (with GCH). It was proved in Erdős *et al.*(1955) that any two real-closed η_α-fields of cardinality $\aleph_\alpha$ are isomorphic as rings.

There are non-commutative analogues of the fields $\mathfrak{F}(\mathbb{R}, \mathbf{G})$; these are the Mal'cev–Neumann constructions of (non-commutative) Laurent series. For an account of this topic, see Lam (1991, §14).

Finally let us mention the amusing totally ordered, real-closed field **No** of Conway (1976). The field **No** has as its domain not a set but a proper class; by Theorems 28 and 29) of Conway(1976), **No** is the unique totally ordered field (or Field) containing all other totally ordered fields.

3

Completions of ordered groups and fields

There is an extensive, and rather technical, theory of various types of completions of ordered groups and fields. (In the literature, the groups are not necessarily either totally ordered or abelian.) We sketch here the aspects of the theory that we shall be concerned with, defining what we shall call the Cauchy completion of an ordered group, to be constructed by using 'Cauchy pregaps'.

Definition 3.1 *Let $\langle A, B\rangle$ be a pregap in a totally ordered group G. Then $\langle A, B\rangle$ is a* Cauchy pregap *if A has no maximum, B has no minimum, and, for each $\varepsilon > 0$ in G, there exists $a \in A$ and $b \in B$ with $b < a + \varepsilon$. The group G is* Cauchy complete *if, for each Cauchy pregap $\langle A, B\rangle$, there exists $x \in G$ with $A << x << B$.*

The element x arising in the above definition is necessarily unique.

Clearly a totally ordered group G is Cauchy complete if and only if there are no Cauchy gaps. For example, the real line $\mathbb{R}$ is certainly Cauchy complete. If G and H are totally ordered groups with $G \simeq H$ and such that G is Cauchy complete, then H is also Cauchy complete.

Definition 3.2 *A totally ordered group G is* discrete *if the set $G^+ \setminus \{0\}$ is empty or has a minimum element, and is* non-discrete *otherwise.*

Thus G is discrete if and only if $G = \{0\}$ or the base number $\delta(G)$ of G is equal to 1. A discrete group is clearly Cauchy complete because it contains no Cauchy pregaps. We shall be concerned with non-discrete groups.

In fact, only certain pregaps have a chance of being a Cauchy pregap.

Proposition 3.3 *Let $\langle A, B\rangle$ be a (κ, λ)-pregap which is a Cauchy pregap in a totally ordered group G. Then $\kappa = \lambda = \delta(G)$.*

Proof Let $(a_\sigma : \sigma < \kappa)$ be a strictly increasing, cofinal subset in A.

First, assume that $\kappa < \delta(G)$. Then there exists $\varepsilon \in G$ such that

$$0 < \varepsilon < a_{\sigma+1} - a_\sigma \quad (\sigma < \kappa)\,.$$

For each $a \in A$, there exists $\sigma < \kappa$ with $a_\sigma > a$, and then

$$a + \varepsilon < a_\sigma + \varepsilon < a_{\sigma+1} << B\,.$$

Thus $\langle A, B\rangle$ is not a Cauchy pregap. But this is a contradiction, and so $\kappa \geq \delta(G)$.

Second, assume that $\kappa > \delta(G)$ and suppose that $(\varepsilon_\rho : \rho < \delta(G))$ is coinitial in $G^+ \setminus \{0\}$. For each $\sigma < \kappa$, choose $\rho_\sigma < \delta(G)$ so that

$$\varepsilon_{\rho_\sigma} < a_{\sigma+1} - a_\sigma \,.$$

Since $\kappa > \delta(G)$, there exists $\rho < \delta(G)$ such that the set $\{a_\sigma : \rho_\sigma = \rho\}$ is cofinal in A. Set $\varepsilon = \varepsilon_\rho$. By relabelling the sequence $(a_\sigma : \sigma < \kappa)$, we may suppose that $\varepsilon < a_{\sigma+1} - a_\sigma \quad (\sigma < \kappa)$, and again this implies that $b > a + \varepsilon \quad (a \in A,\, b \in B)$ and that $\langle A, B\rangle$ is not a Cauchy pregap. Once more, we have reached a contradiction, and so $\kappa \leq \delta(G)$.

By combining these two results, we see that $\kappa = \delta(G)$.

A similar argument applies to show that $\lambda = \delta(G)$ (or apply the result proved to the pregap $\langle -B, -A\rangle$). □

The notion of a Cauchy complete group can also be approached via sequences.

Definition 3.4 *Let G be a totally ordered group, and let σ be an ordinal. A sequence $(x_\tau : \tau < \sigma)$ is:*

(i) *a* Cauchy sequence *if, for each $\varepsilon > 0$ in G, there exists $\sigma_0 < \sigma$ such that*

$$|x_\tau - x_\rho| < \varepsilon \quad (\sigma_0 \leq \tau,\, \rho < \sigma)\,;$$

(ii) convergent *to an element x of G if, for each $\varepsilon > 0$ in G, there exists $\sigma_0 < \sigma$ such that*

$$|x_\tau - x| < \varepsilon \quad (\sigma_0 \leq \tau < \sigma)\,.$$

In the latter case, the element $x \in G$ is uniquely specified; we write

$$\lim x_\tau = x\,.$$

Proposition 3.5 *Let G be a totally ordered group. Then G is Cauchy complete if and only if each Cauchy sequence of length $\delta(G)$ in G is convergent.*

Proof Suppose that G is Cauchy complete, and let (x_τ) be a Cauchy sequence in G. For each $\varepsilon \in G^+ \setminus \{0\}$, there exists σ_ε such that

$$|x_\tau - x_\rho| < \varepsilon \quad (\tau, \rho \geq \sigma_\varepsilon)\,.$$

Set

$$A = \bigcup \{(-\infty, x_{\sigma_\varepsilon} - \varepsilon] : \varepsilon \in G^+ \setminus \{0\}\}\,,$$
$$B = \bigcup \{[x_{\sigma_\varepsilon} + \varepsilon, \infty) : \varepsilon \in G^+ \setminus \{0\}\}\,.$$

Then $\langle A, B\rangle$ is a Cauchy pregap in G, and so there exists an element $x \in G$ with $A << x << B$. Clearly (x_τ) is convergent to x.

Conversely, suppose that each Cauchy sequence of length $\delta = \delta(G)$ is convergent in G, and let $\langle A, B\rangle$ be a Cauchy pregap in G. By Proposition 3.3, $\langle A, B\rangle$ is a (δ, δ)-pregap, and so there is a strictly increasing, cofinal sequence $(a_\sigma : \sigma < \delta)$ in A. Clearly $(a_\sigma : \sigma < \delta)$ is a Cauchy sequence, say it is convergent to $x \in G$. Then $A << x << B$, and so the group G is Cauchy complete. □

Proposition 3.6 *Let G be a totally ordered group. Each Cauchy gap in G is a pseudo-gap. A pseudo-complete group is Cauchy complete.*

Proof We may suppose that G is non-discrete.

Let $\langle A, B\rangle$ be a Cauchy gap in G, and take $x \in G$. Suppose, for example, that there exists $a \in A$ with $a \geq x$. Since $\operatorname{cof} A > 1$, in fact there exists $a \in A$ with $a > x$. Set $\varepsilon = a - x$. There exist $a_1 \in A$ and $b_1 \in B$ with $b_1 < a_1 + \varepsilon$. If $a \leq a_1$, then

$$0 < a_1 - x < b_1 - x < 2(a_1 - x) ,$$

and so v is constant on $[a_1 - x,\ b_1 - x]$. If $a_1 \leq a$, then

$$0 < a - x < b_1 - x < 2(a - x) ,$$

and so v is constant on $[a - x,\ b_1 - x]$. In either case, $\langle A, B\rangle$ is a pseudo-gap.

It follows that G is Cauchy complete in the case where G is a pseudo-complete group. □

Definition 3.7 *Let $(G, +, \leq)$ be a totally ordered group. A totally ordered group $(H, +, \leq)$ is a* Cauchy completion *of $(G, +, \leq)$ if H is Cauchy complete and if G is isomorphic to an order-dense subgroup of H.*

A construction of a Cauchy completion is as follows. It is straightforward to verify the omitted details.

Let G be a totally ordered, non-discrete group, and let $\langle A_1, B_1\rangle$ and $\langle A_2, B_2\rangle$ be two pregaps in G. Then

$$\langle A_1, B_1\rangle \leq \langle A_2, B_2\rangle \tag{3.1}$$

if, for each $a_1 \in A_1$, there exists $a_2 \in A_2$ with $a_1 \leq a_2$. Let $\mathcal{C}$ or $\mathcal{C}_G$ be the family of Cauchy pregaps $\langle A, B\rangle$. For $\langle A_1, B_1\rangle$ and $\langle A_2, B_2\rangle$ in $\mathcal{C}$, set

$$\langle A_1, B_1\rangle \sim \langle A_2, B_2\rangle$$

if $\langle A_1, B_1\rangle \leq \langle A_2, B_2\rangle$ and $\langle A_2, B_2\rangle \leq \langle A_1, B_1\rangle$. Then the relation $\sim$ is an equivalence relation on $\mathcal{C}$. Set

$$H = \mathcal{C}/\sim ,$$

and denote the equivalence class containing $\langle A, B\rangle$ by $[\langle A, B\rangle]$.

Let G' be any totally ordered group containing G as an order-dense subgroup, and, for $x \in G'$, set $A_x = (-\infty, x)$ and $B_x = (x, \infty)$. Since G is non-discrete, $\langle A_x, B_x\rangle \in \mathcal{C}$. The map

$$\iota : x \longmapsto [\langle A_x, B_x\rangle] , \quad G' \to H ,$$

is an injection. In particular $\iota : G \to H$ is an injection.

Define

$$[\langle A_1, B_1\rangle] \leq [\langle A_2, B_2\rangle] \quad \text{if} \quad \langle A_1, B_1\rangle \leq \langle A_2, B_2\rangle .$$

Then $\leq$ is well defined on H, $(H, \leq)$ is a totally ordered set, and the embedding $\iota : G \to H$ is isotonic.

For $\langle A_1, B_1\rangle$, $\langle A_2, B_2\rangle \in \mathcal{C}$, we have $\langle A_1 + A_2, B_1 + B_2\rangle \in \mathcal{C}$. Define

$$[\langle A_1, B_1\rangle] + [\langle A_2, B_2\rangle] = [\langle A_1 + A_2, B_1 + B_2\rangle]$$

in H. Then + is well defined, and $(H, +)$ is an abelian group; the map

$$\iota : (G, +) \to (H, +)$$

is a group morphism. Finally it is clear that $(H, +, \leq)$ is a totally ordered group.

Let $\langle \mathcal{A}, \mathcal{B}\rangle$ be a Cauchy pregap in H, and define

$$A = \bigcup\{A_\alpha : [\langle A_\alpha, B_\alpha\rangle] \in \mathcal{A} \text{ for some } B_\alpha\} ,$$

$$B = \bigcup\{B_\alpha : [\langle A_\alpha, B_\alpha\rangle] \in \mathcal{B} \text{ for some } A_\alpha\} .$$

Then $\langle A, B\rangle \in \mathcal{C}$ and $\mathcal{A} << [\langle A, B\rangle] << \mathcal{B}$ in H. Thus H is Cauchy complete.

We now have the following result.

Theorem 3.8 *Let G be a totally ordered, non-discrete group. Then the group H constructed above is a Cauchy completion of G. If G is divisible, then so is H. If G is a semi-η_1-group, then so is H. Finally $\Gamma_H = \Gamma_G$, and the archimedean valuation on H, when restricted to G, agrees with the archimedean valuation on G.* □

Let G' be any totally ordered group containing G as an order-dense subgroup. Then there is an isotonic morphism from G' into H. This shows that H is essentially unique, and that it is the maximum totally ordered group containing G as an order-dense subgroup.

Definition 3.9 *The Cauchy completion of a totally ordered group G is denoted by $\widetilde{G}$.*

We note that $\delta(\widetilde{G}) = \delta(G)$ and $w(\widetilde{G}) = w(G)$, and hence that

$$\left|\widetilde{G}\right| \leq 2^{w(G)} .$$

Let us present some examples of Cauchy completions.

Definition 3.10 *Let S be a totally ordered set, and let G be a subgroup of $\mathfrak{F}(\mathbb{R}, S)$. An element f of $\mathfrak{F}(\mathbb{R}, S)$ belongs* locally *to G if, for each $s \in S$, there exists $f_s \in G$ such that $f_s(t) = f(t)$ $(t \leq s)$.*

Theorem 3.11 *Let S be a totally ordered set, and let G be a linear subspace of $\mathfrak{F}(\mathbb{R}, S)$, containing $\mathbb{R}\delta_{s_0}$ in the case where S has a maximum element s_0. Then $\widetilde{G}$ is the set of functions of $\mathfrak{F}(\mathbb{R}, S)$ which belong locally to G.*

Proof Write H for the set of functions which belong locally to G.

First, let $f, g \in H$ with $f < g$, and set $s = v(f - g)$, where v is the Hahn valuation on $\mathfrak{F}(\mathbb{R}, S)$. Take $f_s, g_s \in G$ with $f_s(t) = f(t)$ and $g_s(t) = g(t)$ for $t \leq s$, and set

$$h = \tfrac{1}{2}(f_s + g_s) .$$

Then $h \in G$, $f(t) = h(t) = g(t)$ $(t < s)$, and

$$h(s) = \tfrac{1}{2}(f(s) + g(s)),$$

so that $f < h < g$ in $\mathfrak{F}(\mathbb{R}, S)$. Thus G is order-dense in H.

Second, we show that H is Cauchy complete. Let $\langle A, B \rangle$ be a Cauchy pregap in H. For each $s \in S$, there exist $f_s \in A$ and $g_s \in B$ with

$$f_s < g_s < f_s + \delta_s .$$

Then $v(g_s - f_s) \geq s$: set $h_s(t) = f_s(t) = g_s(t)$ $(t < s)$. If $s_1, s_2 \in S$ with $s_1 < s_2$, then necessarily

$$h_{s_1}(t) = h_{s_2}(t) \quad (t < s_1),$$

and so we may consistently define h on $\bigcup\{(-\infty, s) : s \in S\}$ by

$$h(t) = h_s(t) \quad (t < s) .$$

In the case where S has no maximum element, h is already defined on S and $h \in H$. In the case where S has a maximum element s_0, for each $n \in \mathbb{N}$ there exist $p_n \in A$ and $q_n \in B$ with

$$p_n < p_{n+1} < q_{n+1} < q_n < p_n + \frac{1}{n}\delta_{s_0} .$$

Then $p_n(t) = q_n(t)$ $(t \in S \setminus \{s_0\},\ n \in \mathbb{N})$, and

$$\langle \{p_n(s_0) : n \in \mathbb{N}\},\ \{q_n(s_0) : n \in \mathbb{N}\} \rangle$$

is an $(\aleph_0, \aleph_0)$-pregap in $\mathbb{R}$. Hence there is a unique element α of $\mathbb{R}$ such that $p_n(s_0) < \alpha < q_n(s_0)$ $(n \in \mathbb{N})$. Set $h(s_0) = \alpha$; again $h \in H$.

In either case, $A << h << B$ in H, and so H is Cauchy complete.

We have proved that $\widetilde{G} = H$. □

In general, $\widetilde{G} \neq G$. For set $G = \mathfrak{F}_{(1)}(\mathbb{R}, \omega_1)$. Then $\widetilde{G} = \mathfrak{F}(\mathbb{R}, \omega_1) = \mathbb{R}^{\omega_1}$. The theorem does show that $\mathfrak{F}(\mathbb{R}, S)$ itself is always Cauchy complete, a result which also follows from Theorem 1.33 and Proposition 3.6.

The following result gives information about the size of Cauchy complete η_1-groups.

Theorem 3.12 *Let G be a Cauchy complete η_1-group.*

(i) *Suppose that $\delta(G) = \aleph_1$. Then $|G| \geq 2^{\aleph_1}$.*

(ii) *In each case, $|G| \geq \aleph_2$.*

(iii) *Suppose that Γ_G is an α_2-set. Then $|G| = 2^{\aleph_1}$.*

Proof (i) Since $\delta(G) = \aleph_1$, there is a strictly decreasing, coinitial sequence, say $(\varepsilon_\sigma : \sigma < \omega_1)$, in $G^+ \setminus \{0\}$.

Let $\mathcal{T}_{\omega_1} = \{0,1\}^{<\omega_1}$ be the complete $\{0,1\}$-tree of height ω_1, as in Chapter 1. We *claim* that there is a map

$$\alpha \mapsto (x_\alpha, y_\alpha), \quad \mathcal{T}_{\omega_1} \to G \times G,$$

such that:

$$x_\alpha < y_\alpha < x_\alpha + \varepsilon_\sigma \quad \text{for } \alpha \in \{0,1\}^\sigma ;$$
$$x_\alpha < x_\beta < y_\beta < y_\alpha \quad \text{whenever } \alpha < \beta ;$$
$$(x_\alpha, y_\alpha) \cap (x_\beta, y_\beta) = \emptyset \quad \text{whenever } \alpha, \beta \in \mathbf{S}_\sigma \text{ with } \alpha \neq \beta .$$

For, given the map from $\{0,1\}^\sigma$, it is clear that there is a map with the appropriate properties from $\{0,1\}^{\sigma+1}$. Now suppose that $\sigma < \omega_1$ is a limit ordinal and that the map has been defined on $\{0,1\}^{<\sigma}$. Then, for each $\alpha \in \{0,1\}^\sigma$, the existence of the appropriate elements $x_\alpha, y_\alpha \in G$ follows because G is an η_1-set. Thus the claim is established.

Let $\alpha \in \mathbf{S} = \{0,1\}^{\omega_1}$. Then the sequence $(x_{\alpha\restriction\sigma} : \sigma < \omega_1)$ is a Cauchy sequence, and so it converges, say to $X_\alpha \in G$. Clearly, if $\alpha \neq \beta$ in $\mathbf{S}$, then $X_\alpha \neq X_\beta$, and so

$$|G| \geq |\mathbf{S}| = 2^{\aleph_1} .$$

(ii) Since G is an η_1-group, $\delta(G) \geq \aleph_1$, and so the result follows from (i).

(iii) By Theorem 1.40(ii), $\delta(G) = \aleph_1$ and $|G| \leq 2^{\aleph_1}$, and so, $|G| = 2^{\aleph_1}$. □

Definition 3.13 *Let G and H be totally ordered groups, and let I be an interval in G. A function $F : I \to H$ is* uniformly continuous *if, for each $\varepsilon > 0$ in H, there exists $\delta > 0$ in G such that $|F(x) - F(y)| < \varepsilon$ whenever $x, y \in I$ with $|x - y| < \delta$.*

Let G be a totally ordered group, and let I be an interval in G. Then

$$\widetilde{I} = \{x \in \widetilde{G} : a \leq x \leq b \quad \text{for some } a, b \in I\} .$$

Theorem 3.14 *Let G and H be totally ordered groups, let I be an interval in G, and let $F : I \to H$ be a uniformly continuous function. Then there is a uniformly continuous function $\widetilde{F} : \widetilde{I} \to \widetilde{H}$ such that $\widetilde{F} \mid I = F$.*

Proof Let $x \in \widetilde{I}$. Then there is a Cauchy sequence $(x_\tau : \tau < \rho)$ in I with $\lim x_\tau = x$. Clearly $(F(x_\tau) : \tau < \rho)$ is a Cauchy sequence in H, and so it has a unique limit, say $\widetilde{F}(x)$, in $\widetilde{H}$. The element $\widetilde{F}(x)$ is independent of the choice of the sequence $(x_\tau : \tau < \rho)$, and so $\widetilde{F}(x)$ is well defined in $\widetilde{H}$. It is also clear that the function $\widetilde{F} : \widetilde{I} \to \widetilde{H}$ is uniformly continuous. □

We now consider the Cauchy completion of ordered fields. An ordered field K is *Cauchy complete* if the totally ordered group $(K, +)$ is Cauchy complete. A short, alternative proof of the following result is given in the notes, 3.24.

Theorem 3.15 *Let K be an ordered field. Then the Cauchy completion $\widetilde{K}$ of the group $(K, +)$ can be made into an ordered field in such a way that K is a subfield of $\widetilde{K}$. If K is real-closed, then so is $\widetilde{K}$.*

Proof Take $a_1, a_2 \in \widetilde{K}$ with $a_1, a_2 > 0$, say $a_1 = [\langle A_1, B_1 \rangle]$ and $a_2 = [\langle A_2, B_2 \rangle]$. We may suppose that $A_1, A_2 \subset K^+ \setminus \{0\}$. Set

$$A = \{x_1 x_2 : x_1 \in A_1 ,\ x_2 \in A_2\} , \quad B = \{y_1 y_2 : y_1 \in B_1 ,\ y_2 \in B_2\} .$$

Then $\langle A, B\rangle \in \mathcal{C}$. Define

$$a_1a_2 = [\langle A_1, B_1\rangle] \cdot [\langle A_2, B_2\rangle] = [\langle A, B\rangle]\,.$$

The operation $\cdot$ is well defined on $\widetilde{K}$, and $a_1a_2 > 0$.

If $a_1 < 0$ and $a_2 > 0$, set $a_1a_2 = -(-a_1)a_2$, etc.

It is routine to check that the group $\widetilde{K}$, together with this product, is a field with the required properties. □

Let K' be any ordered field containing K as an order-dense subfield. Then there is an isotonic morphism from K' into $\widetilde{K}$, and so $\widetilde{K}$ is the maximum ordered field containing K as an order-dense subfield. The following characterization of Cauchy complete ordered fields is immediate.

Proposition 3.16 *Let K be an ordered field. Then K is Cauchy complete if and only if there is no ordered field L containing K as a proper, order-dense subfield.* □

Proposition 3.17 *Let K be a semi-η_1-field which is not an η_1-field. Then K is Cauchy complete.*

Proof By Proposition 2.20, we have $\delta(K) = \aleph_0$. Let $\langle A, B\rangle$ be a Cauchy pregap in K, so that, by Proposition 3.3, $\langle A, B\rangle$ is an $(\aleph_0, \aleph_0)$-pregap. Since K is a semi-η_1-field, there exists $x \in K$ with $A << x << B$, and so K is Cauchy complete. □

Proposition 3.18 *Let K and L be ordered fields, and let $\theta : K \to L$ be an isotonic isomorphism. Then there is an isotonic isomorphism $\widetilde{\theta} : \widetilde{K} \to \widetilde{L}$ such that $\widetilde{\theta} \mid K = \theta$.*

Proof For $\varepsilon > 0$ in L, there exists $\delta > 0$ in K with $\theta(\delta) = \varepsilon$. Suppose that $x < y < x + \delta$ in K. Then $\theta(x) < \theta(y) < \theta(x) + \varepsilon$ in L, and so $\theta : K \to L$ is uniformly continuous. By Theorem 3.14, there is a uniformly continuous function $\widetilde{\theta} : \widetilde{K} \to \widetilde{L}$ such that $\widetilde{\theta} \mid K = \theta$. Clearly, $\widetilde{\theta}$ is an order-preserving homomorphism. Similarly $\theta^{-1} : L \to K$ has an order-preserving extension to a map from $\widetilde{L}$ into $\widetilde{K}$, and this map is $\widetilde{\theta}^{-1}$. Thus $\widetilde{\theta}$ is an isotonic isomorphism. □

In Definition 2.22, we introduced the fields

$$\mathbf{R} = \mathfrak{F}_{(1)}(\mathbb{R}, \mathbf{G}) \quad \text{and} \quad \widehat{\mathbf{R}} = \mathfrak{F}(\mathbb{R}, \mathbf{G})\,.$$

We know that $\widehat{\mathbf{R}}$ is Cauchy complete; we now consider the Cauchy completion of $\mathbf{R}$. In view of its importance to us, we give a formal definition of this field.

Definition 3.19 *The field $\widetilde{\mathbf{R}}$ is the Cauchy completion of the field $\mathbf{R}$.*

It is immediate from Theorem 3.11 that we have the following explicit description of $\widetilde{\mathbf{R}}$.

Theorem 3.20 *The field $\widetilde{\mathbf{R}}$ is equal to the set of elements $f \in \mathfrak{F}(\mathbb{R}, \mathbf{G})$ such that, for each $x \in \mathbf{G}$, there exists $f_x \in \mathbf{R}$ with $f_x(y) = f(y)$ for $y \leq x$.* □

Theorem 3.21 (i) *The field* $\widetilde{\mathbf{R}}$ *is a real-closed, Cauchy complete* η_1*-field with value group* $\mathbf{G}$.

(ii) $\left|\widetilde{\mathbf{R}}\right| = 2^{\aleph_1}$, $w(\widetilde{\mathbf{R}}) = \mathfrak{c}$, *and* $\operatorname{cof}\widetilde{\mathbf{R}} = \delta(\widetilde{\mathbf{R}}) = \aleph_1$.

(iii) $\widetilde{\mathbf{R}}$ *contains the following types of gap, and no other types:*

$$(0,\aleph_1),\ (\aleph_1,0),\ (1,\aleph_1),\ (\aleph_1,1),\ (\aleph_0,\aleph_1),\ (\aleph_1,\aleph_0),\ (\aleph_1,\aleph_1).$$

(iv) (CH) *Each real-closed, Cauchy complete* η_1*-field of weight* $\mathfrak{c}$ *is isomorphic to* $\widetilde{\mathbf{R}}$.

Proof (i) This follows from Theorem 2.23(i).

(ii) Since $\mathbf{G}$ is an α_2-set, $\left|\widetilde{\mathbf{R}}\right| = 2^{\aleph_1}$ by Theorem 3.12(iii). The remaining results follow from Theorem 2.23(iii).

(iii) The argument giving Theorem 2.23(v) also applies to $\widetilde{\mathbf{R}}$.

(iv) Let L be a real-closed, Cauchy complete η_1-field of weight $\mathfrak{c}$. By Proposition 2.2, L contains an order-dense, real-closed subfield K with $|K| = \mathfrak{c}$ and $\widetilde{K} = L$; since L is an η_1-field, so is K. By Corollary 2.34(ii), there is an isomorphism $\theta : K \to \mathbf{R}$, and so, by Proposition 3.18, there is an extension of θ which is an isomorphism from L onto $\widetilde{\mathbf{R}}$. □

It is apparent from Theorem 3.21 that $\widetilde{\mathbf{R}}$ is the natural analogue of the real line $\mathbb{R}$ 'one cardinality higher'.

Theorem 3.22 *Let* K *be a* β_1*-field. Then there is an embedding of* $\widetilde{K}$ *in* $\widetilde{\mathbf{R}}$.

Proof By Theorem 2.30, there is an embedding of K in $\mathbf{R}$, and, by Proposition 3.18, this map extends to an embedding of $\widetilde{K}$ in $\widetilde{\mathbf{R}}$. □

We summarize some properties of the fields $\mathbf{R}$, $\widetilde{\mathbf{R}}$, and $\widehat{\mathbf{R}}$ in Table 3.1.

Table 3.1

K	$\|K\|$	$w(K)$	$\|\Gamma_K\|$	$\operatorname{cof} K$	Cauchy complete?	Pseudo–complete?
$\mathbf{R}$	$2^{\aleph_0}$	$2^{\aleph_0}$	$2^{\aleph_0}$	$\aleph_1$	No	No
$\widetilde{\mathbf{R}}$	$2^{\aleph_1}$	$2^{\aleph_0}$	$2^{\aleph_0}$	$\aleph_1$	Yes	No
$\widehat{\mathbf{R}}$	$2^{\aleph_1}$	$2^{\aleph_1}$	$2^{\aleph_0}$	$\aleph_1$	Yes	Yes

We conclude this chapter by showing that $\widetilde{\mathbf{R}}$ is not order-isomorphic to either $\mathbf{R}$ or $\widehat{\mathbf{R}}$. This is immediate if we assume CH, for, with CH,

$$w(\widehat{\mathbf{R}}) = \left|\widetilde{\mathbf{R}}\right| = 2^{\aleph_1} > 2^{\aleph_0} = |\mathbf{R}| = w(\widetilde{\mathbf{R}})\,.$$

However, we can also prove this result as a theorem of ZFC.

Theorem 3.23 (i) *There is no isotonic map from* $\widetilde{\mathbf{R}}^+$ *into* $\mathbf{R}$.

(ii) *There is no isotonic map from* $\widehat{\mathbf{R}}$ *into* $\widetilde{\mathbf{R}}$.

Proof (i) This is essentially the same as the proof of Proposition 1.41. Assume towards a contradiction that $\psi : \widetilde{\mathbf{R}}^+ \to \mathbf{R}$ is an isotonic map, and construct elements $f_\sigma \in \widetilde{\mathbf{R}}^+$, intervals $I_\sigma \subset \widetilde{\mathbf{R}}^+$, and $s_\sigma \in \mathbf{G}$ for $\sigma < \omega_1$ as in that proof (with $\psi(I_\sigma) \cap \mathbf{R}_\sigma = \emptyset$). Since $\operatorname{cof}\mathbf{G} = \aleph_1$, we may suppose that $(s_\sigma : \sigma < \omega_1)$ is cofinal in $\mathbf{G}$. But now the element $f : \mathbf{G} \to \mathbb{R}$ constructed from (f_σ) and (s_σ) belongs locally to $\mathbf{R}^+$, and hence, by Theorem 3.20, f belongs to $\widetilde{\mathbf{R}}^+$. Since $f \in I_\sigma$ for each $\sigma < \omega_1$, this is a contradiction, as before.

(ii) To obtain a contradiction, assume that $\psi : \widehat{\mathbf{R}} \to \widetilde{\mathbf{R}}$ is an isotonic map. Since $\widetilde{\mathbf{R}}$ contains the β_1-set $\mathbf{R}$ as an order-dense subset, it follows from Proposition 1.16 that $\psi(\widehat{\mathbf{R}})$ also contains an order-dense β_1-subset. Hence $\widehat{\mathbf{R}}$ contains an order-dense β_1-subset, a contradiction of Theorem 2.23(viii). □

Notes 3.24 The construction of the Cauchy completion of an ordered group was probably first given in Cohen and Goffman (1949), where it is called the *topological completion*, and is carried out via 'Dedekindean lower segments'; an equivalent theory uses equivalence classes of Cauchy, or 'fundamental', sequences. See also Holland (1963) and Priess-Crampe (1983). The idea of using the pregaps that are involved goes back to Baer. Theorem 3.11 seems to be new.

The construction of $\widetilde{K}$ in Theorem 3.15 is essentially that of Baer as expounded in Priess-Crampe (1983, Chapter III). Our pregap is equivalent to the *Dedekindscher Schnitt* of her p.66; our Cauchy pregap is the *eigentlich* cut; our Cauchy completion is the *stetiger Abschluss.* The existence of $\widetilde{K}$ is Satz 1, p.66, and Satz 4, p.68; that $\widetilde{K}$ is real-closed if K is real-closed is Satz 15, p.76.

Our notion of a Cauchy complete ordered field coincides with that of a *complete field* of Scott (1969). The full details of a somewhat different construction of $\widetilde{K}$, and a verification that it has the properties stated in the above theorem, are given by Scott. These fields are the 'Scott complete fields' of Keisler and Schmerl (1991). In this paper, a field K is said to have the *Bolzano–Weierstrass property* if every bounded sequence of length $\delta(K)$ has a convergent subsequence; a field with this property is Cauchy complete, and it is shown in Theorem 3.6 of Keisler and Schmerl (1991) that there is an η_1-field with $\delta(K) = \aleph_2$ and with the Bolzano–Weierstrass property. Other sources where completions of groups or algebras are discussed include Bourbaki (1985), Fuchs (1963), Massaza (1969-70), and Zaharov (1988).

There is a rapid, alternative proof of Theorem 3.15 if one is willing to utilize Kaplansky's embedding theorem. Indeed, let K be an ordered field, and let Γ

be the value group of $\mathcal{R}K$. By Theorem 2.17, K and $\mathcal{R}K$ can be identified with subfields of $\mathfrak{F}(\mathbb{R}, \Gamma)$. Thus, by Theorem 3.11, the Cauchy completions of K and of $\mathcal{R}K$ can be identified with the linear spaces of functions of $\mathfrak{F}(\mathbb{R}, \Gamma)$ which belong locally to K and $\mathcal{R}K$, respectively. Clearly these spaces are themselves subfields of $\mathfrak{F}(\mathbb{R}, \Gamma)$.

4
Algebras of continuous functions

Let X be a non-empty, topological space. Then

$$C(X) \quad \text{and} \quad C(X,\mathbb{C})$$

are the sets of continuous functions on X which are real-valued and complex-valued, respectively. These sets are algebras with respect to the pointwise operations, and we regard $C(X)$ as a real subalgebra of $C(X,\mathbb{C})$. The subalgebras of $C(X)$ and of $C(X,\mathbb{C})$ consisting of bounded functions are denoted by

$$C^b(X) \quad \text{and} \quad C^b(X,\mathbb{C}),$$

respectively. We shall also denote by

$$C(X,\mathbb{I})$$

the set of functions in $C(X)$ whose range is contained in the unit interval $\mathbb{I}$.

Suppose that X is a locally compact space. (Our convention is that a locally compact space is Hausdorff by definition.) A function $f \in C(X,\mathbb{C})$ *vanishes at infinity* if

$$\{x \in X : |f(x)| \geq \varepsilon\}$$

is compact for each $\varepsilon > 0$. The subalgebras of $C(X)$ and of $C(X,\mathbb{C})$ consisting of the functions which vanish at infinity are denoted by

$$C_0(X) \quad \text{and} \quad C_0(X,\mathbb{C}),$$

respectively. Clearly $C_0(X)$ is an ideal in $C^b(X)$.

Finally, $C_{00}(X)$ and $C_{00}(X,\mathbb{C})$ are the ideals in $C(X)$ and $C(X,\mathbb{C})$, respectively, consisting of the functions of compact support.

We have a special notation in the case where $X = \mathbb{N}$: set

$$l^\infty = C^b(\mathbb{N}), \quad c_0 = C_0(\mathbb{N}), \quad c_{00} = C_{00}(\mathbb{N}), \quad c = C(\mathbb{N} \cup \{\infty\}).$$

Thus l^∞ is the algebra of bounded, real-valued sequences with coordinatewise multiplication, c_0 is the ideal in l^∞ consisting of sequences that converge to zero, c_{00} is the ideal in l^∞ of sequences (α_n) such that $\alpha_n = 0$ for all but finitely many $n \in \mathbb{N}$, and c is the subalgebra of l^∞ of sequences converging to some limit. The corresponding algebras of complex-valued functions are $l^\infty(\mathbb{C})$, $c_0(\mathbb{C})$, $c_{00}(\mathbb{C})$, and $c(\mathbb{C})$, respectively.

Before proceeding with a study of these algebras, we record some elementary topological facts and notations that will be used many times. The *closure*, *interior*, and *frontier* of a subset S of a topological space are denoted by

$$\overline{S}, \quad \operatorname{int} S, \quad \text{and} \quad \partial S,$$

respectively. A topological space X is *normal* if it is Hausdorff and if disjoint closed subsets have disjoint closed neighbourhoods. A compact space is certainly normal. Let X be a normal topological space. Then *Urysohn's lemma* asserts that, if F and G are disjoint closed subsets of X, then there exists $f \in C(X, \mathbb{I})$ with $f(F) \subset \{0\}$ and $f(G) \subset \{1\}$, and *Tietze's extension theorem* asserts that, for each closed subset F of X and each $f \in C(F, \mathbb{I})$, there exists $g \in C(X, \mathbb{I})$ with $g \mid F = f$.

A topological space X is *completely regular* if it is a Hausdorff space and if, for each closed subset F of X and each $x \in X \setminus F$, there exists $f \in C(X)$ such that $f(x) = 1$ and $f(F) \subset \{0\}$. Every locally compact space and every metrizable space is completely regular. For each topological space Y, there is a completely regular space X such that $C(X)$ is isomorphic to $C(Y)$, and so, when considering algebraic properties of the algebras $C(X)$, it is sufficient to suppose that X is completely regular.

We shall be particularly concerned with prime ideals in the algebras $C(X)$. Recall that an ideal P in a commutative algebra A is *prime* if $P \neq A$ and if either $a \in P$ or $b \in P$ whenever $a, b \in A$ and $ab \in P$. This is the case precisely when A/P is an integral domain. We shall several times use the following elementary fact about prime ideals. *Let I be an ideal in A, and let S be a non-empty set which is closed under multiplication and disjoint from I. Then there is a prime ideal P in A with $I \subset P$ and $P \cap S = \emptyset$.* Indeed P is an ideal which is maximal in the family of ideals which contain I and are disjoint from S. It follows that the intersection of all the prime ideals in A containing a given ideal I is precisely the set of all elements of which some power belongs to I. Clearly the union and intersection of a chain of prime ideals in a commutative algebra A are prime ideals (save that the union could be the whole algebra when A is non-unital). It follows that every prime ideal contains a minimal prime ideal in A.

Let P and Q be prime ideals in an algebra with $Q \subset P$. Then we say that Q is *below* P and that P is *above* Q.

Let A and B be commutative, unital algebras, and let $\theta : A \to B$ be a unital epimorphism with kernel I. Then the map $P \mapsto \theta(P)$ is a bijection from the set of all prime ideals in A containing I onto the set of all prime ideals in B.

As we stated in Chapter 1, $(\mathbb{R}^X, +, \leq)$ is an ordered group which is a lattice; also $\mathbb{R}^X$ is an ordered algebra. In the case where X is a topological space, $C(X)$ and $C^b(X)$ are subalgebras of $\mathbb{R}^X$, and each of these spaces is a sublattice of $\mathbb{R}^X$. Thus we have the following basic result.

Proposition 4.1 *Let X be a topological space. Then $C(X)$ and $C^b(X)$ are ordered algebras.* □

We have

$$C(X)^+ = \{f \in C(X) : f(x) \geq 0 \quad (x \in X)\}\,.$$

For $f \in C(X)$, set

$$\mathbf{Z}_X(f) = \mathbf{Z}(f) = \{x \in X : f(x) = 0\}\,.$$

Then $\mathbf{Z}(f)$ is the *zero set* of f. We note that, for $f, g \in C(X)$, we have

$$\mathbf{Z}(f) \cap \mathbf{Z}(g) = \mathbf{Z}(f^2 + g^2) \quad \text{and} \quad \mathbf{Z}(f) \cup \mathbf{Z}(g) = \mathbf{Z}(fg)\,.$$

For each ideal I in $C(X)$, set

$$\mathbf{Z}[I] = \{\mathbf{Z}(f) : f \in I\}\,,$$

and write $\mathbf{Z}(X)$ for $\mathbf{Z}[C(X)]$, so that $\mathbf{Z}(X)$ is the family of zero sets of X. In the case where X is a metric space, $\mathbf{Z}(X)$ is the family of closed subsets of X.

Definition 4.2 *Let X be a topological space. Then a non-empty subfamily $\mathcal{F}$ of $\mathbf{Z}(X)$ is a* z-filter *if:*

(i) $\emptyset \notin \mathcal{F}$;

(ii) $F_1 \cap F_2 \in \mathcal{F}$ *whenever* $F_1, F_2, \in \mathcal{F}$;

(iii) $G \in \mathcal{F}$ *whenever* $G \in \mathbf{Z}(X)$ *and* $G \supset F$ *for some* $F \in \mathcal{F}$.

In the case where X is a discrete topological space, of course $C(X) = \mathbb{R}^X$ and $\mathbf{Z}(X) = \mathcal{P}(X)$, the power set of X; we refer to *filters* on X.

For each z-filter $\mathcal{F}$ on X, set

$$\mathbf{Z}^{-1}[\mathcal{F}] = \{f \in C(X) : \mathbf{Z}(f) \in \mathcal{F}\}\,.$$

The first result is straightforward.

Proposition 4.3 *Let X be a completely regular space.*

(i) *For each proper ideal I in $C(X)$, the family $\mathbf{Z}[I]$ is a z-filter, and*

$$I \subset \mathbf{Z}^{-1}[\mathbf{Z}[I]]\,.$$

(ii) *For each z-filter $\mathcal{F}$ on X, the set $\mathbf{Z}^{-1}[\mathcal{F}]$ is a proper ideal in $C(X)$, and*

$$\mathcal{F} = \mathbf{Z}[\mathbf{Z}^{-1}[\mathcal{F}]]\,. \qquad \square$$

Let $\mathcal{F}$ be a z-filter on X. Then we write

$$C(X)/\mathcal{F}$$

for the quotient algebra $C(X)/\mathbf{Z}^{-1}[\mathcal{F}]$; elements of this algebra are denoted by $[f]_{\mathcal{F}}$ or $[f]$, for example.

In general, it is not the case that $I = \mathbf{Z}^{-1}[\mathbf{Z}[I]]$ for a proper ideal I in $C(X)$. For example, let I be the principal ideal $ZC(\mathbb{R})$ generated in $C(\mathbb{R})$ by the coordinate functional $Z : t \mapsto t$ on $\mathbb{R}$. Then

$$\mathbf{Z}^{-1}[\mathbf{Z}[I]] = \{f \in C(\mathbb{R}) : f(0) = 0\} \supsetneq I\,.$$

Ideals I such that $I = \mathbf{Z}^{-1}[\mathbf{Z}[I]]$ play a surprisingly important role in the theory that we are developing.

Definition 4.4 *Let X be a completely regular space. A z-*ideal *in $C(X)$ is an ideal I such that $I = \mathbf{Z}^{-1}[\mathbf{Z}[I]]$.*

Thus an ideal I is a z-ideal if, for each $f \in C(X)$ with $\mathbf{Z}(f) = \mathbf{Z}(g)$ for some $g \in I$, necessarily $f \in I$.

Suppose that $\mathcal{F}$ is a z-filter. Then $\mathbf{Z}^{-1}[\mathcal{F}]$ is a z-ideal, and $\mathbf{Z}^{-1}[\mathbf{Z}[I]]$ is the minimum z-ideal containing a given ideal I.

Definition 4.5 *Let X be a completely regular topological space. A z-filter on X is a z-*ultrafilter *if it is maximal in the family of z-filters on X, the family being ordered by inclusion. A z-ultrafilter is* fixed *if it is the collection of all sets in $\mathbf{Z}(X)$ containing a fixed element of X; otherwise the z-ultrafilter is* free.

In the case where X is discrete, we refer to *ultrafilters* on X.

It follows easily from Zorn's lemma that every z-filter is contained in a z-ultrafilter. It is also easy to see that a filter $\mathcal{F}$ on a discrete space X is an ultrafilter if and only if, for each subset S of X, either $S \in \mathcal{F}$ or $X \setminus S \in \mathcal{F}$. Every family $\mathcal{B}$ of zero sets that has the finite intersection property generates a z-filter $\mathcal{F}$:

$$\mathcal{F} = \{Z \in \mathbf{Z}(X) : Z \supset B_1 \cap \cdots \cap B_n \text{ for some } B_1, \ldots, B_n \in \mathcal{B}\} .$$

Thus $\mathcal{B}$ is contained in a z-ultrafilter.

The following result is an immediate consequence of Proposition 4.3.

Proposition 4.6 *Let X be a completely regular space.*

(i) *Suppose that M is a maximal ideal in $C(X)$. Then $\mathbf{Z}[M]$ is a z-ultrafilter on X.*

(ii) *Suppose that $\mathcal{U}$ is a z-ultrafilter on X. Then $\mathbf{Z}^{-1}[\mathcal{U}]$ is a maximal ideal in $C(X)$.* □

We now turn to our study of prime ideals in the algebras $C(X)$. The following elementary property of $C(X, \mathbb{C})$ will be used many times.

Proposition 4.7 *Let X be a completely regular space.*

(i) *Suppose that $f, g \in C(X, \mathbb{C})$ with $|f| \le |g|$. Then there exists $h \in C(X, \mathbb{C})$ with $f^2 = gh$. In the case where X is discrete, there exists $k \in C(X, \mathbb{C})$ with $f = gk$.*

(ii) *Suppose that P is a prime ideal in $C(X)$ and that $f \in C(X)$ is such that $|f|^n \le M\,|g|$ for some $n \in \mathbb{N}$, $M > 0$, and $g \in P$. Then $f \in P$.*

Proof (i) Set

$$h(x) = \begin{cases} f^2(x)/g(x) & (x \in X \setminus \mathbf{Z}(g)), \\ 0 & (x \in \mathbf{Z}(g)) . \end{cases}$$

Then $|h(x)| \le |g(x)| \quad (x \in X)$, and so $h \in C(X, \mathbb{C})$. Clearly $f^2 = gh$.

In the case where X is discrete, set

$$k(x) = \begin{cases} f(x)/g(x) & (x \in X \setminus \mathbf{Z}(g)), \\ 0 & (x \in \mathbf{Z}(g)) . \end{cases}$$

Then $k \in C(X, \mathbb{C})$ and $f = gk$.

(ii) By (i), there exists $h \in C(X)$ with $f^{2n} = Mgh$, and hence $f^{2n} \in P$ and $f \in P$. □

Let P be a prime ideal in $C(X, \mathbb{C})$. Then $f \in P$ if and only if $|f| \in P$, and so the map $Q \mapsto Q_{\mathbb{C}}$, where

$$Q_{\mathbb{C}} = \{f \in C(X, \mathbb{C}) : |f| \in Q\},$$

is a bijection from the set of prime ideals in $C(X)$ onto the set of prime ideals in $C(X, \mathbb{C})$.

From now on, we shall concentrate on the algebra $C(X)$ of *real*-valued continuous functions on X. Part (i) of the result given below will lead to the quite central fact of our theory that the quotient space $C(X)/P$ is totally ordered.

Proposition 4.8 *Let X be a completely regular space, and let P be a prime ideal in $C(X)$.*

(i) *Suppose that $f \in C(X)$. Then either $f \in f^+ + P$ or $f \in f^- + P$.*

(ii) *The set of prime ideals in $C(X)$ containing P is totally ordered by inclusion.*

(iii) *Suppose that I is a z-ideal containing P. Then I is a prime ideal.*

(iv) *The set $\mathbf{Z}^{-1}[\mathbf{Z}[P]]$ is the minimum prime z-ideal containing P.*

(v) *The ideal P is contained in a unique maximal ideal of $C(X)$.*

Proof (i) We have $f = f^+ + f^-$. Clearly $f^+ f^- = 0$, and so either $f^- \in P$ and $f \in f^+ + P$ or $f^+ \in P$ and $f \in f^- + P$.

(ii) Let Q_1 and Q_2 be prime ideals in $C(X)$ containing P. Assume that $Q_1 \not\subset Q_2$ and that $Q_2 \not\subset Q_1$, and take $f_1 \in Q_1 \setminus Q_2$ and $f_2 \in Q_2 \setminus Q_1$. Suppose that

$$|f_1| - |f_2| \in (|f_1| - |f_2|)^+ + P.$$

Then $|f_2| \leq |f_1| + g$ for some $g \in P$, and $|f_1| + g \in Q_1$, and so $f_2 \in Q_1$, a contradiction. Similarly, if

$$|f_1| - |f_2| \in (|f_1| - |f_2|)^- + P,$$

then $f_1 \in Q_2$, a contradiction. By (i), one of the above two possibilities occurs, and so either $Q_1 \subset Q_2$ or $Q_2 \subset Q_1$.

(iii) For each $f \in C(X)$ and $n \in \mathbb{N}$, we have $\mathbf{Z}(f^n) = \mathbf{Z}(f)$, and so I is the intersection of the prime ideals containing I. By (ii), this set of ideals is a chain, and so I is a prime ideal.

(iv) It is a special case of (iii) that $\mathbf{Z}^{-1}[\mathbf{Z}[P]]$ is prime; the remainder is immediate.

(v) Suppose that M_1 and M_2 are maximal ideals of $C(X)$ which both contain P. Since $M_1 \cap M_2$ is a z-ideal, $M_1 \cap M_2$ is a prime ideal by (iii). But this can only happen if $M_1 = M_2$. Thus P is contained in a unique maximal ideal. □

Definition 4.9 *Let X be a completely regular space. A z-filter $\mathcal{F}$ on X is* prime *if, for each $Z_1, Z_2 \in \mathbf{Z}(X)$ such that $Z_1 \cup Z_2 \in \mathcal{F}$, either $Z_1 \in \mathcal{F}$ or $Z_2 \in \mathcal{F}$.*

Clearly, each z-ultrafilter is a prime z-filter.

Proposition 4.10 *Let X be a completely regular space.*

(i) *Suppose that P is a prime ideal in $C(X)$. Then $\mathbf{Z}[P]$ is a prime z-filter.*

(ii) *Suppose that $\mathcal{F}$ is a prime z-filter on X, then $\mathbf{Z}^{-1}[\mathcal{F}]$ is a prime z-ideal in $C(X)$.*

Proof (i) Set $Q = \mathbf{Z}^{-1}[\,\mathbf{Z}[P]\,]$, so that Q is a prime z-ideal in $C(X)$ by Proposition 4.8(iv). Let $f, g \in C(X)$ be such that $\mathbf{Z}(f) \cup \mathbf{Z}(g) \in \mathbf{Z}[P]$. Then $\mathbf{Z}(fg) \in \mathbf{Z}[Q]$, and so $fg \in Q$, and either $f \in Q$ or $g \in Q$. It follows that either $\mathbf{Z}(f) \in \mathbf{Z}[P]$ or $\mathbf{Z}(g) \in \mathbf{Z}[P]$, and so $\mathbf{Z}[P]$ is a prime z-filter.

(ii) This is immediate. □

The notion introduced in the following definition will be used in Chapter 7 for general completely regular spaces.

Definition 4.11 *Let X be a completely regular space, and let $\mathcal{F}$ be a z-filter on X. Then the cardinal number $\mathfrak{m}(\mathcal{F})$, given by*

$$\mathfrak{m}(\mathcal{F}) = \min\{|F| : \overline{F} \in \mathcal{F}\}\,,$$

is the minimal characteristic *of $\mathcal{F}$. The z-filter $\mathcal{F}$ is* separable *if $\mathfrak{m}(\mathcal{F}) \leq \aleph_0$ and* non-separable *if $\mathfrak{m}(\mathcal{F}) > \aleph_0$.*

Definition 4.12 *Let X be a discrete topological space, and let $\mathcal{F}$ be a filter on X. Then $\mathcal{F}$ is a* uniform *filter if $\mathfrak{m}(\mathcal{F}) = |X|$.*

Thus $\mathcal{F}$ is uniform on X if and only if no set in $\mathcal{F}$ has cardinality strictly less than the cardinality of X. For example, a uniform filter on ω contains no finite sets, and a uniform filter on ω_1 contains no countable sets. Hence, if $\mathcal{U}$ is a uniform ultrafilter on ω_1, then the complement of every countable subset of ω_1 belongs to $\mathcal{U}$.

We are now approaching the central definition of this book: we shall define certain fields K_P, where P is a prime ideal in some $C(X)$. First we define the algebras A_P.

Definition 4.13 *Let X be a completely regular space, and let P be a prime ideal in $C(X)$. Then*

$$A_P = C(X)/P \quad \textit{and} \quad A_P(\mathbb{C}) = C(X,\mathbb{C})/P_{\mathbb{C}}\,.$$

The quotient map from $C(X,\mathbb{C})$ onto $A_P(\mathbb{C})$ is denoted by π_P. Set

$$a \geq 0$$

in A_P if there exists $f \in C(X)^+$ with $a = \pi_P(f)$.

We have $A_P = \pi_P(C(X))$. The algebra A_P is unital; its identity is denoted by 1.

Theorem 4.14 *Let X be a completely regular space, and let P be a prime ideal in $C(X)$. Then A_P is a totally ordered, commutative, unital algebra.*

Proof Certainly A_P is a commutative, unital algebra.

Take $a \in A_P$, say $a = \pi_P(f)$, where $f \in C(X)$. By Proposition 4.8(i), either $f \in f^+ + P$ or $f \in f^- + P$, and so either $a = \pi_P(f^+) \geq 0$ or $a = \pi_P(f^-) \leq 0$. It follows that $(A_P, \leq)$ is a totally ordered set.

Clearly $a + c < b + c$ whenever $a, b, c \in A_P$ with $a < b$, and $ab \geq 0$ whenever $a, b \in A_P$ with $a, b \geq 0$. Thus A_P is a totally ordered algebra. □

In particular, if $\mathcal{F}$ is a prime z-filter on X, then $C(X)/\mathcal{F}$ is a totally ordered algebra.

The quotient map $\pi_P : C(X) \to A_P$ is of course an order-preserving algebra epimorphism. Further, for $f, g \in C(X)$, we have

$$\pi_P(f \vee g) = \pi_P(f) \vee \pi_P(g)\,, \tag{4.1}$$

$$\pi_P(f \wedge g) = \pi_P(f) \wedge \pi_P(g)\,, \tag{4.2}$$

$$\pi_P(|f|) = |\pi_P(f)|\,. \tag{4.3}$$

Let $a \in A_P(\mathbb{C})$. Then we define

$$|a| = \pi_P(|f|), \quad \text{where} \quad a = \pi_P(f)\,.$$

If $f, g \in C(X, \mathbb{C})$ with $\pi_P(f) = \pi_P(g)$, then $\pi_P(|f|) = \pi_P(|g|)$, and so $|a|$ is well defined, given the prime ideal P in $C(X, \mathbb{C})$. However, we caution that the definition of $|a|$ is not intrinsic in the complex algebra $A_P(\mathbb{C})$. The modulus is an additional structure imposed on the algebra $A_P(\mathbb{C})$. We have $|a| = 0$ if and only if $a = 0$, and

$$|a + b| \leq |a| + |b|\,, \qquad |ab| = |a|\,|b| \quad (a, b \in A_P(\mathbb{C}))\,.$$

Note specifically that, if P is a prime ideal in $C(X)$ and $f, g \in C(X)$ with $\pi_P(f) \leq \pi_P(g)$, then there exists $W \in \mathbf{Z}[P]$ with

$$f(x) \leq g(x) \quad (x \in W)\,.$$

Proposition 4.15 *Let X be a completely regular space, and let P be a prime ideal in $C(X)$.*

(i) *Let $a, b \in A_P$ with $a < b$. Then there exist $f, g \in C(X)$ with $\pi_P(f) = a$, $\pi_P(g) = b$, and $f < g$.*

(ii) *Let $a, b, c \in A_P$ with $a < b < c$, and take $f, h \in C(X)$ with $\pi_P(f) = a$, $\pi_P(h) = c$, and $f < h$. Then there exists $g \in C(X)$ with $\pi_P(g) = b$ and $f < g < h$.*

(iii) *Let $a, b \in A_P$ with $|a| \leq |b|$. Then $a^2 \in bA_P$. In the case where X is discrete, $a \in bA_P$.*

Proof (i) Take $f', g \in C(X)$ with $\pi_P(f') = a$ and $\pi_P(g) = b$, and set $f = f' \wedge g$. Then $\pi_P(f') = a \wedge b = a$ by (4.2) and $f < g$.

(ii) Take $g' \in C(X)$ with $\pi_P(g') = b$, and set $g = (g' \vee f) \wedge h$.

(iii) Take $f, g \in C(X)$ with $\pi_P(f) = a$ and $\pi_P(g) = b$; we may suppose that $|f| \leq |g|$. By Proposition 4.7(i), there exists $h \in C(X)$ with $f^2 = gh$, and so $a^2 \in bA_P$. In the case where X is discrete, there exists $k \in C(X)$ with $f = gk$, and so $a \in bA_P$. □

Definition 4.16 *Let X be a completely regular space, and let P be a prime ideal in $C(X)$. Then K_P and $K_P(\mathbb{C})$ are the quotient fields of A_P and $A_P(\mathbb{C})$, respectively.*

An element of K_P has the form a/b, where $a, b \in A_P$ and $b \neq 0$. We may suppose that $b > 0$ in A_P, and then we set $a/b \geq 0$ in K_P if $a \geq 0$. We thus obtain a binary relation $\leq$ on K_P such that

$$(K_P, +, \cdot, \leq)$$

is an ordered field. The set of infinitesimals and the set of finite elements in K_P are denoted by

$$K_P^0 \quad \text{and} \quad K_P^{\#},$$

respectively. The value group of K_P is denoted by Γ_P, and the archimedean valuation is denoted by v_P.

An element of $K_P(\mathbb{C})$ has the form $x = a/b$, where $a, b \in A_P(\mathbb{C})$ and $b \neq 0$. We set

$$|x| = |a| \, / \, |b| \, .$$

Then $|x|$ is well defined, $|x| = 0$ if and only if $x = 0$, and

$$|x + y| \leq |x| + |y| \, , \qquad |xy| = |x| \, |y| \quad (x, y \in K_P(\mathbb{C})) \, .$$

We define

$$K_P(\mathbb{C})^0 = \{x \in K_P(\mathbb{C}) : |x| = \mathbf{o}(1)\}, \quad K_P(\mathbb{C})^{\#} = \{x \in K_P(\mathbb{C}) : |x| = \mathbf{O}(1)\};$$

these are the sets of *infinitesimal* and *finite* elements of $K_P(\mathbb{C})$, respectively. As in the case of $A_P(\mathbb{C})$, the modulus is an additional structure imposed on the complex field $K_P(\mathbb{C})$; in contrast to the situation for K_P, the modulus and the subfields $K_P(\mathbb{C})^0$ and $K_P(\mathbb{C})^{\#}$ are not intrinsic to the field.

In the particular case where M is a maximal ideal in $C(X)$, of course A_M itself is already a field; if $A_M \cong \mathbb{R}$, then M is a *real ideal* in $C(X)$, and, if $A_M \not\cong \mathbb{R}$, then M is a *hyper-real ideal* in $C(X)$.

Definition 4.17 *Let K be an ordered field. Then:*

(i) *K is a* super-real field *if $K \not\cong \mathbb{R}$ and if K is isomorphic to an ordered field of the form K_P, where P is a prime ideal in $C(X)$ for some completely regular space X;*

(ii) *K is a* hyper-real field *if K is isomorphic to an ordered field $C(X)/M$, where M is a hyper-real maximal ideal in $C(X)$ for some completely regular space X.*

Clearly every hyper-real field is a super-real field. We shall see later in this book that there are real-closed η_1-fields that are not super-real fields, thus answering an open question.

We again refer to the discussion in Chapter 2 on our meaning for the word 'field'. We think it likely that there exists an ordered field K such that K is isomorphic as a ring to a super-real field, but such that K is not itself a super-real field.

We have introduced the terminology 'super-real' fields because we believe that this class of fields has the appropriate properties to make its members the natural candidates to play the role of a 'bigger' version of the real field $\mathbb{R}$. We hope to convince the reader in due course that this is a reasonable thesis.

We shall determine some of the properties of super-real fields; we began our study with the intention of characterizing the super-real fields algebraically in the class of all totally ordered fields, but the possibility that there is an intrinsic characterization of these fields disappeared over the horizon as our work progressed. We shall discuss in Chapter 11 the more modest possibility that there is a characterization of the super-real subfields of the special example $\widetilde{\mathbf{R}}$; even this seems unattainable (see Theorem 11.28).

We finally introduce a further special subclass of the hyper-real fields: we take X in Definition 4.17(ii) to be discrete, and hence we regard X as a cardinal number κ with the discrete topology.

Definition 4.18 *Let K be an ordered field. Then K is an* ultrapower *if K is isomorphic to an ordered field $\mathbb{R}^\kappa/M$, where κ is a cardinal number and M is a hyper-real maximal ideal in $\mathbb{R}^\kappa$.*

The maximal ideals of $\mathbb{R}^\kappa$ correspond to ultrafilters on κ, and so ultrapowers have the form $\mathbb{R}^\kappa/\mathcal{U}$, where $\mathcal{U}$ is a free ultrafilter on κ; this is the standard form in which ultrapowers are presented. We give an *ab ovo* description of these fields.

Let κ be a cardinal, and let $\mathcal{U}$ be a free ultrafilter on κ. For $f, g \in \mathbb{C}^\kappa$, set

$$f \sim_{\mathcal{U}} g \quad \text{if } \{\sigma < \kappa : f(\sigma) = g(\sigma)\} \in \mathcal{U} .$$

Then $\sim_{\mathcal{U}}$ is an equivalence relation on $\mathbb{C}^\kappa$; the equivalence class containing f is $[f]_{\mathcal{U}}$ or $[f]$, and the collection of equivalence classes is $\mathbb{C}^\kappa/\mathcal{U}$. For $[f], [g] \in \mathbb{C}^\kappa/\mathcal{U}$ and $\alpha \in \mathbb{C}$, define

$$[f] + [g] = [f + g], \quad \alpha[f] = [\alpha f], \quad [f][g] = [fg] .$$

Then $\mathbb{C}^\kappa/\mathcal{U}$ is a complex algebra with identity $[1]$ with respect to these operations, and we regard $\mathbb{R}^\kappa/\mathcal{U}$ as a real subalgebra of $\mathbb{C}^\kappa/\mathcal{U}$. It is directly clear that $\mathbb{C}^\kappa/\mathcal{U}$ is an algebraically closed complex field and that $\mathbb{R}^\kappa/\mathcal{U}$ is real-closed.

For $f, g \in \mathbb{R}^\kappa$, set

$$f \leq_{\mathcal{U}} g \quad \text{if} \quad \{\sigma \leq \kappa : f(\sigma) \leq g(\sigma)\} \in \mathcal{U} ,$$

and set

$$[f] \leq [g] \quad \text{if} \quad f \leq_{\mathcal{U}} g .$$

Then $\leq$ is well defined on $\mathbb{R}^\kappa/\mathcal{U}$, and $(\mathbb{R}^\kappa, +, \cdot, \leq)$ is an ordered field. The fields $\mathbb{C}^\kappa/\mathcal{U}$ and $\mathbb{R}^\kappa/\mathcal{U}$ are *ultrapowers* of $\mathbb{C}$ and $\mathbb{R}$, respectively. The map $[f] \to [\,|f|\,]$ is a well-defined map from $\mathbb{C}^\kappa/\mathcal{U}$ into $\mathbb{R}^\kappa/\mathcal{U}$; we have

$$(\mathbb{C}^\kappa/\mathcal{U})^0 = \left\{[f] \in \mathbb{C}^\kappa/\mathcal{U} : [\,|f|\,] \in (\mathbb{R}^\kappa/\mathcal{U})^0\right\},$$

$$(\mathbb{C}^\kappa/\mathcal{U})^\# = \left\{[f] \in \mathbb{C}^\kappa/\mathcal{U} : [\,|f|\,] \in (\mathbb{R}^\kappa/\mathcal{U})^\#\right\}.$$

This completes our description of the ultrapowers $\mathbb{R}^\kappa/\mathcal{U}$.

Before proceeding, we recall some remarks about the character space of a Banach algebra. (For the definition of a Banach algebra, see Chapter 5.)

As in Chapter 2, a *character* on a complex algebra A is a non-zero homomorphism $\varphi : A \to \mathbb{C}$; the space of characters on A is denoted by Φ_A. In the case where A is a Banach algebra, each character φ on A is continuous with $\|\varphi\| \leq 1$, and so the space Φ_A may be identified with a subset of the closed unit ball of A', where A' is the Banach space of all continuous linear functionals on A; Φ_A is a locally compact space with respect to the relative weak $*$-topology from A'. (In the weak $*$-topology on A', a net (λ_ν) converges to λ if and only if $\lambda_\nu(a) \to \lambda(a)$ for each $a \in A$.) If A is commutative and has an identity, the topological space Φ_A is non-empty and compact, and the set of maximal ideals in A is exactly the set

$$\{\ker\varphi : \varphi \in \Phi_A\};$$

for this reason, Φ_A is often called the *maximal ideal space* of A. For $a \in A$, define

$$\widehat{a}(\varphi) = \varphi(a) \quad (\varphi \in \Phi_A).$$

Then $\widehat{a} \in C(\Phi_A)$, and the map $a \mapsto \widehat{a}$, $\;A \to C(\Phi_A)$ is a homomorphism, called the *Gelfand transform* of A.

Let X be a completely regular space, and set

$$|f|_X = \sup\{|f(x)| : x \in X\} \quad (f \in C^b(X, \mathbb{C})).$$

Then $|\cdot|_X$ is the *uniform norm* on X, and $(C^b(X,\mathbb{C}), |\cdot|_X)$ is a complex, commutative, unital Banach algebra. In particular, for each compact space Ω, $(C(\Omega,\mathbb{C}), |\cdot|_\Omega)$ is a complex, commutative, unital Banach algebra.

We first make some remarks about general ideals in $C(\Omega)$.

Definition 4.19 *Let Ω be a compact space.*

(i) *For a closed subset F of Ω,*

$$I(F) = \{f \in C(\Omega) : f \mid F = 0\},$$
$$J(F) = \{f \in C(\Omega) : \mathbf{Z}(f) \text{ is a neighbourhood of } F\}.$$

(ii) *For an ideal I of $C(\Omega)$, the* hull *of I is*

$$h(I) = \bigcap\{\mathbf{Z}(f) : f \in I\}.$$

(iii) *For $x_0 \in \Omega$, $M_{x_0} = I(\{x_0\})$ and $J_{x_0} = J(\{x_0\})$.*

Let F be a closed subset of Ω. Clearly both $J(F)$ and $I(F)$ are ideals in $C(\Omega)$, the ideal $I(F)$ is closed, and $J(F) \subset I(F)$. The hull of an ideal is a closed subset of Ω.

Proposition 4.20 *Let Ω be a compact space.*

(i) *For each proper ideal I in $C(\Omega)$, $h(I) \neq \emptyset$.*

(ii) *$\{M_x : x \in \Omega\}$ is the family of maximal ideals in $C(\Omega)$.*

(iii) *Each character on $C(\Omega)$ has the form $f \mapsto f(x)$ for some $x \in \Omega$.*

(iv) *For each closed subset F of Ω, $J(F)$ is dense in $I(F)$.*

(v) *Let I be an ideal in $C(\Omega)$, and set $F = h(I)$. Then*

$$J(F) \subset I \subset I(F).$$

In the case where I is closed, $I = I(F)$.

(vi) *Let P be a prime ideal in $C(\Omega)$. Then $h(P)$ is a singleton.*

Proof (i) Take $f_1, \dots, f_n \in I$. We *claim* that $\bigcap_{j=1}^n \mathbf{Z}(f_j) \neq \emptyset$. For assume towards a contradiction that $\bigcap_{j=1}^n \mathbf{Z}(f_j) = \emptyset$, and set

$$g = \sum_{j=1}^n f_j^2 .$$

Then $g \in C(\Omega)$ and $g(x) > 0 \quad (x \in \Omega)$, so that $1/g \in C(\Omega)$ and

$$1 = \sum_{j=1}^n f_j^2/g \in I,$$

a contradiction of the fact that I is a proper ideal.

It follows that the family $\mathbf{Z}[I]$ has the finite intersection property, and so $h(I)$ is non-empty.

(ii) For $x \in \Omega$, M_x has codimension 1 in $C(\Omega)$, and so M_x is a maximal ideal. Let M be a maximal ideal in $C(\Omega)$. Then $h(M) \neq \emptyset$ by (i). Take $x \in h(M)$. Then $M \subset M_x$, and so $M = M_x$.

(iii) This is immediate from (ii).

(iv) For each $f \in I(F)$ and $\varepsilon > 0$, take

$$U = \{x \in \Omega : |f(x)| < \varepsilon\}.$$

By Urysohn's lemma, there exists $g \in J(F) \cap C(\Omega, \mathbb{I})$ with $g \mid (\Omega \setminus U) = 1$, and then $fg \in J(F)$ and $|f - fg|_\Omega < \varepsilon$. Thus $J(F)$ is dense in $I(F)$.

(v) Certainly $I \subset I(F)$. Let K be a closed subset of Ω with $F \cap K = \emptyset$. For each $x \in K$, there exists $f_x \in I$ with $f_x(x) = 1$, and there is a neighbourhood U_x of x such that

$$f_x(y) > 1/2 \quad (y \in U_x).$$

Since K is compact, there exist $f_1, \ldots, f_n$ in the family $\{f_x : x \in I\}$ such that, for each $x \in K$, there exists $j \in \{1, \ldots, n\}$ with $f_j(x) > 1/2$. Set

$$g = \sum_{j=1}^{n} f_j^2 \in I .$$

Then $g(x) \geq 1/4 \quad (x \in K)$, and so there exists a function $h \in C(\Omega)$ with $g(x)h(x) = 1 \quad (x \in K)$, and $gh \in I$.

Now take $f \in J(F)$, and set $K = \overline{\Omega \setminus \mathbf{Z}(f)}$, so that $F \cap K = \emptyset$. Then there exists $k \in I$ with $k(x) = 1 \quad (x \in K)$. We have $f = fk$, and so $f \in I$ and $J(F) \subset I$.

It follows from (iv) that $I = I(F)$ in the case where I is closed.

(vi) Assume towards a contradiction that there exist $x_1, x_2 \in h(P)$ with $x_1 \neq x_2$. Then there are neighbourhoods U_1 and U_2 of x_1 and x_2, respectively, with $\overline{U}_1 \cap \overline{U}_2 = \emptyset$, and there exist $f_1, f_2 \in C(\Omega)$ with $f_1(x_1) = f_2(x_2) = 1$ and $f_1 \,|(\Omega \setminus U_1) = f_2|\,(\Omega \setminus U_2) = 0$. We have $f_1 f_2 = 0$, but $f_1 \notin P$ and $f_2 \notin P$, a contradiction. □

In fact, (vi), above, already follows from (ii) and Proposition 4.8(v).

Now let P be a prime ideal in $C(\Omega)$. Then the unique element $x \in \Omega$ such that $h(P) = \{x\}$ is denoted by x_P. By Proposition 4.20(iv),

$$J_{x_P} \subset P \subset M_{x_P} ,$$

and so, by (iii), P is closed in $C(\Omega)$ if and only if $P = M_{x_P}$. In particular, if Q is a minimal prime ideal contained in P, then $J_{x_P} \subset Q$. The ideal J_{x_P} may or may not be a prime ideal; if it is a prime ideal, then it is the unique minimal prime ideal contained in P, and the family of prime ideals in $C(\Omega)$ with hull equal to $\{x_P\}$ is a chain with maximum element M_{x_P} and minimum element J_{x_P}.

Again set $A_P = C(\Omega)/P$. Then A_P has a unique maximal ideal, namely $\pi_P(M_{x_P})$; this maximal ideal is denoted by M_P.

Proposition 4.21 *Let Ω be a compact space, and let P be a non-maximal, prime ideal in $C(\Omega)$. Then:*

(i) $A_P \subset K_P^{\#}$ *and* $M_P = A_P \cap K_P^0 = \{a \in A_P : n\,|a| \leq 1 \quad (n \in \mathbb{N})\}$;

(ii) $A_P^+ \setminus \{0\}$ *is coinitial in* $K_P^+ \setminus \{0\}$;

(iii) M_P *is cofinal in* K_P^0.

Proof (i) Let $a \in A_P$, and take $f \in C(\Omega)^+$ with $\pi_P(f) = |a|$. Then there exists $n \in \mathbb{N}$ with $|f|_\Omega \leq n$. We have $f \leq n$ in $C(\Omega)$ and $|a| \leq n$ in A_P, whence $a \in K_P^{\#}$.

If $a \in M_P$, then $f \in M_{x_P}$, and, for each $n \in \mathbb{N}$, $f(x) = f(x) \wedge (1/n)$ on a neighbourhood of x_P. Since $J_{x_P} \subset P$, this implies that

$$\pi_P(f) = \pi_P(f) \wedge (1/n) ,$$

and so $n\,|a| \leq 1$. Thus $a \in K_P^0$, and hence $M_P = A_P \cap K_P^0$.

(ii) Take $a, b \in A_P^+ \setminus \{0\}$ with $a/b < 1$; we may suppose that $a < 1$. By Proposition 4.15(iii), there exists $c \in A_P$ with $a^2 = bc$, and then $0 < c < a/b$ in K_P^+.

(iii) Take $a \in K_P^0$, and choose $f, g \in M_{x_P}^+$ with $\pi_P(f) = a\pi_P(g)$. Set

$$F_n = \{x \in \Omega : (n+1)f(x) \leq g(x)\} \quad (n \in \mathbb{N}).$$

Then each F_n is a neighbourhood of x_P. Define

$$h(x) = \frac{1}{n} \quad (x \in \partial F_n,\, n \in \mathbb{N}).$$

Then h has a continuous extension to $h \in M_{x_P}$ with

$$\frac{1}{(n+1)} \leq h(x) \leq \frac{1}{n} \quad (x \in F_n \setminus \operatorname{int} F_{n+1},\, n \in \mathbb{N}).$$

Clearly $f \leq hg$ in $C(\Omega)$, and so $a \leq \pi_P(h) \in M_P$. □

We shall utilize the Stone–Čech compactification βX of a completely regular topological space X in a minor way in this book; we recall the definition of βX. Let X be a completely regular space. Then βX is a compact space containing X as a dense subspace and βX has the property that each continuous map from X into a compact space K has an extension to a continuous map from βX into K. The space βX is unique in the sense that, if T is a compact space satisfying the same conditions, then there is a homeomorphism from βX onto T leaving X pointwise fixed. There is a considerable number of constructions of βX; we shall briefly recall one such approach which is natural in the context of Banach algebra theory.

The character space of $C^b(X, \mathbb{C})$ is easily seen to have the characteristic properties of the Stone–Čech compactification of X, and so it can be identified with the space βX; we regard X as a dense subset of βX. The Gelfand transform

$$f \mapsto \widehat{f}, \quad C^b(X, \mathbb{C}) \to C(\beta X, \mathbb{C}),$$

is an isometric isomorphism, and we shall identify $C^b(X)$ and $C^b(X, \mathbb{C})$ with $C(\beta X)$ and $C(\beta X, \mathbb{C})$, respectively. For example, we identify l^∞ with $C(\beta\mathbb{N})$, and so

$$c_0 = \{f \in C(\beta\mathbb{N}) : f \mid (\beta\mathbb{N} \setminus \mathbb{N}) = 0\},$$
$$c = \{f \in C(\beta\mathbb{N}) : f \mid (\beta\mathbb{N} \setminus \mathbb{N}) \text{ is constant}\}.$$

For a subset S of X, cl S denotes the closure of S in βX. It is a standard property of βX that, if $Z_1, Z_2 \in \mathbf{Z}(X)$, then

$$\operatorname{cl}(Z_1 \cap Z_2) = \operatorname{cl} Z_1 \cap \operatorname{cl} Z_2.$$

Let $\mathcal{U}$ be a z-ultrafilter on X. Then $\bigcap\{\operatorname{cl} Z : Z \in \mathcal{U}\}$ is a singleton in βX. Conversely, for each $\mathfrak{p} \in \beta X$, $\{Z \in \mathbf{Z}(X) : \mathfrak{p} \in \operatorname{cl} Z\}$ is a z-ultrafilter on X. This correspondence between $\mathcal{U}$ and $\mathfrak{p}$ is a bijection, and so the points of βX correspond to the maximal ideals of $C(X)$; this is the *Gelfand–Kolmogoroff theorem*. For example, $\beta\mathbb{N}$ can be identified with (and is often defined as) the set of ultrafilters on $\mathbb{N}$ with the so-called *Stone topology*; in this setting, it is the *Stone space* of $\mathbb{N}$.

In summary, the Stone–Čech compactification βX of a completely regular space X can be identified with each of the following sets: (i) the character space of $C^b(X)$; (ii) the maximal ideal space of $C^b(X)$; (iii) the set of maximal ideals of $C(X)$; (iv) the set of z-ultrafilters on X.

In fact every super-real field has the form K_P for a prime ideal P in $C(\Omega)$ for some compact space Ω. For let Q be a prime ideal in $C(X)$, where X is a completely regular space, and set $P = C^b(X) \cap Q = C(\beta X) \cap Q$. Then P is a prime ideal in $C(\beta X)$, and $K_P \cong K_Q$ because each continuous function $f \in C(X)$ can be expressed as $f = g/h$, where $g, h \in C^b(X)$.

Now suppose that M is a maximal ideal in $C(X)$. Then $P = C^b(X) \cap M$ is given by

$$P = \{f \in C^b(X) : \mathfrak{p} \in \mathrm{cl}_{\beta X} \mathbf{Z}_X(f)\} , \tag{4.4}$$

where $\mathfrak{p}$ is the point of βX corresponding to the maximal ideal M, and so P is a z-ideal. Thus every hyper-real field is a super-real field of the form K_P where P is a prime z-ideal in some algebra $C(\Omega)$, with Ω compact.

Finally, suppose further that X is discrete. For each $f \in C^b(X)$,

$$\{\mathbf{Z}(f),\ X \setminus \mathbf{Z}(f)\}$$

is a partition of X into disjoint closed sets, and so

$$\{\mathrm{cl}\,\mathbf{Z}(f),\ \mathrm{cl}\,(X \setminus \mathbf{Z}(f))\}$$

is a partition of βX into disjoint closed sets. Thus, if $f \in C^b(X) \cap M$, then $\mathrm{cl}\,\mathbf{Z}(f)$ is a neighbourhood of $\mathfrak{p}$ in βX, and so $C^b(X) \cap M = J_{\mathfrak{p}}$. Hence an ultrapower has the form K_P, where $P = J_{\mathfrak{p}}$ is a minimal prime ideal in the algebra $C(\beta X)$.

For example, let $\mathcal{U}$ be a free ultrafilter on $\mathbb{N}$, corresponding to the point $\mathfrak{p} \in \beta\mathbb{N} \setminus \mathbb{N}$, say. Then we set

$$A_{\mathfrak{p}} = C(\beta\mathbb{N})/J_{\mathfrak{p}} ,$$

and we write

$$\pi_{\mathfrak{p}} : C(\beta\mathbb{N}) \to A_{\mathfrak{p}}$$

for the quotient map; the super-real field which is the quotient field of $A_{\mathfrak{p}}$ is denoted by $K_{\mathfrak{p}}$ and is identified with the ultrapower $\mathbb{R}^{\mathbb{N}}/\mathcal{U}$.

Proposition 4.22 *Let Ω be a compact space, and let P be a prime ideal in $C(\Omega)$. Set*

$$\mathcal{F} = \{Z \in \mathbf{Z}[P] : I(Z) \subset P\} .$$

Then $\mathcal{F}$ is a prime z-filter.

Proof Since $I(\emptyset) = C(\Omega)$, we have $\emptyset \notin \mathcal{F}$.

Suppose that $Z_1, Z_2 \in \mathcal{F}$, and take $h \in I(Z_1 \cap Z_2)$. Define $h_1 \in C(Z_1 \cup Z_2)$ by the requirements that $h_1 \mid Z_1 = 0$ and $h_1 \mid Z_2 = h$, and extend h_1 to be an element of $C(\Omega)$. Set $h_2 = h - h_1$. Then $h_1 \in I(Z_1)$ and $h_2 \in I(Z_2)$, and so $h_1, h_2 \in P$. Hence $h = h_1 + h_2 \in P$; this shows that $I(Z_1 \cap Z_2) \subset P$, and hence that $Z_1 \cap Z_2 \in \mathcal{F}$.

Suppose that $Z \in \mathcal{F}$ and that $W \in \mathbf{Z}(\Omega)$ with $W \supset Z$. Since $I(W) \subset I(Z)$, we have $W \in \mathcal{F}$.

Suppose that $Z_1, Z_2 \in \mathbf{Z}(\Omega)$ with $Z_1 \cup Z_2 \in \mathcal{F}$. Assume that there exist $h_1 \in I(Z_1) \setminus P$ and $h_2 \in I(Z_2) \setminus P$. Then $h_1 h_2 \in I(Z_1 \cup Z_2) \subset P$, and so either $h_1 \in P$ or $h_2 \in P$, a contradiction. Thus either $I(Z_1) \subset P$ or $I(Z_2) \subset P$, and either $Z_1 \in \mathcal{F}$ or $Z_2 \in \mathcal{F}$.

We have shown that $\mathcal{F}$ is a prime z-filter. □

Proposition 4.23 *Let P be a prime ideal in $C(\Omega)$. Then there is a maximum z-ideal Q with Q below P, and Q is a prime ideal.*

Proof Set $Q = \mathbf{Z}^{-1}[\mathcal{F}]$, where $\mathcal{F}$ is the prime z-filter specified in Proposition 4.22. Then Q has the required properties. □

Corollary 4.24 *Let Ω be a compact space. Then every minimal prime in $C(\Omega)$ is a z-ideal.* □

The following minor technicality about prime z-ideals will be required several times.

Proposition 4.25 *Let P be a prime z-ideal in $C(\Omega)$ with P below M_{x_0}, let $f_1, f_2 \in C(\Omega)$, and let W be a neighbourhood of x_0. Suppose that there exists $g \in P$ such that*

$$f_1(x) \leq f_2(x) \quad (x \in W \cap \mathbf{Z}(g)).$$

Then $\pi_P(f_1) \leq \pi_P(f_2)$ in A_P.

Proof Take $h \in J_{x_0}$ with $\mathbf{Z}(h) \subset W$. Then $h \in P$ and $\mathbf{Z}(g) \cap \mathbf{Z}(h) \in \mathbf{Z}[P]$. Set $f = (f_1 \vee f_2) - f_2$. Then $\mathbf{Z}(g) \cap \mathbf{Z}(h) \subset \mathbf{Z}(f)$, and so $\mathbf{Z}(f) \in \mathbf{Z}[P]$. Since P is a z-ideal, we have $f \in P$, and so $\pi_P(f_1) \vee \pi_P(f_2) = \pi_P(f_2)$ in A_P. Thus $\pi_P(f_1) \leq \pi_P(f_2)$. □

We shall prove later that various sets are order-dense in K_P; the following result gives an easy way of checking this.

Proposition 4.26 *Let P be a prime ideal in $C(\Omega)$. Suppose that S is a subgroup of $(K_P^+ \setminus \{0\}, \cdot)$ such that $S \cap [c, d] \neq \emptyset$ whenever $c < d$ in $M_P^+ \setminus \{0\}$. Then S is order-dense in K_P^+.*

Proof It is sufficient to show that, given $a_1, a_2, b_1, b_2 \in M_P^+ \setminus \{0\}$ such that $a_1/b_1 < a_2/b_2$ in K_P^+, there exist $a, b \in S$ with

$$\frac{a_1}{b_1} < \frac{a}{b} < \frac{a_2}{b_2}.$$

Thus, take such a_1, a_2, b_1, b_2, as specified, and set $\varepsilon = a_2 b_1 - a_1 b_2$, so that $\varepsilon \in K_P^+ \setminus \{0\}$. By Proposition 4.21(ii), there exists $\eta \in M_P^+ \setminus \{0\}$ such that

$$\eta < \frac{\varepsilon b_2}{3a_2} < \frac{\varepsilon b_1}{3a_1}$$

in K_P. By hypothesis, there exist $a, b \in S$ with

$$a_1b_2 + \frac{\varepsilon}{3} \leq a \leq a_2b_1 - \frac{\varepsilon}{3} \quad \text{and} \quad b_1b_2 - \eta \leq b \leq b_1b_2 + \eta\,.$$

Then

$$ab_1 - a_1b \geq (a_1b_2 + \frac{\varepsilon}{3})\,b_1 - a_1(b_1b_2 + \eta) = \frac{\varepsilon b_1}{3} - a_1\eta > 0\,,$$

and, similarly, we have $a_2b - ab_2 > 0$, and so $a_1/b_1 < a/b < a_2/b_2$, giving the result. □

We now prove the key fact that super-real fields are real-closed.

Theorem 4.27 *Let Ω be a compact space, and let P be a non-maximal, prime ideal in $C(\Omega)$. Then K_P is a real-closed field.*

Proof This result is equivalent to the statement that the quotient field $K_Q(\mathbb{C})$ of $A_Q(\mathbb{C})$ is algebraically closed for an appropriate prime ideal $Q = P_{\mathbb{C}}$ in $C(\Omega, \mathbb{C})$, and hence to the statement that each monic polynomial over $A_Q(\mathbb{C})$ has a root in $A_Q(\mathbb{C})$. We shall prove this form of the statement.

Let p be a monic polynomial over $A_Q(\mathbb{C})$, say

$$p = a_0 + a_1X + \cdots + a_{n-1}X^{n-1} + X^n\,,$$

where $n \in \mathbb{N}$ and $a_0, \ldots, a_{n-1}$ belong to $A_Q(\mathbb{C})$. For $j = 0, \ldots, n-1$, choose $f_j \in C(\Omega, \mathbb{C})$ such that $\pi_P(f_j) = a_j$, and set

$$p_x = f_0(x) + f_1(x)X + \cdots + f_{n-1}(x)X^{n-1} + X^n \quad (x \in \Omega)\,,$$

so that $p_x \in \mathbb{C}[X]$. Let $z_1(x), \ldots, z_n(x)$ be the roots of p_x in $\mathbb{C}$, and set

$$r_j(x) = \operatorname{Re} z_j(x) \quad (j = 1, \ldots, n)\,,$$

ordering the roots so that

$$r_1(x) \leq \cdots \leq r_n(x)\,.$$

Take $x \in \Omega$, and let $\rho_1, \ldots, \rho_m$ be the distinct values in the set

$$\{r_1(x), \ldots, r_n(x)\}\,,$$

with $\rho_1 < \cdots < \rho_m$. Fix $\varepsilon > 0$, and take $\eta \in (0, \varepsilon)$ and $C > 0$ such that the rectangles $\overline{W}_1, \ldots, \overline{W}_m$ in $\mathbb{R}^2$ are disjoint, where

$$W_j = (\rho_j - \eta, \rho_j + \eta) \times (-C, C) \qquad (j = 1, \ldots, m)\,,$$

and such that

$$\{z_1(x), \ldots, z_n(x)\} \subset \bigcup_{j=1}^{m} W_j\,.$$

Set $W = \bigcup_{j=1}^{m} W_j$ and $\delta_x = \inf\{|p_x(w)| : w \in \partial W\}$. Since ∂W is compact and $p_x(w) \neq 0$ for $w \in \partial W$, necessarily $\delta_x > 0$. The map

$$(y, w) \mapsto p_y(w)\,, \quad \Omega \times \partial W \to \mathbb{C}\,,$$

is continuous, and so there is a neighbourhood U of x in Ω such that

$$|p_y(w) - p_x(w)| < \delta_x \leq p_x(w) \qquad (y \in U,\ w \in \partial W)\ .$$

By Rouché's theorem, for each $y \in U$ and each $j \in \{1, \dots, m\}$, p_y and p_x have the same number of zeros in W_j, and so

$$|r_j(y) - r_j(x)| < \varepsilon \quad (y \in U,\ j = 1, \dots, n)\,.$$

Thus the functions $r_1, \dots, r_n$ are continuous on Ω. Clearly, if $(y, w) \in \Omega \times \mathbb{C}$ and $p_y(w) = 0$, then $\operatorname{Re} w = r_j(y)$ for some $j \in \{1, \dots, n\}$.

The real parts of the roots of the monic polynomial

$$\mathrm{i}^{-n}(f_0(x) + \mathrm{i}f_1(x)X + \cdots + \mathrm{i}^n X^n)$$

are exactly the imaginary parts of the roots of p_x, and so there exist continuous, real-valued functions $s_1, \dots, s_n$ on Ω such that, if $(y, w) \in \Omega \times \mathbb{C}$ and $p_y(w) = 0$, then $\operatorname{Im} w = s_j(y)$ for some $j \in \{1, \dots, n\}$.

For $j, k \in \{1, \dots, n\}$ and $x \in \Omega$, set

$$u_{j,k}(x) = p_x(r_j(x) + \mathrm{i}s_k(x))\,.$$

Then $u_{j,k} \in C(\Omega, \mathbb{C})$. If $x \in \Omega$, then there exists $z \in \mathbb{C}$ with $p_x(z) = 0$, and so there exist $j, k \in \{1, \dots, n\}$ such that $\operatorname{Re} z = r_j(x)$ and $\operatorname{Im} z = s_k(x)$. We have $u_{j,k}(x) = 0$, and so

$$\prod\nolimits_{j,k} u_{j,k} = 0 \quad \text{in} \quad C(\Omega, \mathbb{C})\,.$$

Since Q is a prime ideal in $C(\Omega, \mathbb{C})$, there exist $j_0, k_0 \in \{1, \dots, n\}$ such that $u_{j_0,k_0} \in Q$. Set $a = \pi_P(r_{j_0} + \mathrm{i}s_{k_0})$. Then $a \in A_Q(\mathbb{C})$ and $p(a) = 0$, proving that p has a root in $A_Q(\mathbb{C})$, as required. □

Next we examine the separation properties of super-real fields. A totally ordered algebra $(A, +, \cdot, \leq)$ is a *semi-η_1-algebra* (respectively, an *η_1-algebra*) if $(A, \leq)$ is a semi-η_1-set (respectively, an η_1-set).

Theorem 4.28 *Let Ω be a compact space, and let P be a non-maximal prime ideal in $C(\Omega)$. Then:*

(i) *A_P is a semi-η_1-algebra;*

(ii) *K_P is an η_1-field if and only if $\delta(K_P) \geq \aleph_1$;*

(iii) *$\delta(\Gamma_P) \geq \aleph_1$.*

Proof (i) To prove that $(A_P, \leq)$ is a semi-η_1-set, we must show that, whenever (a_n) and (b_n) are sequences in A_P with

$$a_m < a_n < b_n < b_m$$

for each $m, n \in \mathbb{N}$ with $m < n$, there exists $a \in A_P$ with

$$a_m < a < b_n \quad (m, n \in \mathbb{N})\,. \tag{4.5}$$

It follows from Proposition 4.15, (i) and (ii), that there exist sequences (f_n) and (g_n) in $C(\Omega)$ such that $\pi_P(f_n) = a_n$ and $\pi_P(g_n) = b_n$ for $n \in \mathbb{N}$ and such that

$$f_m < f_n < g_n < g_m \quad (m, n \in \mathbb{N},\ m < n)\,.$$

We now obtain the element a in A_P. Set $x = x_P$. Then

$$f_m(x) \leq g_n(x) \quad (m, n \in \mathbb{N}),$$

and so we may suppose that $f_m(x) \leq 0 \leq g_n(x) \quad (m, n \in \mathbb{N})$.

Suppose first that $f_m(x) < 0 < g_n(x)$ for each $m, n \in \mathbb{N}$. For $n \in \mathbb{N}$,

$$g_{n+1} - g_{n+1}^{+} \in J_x \subset P,$$

and so $b_n > b_{n+1} = b_{n+1}^{+} \geq 0$. Similarly, $a_m < 0 \quad (m \in \mathbb{N})$, and so we may take $a = 0$ in this case.

Now suppose that $f_m(x) = 0$ eventually; by relabelling the sequence, we may suppose that $f_m(x) = 0 \quad (m \in \mathbb{N})$. Clearly, we have

$$(a_{m+1} - a_m) \wedge 2^{-m} = a_{m+1} - a_m \quad (m \in \mathbb{N}).$$

Set

$$f = f_1 + \sum_{k=1}^{\infty}(f_{k+1} - f_k) \wedge 2^{-k}.$$

The series is uniformly convergent on Ω, and $f \in M_x$. Set $a = \pi_P(f)$, so that $a \in M_P$. For each $m \in \mathbb{N}$,

$$f \geq f_1 + \sum_{k=1}^{m}(f_{k+1} - f_k) \wedge 2^{-k},$$

and so

$$a \geq a_1 + \sum_{k=1}^{m}(a_{k+1} - a_k) = a_{m+1} > a_m.$$

Also, for each $y \in \Omega$ and $n \in \mathbb{N}$,

$$\begin{aligned} f(y) &\leq f_1(y) + \limsup_{m\to\infty} \sum_{k=1}^{m}(f_{k+1}(y) - f_k(y)) \\ &\leq \limsup_{m\to\infty} f_{m+1}(y) \leq g_{n+1}(y), \end{aligned}$$

and so $a \leq b_{n+1} < b_n$. A similar argument applies in the case where $g_n(x) = 0$ eventually.

We have obtained the element $a \in A_P$ to satisfy (4.5).

(ii) Certainly $\delta(K_P) \geq \aleph_1$ in the case where K_P is an η_1-field.

Now suppose that $\delta(K_P) \geq \aleph_1$. By Proposition 2.20, it suffices to show that K_P is a semi-η_1-field. For this, take sequences (a_n), (b_n), (c_n), and (d_n) in $A_P^+ \setminus \{0\}$ with

$$\frac{a_m}{b_m} < \frac{a_n}{b_n} < \frac{c_n}{d_n} < \frac{c_m}{d_m} \quad (m, n \in \mathbb{N},\ m < n). \tag{4.6}$$

Since $\delta(K_P) \geq \aleph_1$, by Proposition 4.21(ii) there exists $b \in A_P^+ \setminus \{0\}$ with

$$b << \{b_m, d_n : m, n \in \mathbb{N}\}.$$

By Proposition 4.15(iii), there exist sequences (x_m) and (y_n) in A_P such that $b^2 = b_m x_m \quad (m \in \mathbb{N})$ and $b^2 = d_n y_n \quad (n \in \mathbb{N})$. From (4.6), we have

$$a_m x_m < a_n x_n < c_n y_n < c_m y_m \quad (m, n \in \mathbb{N},\ m < n).$$

By (i), there exists $c \in A_P$ with

$$a_m x_m < c < c_n y_n \quad (m, n \in \mathbb{N}).$$

But now

$$\frac{a_m}{b_m} < \frac{c}{b^2} < \frac{c_n}{d_n} \quad (m, n \in \mathbb{N}),$$

and so K_P is a semi-η_1-field.

(iii) Suppose that (s_n) is a strictly decreasing sequence in $\Gamma_P^+ \setminus \{0\}$. By Proposition 4.21(iii), there exists $(f_n) \subset M_P^+$ with

$$v_P(f_n) \leq s_n \quad (n \in \mathbb{N}),$$

and, by (i), there exists $f \in M_P$ with $f > f_n \quad (n \in \mathbb{N})$. We have

$$v(f) << \{s_n : n \in \mathbb{N}\},$$

and so (s_n) is not coinitial in $\Gamma_P^+ \setminus \{0\}$. Thus $\delta(\Gamma_P) \geq \aleph_1$. □

In due course, we shall see that there are super-real fields which are semi-η_1-fields, but not η_1-fields, and that there are super-real fields which are not even semi-η_1-fields. However, before this, we shall prove that in many important cases a super-real field is a semi-η_1-field and even an η_1-field; the first theorem covers all hyper-real fields.

Theorem 4.29 *Let K be a hyper-real field. Then K is a real-closed η_1-field.*

Proof Suppose that $K = C(X)/M$ for a completely regular space X and a hyper-real maximal ideal M. By Theorem 4.28(ii), it suffices to prove that $\delta(K) \geq \aleph_1$. We shall prove the equivalent fact that $\operatorname{cof} K^+ \geq \aleph_1$.

Let (a_n) be a strictly increasing sequence in K^+, and take (f_n) in $C(X)$ with $f_n + M = a_n$ and

$$0 < f_n < f_{n+1} \quad (n \in \mathbb{N}).$$

Since M is hyper-real, there exists $g \in C(X)$ such that $g + M \geq k$ for each $k \in \mathbb{N}$. Define auxiliary functions $\varphi_k \in C(\mathbb{R}^+)$ for $k \in \mathbb{N}$ such that, for each $k \in \mathbb{N}$, $\varphi_k(k-1) = \varphi_k(k+1) = 0$, $\varphi_k(k) = 1$, φ_k is linear on $[k-1, k]$ and on $[k, k+1]$, and $\varphi_k = 0$ on $\mathbb{R}^+ \setminus [k-1, k+1]$. Then

$$\sum_{k=1}^{\infty} \varphi_k(t) = 1 \quad (t \geq 1),$$

and, for each $t \in \mathbb{R}^+$, $\varphi_k(t) = 0$ for all but at most two values of k. Set

$$h(x) = \sum_{k=1}^{\infty} \varphi_k(g(x)) f_k(x) \qquad (x \in X).$$

Then h is defined on X. On each open set of the form

$$U_n = \{x \in X : n - 1 < g(x) < n + 1\},$$

we have

$$h(x) = \sum_{k=n-1}^{n+1} \varphi_k(g(x)) f_k(x),$$

and so $h \mid U_n$ is continuous. It follows that $h \in C(X)$. We define $a = h + M$, an element of K^+.

Let $n \in \mathbb{N}$, and set $Z_n = \mathbf{Z}(g - (n+1))$. For $x \in Z_n$, we have

$$h(x) = \sum_{k=n+1}^{\infty} \varphi_k(g(x)) f_k(x) \geq f_{n+1}(x),$$

and so $a \geq a_{n+1} > a_n$. Thus (a_n) is not cofinal in K, and so $\operatorname{cof} K^+ \geq \aleph_1$, as required. □

Corollary 4.30 (CH) *Let $\mathcal{U}$ and $\mathcal{V}$ be free ultrafilters on* $\mathbb{N}$. *Then*

$$\mathbb{R}^{\mathbb{N}}/\mathcal{U} \cong \mathbb{R}^{\mathbb{N}}/\mathcal{V} \cong \mathbf{R} \quad \text{and} \quad (\mathbb{C}^{\mathbb{N}}/\mathcal{U})^{\#} \cong (\mathbb{C}^{\mathbb{N}}/\mathcal{V})^{\#} \cong \mathbf{C}^{\#}.$$

Proof By Theorem 4.29, $\mathbb{R}^{\mathbb{N}}/\mathcal{U}$ and $\mathbb{R}^{\mathbb{N}}/\mathcal{V}$ are real-closed η_1-fields, and these fields have cardinality $\mathfrak{c}$. Thus the result for $\mathbf{R}$ follows from Corollary 2.34(ii). The result for the complex algebra $\mathbf{C}$ is an immediate consequence of the result for $\mathbf{R}$. □

We have proved that, with CH, $\mathbf{R}$ is an ultrapower.

Note that it does not follow obviously from the fact that $\mathbb{C}^{\mathbb{N}}/\mathcal{U} \cong \mathbf{C}$ that $(\mathbb{C}^{\mathbb{N}}/\mathcal{U})^{\#} \cong \mathbf{C}^{\#}$; indeed, it is not difficult to see that it is always true that $\mathbb{C}^{\mathbb{N}}/\mathcal{U} \cong \mathbf{C}$ because these two algebraically closed fields both have transcendence degree $\mathfrak{c}$ over $\mathbb{C}$.

Definition 4.31 *Let Ω be a compact space, and let P be a prime ideal in $C(\Omega)$. Then P is a* valuation prime *if A_P is a valuation algebra.*

Proposition 4.32 *Let P be a non-maximal, prime ideal in $C(\Omega)$. Then P is a valuation prime if and only if $K_P^{\#} = A_P$.*

Proof By Proposition 2.6, $K_P^{\#}$ is a valuation algebra, and so P is a valuation prime in the case where $A_P = K_P^{\#}$.

Now suppose that A_P is a valuation algebra, and take $a \in K_P^{\#} \setminus \{0\}$, say $a = bc^{-1}$, where $b, c \in A_P \setminus \{0\}$. Then $v_P(b) \geq v_P(c)$, and so $|b| \leq n\,|c|$ for some $n \in \mathbb{N}$. Since A_P is a valuation algebra, either $b \in cA_P$ or $c \in bA_P$. In the former case, $a \in A_P$. In the latter case, there exists $d \in A_P$ with $c = bd$, and $|d| \geq 1/n$, so that $d \notin M_P$. Hence $d \in \operatorname{Inv} A_P$ and $a = d^{-1} \in A_P$. Thus, in either case, $a \in A_P$, and so $A_P = K_P^{\#}$. □

Proposition 4.33 *Let $K = C(X)/M$ be a hyper-real field, where M is a hyper-real maximal ideal in $C(X)$. Then $K \cong K_P$ for the valuation prime z-ideal $P = C^b(X) \cap M$ in $C(\beta X)$.*

Proof Set $P = C^b(X) \cap M$. Then P is a prime z-ideal in $C(\beta X)$ and $K \cong K_P$. An element of $K_P^\#$ has the form $f + M$, where $f \in C^b(X)$, and so has the form $\widehat{f} + P$, where $\widehat{f} \in C(\beta X)$. Thus $K_P^\# = A_P$ and P is a valuation prime. □

There are examples of spaces Ω such that each prime ideal in $C(\Omega)$ is a valuation prime; the following result covers the algebra l^∞. Since there are prime ideals in l^∞ which are not z-ideals, it shows that a valuation prime is not necessarily a z-ideal. We shall shortly give an example of a prime z-ideal which is not a valuation prime.

Proposition 4.34 *Let X be a discrete space, and let P be a prime ideal in $C(\beta X)$. Then P is a valuation prime.*

Proof Take $a, b \in A_P$. Then either $|a| \le |b|$ or $|b| \le |a|$. By Proposition 4.15(iii), either $a \in bA_P$ or $b \in aA_P$. □

We next present a variant of the notion of a valuation prime. We have noted that always $M_P \subset K_P^0$ and that P is a valuation prime if and only if $M_P = K_P^0$.

Definition 4.35 *Let Ω be a compact space, and let P be a prime ideal in $C(\Omega)$. Then P is a* strongly convex prime *if M_P is an interval in K_P.*

Thus P is a strongly convex prime if $b \in M_P$ whenever $b \in K_P^+$ and $b \le a$ for some $a \in M_P^+$. Equivalently, P is a strongly convex prime if and only if M_P is an ideal in $K_P^\#$.

Certainly each valuation prime is a strongly convex prime; we shall shortly observe that the converse is not true. In Theorem 4.40, we shall exhibit a prime z-ideal which is not a strongly convex prime.

Proposition 4.36 *Let P be a prime ideal in c. Then P is a strongly convex prime.*

Proof This is immediate from Proposition 4.34 since c_0 is an ideal in l^∞. □

Proposition 4.37 *Let P be a strongly convex prime in $C(\Omega)$. Then K_P is a semi-η_1-field.*

Proof Take sequences (a_n) and (b_n) in K_P with

$$0 < a_m < a_n < b_n < b_m \quad (m, n \in \mathbb{N},\ m < n)\,.$$

There exists $c \in K_P^+ \setminus \{0\}$ such that $b_1 c \in K_P^\#$. Take $d \in M_P^+ \setminus \{0\}$. Then $(a_n cd), (b_n cd) \subset M_P$ because M_P is an ideal in $K_P^\#$. By Theorem 4.28(i), there exists $a \in M_P$ with $a_m cd < a < b_n cd \quad (m, n \in \mathbb{N})$, and then

$$a_m < ac^{-1}d^{-1} < b_n \quad (m, n \in \mathbb{N})\,,$$

so that K_P is a semi-η_1-field. □

Corollary 4.38 *Let X be a discrete space, and let P be a prime ideal in $C(\beta X)$. Then K_P is a semi-η_1-field.*

Proof By Proposition 4.34, P is a valuation prime, and hence P is a strongly convex prime. □

Suppose that P and Q are prime ideals in $C(X)$ with Q below P. Then we have the canonical map

$$\pi_{PQ} : f + Q \mapsto f + P, \quad A_Q \to A_P .$$

The map π_{PQ} is a unital epimorphism, and π_{PQ} is order-preserving. Of course $\pi_{PQ} \circ \pi_Q = \pi_P$. In the case where Q is a valuation prime, so is P, for the image of a valuation algebra by a unital epimorphism is a valuation algebra. Theorem 4.40, below, shows that the converse of this result is false in general, even in the case where K_P is an ultrapower. Indeed Q is not necessarily a strongly convex prime. This is an important point, for it is very tempting in the middle of a proof to assume that an arbitrary prime ideal in $C(\Omega)$ is a strongly convex prime; because this is not the case, certain later proofs require rather long and delicate arguments.

We shall first show that a prime above a strongly convex prime is also strongly convex, and then we shall give an example to show that a prime below a strongly convex prime is not necessarily strongly convex.

Proposition 4.39 *Let Q and P be prime ideals in $C(\Omega)$ with Q below P. Suppose that Q is a strongly convex prime. Then P is a strongly convex prime.*

Proof Take $a \in M_P$ and $b \in K_P^{\#}$, say $a = \pi_P(f)$ and $b = \pi_P(g)/\pi_P(h)$, where $f \in M_{x_Q}$, $g \in C(\Omega)$, and $h \in C(\Omega) \setminus P$; we may suppose that $|g| \leq |h|$ in $C(\Omega)$. Then $\pi_Q(f) \in M_Q$ and $\pi_Q(g)/\pi_Q(h) \in K_Q^{\#}$. Since Q is strongly convex,

$$\pi_Q(f)\pi_Q(g)/\pi_Q(h) \in M_Q ,$$

and so

$$ab = \pi_P(f)\pi_P(g)/\pi_P(h) \in M_P .$$

Thus M_P is an ideal in $K_P^{\#}$, and hence P is strongly convex. □

Let Ω be a compact space, and take $x \in \Omega$. Then x is a *P-point* of Ω if $M_x = J_x$; this is equivalent to the condition that every G_δ-set containing x is a neighbourhood of x. Clearly each infinite compact space contains a point which is not a P-point, and so the compact space $\beta\mathbb{N} \setminus \mathbb{N}$ contains such a point. It can be shown that, with CH, $\beta\mathbb{N} \setminus \mathbb{N}$ contains a P-point; however, this is not a theorem of ZFC.

Let $\mathcal{U}$ be a free ultrafilter on $\mathbb{N}$ corresponding to the point $\mathfrak{p} \in \beta\mathbb{N} \setminus \mathbb{N}$, and regard $c_0/\mathcal{U}$ as a subset of $(l^\infty/\mathcal{U})^0$. Then $c_0/\mathcal{U} = (l^\infty/\mathcal{U})^0$ if and only if $\mathfrak{p}$ is a P-point of the compact space $\beta\mathbb{N} \setminus \mathbb{N}$. For clearly $c_0/\mathcal{U} = (l^\infty/\mathcal{U})^0$ if and only if $c_0/(c_0 \cap J_\mathfrak{p}) = M_\mathfrak{p}/J_\mathfrak{p}$, where $J_\mathfrak{p}$ and $M_\mathfrak{p}$ are taken to be ideals in $C(\beta\mathbb{N})$. Suppose that $\mathfrak{p}$ is a P-point of $\beta\mathbb{N} \setminus \mathbb{N}$, and take $f \in M_\mathfrak{p}$. Then there is an open and closed neighbourhood U of $\mathfrak{p}$ in $\beta\mathbb{N}$ such that $f \mid (U \cap (\beta\mathbb{N} \setminus \mathbb{N})) = 0$. Set

$g(x) = f(x) \quad (x \in U)$ and $g(x) = 0 \quad (x \in \beta\mathbb{N} \setminus U)$. Then $g \in c_0$ and $f - g \in J_\mathfrak{p}$, and so $f \in c_0 + J_\mathfrak{p}$, showing that $c_0/(c_0 \cap J_\mathfrak{p}) = M_\mathfrak{p}/J_\mathfrak{p}$. The converse is similarly easy.

Theorem 4.40 *There are a compact space Ω and prime z-ideals P and Q in $C(\Omega)$ such that:*

(i) *P is a valuation prime and K_P is an ultrapower;*

(ii) *$Q \subsetneq P$;*

(iii) *Q is not a strongly convex prime;*

(iv) *K_Q is an η_1-field.*

Proof Set $S = \{0\} \cup \{1/k : k \in \mathbb{N}\}$ (with the relative topology from $\mathbb{R}$), set $X = \mathbb{N} \times S$, and set $\Omega = \beta X$. Take $\mathcal{U}$ to be any free ultrafilter on $\mathbb{N}$, and take $\mathcal{V}$ to be a free ultrafilter on $\mathbb{N}$ corresponding to a point $\mathfrak{p}$ in $\beta\mathbb{N} \setminus \mathbb{N}$ such that $\mathfrak{p}$ is not a P-point of $\beta\mathbb{N} \setminus \mathbb{N}$.

Set

$$P = \left\{F \in C^b(X) : \{n \in \mathbb{N} : F(n,0) = 0\} \in \mathcal{U}\right\}.$$

Then P is a prime z-ideal in the algebra $C^b(X) = C(\Omega)$, and $A_P = C(\Omega)/P$ is isomorphic to $l^\infty/\mathcal{U}$, whose quotient field K_P is the ultrapower $\mathbb{R}^\mathbb{N}/\mathcal{U}$.

Take $F, G \in C^b(X)$ with $|F| \leq |G|$. Then there exists $h \in l^\infty$ such that $F(n,0) = G(n,0)h(n) \quad (n \in \mathbb{N})$. Set

$$H(n, 1/k) = H(n,0) = h(n) \quad (n, k \in \mathbb{N}).$$

Then $H \in C^b(X)$ and $\pi_P(F) = \pi_P(G)\pi_P(H)$. Thus $K_P^\# = A_P$, and so, by Proposition 4.32, P is a valuation prime.

For $F \in C^b(X)$ and $n \in \mathbb{N}$, set

$$\sigma_{F,n} = \{k \in \mathbb{N} : F(n, 1/k) = 0\}, \quad \sigma_F = \{n \in \mathbb{N} : \sigma_{F,n} \in \mathcal{V}\}.$$

Now define

$$Q = \{F \in C^b(X) : \sigma_F \in \mathcal{U}\}.$$

It is clear that Q is a z-ideal in $C^b(X)$ and that $Q \subsetneq P$. Take F, G in $C^b(X)$. Since $\sigma_{FG,n} = \sigma_{F,n} \cup \sigma_{G,n}$ for each $n \in \mathbb{N}$, it follows that $\sigma_{FG} = \sigma_F \cup \sigma_G$ because $\mathcal{U}$ is an ultrafilter. Thus, if $FG \in Q$, then either $F \in Q$ or $G \in Q$, again because $\mathcal{U}$ is an ultrafilter. This shows that Q is a prime ideal in $C^b(X)$.

We have

$$K_Q \cong (l^\infty/\mathcal{U})^\mathbb{N}/\mathcal{V},$$

and so $\delta(K_Q) = \delta(K_P) \geq \aleph_1$. By Theorem 4.28(ii), K_Q is an η_1-field.

It remains to prove that Q is not strongly convex. Define

$$F_0(n, s) = 1/n \quad (n \in \mathbb{N},\ s \in S),$$

so that $F_0 \in C^b(X)^+ \setminus Q$, and set $a_0 = \pi_Q(F_0) \in A_Q^+ \setminus \{0\}$. For each $m \in \mathbb{N}$, we have $F_0 - F_0 \wedge (1/m) \in Q$, and so $a_0 \in M_Q^+$.

Since $\mathfrak{p}$ is not a P-point of $\beta\mathbb{N} \setminus \mathbb{N}$, we have $c_0/\mathcal{V} \subsetneq (l^\infty/\mathcal{V})^0$, and so there exist $g, h \in c_0$ such that $[h]_\mathcal{V} \neq 0$ and

$$[g]_\mathcal{V}/[h]_\mathcal{V} \in (l^\infty/\mathcal{V})^0 \setminus (c_0/\mathcal{V}). \tag{4.7}$$

We may suppose that $g(m), h(m) > 0 \quad (m \in \mathbb{N})$.

For each $n \in \mathbb{N}$, take $S_n \in \mathcal{V}$ such that

$$0 < g(k)/h(k) \leq 1/n \quad (k \in S_n),$$

and set $T_n = \mathbb{N}\setminus S_n$, so that $T_n \notin \mathcal{V}$. Next, define G and H on X by the formulae

$$\left.\begin{aligned} G(n, 1/k) &= \chi_{S_n}(k)g(k) \\ H(n, 1/k) &= \chi_{S_n}(k)h(k) \\ G(n, 0) &= H(n, 0) = 0 \end{aligned}\right\}, \quad (n, k \in \mathbb{N}).$$

Then $G, H \in C^b(X)$. For each $n \in \mathbb{N}$, we have $\sigma_{G,n} = \sigma_{H,n} = T_n$, and so $\sigma_G = \sigma_H = \emptyset$. Thus $G, H \notin Q$. Set $b = \pi_Q(G)$ and $c = \pi_Q(H)$ in $A_Q \setminus \{0\}$. Since

$$G(x) \leq F_0(x)H(x) \quad (x \in X),$$

we have $b \leq a_0c$ in A_Q, and so $b(a_0c)^{-1} \in K_Q^\#$.

To prove that Q is not a strongly convex prime, it suffices to show that $bc^{-1} \notin M_Q$. Assume towards a contradiction that $bc^{-1} \in M_Q$. Then there exists $F \in C^b(X)$ with $G - FH \in Q$. In particular, there exists $n \in \mathbb{N}$ such that

$$\{k \in \mathbb{N} : (G - FH)(n, 1/k) = 0\} \in \mathcal{V},$$

and so there exists $f \in c_0$ with $[g]_\mathcal{V} = [f]_\mathcal{V}[h]_\mathcal{V}$, a contradiction of (4.7). Thus Q is not a strongly convex prime.

This completes the proof of the theorem. □

In fact the above procedure can be iterated to produce a prime z-ideal Q in some $C(\Omega)$ such that M_Q does not contain any interval of the form $[-a, a]$ for $a \in K_P^+ \setminus \{0\}$.

Although it is not true that every prime ideal in $C(\Omega)$ is strongly convex, the following result, which we shall require in Chapters 5 and 6, shows that, for an arbitrary prime ideal P in $C(\Omega)$, there is a (necessarily strongly convex) prime ideal Q in c_0 and an embedding of c_0/Q into $C(\Omega)/P$. An ideal I in an algebra A is *modular* if A/I has an identity.

Theorem 4.41 *Let Ω be a compact space, and let P be a non-maximal, prime ideal in $C(\Omega)$. Then there is a non-modular, prime ideal Q in c_0 such that $c_{00} \subset Q$ and such that there is an embedding of c_0/Q into M_P.*

Proof Set $M = M_{x_P}$, and choose $f \in M^+ \setminus P$. There is a sequence (x_n) in $\Omega \setminus \{x_P\}$ such that

$$f(x_{n+1}) < f(x_n) < 1/n \quad (n \in \mathbb{N}).$$

Take sequences (r_n), (s_n), (t_n), (u_n), and (v_n) in $\mathbb{R}$ such that

$$f(x_{n+1}) < r_n < s_n < t_n < u_n < v_n < f(x_n) \quad (n \in \mathbb{N}),$$

and let g be a continuous function on $\mathbb{R}^+$ such that $0 \le g(t) \le t \quad (t \in \mathbb{R}^+)$ and

$$g(t) = t \quad (t \in \bigcup [s_n, u_n]), \quad g(t) = 0 \quad (t \in \bigcup [v_{n+1}, r_n]) .$$

Set $h(t) = t - g(t) \quad (t \in \mathbb{R}^+)$. Then $g \circ f,\ h \circ f \in M$ and $g \circ f + h \circ f = f$, and so either $g \circ f \notin P$ or $h \circ f \notin P$: in the former case, write f_1 for $g \circ f$, and set $I_n = [r_n, v_n]$ and $J_n = (f(x_{n+1}), f(x_n))$, and, in the latter case, write f_1 for $h \circ f$, and set $I_n = [u_{n+1}, s_n]$ and $J_n = (t_{n+1}, t_n)$. Then $f_1 \in M \setminus P$ and

$$\operatorname{supp} f_1 \subset f^{-1}(\{0\} \cup \bigcup I_n) .$$

For each $n \in \mathbb{N}$, choose $e_n \in C(\mathbb{R}^+, \mathbb{I})$ with $e_n(I_n) = \{1\}$ and $\operatorname{supp} e_n \subset J_n$, and then, for $\alpha = (\alpha_n) \in c_0$, set

$$\psi(\alpha) = \sum_{n=1}^{\infty} \alpha_n (e_n \circ f) .$$

Each point $x \in \Omega$ has a neighbourhood V such that $(e_n \circ f) \mid V = 0$ for all save at most one value of n, and so $\psi(\alpha)$ is continuous on Ω. Thus $\psi(\alpha) \in M$.

The map $\psi : c_0 \to M$ is clearly linear. For $m, n \in \mathbb{N}$, we have

$$(e_m - e_m^2) \mid I_n = 0 \quad \text{and} \quad e_m e_n = \delta_{m,n} e_n^2 ,$$

and so

$$(\psi(\alpha\beta) - \psi(\alpha)\psi(\beta)) f_1 = 0 \qquad (\alpha, \beta \in c_0) .$$

Since $f_1 \notin P$, we have $\psi(\alpha\beta) - \psi(\alpha)\psi(\beta) \in P$, and thus

$$\pi_P \circ \psi : c_0 \to M_P$$

is a homomorphism.

Set $Q = \ker(\pi_P \circ \psi)$, so that there is an embedding of c_0/Q into M_P. We define $\alpha = (1/n) \in c_0$, and take $x \in \Omega$, with $f(x) \in \overline{J}_n$, say. Then clearly $f(x) \le f(x_n) < 1/n$, and so $(1/n)(e_n \circ f) \ge f_1(x)$. Thus $\psi(\alpha)(x) \ge f_1(x)$ for each x in a neighbourhood of φ. It follows that $\psi(\alpha) \notin P$, and so $\alpha \notin Q$. Hence $Q \ne c_0$, and Q is a prime ideal in c_0.

For each $n \in \mathbb{N}$, we have $e_n \circ f \in J_\varphi$, and so $\psi(c_{00}) \subset J_\varphi \subset P$. Thus $c_{00} \subset Q$, and so Q is non-modular. □

A rather curious class of fields will arise naturally in Chapter 9. They are determined by prime ideals which we shall call 'cut primes'; the reason for this terminology will emerge in Chapter 9, and we shall see that the super-real fields of the form K_P, where P is a cut prime, form an important class.

Definition 4.42 *Let P be a non-maximal, prime ideal in an algebra $C(\Omega)$. Then P is a* cut prime *if P strictly contains a strongly convex prime.*

By Proposition 4.39, a cut prime is a strongly convex prime, and so Theorem 4.40 shows that there is a prime z-ideal which is not a cut prime. The first theorem below shows that the above-mentioned class at least contains all the ultrapowers. The second theorem, which will be required in Chapter 8, shows that every super-real field is isomorphic to a super-real field K_P, where P is a prime ideal strictly containing some other prime z-ideal Q (not necessarily strongly convex) such that K_Q is an η_1-field.

Theorem 4.43 *Let K be an ultrapower. Then there are a compact space Ω and prime z-ideals P and Q in $C(\Omega)$ such that $K \cong K_P$, $Q \subsetneq P$, and Q is a valuation prime.*

Proof Take κ to be an infinite cardinal and $\mathcal{U}$ to be a free ultrafilter on κ such that $K \cong \mathbb{R}^\kappa/\mathcal{U}$. Set $X = \kappa \times \mathbb{N}$, where κ has the discrete topology, and set $\Omega = \beta X$. Let $\mathcal{V}$ be a free ultrafilter on $\mathbb{N}$, corresponding to $\mathfrak{p} \in \beta\mathbb{N} \setminus \mathbb{N}$, say.

Set

$$P = \{F \in C^b(X) : \{\zeta < \kappa : F(\zeta, \mathfrak{p}) = 0\} \in \mathcal{U}\}\,, \quad Q = \{F \in C^b(X) : \sigma_F \in \mathcal{U}\}\,,$$

where $\sigma_F = \{\zeta < \kappa : \sigma_{F,\zeta} \in \mathcal{V}\}$ and $\sigma_{F,\zeta} = \{k \in \mathbb{N} : F(\zeta, k) = 0\}$. Then, essentially as in Theorem 4.40, P is a prime z-ideal in $C(\Omega)$, $K_P \cong \mathbb{R}^\kappa/\mathcal{U}$, and Q is a prime z-ideal in $C(\Omega)$ with $Q \subsetneq P$. Since X is a discrete space, Q is a valuation prime by Proposition 4.34. □

Theorem 4.44 *Let K be a super-real field. Then there are a compact space Ω, a prime ideal P in $C(\Omega)$ such that $K \cong K_P$, and a prime z-ideal Q in $C(\Omega)$ such that $Q \subsetneq P$ and K_Q is an η_1-field.*

Proof We first make the following *claim.* Let Y be a compact space, let $X \in \mathbf{Z}(Y)$ be such that $\mathrm{int}_Y X = \emptyset$, and suppose that I is a prime ideal in $C(X)$. Then there is a prime z-ideal J in $C(Y)$ such that

$$J \subsetneq \{f \in C(Y) : f \mid X \in I\}\,.$$

Indeed, set $I' = \{f \in C(Y) : f \mid X \in I\}$, take $f_0 \in C(Y)$ with $\mathbf{Z}(f_0) = X$, and consider

$$S = \{f_0^n g : n \in \mathbb{Z}^+,\ g \in C(Y) \setminus I'\}\,.$$

The set S is closed under multiplication. Assume that $0 \in S$, say $0 = f_0^n g$. Then $g \mid (Y \setminus X) = 0$, and so $g = 0$ because $\mathrm{int}_Y X = \emptyset$, a contradiction. Thus $0 \notin S$, and so there is a prime ideal J in $C(Y)$ with $J \cap S = \emptyset$. By Proposition 4.23, we may suppose that J is a z-ideal. We have $J \subset I'$, and $J \neq I'$ because $f_0 \in I' \setminus J$. This establishes the claim.

We now begin the main part of the proof by taking a compact space Ω_0 and a prime ideal P_0 in $C(\Omega_0)$ such that $K \cong K_{P_0}$.

We shall inductively define sequences $(\Omega_\sigma : 1 \leq \sigma \leq \omega_1)$ of compact spaces and $(P_\sigma : 1 \leq \sigma \leq \omega_1)$ of prime z-ideals in $C(\Omega_\sigma)$ such that, for each $\tau < \sigma \leq \omega_1$:

(i) Ω_τ is a closed subspace of Ω_σ;

(ii) $P_\sigma \subsetneq \{f \in C(\Omega_\tau) : f \mid \Omega_\tau \in P_\tau\}$.

Thus, first suppose that $\sigma < \omega_1$, and that Ω_τ and P_τ have been defined for $\tau \leq \sigma$. Define $S = \{0\} \cup \{1/k : k \in \mathbb{N}\}$, and $\Omega_{\sigma+1} = \Omega_\sigma \times S$, identifying Ω_σ with

$\Omega_\sigma \times \{0\}$ in $\Omega_{\sigma+1}$. By the claim, there is a prime z-ideal $P_{\sigma+1}$ in $C(\Omega_{\sigma+1})$ such that

$$P_{\sigma+1} \subsetneq \{f \in C(\Omega_{\sigma+1}) : f \mid \Omega_\sigma \in P_\sigma\}.$$

Now suppose that $\sigma \leq \omega_1$ is a limit ordinal and that Ω_τ and P_τ have been defined for $\tau < \sigma$. Set $X = \bigcup\{\Omega_\tau : \tau < \sigma\}$ with the topology such that $U \subset X$ is open if $U \cap \Omega_\tau$ is open in Ω_τ for each $\tau < \sigma$. Then X is a completely regular space, and the map

$$f \mapsto f \mid \Omega_\tau, \quad C^b(X) \to C(\Omega_\tau),$$

is an epimorphism for each $\tau < \sigma$. Define $\Omega_\sigma = \beta X$ and

$$P_\sigma = \{f \in C(\Omega_\sigma) : f \mid \Omega_\tau \in P_\tau \quad (\tau < \sigma)\}.$$

Clearly P_σ is a z-ideal in $C(\Omega_\sigma)$. Take $f, g \in C(\Omega_\sigma)$ with $fg \in P_\sigma$. Then we may suppose that $\{\tau < \sigma : f \mid \Omega_\tau \in P_\tau\}$ is cofinal in σ, and hence equal to σ. Thus $f \in P_\sigma$, and so P_σ is a prime ideal. Clearly condition (ii), above, is satisfied.

This completes the inductive construction.

Define $\Omega = \Omega_{\omega_1}$, $P'_\sigma = \{f \in C(\Omega) : f \mid \Omega_\sigma \in P_\sigma\} \quad (\sigma < \omega_1)$, $P = P'_0$, and $Q = P_{\omega_1}$. Then $K_P \cong K_{P_0} \cong K$, and Q is a z-ideal in $C(\Omega)$ with $Q \subsetneq P$. Further $Q \subsetneq P_\sigma \subsetneq P_\tau \subset P \quad (\tau < \sigma < \omega_1)$, and so $\delta(K_Q) \geq \aleph_1$. By Theorem 4.28(ii), K_Q is an η_1-field. □

There is one final special type of prime ideal that we shall consider later. The reason for the terminology we use will become apparent in Chapter 6.

The functions E and L are defined on $\mathbb{R}^+$ and $[0,1)$, respectively, by the formulae:

$$E(t) = \begin{cases} \exp(-1/t) & (t > 0), \\ 0 & (t = 0); \end{cases} \tag{4.8}$$

$$L(t) = \begin{cases} 1/(\log(1/t)) & (0 < t < 1), \\ 0 & (t = 0). \end{cases} \tag{4.9}$$

Clearly $E : \mathbb{R}^+ \to [0,1)$ and $L : [0,1) \to \mathbb{R}^+$ are continuous, strictly increasing functions, and these functions are the inverses of each other. Note that

$$E(2t)^2 = E(t) \quad (t \in \mathbb{R}^+)$$

and that, for each $n \in \mathbb{N}$, there exists $\varepsilon_n > 0$ such that

$$L(t)^n > t \quad (t \in (0, \varepsilon_n)). \tag{4.10}$$

Now let Ω be a compact space, and take $f, g \in C(\Omega)^+$ with $g < 1$. Then we define

$$\exp(-1/f) = E \circ f \quad \text{and} \quad 1/(\log(1/g)) = L \circ g$$

pointwise on Ω to obtain further functions in $C(\Omega)^+$ with

$$\mathbf{Z}(\exp(-1/f)) = \mathbf{Z}(f) \quad \text{and} \quad \mathbf{Z}(1/(\log(1/g))) = \mathbf{Z}(g).$$

Definition 4.45 *Let P be a prime ideal in an algebra $C(\Omega)$. Then P is an* exponential prime *if $\exp(-1/f) \notin P$ whenever $f \in C(\Omega)^+ \setminus P$.*

Equivalently, P is an exponential prime if and only if $1/\log(1/g) \in P$ whenever $g \in P^+$ with $g < 1$. Clearly a prime z-ideal is an exponential prime; we shall shortly give examples of exponential primes which are not z-ideals and of valuation primes which are not exponential primes.

The following construction is a useful method for producing prime ideals above a given prime.

Theorem 4.46 *Let Q be a non-maximal, prime ideal in $C(\Omega)$, and let (a_k) be a sequence in $M_Q^+ \setminus \{0\}$.*

(i) *Suppose that $a_{k+1} \le a_k^2 \quad (k \in \mathbb{N})$, and set*

$$P = \{f \in C(\Omega) : \pi_Q(|f|) \le a_k \quad (k \in \mathbb{N})\}.$$

Then P is a non-maximal, prime ideal in $C(\Omega)$ with P above Q, and $(\pi_{PQ}(a_k))$ is coinitial in $A_P^+ \setminus \{0\}$.

(ii) *Suppose that $a_{k+1}^2 \ge a_k \quad (k \in \mathbb{N})$, and set*

$$P = \{f \in C(\Omega) : \pi_Q(|f|) \le a_k \text{ for some } k \in \mathbb{N}\}.$$

Then P is a non-maximal, prime ideal in $C(\Omega)$ with P above Q, and K_P is an η_1-field.

Proof (i) It is straightforward to check that P is a non-maximal ideal in $C(\Omega)$. Take $f, g \in C(\Omega) \setminus P$. There exists $k \in \mathbb{N}$ with $\pi_Q(|f|) > a_k$ and $\pi_Q(|g|) > a_k$, and then $\pi_Q(|fg|) > a_k^2 \ge a_{k+1}$, and so $fg \notin P$. Hence P is a prime ideal. Since (a_k) is coinitial in $M_Q^+ \setminus \pi_Q(P)$, the sequence $(\pi_{PQ}(a_k))$ is coinitial in $A_P^+ \setminus \{0\}$.

(ii) Take $f \in P$ and $g \in C(\Omega)$, say $\pi_Q(|f|) \le a_k$ and $\pi_Q(|g|) \le n$, where $n \in \mathbb{N}$. Then

$$\pi_Q(|fg|) \le na_k \le na_{k+1}^2 \le a_{k+1}$$

because $a_{k+1} \le 1/n$, and so $fg \in P$. It is now easily seen that P is a non-maximal ideal in $C(\Omega)$. Take $f, g \in C(\Omega) \setminus P$ and $k \in \mathbb{N}$. Then $\pi_Q(|f|) > a_{k+1}$ and $\pi_Q(|g|) > a_{k+1}$, and so $\pi_Q(|fg|) > a_{k+1}^2 \ge a_k$, whence $fg \notin P$. Thus P is a prime ideal.

Assume towards a contradiction that $\delta(K_P) = \aleph_0$. By Proposition 4.21(ii), there is a strictly decreasing, coinitial sequence (b_k) in $A_P^+ \setminus \{0\}$. For $k \in \mathbb{N}$, choose $c_k \in A_Q$ with $\pi_{PQ}(c_k) = b_k$; the sequence (c_k) is strictly decreasing in A_Q and $a_j < c_k \quad (j, k \in \mathbb{N})$. By Theorem 4.28(i), there exists $a \in A_Q$ with $a_j < a < c_k \quad (j, k \in \mathbb{N})$, and then $0 < \pi_{PQ}(a) < b_k \quad (k \in \mathbb{N})$, a contradiction. Thus $\delta(K_P) \ge \aleph_1$, and so, by Theorem 4.28(ii), K_P is an η_1-field. □

A simple way of obtaining sequences (a_k) to satisfy (i) and (ii), above, is to take $a_1 \in M_Q^+ \setminus \{0\}$ and inductively define $a_{k+1} = a_k^2$ and $a_{k+1} = a_k^{1/2}$, respectively. Part (ii) of the above result shows that, for each non-maximal, prime ideal Q, there is a prime ideal P above Q such that K_P is an η_1-field. Using the above construction, we can give some examples that were referred to earlier.

Let Q be a non-maximal, prime z-ideal in $C(\Omega)$, and take $f_1 \in M_{x_Q}^+ \setminus \{0\}$ with $f_1 < 1$. Inductively define (f_k) in $M_{x_Q}^+ \setminus \{0\}$ by the formula

$$f_{k+1}(x) = (L \circ f_k)(x) \wedge \varepsilon_2 \quad (k \in \mathbb{N},\, x \in \Omega),$$

where ε_2 is as in (4.10); set $a_k = \pi_Q(f_k)$, so that $a_{k+1}^2 \geq a_k \quad (k \in \mathbb{N})$. The corresponding prime ideal P defined in Theorem 4.46(ii) is easily seen to be an exponential prime with K_P an η_1-field, but in general P is not a z-ideal.

Again take $Q = J_\mathfrak{p}$ to be a minimal prime in $C(\beta\mathbb{N})$, where $\mathfrak{p} \in \beta\mathbb{N} \setminus \mathbb{N}$, and set

$$a_k = \left((n+1)^{-1/2^k} : n \in \mathbb{N}\right) + Q \in A_Q$$

for $k \in \mathbb{N}$. Then $a_{k+1}^2 = a_k \ (k \in \mathbb{N})$, and the corresponding prime ideal P is such that K_P is an η_1-field. The function

$$g : n \mapsto \frac{1}{n+1}, \quad \mathbb{N} \to \mathbb{R}$$

belongs to P^+ with $g < 1$, but

$$1/\log(1/g) : n \mapsto \frac{1}{\log(n+1)}, \quad \mathbb{N} \to \mathbb{R}$$

does not belong to P because the sequence $\left(n/(\log(n+1))^{2^k} : n \in \mathbb{N}\right)$ is unbounded for each $k \in \mathbb{N}$. By Proposition 4.34, P is a valuation prime.

In Corollary 2.35, we showed that there are only four possible isomorphism classes of real-closed semi-η_1-fields which are β_1-fields. We shall first show that each of possibilities (2)–(4) of that result actually occurs in the class of super-real fields under the assumption that there is a free ultrafilter $\mathcal{U}$ on $\mathbb{N}$ such that $\mathbb{R}^\mathbb{N}/\mathcal{U}$ is a β_1-field (this assumption is certainly true in the theory ZFC + CH).

Indeed, let $\mathcal{U}$ be such an ultrafilter, and set $\mathbb{R}^\mathbb{N}/\mathcal{U} = K_\mathfrak{p}$, where $\mathfrak{p} \in \beta\mathbb{N} \setminus \mathbb{N}$ corresponds to $\mathcal{U}$. Then $K_\mathfrak{p}$ is a super-real field which is a β_1-η_1-field, giving possibility (2) of Corollary 2.35.

Now take $f_0 \in M_\mathfrak{p}^+ \setminus J_\mathfrak{p}$, and set

$$P_1 = \{f \in C(\beta\mathbb{N}) : \pi_\mathfrak{p}(|f|) \leq \pi_\mathfrak{p}(f_0^k) \quad (k \in \mathbb{N})\}.$$

Also take a sequence (g_k) in $M_\mathfrak{p}^+ \setminus J_\mathfrak{p}$ such that $g_{k+1} \leq g_k^n \quad (n, k \in \mathbb{N})$, and set

$$P_2 = \{f \in C(\beta\mathbb{N}) : \pi_\mathfrak{p}(|f|) \leq \pi_\mathfrak{p}(g_k) \quad (k \in \mathbb{N})\}.$$

By Theorem 4.46(i), each of P_1 and P_2 is a non-maximal, prime ideal in $C(\beta\mathbb{N})$ above Q, and, by Corollary 4.38, each of the super-real fields K_{P_1} and K_{P_2} is a semi-η_1-field. The sequence $(\pi_{P_1}(f_0^k))$ is coinitial in $A_{P_1}^+ \setminus \{0\}$, and so

$$\operatorname{cof} \Gamma_{P_1} = (\aleph_0, 1)$$

in the extended sense. However there is no sequence (a^n) which is coinitial in $A_{P_2}^+ \setminus \{0\}$, and so

$$\operatorname{cof} \Gamma_{P_2} = (\aleph_0, \aleph_0)$$

in the extended sense. Thus we have obtained possibilities (3) and (4) of Corollary 2.35 in our class.

Note that both K_{P_1} and K_{P_2} are semi-η_1-fields which are not η_1-fields.

So far, we have not exhibited a super-real field which is not even a semi-η_1-field; we remedy this now. The existence of such fields is rather subtle. At this stage, our classifications do not encompass such fields.

Theorem 4.47 *There is a prime z-ideal P in an algebra $C(\Omega)$ such that the super-real field K_P is not a semi-η_1-field.*

Proof We begin by defining a certain free ultrafilter $\mathcal{U}$ on $\mathbb{N}$ and a certain function Φ on $\mathbb{N}$.

Indeed, let $\{T_m : m \in \mathbb{N}\}$ be a partition of $\mathbb{N}$ into subsets each of which is infinite, and let $\mathcal{F}$ be the family of subsets S of $\mathbb{N}$ such that $T_m \setminus S$ is finite for all but finitely many values of $m \in \mathbb{N}$. Clearly $\mathcal{F}$ is a filter on $\mathbb{N}$. We take $\mathcal{U}$ to be any ultrafilter containing $\mathcal{F}$; certainly $\mathcal{U}$ is a free ultrafilter on $\mathbb{N}$ because $\mathbb{N}\setminus\{n_0\} \in \mathcal{F}$ for each $n_0 \in \mathbb{N}$. (In fact, $\mathcal{U}$ corresponds to a point of $\beta\mathbb{N}\setminus\mathbb{N}$ which is not a P-point.)

The function $\Phi : \mathbb{N} \to \mathbb{Q}$ is defined by the rule

$$\Phi \mid T_m = \frac{1}{m+1} \quad (m \in \mathbb{N}).$$

Thus $\Phi \in l^\infty$, and $0 < \Phi(n) < 1$ $(n \in \mathbb{N})$. For each $S \in \mathcal{U}$, we have $\mathbb{N} \setminus S \notin \mathcal{F}$, and so, for infinitely many values of m, $T_m \cap S$ is infinite. Thus S has two infinite subsets, say S_1 and S_2, such that $\Phi \mid S_1$ and $\Phi \mid S_2$ take different constant values.

We next inductively define ultrafilters on $\mathbb{N}^k$ for each $k \in \mathbb{N}$. We start with $\mathcal{U}_1 = \mathcal{U}$ on $\mathbb{N}$. Now suppose that $\mathcal{U}_k$ has been defined as an ultrafilter on $\mathbb{N}^k$. We define $\mathcal{U}_{k+1}$ on $\mathbb{N}^{k+1}$ by the following rule: for a subset X of $\mathbb{N}^{k+1}$, we say that $X \in \mathcal{U}_{k+1}$ if

$$\{m \in \mathbb{N}^k : \{n \in \mathbb{N} : m^\frown n \in X\} \in \mathcal{U}\} \in \mathcal{U}_k .$$

It is immediately verified that $\mathcal{U}_{k+1}$ is indeed an ultrafilter on $\mathbb{N}^{k+1}$.

The third preliminary task is to define functions $\varphi_k, \psi_k : \mathbb{N}^k \to \mathbb{R}$ for each $k \in \mathbb{N}$. We start by choosing $\varphi_1, \psi_1 : \mathbb{N} \to \mathbb{R}$ such that, for each $n \in \mathbb{N}$:

$$\tfrac{1}{2}(\varphi_1(n) + \psi_1(n)) = \Phi(n) ;$$

$$0 < \varphi_1(n) < \psi_1(n) < 1 ;$$

$$0 < \psi_1(n) - \varphi_1(n) < 1/n .$$

Now suppose that $\varphi_k, \psi_k : \mathbb{N}^k \to \mathbb{R}$ have been defined, and choose functions $\varphi_{k+1}, \psi_{k+1} : \mathbb{N}^{k+1} \to \mathbb{R}$ such that, for each $m \in \mathbb{N}^k$ and $n \in \mathbb{N}$:

$$\tfrac{1}{2}(\varphi_{k+1}(m^\frown n) + \psi_{k+1}(m^\frown n)) = \varphi_k(m) + (\psi_k(m) - \varphi_k(m))\Phi(n) ; \tag{4.11}$$

$$0 < \varphi_k(m) < \varphi_{k+1}(m^\frown n) < \psi_{k+1}(m^\frown n) < \psi_k(m) < 1 ; \tag{4.12}$$

$$0 < \psi_{k+1}(m^\frown n) - \phi_{k+1}(m^\frown n) < 1/n . \tag{4.13}$$

The compact set on which we shall base our construction is the Cantor set K: K is the compact subset of $\mathbb{I}$ defined by successively deleting the open intervals $(1/3, 2/3), (1/9, 2/9), (7/9, 8/9), \ldots$ from $\mathbb{I}$. Thus K is a compact, perfect, totally disconnected subset of $\mathbb{I}$ (and every metric space with these properties is homeomorphic to K). The space K is homeomorphic to the product space $\{0,1\}^{\omega}$.

We shall next define a sequence (F_k) of closed subsets of K and a sequence $(\mathcal{I}_k)$, where

$$\mathcal{I}_k = \{I_m : m \in \mathbb{N}^k\}$$

is a family of pairwise disjoint, closed subintervals of $\mathbb{I}$ related to K.

We start with $F_1 = \{0\}$. We then choose $(x_n) \subset K$ and $(y_n) \subset \mathbb{I}$ such that:

$$0 < x_{n+1} < y_{n+1} < x_n < y_n < 1/n \quad (n \in \mathbb{N});$$

$$K \subset F_1 \cup \bigcup \{[x_n, y_n] : n \in \mathbb{N}\}.$$

(For example take $x_n = 2/3^{n+1}$ and $y_n = 1/3^n$.) Then $\mathcal{I}_1 = \{I_n : n \in \mathbb{N}\}$, where $I_n = [x_n, y_n]$, is the first *generation* of subintervals. We set

$$L_1 = \{x_n : n \in \mathbb{N}\} \quad \text{and} \quad F_2 = F_1 \cup L_1,$$

so that F_2 is a closed subset of K and the set of accumulation points, or *derived set*, of F_2 is exactly F_1.

We inductively define the families $\mathcal{I}_k = \{I_m : m \in \mathbb{N}^k\}$ of closed subintervals to satisfy the following conditions for each $k \in \mathbb{N}$:

(i) I_m for $m \in \mathbb{N}^k$ are pairwise disjoint, closed subintervals of $\mathbb{I}$;

(ii) $I_{m^\frown n} \subset I_m \quad (m \in \mathbb{N}^k, n \in \mathbb{N})$;

(iii) the set L_k of left-hand end-points of the intervals of $\mathcal{I}_k$ has as its derived set the closed subset F_k of K;

(iv) $F_{k+1} = F_k \cup L_k$;

(v) $K \subset F_k \cup \bigcup \{I_m : m \in \mathbb{N}^k\}$ and $F_k \cap \bigcup \{I_m : m \in \mathbb{N}^k\} = \emptyset$.

For the inductive step of this construction, choose $x \in L_k$. Then there is a unique $m \in \mathbb{N}^k$ such that x is the left-hand end-point of $I_m \in \mathcal{I}_k$. Choose sequences $(s_n) \subset K \cap I_m$ and $(t_n) \subset I_m$ such that:

$$x < s_{n+1} < t_{n+1} < s_n < t_n < 1/n \quad (n \in \mathbb{N});$$

$$K \cap I_m \subset \{x\} \cup \bigcup \{[s_n, t_n] : n \in \mathbb{N}\}.$$

Then we define $I_{m^\frown n} = [s_n, t_n]$ and

$$\mathcal{I}_{k+1} = \{I_{m^\frown n} : m \in \mathbb{N}^k,\ n \in \mathbb{N}\};$$

we say that $\mathcal{I}_{k+1}$ is the $(k+1)$ st *-generation* of subintervals.

This completes the definition of the sequences (F_k) and $(\mathcal{I}_k)$. It is clear that the sequences have the required properties.

Note that, for each $x \in L_k$ with $x \in I_m$ and each $\varepsilon > 0$, $I_{m^\frown n} \subset [x, x+\varepsilon]$ for all but finitely many $n \in \mathbb{N}$.

We now move to the definition of certain subsets R_k of $C(K)$ for $k \in \mathbb{N}$. Indeed, R_k is the set of functions $f \in C(K)$ such that

$$\{m \in \mathbb{N}^k : f \mid (K \cap I_m) \text{ is constant with range in } \mathbb{R}^+ \setminus \{0\}\} \in \mathcal{U}_k .$$

Note that, for each $k \in \mathbb{N}$, we have $0 \notin R_k$, R_k is closed under multiplication, and $R_k \subset R_{k+1}$. Set

$$R = \bigcup \{R_k : k \in \mathbb{N}\} .$$

Then R is a non-empty set which is closed under multiplication, and $0 \notin R$, and so there is a prime ideal Q in $C(K)$ with $Q \cap R = \emptyset$. Since every prime ideal contains a prime z-ideal, we may suppose that Q is a z-ideal.

Let $f \in C(K)$ and $k \in \mathbb{N}$ be such that

$$\{m \in \mathbb{N}^k : f \mid (K \cap I_m) = 0\} \in \mathcal{U}_k .$$

Then there exists $g \in R_k$ such that $fg = 0$; since $g \notin Q$, it follows that $f \in Q$. In particular, if S_m is a finite subset of $\mathbb{N}$ for each $m \in \mathbb{N}^k$, then

$$F_k \cup \bigcup \{I_{m^\frown n} : m \in \mathbb{N}^k, n \in \mathbb{N} \setminus S_m\} \in \mathbf{Z}[Q] . \tag{4.14}$$

The final object that we shall define is a specific sequence (f_k) of functions in $C(K) \setminus Q$ with $f_k(0) = 0 \quad (k \in \mathbb{N})$. Thus, take $k \in \mathbb{N}$ and $x \in K$. Then

$$f_k(x) = \begin{cases} \dfrac{1}{n_1 \cdots n_k} & (x \in I_{(n_1,\dots,n_k)}) , \\ 0 & (x \in F_k) . \end{cases}$$

We note that $f_k \in R_k \subset C(K)$, so that $f_k \notin Q$, and that $\mathbf{Z}(f_k) = F_k$.

Take $k \in \mathbb{N}$ and $m = (n_1, \dots, n_k) \in \mathbb{N}^k$. Then

$$\frac{1}{n_1 \dots n_k n} \le \left(\frac{1}{n_1 \dots n_k} \right)^2$$

for all save finitely many values of n, and so, by (4.14),

$$\{x \in K : f_{k+1}(x) \le f_k(x)^2\} \in \mathbf{Z}[Q] .$$

Thus $\pi_Q(f_{k+1}) \le \pi_Q(f_k^2)$. Similarly, we have $\pi_Q(h) \ge \pi_Q(f_{k+1})$ for each $h \in R_k$.

We *claim* that, for each $W \in \mathbf{Z}[Q]$ and each $k \in \mathbb{N}$, necessarily

$$F_k \cap W \subsetneq F_{k+1} \cap W . \tag{4.15}$$

Certainly $F_k \cap W \subset F_{k+1} \cap W$. For $m \in \mathbb{N}^k$, set

$$S_m = \{n \in \mathbb{N} : I_{m^\frown n} \cap W \neq \emptyset\} .$$

Assume that S_m is finite for each $m \in \mathbb{N}^k$. Then there exists $f \in R_{k+1}$ with $\mathbf{Z}(f) \supset W$, and so $\mathbf{Z}(f) \in \mathbf{Z}[Q]$. But then $f \in Q \cap R_{k+1}$, a contradiction. Thus S_m is infinite for some $m \in \mathbb{N}^k$, and so there is a sequence in W converging to a point $t \in L_k$. Since W is a closed set, we have $t \in W$, and so

$$t \in (F_{k+1} \cap W) \setminus (F_k \cap W) ,$$

giving (4.15), as claimed.

Define

$$P = \{h \in C(K) : \pi_Q(|h|) \leq \pi_Q(f_k) \quad (k \in \mathbb{N})\}.$$

By Proposition 4.46(i), P is a prime ideal in $C(K)$, and $Q \subset P$. We show that P is a z-ideal. For assume that there exists $h \in C(K)^+ \setminus P$ and $g \in P^+$ with $\mathbf{Z}(h) = \mathbf{Z}(g)$. Then there exists $k \in \mathbb{N}$ and $W \in \mathbf{Z}[Q]$ with

$$f_k(x) \leq h(x) \quad (x \in W).$$

By reducing W if necessary, we may suppose that

$$g(x) \leq f_{k+1}(x) \quad (x \in W).$$

But now

$$F_{k+1} \cap W = \mathbf{Z}(f_{k+1}) \cap W \subset \mathbf{Z}(g) = \mathbf{Z}(h) \subset \mathbf{Z}(f_k) \cap W = F_k \cap W,$$

a contradiction of (4.15). Hence P is a prime z-ideal in $C(K)$. Since

$$\pi_Q(h) \geq \pi_Q(f_{k+1}) \quad (h \in R_k),$$

we have $P \cap R = \emptyset$.

Define $c_k = \pi_P(f_k) \quad (k \in \mathbb{N})$. By Theorem 4.46(i), (c_k) is coinitial in $A_P^+ \setminus \{0\}$, and so $\delta(K_P) = \aleph_0$. (This shows that K_P is not an η_1-field.)

We shall show that the super-real field K_P is not a semi-η_1-field. To do this, we must first recall the functions $\varphi_k, \psi_k : \mathbb{N}^k \to \mathbb{R}$, which were defined earlier. For each $k \in \mathbb{N}$, we define functions $\Phi_k, \Psi_k \in C(K)$ by the rules

$$\Phi_k \mid (K \cap I_m) = \varphi_k(m)\, f_k \mid (K \cap I_m),$$
$$\Psi_k \mid (K \cap I_m) = \psi_k(m)\, f_k \mid (K \cap I_m),$$

for each $m \in \mathbb{N}^k$, and $\Phi_k \mid F_k = \Psi_k \mid F_k = 0$. The functions Φ_k and Ψ_k are continuous on each $K \cap I_m$, and we have $\mathbf{Z}(\Phi_k) = \mathbf{Z}(\Psi_k) = \mathbf{Z}(f_k) = F_k$. Also

$$0 \leq \Phi_k(x) \leq \Psi_k(x) \leq f_k(x) \quad (x \in K)$$

by (4.12), and so $\Phi_k, \Psi_k \in C(K)^+ \setminus P$.

Define $a_k = \pi_P(\Phi_k)$ and $b_k = \pi_P(\Psi_k)$, so that $a_k, b_k \in A_P^+ \setminus \{0\}$ with $0 < a_k < b_k < c_k$. It follows from (4.12) that

$$0 < \frac{a_j}{c_j} < \frac{a_k}{c_k} < \frac{b_k}{c_k} < \frac{b_j}{c_j} < 1$$

in $K_P^+ \setminus \{0\}$ whenever $j < k$ in $\mathbb{N}$.

To show that K_P is not a semi-η_1-field, we shall show that there is no element $a \in K_P$ with

$$\frac{a_j}{c_j} < a < \frac{b_k}{c_k} \quad (j, k \in \mathbb{N}). \tag{4.16}$$

To obtain a contradiction, assume that such an element a exists, say

$$a = \pi_P(f)/\pi_P(g),$$

where $f, g \in C(K)^+ \setminus P$. We may suppose that $f < g$. There exists $k \in \mathbb{N}$ such that $\pi_Q(f) \geq \pi_Q(f_k)$, and so there exists $W \in \mathbf{Z}[Q]$ such that

$$f(x) \geq f_k(x) \quad (x \in W).$$

Since $a_{k+1} < ac_{k+1} < b_{k+1}$, we may suppose, by reducing W, that

$$\Phi_{k+1}(x)g(x) \leq f_{k+1}(x)f(x) \leq \Psi_{k+1}(x)g(x) \quad (x \in W).$$

We thus have

$$\frac{\Phi_{k+1}(x)}{f_{k+1}(x)} \leq \frac{f(x)}{g(x)} \leq \frac{\Psi_{k+1}(x)}{f_{k+1}(x)} \quad (x \in W \setminus F_{k+1}). \tag{4.17}$$

The set $\{m \in \mathbb{N}^k : I_m \cap W \neq \emptyset\}$ belongs to $\mathcal{U}_k$ because $P \cap R = \emptyset$, and so there exists $m \in \mathbb{N}^k$ and $x_0 \in L_k$ such that $S \in \mathcal{U}$, where

$$S = \{n \in \mathbb{N} : I_{m^\frown n} \cap W \neq \emptyset\}$$

and x_0 is the left-hand end-point of I_m. We have $x_0 \in W \setminus F_k$. By the definition of the function Φ, there exist infinite subsets S_1 and S_2 of S such that

$$\Phi(n) = \delta_j \quad (n \in S_j)$$

for $j = 1, 2$, where δ_1 and δ_2 are distinct numbers in $\mathbb{R}^+ \setminus \{0\}$. Set

$$\Delta_j = \varphi_k(m) + (\psi_k(m) - \varphi_k(m))\delta_j \quad (j = 1, 2),$$

so that Δ_1 and Δ_2 are distinct numbers, and take disjoint neighbourhoods V_1 and V_2 of Δ_1 and Δ_2, respectively. By (4.11) and (4.13), we have

$$[\varphi_{k+1}(m^\frown n), \psi_{k+1}(m^\frown n)] \subset V_j \quad (n \in S_j') \tag{4.18}$$

for $j = 1, 2$, where S_j' is an infinite subset of S_j. For $x \in I_{m^\frown n} \cap W$, where $n \in \mathbb{N}$, we have

$$\frac{\Phi_{k+1}(x)}{f_{k+1}(x)} = \varphi_{k+1}(m^\frown n), \quad \frac{\Psi_{k+1}(x)}{f_{k+1}(x)} = \psi_{k+1}(m^\frown n),$$

and so, by (4.17) and (4.18),

$$\frac{f(x)}{g(x)} \in V_j \quad (x \in I_{m^\frown n} \cap W, \ n \in S_j').$$

Since the sets of intervals $I_{m^\frown n}$ for $n \in S_j'$ both accumulate at x_0, this shows that

$$\frac{f(x_0)}{g(x_0)} \in V_1 \cap V_2.$$

But $V_1 \cap V_2 = \emptyset$, and so finally we have reached a contradiction.

Thus K_P is not a semi-η_1-field. □

We are now in a position to summarize the relationship between some of the classes of fields that we have described. We first list, with short title, the main classes that we are considering. Throughout we shall restrict ourselves to consideration of η_1-fields, although there are interesting questions if we step outside this class. (We recall our standing convention that all fields have ground field $\mathbb{R}$.) The classes are as follows:

- **real-closed η_1-fields** – the class of all real-closed η_1-fields;
- **super-real η_1** – the class of all super-real η_1-fields;
- **exponential η_1** – the class of all super-real η_1-fields isomorphic to K_P, where P is an exponential prime in some space $C(\Omega)$;
- **z-ideal η_1** – the class of all super-real η_1-fields isomorphic to K_P, where P is a prime z-ideal in some space $C(\Omega)$;
- **strongly convex η_1** – the class of all super-real η_1-fields isomorphic to K_P, where P is a strongly convex prime ideal in some space $C(\Omega)$;
- **strongly convex z-ideal η_1** – the class of all super-real η_1-fields isomorphic to K_P, where P is a strongly convex prime z-ideal in some space $C(\Omega)$;
- **valuation η_1** – the class of all super-real η_1-fields isomorphic to K_P, where P is a valuation prime ideal in some space $C(\Omega)$;
- **cut prime η_1** – the class of all super-real η_1-fields isomorphic to K_P, where P is a cut prime ideal in some space $C(\Omega)$;
- **valuation z-ideal η_1** – the class of all super-real η_1-fields isomorphic to K_P, where P is a valuation prime z-ideal in some space $C(\Omega)$;
- **hyper-real** – the class of all hyper-real fields;
- **cut prime z-ideal η_1** – the class of all super-real η_1-fields isomorphic to K_P, where P is a cut prime z-ideal in some space $C(\Omega)$;
- **ultrapowers** – the class of all ultrapowers.

Note that we have inserted 'η_1' in the short title of the classes just where this is necessary; for example, hyper-real fields are automatically η_1-fields.

We are concerned with the relationships between the above classes. Note first that the question whether **cut prime η_1** is equal to **strongly convex η_1**, for example, is much more substantial than the rather trivial question whether there is a strongly convex prime giving an η_1-field that is not a cut prime. Indeed, the question of the equality of any two of the above classes can be interpreted in different ways. Each class consists of real-closed fields. We could say that a class $\mathcal{C}$ is contained in a class $\mathcal{D}$ if each member of $\mathcal{C}$ is, respectively:

- order-isomorphic to a member of $\mathcal{D}$;
- isomorphic as a totally ordered group to a member of $\mathcal{D}$;
- isomorphic as a ring to a member of $\mathcal{D}$;
- isomorphic as a field, and hence as an ordered field, to a member of $\mathcal{D}$.

Naturally we shall seek to prove the strongest possible results, both in our positive theorems and in our counter-examples.

Finally, the equality or non-equality of two classes could be proved in various theories. We shall strive to prove results in the theory ZFC; sometimes the results only hold, or are only known to hold, in a stronger theory than ZFC, for

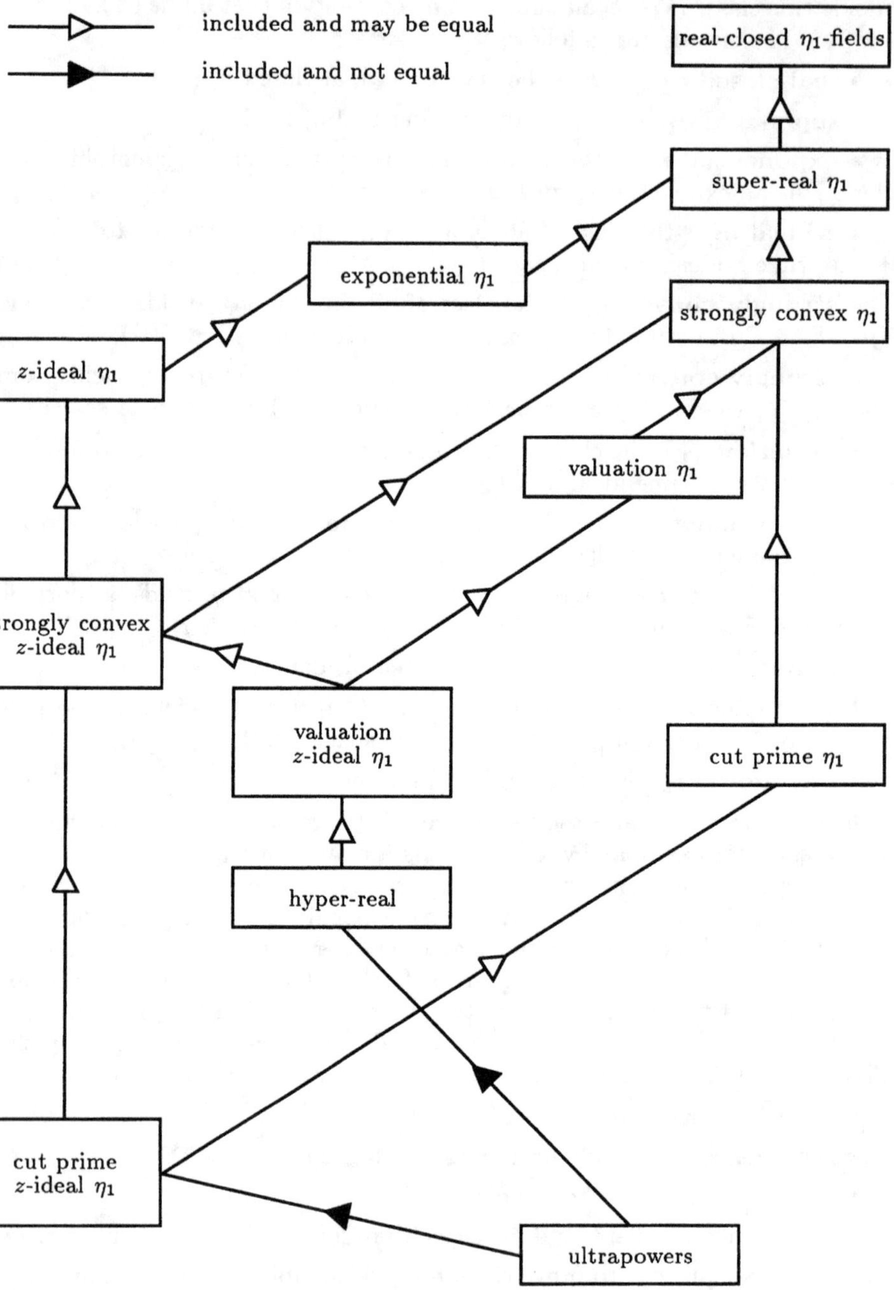

Fig. 4.1. The relationships between classes

example in ZFC + CH or ZFC + GCH. Indeed, in some cases, we can just prove that it is relatively consistent with ZFC that a result holds.

We present a diagram (Figure 4.1) that shows the inclusions between the above classes that are either trivial or have been proved so far. The lines with arrows indicate that it is a theorem of ZFC that every member of the lower class is isomorphic as an ordered field to a member of the higher class. Later, we shall present more complete versions of this same diagram; the successive versions will incorporate the extra information that we have acquired as our work progresses. It will become apparent at the end of Chapter 11 that some questions of containment between various classes remain open; these open questions will be discussed in Chapter 12.

The inclusions that inform this diagram have been proved as follows:

(i) by Theorem 4.27, every super-real field is real-closed;

(ii) by Proposition 4.39, a cut prime is a strongly convex prime;

(iii) by Theorem 4.29, a hyper-real field is an η_1-field, and, by Proposition 4.33, a hyper-real field has the form K_P with P a valuation prime z-ideal;

(iv) by Theorem 4.43, an ultrapower has the form K_P with P a cut prime z-ideal.

The other inclusions are trivial.

In the paper by Antonovskij *et al.* (1981), the class of ultrapowers is distinguished from the class of hyper-real fields on cardinality grounds. Let κ be an infinite cardinal, let $\mathcal{U}$ be a free ultrafilter on κ, and set $K = \mathbb{R}^\kappa/\mathcal{U}$. Then, by Theorem 4.11 of Antonovskij *et al.* (1981), $|K| = \aleph_1$ or $|K| = |K|^{\aleph_1}$. On the other hand there exists a hyper-real field L such that $\aleph_1 < |L| < |L|^{\aleph_1}$. Hence L is a hyper-real field that is not order-isomorphic to any ultrapower. A similar argument shows that there is a field in **cut prime z-ideal η_1** which is not an ultrapower. (In Chapter 9, we shall give an example in the theory ZFC+GCH of a hyper-real field K with $|K| = |K|^{\aleph_1}$ which is not isomorphic as a totally ordered group to any ultrapower; the example has weight $\aleph_1$.) We believe that this was the only previous knowledge about the equality of various of the above classes. Indeed Antonovskij *et al.* raised the question whether or not the class **hyper-real** is equal to the class **real-closed η_1-fields**; we shall show that this is not the case.

Various examples we have given suggest, but do not prove, that no other inclusions in the above diagram will be easily found.

Notes 4.48 As we have noted, the starting point of our work in this chapter is the text of Gillman and Jerison (1960), where most of the material in the first part of the chapter can be found; for prime ideals in $C(X)$, see Chapter 14 of this text, which is largely an exposition of the work of Kohls (1958). There are several constructions of the Stone–Čech compactification βX in this book. The classic Gelfand–Kolmogoroff theorem is given in §7.3 of Gillman and Jerison (1960).

A purely algebraic characterization of commutative, unital algebras which have the form $C(\Omega)$ for some compact space Ω is given by Delfosse (1975).

There is an enormous corpus of work on ultrafilters: see, for example, the monograph of Comfort and Negrepontis (1974). Uniform ultrafilters, and more technical versions of this notion, are studied in §7 of this work; it is proved, for example, that there is a uniform ultrafilter on each infinite cardinal.

Definition 4.11 is Definition 4.1 of Antonovskij *et al.* (1981).

For the general theory of Banach algebras, see Bonsall and Duncan (1973), Rickart (1960), and Rudin (1991), for example. In the case where A is the unital Banach algebra $C^b(X, \mathbb{C})$ for a completely regular space X, the identification of Φ_A with βX is given in detail in Dales (1997), Gamelin (1969), and Stout (1971). Properties of $C(\Omega, \mathbb{C})$ as a Banach algebra are given in Dales (1997, §3.2); for a study of prime ideals in commutative Banach algebras, see Dales (1997, §3.5).

The terminology 'super-real' field is new. Hyper-real fields are discussed by Gillman and Jerison (1960, Chapter 13), where it is shown that each such field is a real-closed η_1-field. As we said, the study of hyper-real fields has been taken up more recently by Antonovskij *et al.* (1981), where hyper-real fields are termed *H-fields*. Proposition 4.22 may be new, but Proposition 4.24 is 14.7 of Gillman and Jerison (1960).

The proofs of Theorems 4.27 and 4.28 are small variations of the arguments given in Gillman and Jerison (1960): see 13.4, 14.15(a), and 14.16(b), of that book.

Let $\mathcal{U}$ be a free ultrafilter on $\mathbb{N}$. Then the ultrapower $\mathbb{R}^{\mathbb{N}}/\mathcal{U}$ is an η_1-field; this is a special case of Theorem 4.29, and is also easily seen directly. With CH, each of these ultrapowers is a β_1-field by Corollary 2.28; the situation in the theory ZFC + $\neg$CH will be discussed in Notes 5.29.

Let P be a non-maximal, prime ideal in $C(\Omega)$. The family of prime ideals in A_P is described by Gillman and Jerison (1960, 14.19): the chain of all prime ideals is Dedekind complete (in the sense that every non-empty subfamily bounded above has a supremum) and contains at least $2^{\aleph_1}$ members. Much of our later theory (see, in particular, Chapter 8) is much easier for prime z-ideals. To justify our consideration of the case of general prime ideals, note that, whenever $J_x \neq M_x$ in $C(\Omega)$, there exists a prime ideal P with $J_x \subset P \subset M_x$ which is not a z-ideal, and there are easy examples (Gillman and Jerison 1960, 4M) such that none of the $2^{\aleph_1}$ prime ideals between J_x and M_x is a z-ideal.

As we stated, an ideal J_x in $C(\Omega)$ may or may not be a prime ideal. Indeed the compact space Ω is such that every ideal J_x in $C(\Omega)$ is a prime ideal if and only if Ω is an F-space (Gillman and Jerison 1960, 14.25). (A topological space X is an F-space if $\overline{U} \cap \overline{V} = \emptyset$ whenever U and V are open F_σ-sets with $U \cap V = \emptyset$; for example $\beta\mathbb{N}$ and $\beta\mathbb{N} \setminus \mathbb{N}$ are F-spaces.)

An example (Kohls 1958, Example 4.2) shows that even in the case where $P = J_x$, K_P is not necessarily an η_1-field: there is a completely regular space X and a point $\mathfrak{p}$ in βX such that $(C(\beta X)/J_{\mathfrak{p}})^+ \setminus \{0\}$ has a countable, coinitial subset.

The concept of a valuation prime has also been studied by Cherlin and Dickmann (1986) using a different terminology. We remark that the argument of

Proposition 4.34 shows that each prime ideal in $C(\Omega)$ is a valuation prime in the more general case where Ω is an F-space. However this property does not characterize F-spaces, as is shown by Example 10 of §2.1 of Cherlin and Dickmann (1986).

The characterization of ultrafilters $\mathcal{U}$ such that $c_0/\mathcal{U} = (l^\infty/\mathcal{U})^0$ given before Theorem 4.40 is from (Dales and Woodin 1987, Theorem 2.24), where other equivalent conditions are also given. See also Cherlin and Dickmann (1986, 3.1). The fact that, with CH, $\beta\mathbb{N} \setminus \mathbb{N}$ contains P-points is proved with other results in Walker (1974, Chapter 4); that the existence of such P-points is not a theorem of ZFC was proved by Shelah (1982, VI, §4).

It is of interest to specifically describe the situation for the algebras l^∞ and c. Take $\mathfrak{p} \in \beta\mathbb{N} \setminus \mathbb{N}$. Then the ideal $J_\mathfrak{p}$ is a non-maximal, prime z-ideal in $C(\beta\mathbb{N})$, and the set of prime ideals contained in $M_\mathfrak{p}$ is a chain of cardinality $\aleph_2$ (with GCH). The quotient field $K_\mathfrak{p}$ of $C(\beta\mathbb{N})/J_\mathfrak{p}$ is the ultrapower $\mathbb{R}^\mathbb{N}/\mathcal{U}_\mathfrak{p}$, where $\mathcal{U}_\mathfrak{p}$ is the ultrafilter on $\mathbb{N}$ corresponding to $\mathfrak{p}$, and $K_\mathfrak{p}$ is a real-closed η_1-field of cardinality $\mathfrak{c}$. Let P be a non-maximal, prime ideal in l^∞. Then $c \cap P$ is a non-maximal, prime ideal in c, and every prime ideal in c has this form: if Q is a prime ideal in c, set

$$P = \{f \in l^\infty : fc_0 \subset Q\} ,$$

and verify that P is the unique prime ideal in l^∞ such that $c \cap P = Q$. Clearly $K_Q \cong K_P$. Although $c \cap P$ is always a strongly convex prime in c (by Proposition 4.36), it is only a valuation prime if $\mathfrak{p}$ is a P-point of $\beta\mathbb{N} \setminus \mathbb{N}$.

Let P and Q be non-maximal, prime ideals in c. Conditions for the fields K_P and K_Q to be isomorphic (in the case where CH holds) are given by Moloney (1990, Theorem 2.2.13). In fact the conditions divide these fields K_P into exactly nine classes up to isomorphism.

The final example (Theorem 4.47) of a super-real field which is not a semi-η_1-field is new. It is perhaps surprising that the question of the existence of such fields is not mentioned in the text of Gilman and Jerison (1960).

5

Normability and universality

One of our motives for beginning our investigations of super-real fields was a certain *normability* question. Let us explain this question.

Definition 5.1 *Let A be an algebra over a field $\mathbb{F}$, where $\mathbb{F}$ is $\mathbb{R}$ or $\mathbb{C}$. An* algebra norm *on A is a map $\|\cdot\| : A \to \mathbb{R}$ such that:*

(i) $\|a\| \geq 0 \quad (a \in A)$, *and* $\|a\| = 0$ *if and only if* $a = 0$;

(ii) $\|a + b\| \leq \|a\| + \|b\| \quad (a, b \in A)$;

(iii) $\|\alpha a\| = |\alpha|\,\|a\| \quad (\alpha \in \mathbb{F}, a \in A)$;

(iv) $\|ab\| \leq \|a\|\,\|b\| \quad (a, b \in A)$.

The pair $(A, \|\cdot\|)$ is a normed algebra. *The algebra A is* normable *if there is an algebra norm on A.*

Of course, a norm $\|\cdot\|$ defines a metric, say d, on A, where

$$d(a, b) = \|a - b\| \quad (a, b \in A).$$

The norm is *complete* if the metric space (A, d) is complete. An algebra with a complete algebra norm is a *Banach algebra*. It is easy to see that, for each normed algebra A, there is an isometric isomorphism of A onto a dense subalgebra of a Banach algebra, called the *completion* of A, and so an algebra A is normable if and only if there is an embedding of A into a Banach algebra. We also note that, if $\|\cdot\|$ is an algebra norm on a real algebra A, then there is an algebra norm on the complexification $A_{\mathbb{C}}$ of A such that the embedding of A into $A_{\mathbb{C}}$ is an isometry.

For example, as we stated in Chapter 4, the uniform norm $|\cdot|_X$ is an algebra norm on $C^b(X, \mathbb{C})$ such that $(C^b(X, \mathbb{C}), |\cdot|_X)$ is a commutative, unital Banach algebra.

A question we wish to consider is the following.

Question 1 For which ordered fields K is the algebra $K^{\#}$ of finite elements of K normable?

We remark that the only real fields which are themselves normable are $\mathbb{R}$ and $\mathbb{C}$; this is the famous *Gelfand–Mazur* theorem for normed algebras.

A closely related question is the following.

Question 2 Let Ω be a compact space. For which non-maximal, prime ideals P in $C(\Omega)$ are the algebras $A_P = C(\Omega)/P$ and/or $K_P^{\#}$ normable?

Let P be a non-maximal, prime ideal in $C(\Omega)$. Certainly, if $K_P^{\#}$ is normable, then so is A_P because $A_P \subset K_P^{\#}$. We do not know whether or not the converse is true, even in the theory ZFC + CH.

The interest in Question 2 arose from a famous *problem of Kaplansky*. Let Ω be a compact space, and let $(C(\Omega, \mathbb{C}), |\cdot|_\Omega)$ be the commutative, unital Banach algebra discussed earlier. Now suppose that $\|\cdot\|$ is an arbitrary algebra norm on $C(\Omega, \mathbb{C})$. It was shown by Kaplansky (1949) that necessarily

$$|f|_\Omega \leq \|f\| \quad (f \in C(\Omega, \mathbb{C})),$$

and so the norm $\|\cdot\|$ is equivalent to the uniform norm $|\cdot|_\Omega$ if and only if there is a constant M such that

$$\|f\| \leq M\,|f|_\Omega \quad (f \in C(\Omega, \mathbb{C})).$$

It is easy to see that this is indeed the case whenever $\|\cdot\|$ is a complete norm. The problem of Kaplansky was whether or not an arbitrary norm $\|\cdot\|$, not *a priori* supposed to be complete, on $C(\Omega, \mathbb{C})$ is necessarily equivalent to the uniform norm $|\cdot|_\Omega$; the question was open for each infinite compact space Ω. It is also easy to see that Kaplansky's problem is answered affirmatively if and only if every homomorphism from the Banach algebra $(C(\Omega, \mathbb{C}), |\cdot|_\Omega)$ into an arbitrary Banach algebra is automatically continuous, and this is the form in which the question was studied. This study eventually led to a further equivalent reformulation of the problem, given as (c), below.

Theorem 5.2 *Let Ω be a compact space. Then the following conditions are equivalent:*

(a) *there is a discontinuous homomorphism from $C(\Omega, \mathbb{C})$ into a Banach algebra;*

(b) *there is an algebra norm on $C(\Omega, \mathbb{C})$ which is not equivalent to the uniform norm $|\cdot|_\Omega$;*

(c) *there is a non-maximal, prime ideal P in $C(\Omega)$ such that $C(\Omega)/P$ is normable.* □

The problem of Kaplansky was eventually shown to be independent of the theory ZFC, a remarkable event in the history of analysis. It was proved independently by Dales and by Esterle that, in the theory ZFC + CH, there exists a non-maximal, prime ideal P in $C(\Omega, \mathbb{C})$ for each infinite compact space Ω such that the algebra $C(\Omega, \mathbb{C})/P$ is normable, and hence that there is a discontinuous homomorphism from $C(\Omega, \mathbb{C})$ into a Banach algebra. However, the existence of such discontinuous homomorphisms is not a theorem of ZFC: it was proved by Solovay and Woodin that it is relatively consistent with ZFC that all homomorphisms from $C(\Omega, \mathbb{C})$ into every Banach algebra are automatically continuous.

In the light of these results, we address Questions 1 and 2 in the theory ZFC + CH. A full solution to Question 2 in this setting would complete one aspect of the work of Dales and Esterle on Kaplansky's problem; Question 1 seems to be an interesting and important generalization of Question 2.

We first present a necessary condition for an algebra $K^{\#}$ to be normable.

Theorem 5.3 *(Esterle) Let K be an ordered field, and suppose that $K^{\#}$ is normable. Then $|\Gamma_K| \le \mathfrak{c}$.*

Proof Let $\|\cdot\|$ be an algebra norm on $K^{\#}$.

First, suppose that $a, b \in K^0 \backslash \{0\}$ with $v(a) > v(b)$, where v is the archimedean valuation on K. Then there exists $c \in K^0$ such that $a = bc$ in K^0. We have

$$\|a^n\| \le \|b^n\| \, \|c^n\| \quad (n \in \mathbb{N}). \tag{5.1}$$

Since the element $z - c$ is invertible in the complexification $K^0_{\mathbb{C}}$ of K^0 for each $z \neq 0$, it follows from the spectral radius formula for Banach algebras that $\|c^n\|^{1/n} \to 0$ as $n \to \infty$. Since $\|b^n\| > 0 \quad (n \in \mathbb{N})$, we see from (5.1) that

$$\|a^n\| < \|b^n\|$$

for all sufficiently large n.

Now define

$$\psi : a \mapsto (\|a^n\|), \quad K^0 \setminus \{0\} \to \mathbb{R}^{\mathbb{N}}.$$

We have shown that, if $\psi(a) = \psi(b)$, then $v(a) = v(b)$, and so ψ induces an injection from Γ_K into $\mathbb{R}^{\mathbb{N}}$. Since $\left|\mathbb{R}^{\mathbb{N}}\right| = \mathfrak{c}$, it follows that $|\Gamma_K| \le \mathfrak{c}$. □

Theorem 5.3 leads to an upper bound on the cardinality of a field K for which $K^{\#}$ is normable.

Theorem 5.4 *Let K be an ordered field such that $K^{\#}$ is normable. Then:*

(i) $|K| \le 2^{\mathfrak{c}}$;

(ii) (GCH) $|K| \le \aleph_2$.

Proof (i) By Theorem 5.3, $|\Gamma_K| \le \mathfrak{c}$, and so, by Theorem 1.28(ii), we have $|K| \le 2^{\mathfrak{c}}$.

(ii) With GCH, $2^{\mathfrak{c}} = \aleph_2$. □

We note that the converse of Theorem 5.4(ii) does not hold: there are fields of cardinality $\aleph_2$ which are not normable. For set $H = \widehat{\mathbf{R}}$, regarded as a divisible group. Then, with GCH, $|H| = \aleph_2$ by Theorem 2.23(iv). Set

$$K = \mathfrak{F}_{(1)}(\mathbb{R}, H).$$

Then $|K| = \aleph_2$ by Proposition 1.28(i), but $|\Gamma_K| = |H| = \aleph_2$, and so, by Theorem 5.3, K is not normable.

The criterion of Theorem 5.3 allows us to exhibit ultrapowers $\mathbb{R}^{\kappa}/\mathcal{U}$ such that $(\mathbb{R}^{\kappa}/\mathcal{U})^{\#}$ is not normable.

Proposition 5.5 *Let κ be a cardinal with $\kappa \ge \mathfrak{c}$. Then there is an ultrafilter $\mathcal{U}$ on κ such that $(\mathbb{R}^{\kappa}/\mathcal{U})^{\#}$ is not normable.*

Proof Let $\mathcal{F}$ be the family of finite subsets of κ, so that $|\mathcal{F}| = \kappa$; we set

$$\mathcal{F} = \{F_x : x \in \kappa\}.$$

For $x \in \kappa$, set $Z_x = \{y \in \kappa : x \in F_y\}$. Then $\{Z_x : x \in \kappa\}$ has the finite intersection property: the intersection $Z_{x_1} \cup \cdots \cup Z_{x_k}$ contains the point y for which $F_y = \{x_1, \ldots, x_k\}$. So $\{Z_x : x \in \kappa\}$ is contained in an ultrafilter, say $\mathcal{U}$, on κ. Set

$$K = \mathbb{R}^\kappa / \mathcal{U} .$$

Let S be any subset of $C(\beta\kappa)$ with $|S| \leq \mathfrak{c}$, and set

$$T = \{|g|/n : g \in S, \quad n \in \mathbb{N}\},$$

so that $|T| \leq \mathfrak{c}$. Since $\kappa \geq \mathfrak{c}$, there is a surjection $x \mapsto h_x$, $\kappa \to T$. Define

$$f(y) = \min\{h_x(y) : x \in F_y\} \qquad (y \in \kappa) .$$

Since F_y is finite and not empty, $f(y)$ is well defined; the function f defines the element $[f]$ of K. For each $g \in S$ and $n \in \mathbb{N}$, there exists $x \in \kappa$ with $h_x = |g|/n$. For $y \in Z_x$, we have $x \in F_y$, and so $f(y) \leq h_x(y)$. Thus

$$f(y) \leq |g|/n \quad (y \in Z_x) ,$$

and this shows that

$$v([f]) << \{v([g]) : g \in S\} .$$

It follows that $|\Gamma_K| > \mathfrak{c}$, and so, by Theorem 5.3, the algebra $K^\#$ is not normable. □

In the other direction to Theorem 5.3, we have the following very deep and important theorem of Esterle. (Note that this theorem does not require CH.)

Theorem 5.6 *(Esterle) The algebras* $\mathbf{C}^\#$ *and* $\mathbf{R}^\#$ *of finite elements of the fields* $\mathbf{C}$ *and* $\mathbf{R}$ *are normable.* □

It follows immediately from this theorem together with Theorem 2.30 and Corollary 2.28 that we have the following positive answer to Question 1.

Theorem 5.7 *Let K be a β_1-field. Then the algebra $K^\#$ of finite elements in K is normable. In particular, $K^\#$ is normable in the case where K has transcendence degree at most $\aleph_1$ over $\mathbb{R}$.* □

Necessarily, a field (by our convention, with ground field $\mathbb{R}$) has cardinality at least $\mathfrak{c}$. The above theorem shows that, if the *continuum hypothesis* (CH) *be assumed*, then, for each field K with $|K| = \mathfrak{c}$, $K^\#$ is normable. The fact that $K^\#$ is normable for each such field K is not, however, a theorem of ZFC.

A purpose of the present book is to investigate the gap between the conditions on K given in Theorem 5.3 and in Theorem 5.7, where we are working in the theory ZFC + GCH. In this theory, Theorem 5.3 contains the only constraint that we know for the normability of $K^\#$. It is possible therefore to make the following conjecture.

Conjecture 1 Let K be an ordered field such that $|\Gamma_K| = \aleph_1$. Then $K^\#$ is normable.

The following weaker conjecture would still amount to a very attractive strengthening of Theorem 5.7.

Conjecture 2 Let K be an ordered field such that $w(K) = \aleph_1$. Then $K^\#$ is normable.

Let K be an ordered field, and assume that GCH holds. Then we know whether or not $K^\#$ is normable in every case except where

$$|K| = \aleph_2 \quad \text{and} \quad |\Gamma_K| = \aleph_1 . \tag{5.2}$$

We can reduce to the case (5.2) for the following reasons: if $|K| > \aleph_2$ or if $|\Gamma_K| > \aleph_1$, then $K^\#$ is not normable by Theorems 5.3 and 5.4; if $|K| < \aleph_2$, then $|K| = \aleph_1$, and so $K^\#$ is normable by Theorem 5.7; and, by Theorem 1.28(ii), the case where $|\Gamma_K| < \aleph_1$ and $|K| = \aleph_2$ does not arise. We are thus led to investigate the normability of $K^\#$ for fields K satisfying (5.2). We confess total failure in this respect: we can give no information at all about whether or not the algebra $K^\#$ is normable for any ordered field K such that K satisfies (5.2).

One of the questions that we shall investigate is whether there *are* ordered fields K satisfying (5.2), and whether there are examples occurring in the various classes of Chapter 4. In particular, it is not trivial to exhibit examples of super-real fields K_P which satisfy condition (5.2); if there were no such fields, Question 2 would already be fully resolved (at least in the theory ZFC + GCH), without any further work. We would particularly like to inspect a super-real η_1-field satisfying condition (5.2).

We first give an example of a real-closed field satisfying (5.2).

Definition 5.8 *Let σ be an infinite ordinal. Then*

$$\Delta_\sigma = \mathfrak{F}_{(0)}(\mathbb{Q}, \sigma) = \{u \in \mathbb{Q}^\sigma : \operatorname{supp} u \text{ is finite}\} \quad \textit{and} \quad L_\sigma = \mathfrak{F}(\mathbb{R}, \Delta_\sigma) .$$

As in Chapter 1, Δ_σ is a divisible, totally ordered group, and so L_σ can be defined. Clearly $|\Delta_\sigma| = |\sigma|$.

Theorem 5.9 *Let σ be an infinite ordinal. Then L_σ is a real-closed field with $|L_\sigma| = w(L_\sigma) = 2^{|\sigma|}$ and $|\Gamma_{L_\sigma}| = \delta(L_\sigma) = |\sigma|$.*

Proof Set $\kappa = |\sigma|$. By Theorem 2.15(iv), L_σ is a real-closed field. Its value group Γ_{L_σ} is Δ_σ, and so $|\Gamma_{L_\sigma}| = \kappa$. Using Proposition 1.28, it is easy to see that $|L_\sigma| = 2^\kappa$ and that $\delta(L_\sigma) = \kappa$. Since Δ_σ contains the increasing sequence $(-\delta_\alpha : \alpha < \sigma)$ bounded above by δ_0, $w(L_\sigma) = 2^\kappa$. □

Corollary 5.10 (GCH) *Let $K = L_{\aleph_1}$. Then K is a real-closed field, with*

$$|K| = w(K) = \aleph_2 \quad \text{and} \quad |\Gamma_K| = \delta(K) = \aleph_1 . \qquad \square$$

Thus $L_{\aleph_1}$ satisfies (5.2), above. However, L_σ is not a semi-η_1-field for any infinite ordinal σ. For let u and v be the characteristic functions of $\{0\}$ and $\{1\}$, respectively, in Δ_σ. Then

$$mv < (m+1)v < \frac{1}{n+1}u < \frac{1}{n}u \quad (m, n \in \mathbb{N}),$$

but there is no $h \in \Delta_\sigma$ such that $mv < h < (1/n)u \quad (m, n \in \mathbb{N})$. Thus, Δ_σ is not a semi-η_1-set, and so L_σ is not a semi-η_1-field. It follows from Theorem

4.28(ii) that $L_{\aleph_1}$ is not a super-real field, and so we must look elsewhere for super-real η_1-fields satisfying (5.2). We do not know whether or not the algebra $L^{\#}_{\aleph_1}$ is normable.

The first obvious examples of η_1-fields which do satisfy (5.2) are the fields $\widehat{\mathbf{R}}$ and $\widetilde{\mathbf{R}}$, which were introduced earlier.

Proposition 5.11 *The field* $\widehat{\mathbf{R}}$ *is a real-closed* η_1*-field of cardinality and weight* $2^{\aleph_1}$ *whose value group* $\mathbf{G}$ *has cardinality* $\mathfrak{c}$.

Proof This is part of Theorem 2.23. □

Corollary 5.12 (GCH) *The field* $\widehat{\mathbf{R}}$ *is a real-closed* η_1*-field of cardinality and weight* $\aleph_2$ *whose value group* $\mathbf{G}$ *has cardinality* $\aleph_1$. □

In fact, the field $\widehat{\mathbf{R}}$ is a 'test-case' for Conjecture 1, above.

Theorem 5.13 *The algebra* $K^{\#}$ *is normable for each ordered field* K *such that* $|\Gamma_K| = \aleph_1$ *if and only if* $\widehat{\mathbf{R}}^{\#}$ *is normable.*

Proof Let K be an ordered field with $|\Gamma_K| = \aleph_1$. Then Γ_K is a β_1-group, and so, by Theorem 2.24, there is an embedding of K in $\widehat{\mathbf{R}}$. Hence there is an embedding of $K^{\#}$ in $\widehat{\mathbf{R}}^{\#}$; if $\widehat{\mathbf{R}}^{\#}$ is normable, then so is $K^{\#}$. □

The field $\widetilde{\mathbf{R}}$ also satisfies (5.2). It is a 'test-case' for Conjecture 2, above.

Proposition 5.14 (GCH) *The field* $\widetilde{\mathbf{R}}$ *is a real-closed* η_1*-field of cardinality* $\aleph_2$ *and weight* $\aleph_1$ *whose value group* $\mathbf{G}$ *has cardinality* $\aleph_1$.

Proof This is part of Theorem 3.21. □

Theorem 5.15 *The algebra* $K^{\#}$ *is normable for each ordered field* K *with weight* $\aleph_1$ *if and only if* $\widetilde{\mathbf{R}}^{\#}$ *is normable.*

Proof By Proposition 2.2, K contains an order-dense subfield L of cardinality $\aleph_1$, and L is a β_1-field. By Theorem 3.22, there is an embedding of $\widetilde{L}$ in $\widetilde{\mathbf{R}}$. Also, there is an embedding of K into $\widetilde{L}$, and hence there is an embedding of K in $\widetilde{\mathbf{R}}$. The result follows, as in Theorem 5.13. □

The truth is that we have not been able to determine whether or not either of the algebras $\widetilde{\mathbf{R}}^{\#}$ or $\widehat{\mathbf{R}}^{\#}$ is normable, so we cannot resolve Question 1. It is clearly relevant to Question 2 whether or not the fields $\widehat{\mathbf{R}}$ and $\widetilde{\mathbf{R}}$ are super-real fields. This problem will be taken up in Chapter 6, where we shall prove that $\widehat{\mathbf{R}}$ is not a super-real field. Examples which will show that Question 2 is open will be given in Chapter 7. Finally, we shall prove in Theorem 9.26 that (with GCH) $\widetilde{\mathbf{R}}$ is a super-real field, and so Question 2 cannot be resolved until it is known whether or not $\widetilde{\mathbf{R}}^{\#}$ is normable. An example to be given in Theorem 9.60 will show that, even if $\widetilde{\mathbf{R}}^{\#}$ were known to be normable, we would not have fully resolved Question 2.

Our second topic in the present chapter is the universality of certain algebras in various classes of algebras.

Definition 5.16 *Let C be a class of algebras. A subclass C' of C is* universal *for C if each member of C is isomorphic (by a mapping appropriate to the class) to a subalgebra of some member of C'. An individual algebra A is* universal *for C if A belongs to C and if each member of C is isomorphic to a subalgebra of A.*

Thus, for example, as we noted after Theorem 2.30, $\mathbf{R}$ is universal in the class of real-closed β_1-fields; with CH, $\mathbf{R}$ is universal in the class of real-closed fields of cardinality $\aleph_1$.

Clearly, if C' is a universal subclass of a class C and if each member of C' is normable, then each member of C is also normable.

Consider the class of non-unital integral domains of cardinality $\mathfrak{c}$. We shall first show that the subclass of integral domains $(\mathbb{C}^{\mathbb{N}}/\mathcal{U})^0$, where $\mathcal{U}$ is a free ultrafilter on $\mathbb{N}$, is a universal subclass. This will imply that, with CH, each non-unital integral domain of cardinality $\mathfrak{c}$ is normable. In fact, we shall prove the stronger result that the family $c_0(\mathbb{C})/\mathcal{U}$ is universal in the class of non-unital, complex integral domains of cardinality $\mathfrak{c}$; we do not need this stronger form of the result to obtain the mentioned corollary, and in fact the corollary could be obtained with considerably less technical effort. Note that our main result (Theorem 5.25) does not require the Continuum Hypothesis.

We shall require some preliminary lemmas. In these lemmas, we write $\|\cdot\|_n$ for the supremum norm in $\mathbb{R}^n$, so that

$$\|x\|_n = \max\{|x_1|, \dots, |x_n|\} \quad (x = (x_1, \dots, x_n) \in \mathbb{R}^n),$$

and, for $x \in \mathbb{R}^n$ and $\varepsilon > 0$, we set

$$B_n(x;\varepsilon) = \{y \in \mathbb{R}^n : \|x-y\|_n < \varepsilon\},$$
$$S_n(x;\varepsilon) = \{y \in \mathbb{R}^n : \|x-y\|_n = \varepsilon\},$$

for the ball and sphere in $\mathbb{R}^n$ with centre x and radius ε, respectively.

In the first lemma, we regard $[a,b]$ as a subset of $\mathbb{N}$ for $a, b \in \mathbb{N}$.

Lemma 5.17 *Suppose that, for each $m, n \in \mathbb{N}$, $\mathbf{U}_{m,n}$ is a finite collection of open subsets of $\mathbb{R}^n$ such that $U \subset B_n(0;1/m)$ for each $U \in \mathbf{U}_{m,n}$. Then there exists $g_0 \in c_0^+$ and a subset S of $\mathbb{Q}^{\mathbb{N}}$ with $|S| = \mathfrak{c}$ such that:*

(i) *for each $k \in \mathbb{N}$ and $f \in S$, we have $|f(k)| \leq g_0(k)$;*

(ii) *for each $n \in \mathbb{N}$, for each finite subset $\{f_1, \dots, f_n\}$ of S, and for each infinite sequence $(V_m : m \in \mathbb{N})$ such that $V_m \in \mathbf{U}_{m,n}$ $(m \in \mathbb{N})$, the set*

$$\left\{k \in \mathbb{N} : (f_1(k), \dots, f_n(k)) \in \dot{\bigcup}\{V_m : m \in \mathbb{N}\}\right\}$$

is infinite.

Proof We define by induction sequences (m_i) and (P_i) such that, for each $i \in \mathbb{N}$:

(1) $m_i \in \mathbb{N}$ and $P_i \subset \mathbb{Q}^{m_i}$;

(2) $m_i < m_{i+1}$ and $\{s \mid [1, m_i] : s \in P_{i+1}\} = P_i$;

(3) for $s \in P_i$, there are exactly two elements $t \in P_{i+1}$ with $t \mid [1, m_i] = s$;

(4) for $s \in P_{i+1}$ and $k \in [m_{i+1}, m_{i+1}]$, we have $|s(k)| \le 1/i$;

(5) for each set $\{s^{(1)}, \ldots, s^{(n)}\} \subset P_{i+1}$ with $n \le i$ and each $U \in \mathbf{U}_{i,n}$, there exists $k \in [m_i + 1, m_{i+1}]$ with $(s^{(1)}(k), \ldots, s^{(n)}(k)) \in U$.

Indeed, set $m_1 = 1$ and take P_1 to consist of the single 'sequence' of length 1 and taking the value 1.

Now assume that $m_1, \ldots, m_i$ and $P_1, \ldots, P_i$ have been defined. Set

$$\widetilde{P}_{i+1} = \left\{ s \in \mathbb{Q}^{m_i+1} : s \mid [1, m_i] \in P_i, \quad s_{m_i+1} \in \left\{ \frac{1}{i+1}, \frac{1}{i+2} \right\} \right\}.$$

Thus, for each $s \in P_i$, there are exactly two elements $t \in \widetilde{P}_{i+1}$ with $t \mid [1, m_i] \in P_i$.

Consider the family of all pairs $(U, \mathbf{s})$ such that there exists $n \in \{1, \ldots, i\}$ with $U \in \mathbf{U}_{i,n}$ and such that $\mathbf{s} = (s^{(1)}, \ldots, s^{(n)})$ belongs to $(\widetilde{P}_{i+1})^n$, where $s^{(j_1)}$ and $s^{(j_2)}$ are distinct elements of $\widetilde{P}_{i+1}$ whenever $j_1, j_2 \in \{1, \ldots, n\}$ with $j_1 \ne j_2$. This family is finite, and can be enumerated as

$$(U_1, \mathbf{s}_1), \ \ldots \ , (U_M, \mathbf{s}_M).$$

For each $j \in \{1, \ldots, M\}$, choose $x^{(j)} \in U_j$; since there exists $n \in \{1, \ldots, i\}$ with $U_j \in \mathbf{U}_{i,n}$, we have $x^{(j)} \in \mathbb{R}^n$, say $x^{(j)} = (x_1^{(j)}, \ldots, x_n^{(j)})$, and $\left\|x^{(j)}\right\|_n < 1/i$, so that

$$\left|x_k^{(j)}\right| < 1/i \quad (k = 1, \ldots, n).$$

Define $m_{i+1} = m_i + 1 + M$. We extend each $s \in \widetilde{P}_{i+1}$ to a unique element $\overline{s} \in \mathbb{Q}^{m_{i+1}}$ in such a way that, for each $k \in [m_i + 2, m_{i+1}]$, we have $|\overline{s}(k)| < 1/i$, and such that, for each $j \in \{1, \ldots, M\}$,

$$\overline{s}(m_i + 1 + j) = x_t^{(j)}$$

whenever $\mathbf{s}_j^{(t)} = s$. (For each j, there is at most one t with $\mathbf{s}_j^{(t)} = s$; if there is no such t, there is no constraint on the choice of $\overline{s}(m_i + 1 + j)$ in $\mathbb{Q}$ beyond the requirement that $|\overline{s}(m_i + 1 + j)| < 1/i$.) The subset P_{i+1} of $\mathbb{Q}^{m_{i+1}}$ is the collection of all these extensions $\overline{s}$.

It is clear that the appropriate versions of conditions (1)–(4) are satisfied at this stage of the induction. Now take $\{s^{(1)}, \ldots, s^{(n)}\} \subset P_{i+1}$ with $n \le i$, and take $U \in \mathbf{U}_{i,n}$. Then there exists $j \in \{1, \ldots, M\}$ with $U = U_j$ and with $\mathbf{s}_j = (s^{(1)}, \ldots, s^{(n)})$. Consider the number

$$k = m_i + 1 + j \in [m_i + 1, m_{i+1}].$$

Then

$$(s^{(1)}(k), \ldots, s^{(n)}(k)) = (x_1^{(j)}, \ldots, x_n^{(j)}) = x^{(j)} \in U_j = U,$$

and so the appropriate version of condition (5) is also satisfied.

This shows that the inductive construction continues, and thus we may suppose that we have obtained $(m_i : i \in \mathbb{N})$ and $(P_i : i \in \mathbb{N})$ such that the above conditions are satisfied for all $i \in \mathbb{N}$.

We now define $g_0 \in c_0^+$ by the requirement that

$$g_0(k) = \frac{1}{i} \qquad (k \in [m_i + 1, m_{i+1}]),$$

and we set

$$S = \{f \in \mathbb{Q}^{\mathbb{N}} : f \mid [1, m_i] \in P_i \qquad (i \in \mathbb{N})\}.$$

It follows from condition (3), above, that $|S| = \mathfrak{c}$, and from condition (4) that

$$|f(k)| < g_0(k) \quad (k \in \mathbb{N})$$

for each $f \in S$.

Now suppose that $n \in \mathbb{N}$, that $\{f_1, \ldots, f_n\}$ is a finite subset of S, and that (V_m) is an infinite sequence with $V_m \in \mathbf{U}_{m,n} \quad (m \in \mathbb{N})$. Choose $i_0 \in \mathbb{N}$ such that $i_0 > n$ and such that

$$f_{j_1} \mid [1, i_0] \neq f_{j_2} \mid [1, i_0]$$

whenever $j_1, j_2 \in \{1, \ldots, n\}$ with $j_1 \neq j_2$. By condition (5), for each $i \geq i_0$, there exists $k \in [m_i + 1, m_{i+1}]$ with $(f_1(k), \ldots, f_n(k)) \in V_i$. Thus

$$\left\{k \in \mathbb{N} : (f_1(k), \ldots, f_n(k)) \in \bigcup\{V_m : m \in \mathbb{N}\}\right\}$$

is infinite, so establishing clause (ii) of the lemma. □

We now need to call in aid some facts about semi-algebraic sets and the Tarski–Seidenberg theorem. (Note again that the theorems at the end of this chapter only require that the family of integral domains $(\mathbb{C}^{\mathbb{N}}/\mathcal{U})^0$ be universal in the class of non-unital integral domains of cardinality $\mathfrak{c}$, rather than the stronger theorem that we prove in Theorem 5.25, and this weaker theorem does not require the Tarski–Seidenberg theorem.)

Let k be a field. As in Chapter 2, the algebra of polynomials in n indeterminates over k is denoted by $k[X_1, \ldots, X_n]$.

A family $\mathcal{S}$ of subsets of $\mathbb{R}^n$ is a *Boolean ring* of subsets if:

(i) $\emptyset \in \mathcal{S}$;

(ii) $S \cup T \in \mathcal{S}$ and $S \cap T \in \mathcal{S}$ whenever $S, T \in \mathcal{S}$;

(iii) $\mathbb{R}^n \setminus S \in \mathcal{S}$ whenever $S \in \mathcal{S}$.

The intersection of a collection of Boolean rings is a Boolean ring, and so each family $\mathcal{F}$ of subsets of $\mathbb{R}^n$ generates a Boolean ring; this is the minimum Boolean ring containing $\mathcal{F}$.

Definition 5.18 *Let $n \in \mathbb{N}$. The* semi-algebraic sets *in $\mathbb{R}^n$ are the members of the Boolean ring of subsets of $\mathbb{R}^n$ generated by sets of the form*

$$\{x = (x_1, \ldots, x_n) \in \mathbb{R}^n : p(x_1, \ldots, x_n) > 0\},$$

where $p \in \mathbb{R}[X_1, \ldots, X_n]$.

For example, it is easy to see that a semi-algebraic subset E of $\mathbb{R}$ is either empty or a finite union of intervals; if $E \subset \mathbb{R}^+$ and $0 \in \overline{E} \setminus E$, then there exists $\varepsilon_0 > 0$ such that $(0, \varepsilon_0) \subset E$. It is rather easy to show that a semi-algebraic set in $\mathbb{R}^n$ has a finite number of connected components, and that each such component is itself a semi-algebraic set.

Theorem 5.19 *(Tarski–Seidenberg). Let E be a semi-algebraic set in $\mathbb{R}^{m+n}$, and let $\pi : \mathbb{R}^{m+n} \to \mathbb{R}^m$ be the projection onto the first m coordinates. Then $\pi(E)$ is a semi-algebraic set in $\mathbb{R}^m$.* □

The proof of the Tarski–Seidenberg theorem is not difficult: it involves what is called 'elimination theory' in algebra. The theorem is also well known in logic, where it is regarded as a statement about elimination of quantifiers; it is related to the decision procedure for deciding the truth of an elementary sentence built up from finitely many inequalities of the form

$$p_i(x_1, \ldots, x_n) > 0\,,$$

where the p_i are polynomials in $k[X_1, \ldots, X_n]$ for a real-closed field k, by using the basic logical connectives '$\wedge$', '$\vee$', '$\neg$', '$\exists$', and '$\forall$'. Applications of the theorem basically say that a set defined in terms of semi-algebraic sets by elementary sentences is still a semi-algebraic set.

For example, it follows from the Tarski–Seidenberg theorem that the closure and interior of a semi-algebraic set are also semi-algebraic.

Let E and F be semi-algebraic sets in $\mathbb{R}^m$ and $\mathbb{R}^n$, respectively. A map $f : E \to F$ is *semi-algebraic* if its graph is a semi-algebraic set in $\mathbb{R}^{m+n}$. For example, let H be a semi-algebraic set in $\mathbb{R}^2$ contained in $\{(x, y) : y \leq 1\}$, and set $h(x) = \sup\{y : (x, y) \in H\}$ for $x \in \mathbb{R}$. Then the Tarski–Seidenberg theorem implies that $h : \mathbb{R} \to \mathbb{R}$ is a semi-algebraic function.

Theorem 5.20 *(Łojasiewicz's inequality) Let $f, g : [0, 1] \to \mathbb{R}$ be semi-algebraic functions such that $\mathbf{Z}(f) \subset \mathbf{Z}(g)$. Then there exists $\varepsilon_0 > 0$, $n \in \mathbb{N}$, and $c > 0$ such that*

$$|g(x)|^n < c\,|f(x)| \qquad (x \in [0, \varepsilon_0])\,.$$ □

We can now prove the technical lemma that we require.

Lemma 5.21 *Let U be an open subset of $\mathbb{R}^n$ such that U is a semi-algebraic set. Suppose that $0 \in \overline{U}$. Then there exists $\varepsilon_0 > 0$ and $m_0 \in \mathbb{N}$ such that, for each ε with $0 < \varepsilon < \varepsilon_0$, there exists $x \in S_n(0; \varepsilon)$ with $B_n(x; \varepsilon^{m_0}) \subset U$.*

Proof We define the following subsets of $\mathbb{R}^{2n+2}$, where $(\varepsilon, \delta, x, y)$ with $\varepsilon \in \mathbb{R}$, $\delta \in \mathbb{R}$, $x \in \mathbb{R}^n$, and $y \in \mathbb{R}^n$ is a generic point of $\mathbb{R}^{2n+2}$:

$$\begin{aligned}
E_1 &= \{(\varepsilon, \delta, x, y) : \varepsilon \leq 0\}\,;\\
E_2 &= \{(\varepsilon, \delta, x, y) : \delta \leq 0\}\,;\\
E_3 &= \{(\varepsilon, \delta, x, y) : \varepsilon > 0 \ \text{ and } x \notin S_n(0; \varepsilon)\}\,;\\
E_4 &= \{(\varepsilon, \delta, x, y) : \delta > 0, y \in B_n(x; \delta), \text{and } y \notin U\}\,;\\
E &= E_1 \cup E_2 \cup E_3 \cup E_4\,.
\end{aligned}$$

It is clear that each of the sets E_1, E_2, E_3, and E_4 is semi-algebraic, and so E is a semi-algebraic set.

Define

$$F = \{(\varepsilon, \delta, x) \in \mathbb{R}^{n+2} : (\varepsilon, \delta, x, y) \in E \text{ for some } y \in \mathbb{R}^n\}, \quad G = \mathbb{R}^{n+2} \setminus F.$$

By the Tarski–Seidenberg theorem, F is a semi-algebraic set in $\mathbb{R}^{n+2}$, and so G is also semi-algebraic. Note that

$$G = \{(\varepsilon, \delta, x) \in \mathbb{R}^{n+2} : \varepsilon > 0,\ \delta > 0,\ x \in B_n(0;\varepsilon), \text{ and } B_n(0;\varepsilon) \subset U\}.$$

Now define

$$H = \{(\varepsilon, \delta) \in \mathbb{R}^2 : \delta \leq 1 \text{ and } (\varepsilon, \delta, x) \in G \text{ for some } x \in \mathbb{R}^n\}.$$

Again by the Tarski–Seidenberg theorem, H is a semi-algebraic set in $\mathbb{R}^2$.

Next define

$$D = \{\varepsilon \in \mathbb{R} : (\varepsilon, \delta) \in H \text{ for some } \delta \in \mathbb{R}\}.$$

Then D is a semi-algebraic subset of $\mathbb{R}$, with $D \subset \mathbb{R}^+$. The hypothesis that $0 \in \overline{U}$ implies that $0 \in \overline{D}$, and so there exists $\varepsilon_0 > 0$ such that $(0, \varepsilon_0) \subset D$.

Finally define $h : D \to \mathbb{R}^+$ by

$$h(\varepsilon) = \sup\{\delta \in \mathbb{R} : (\varepsilon, \delta) \in H\}.$$

Since H is a semi-algebraic set, h is a semi-algebraic function, and so, by the Łojasiewicz inequality, Theorem 5.20, there exists $n \in \mathbb{N}$ and $c > 0$ such that $\varepsilon^n < ch(\varepsilon)$ for $0 < \varepsilon < \varepsilon_0$.

The result follows. □

Lemma 5.22 *There exists $g_0 \in c_0^+$ and $S \subset \mathbb{Q}^{\mathbb{N}}$ with $|S| = \mathfrak{c}$ such that:*

(i) *for each $k \in \mathbb{N}$ and $f \in S$, we have*

$$|f(k)| < g_0(k);$$

(ii) *for each $n \in \mathbb{N}$, for each finite subset $\{f_1, \ldots, f_n\}$ of S, and for each open, semi-algebraic set U in $\mathbb{R}^n$ with $0 \in \overline{U}$, the set*

$$\{k \in \mathbb{N} : (f_1(k), \ldots, f_n(k)) \in U\}$$

is infinite.

Proof For each $m, n \in \mathbb{N}$, choose a finite set $S_{m,n}$ in $S_n(0; 1/2^m)$ such that

$$S_n(0; 1/2^m) \subset \bigcup\{B_n(x; 1/2^{m^2}) : x \in S_{m,n}\},$$

and set

$$\mathbf{U}_{m,n} = \{B_n(x; 1/2^{m^2}) : x \in S_{m,n}\}.$$

Suppose that $n \in \mathbb{N}$ and that $U \subset \mathbb{R}^n$ is an open, semi-algebraic set with $0 \in \overline{U}$. By Lemma 5.21, there exists $\varepsilon_0 > 0$ and $m_0 \in \mathbb{N}$ such that, for all ε with $0 < \varepsilon < \varepsilon_0$, there exists $x \in S_n(0;\varepsilon)$ such that

$$B_n(x; \varepsilon^{m_0}) \subset U.$$

Choose $m_1 \in \mathbb{N}$ with $m_1 > m_0$ and with $1/2^{m_1} < \varepsilon_0$. For each $m > m_1$, there exists $x_m \in S_n(0; 1/2^m)$ such that

$$B_n(x_m; 1/2^{mm_0}) \subset V .$$

However,

$$x_m \in B_n(x'_m; 1/2^{m^2})$$

for some $x'_m \in S_{m,n}$, and we have

$$B_n(x'_m; 1/2^{m^2}) \subset U \cap B_n(0; 1/m) .$$

By Lemma 5.17, there exists $g_0 \in c_0^+$ and $S \subset \mathbb{Q}^{\mathbb{N}}$ with $|S| = \mathfrak{c}$ such that $|f(k)| < g_0(k)$ for $f \in S$ and $k \in \mathbb{N}$, and such that, for each $n \in \mathbb{N}$ and $\{f_1, \ldots, f_n\} \subset S$, the set

$$\left\{k \in \mathbb{N} : (f_1(k), \ldots, f_n(k)) \subset \bigcup\{B_n(x'_m; 1/2^{m^2}) : m \in \mathbb{N}\}\right\}$$

is infinite. It follows that $\{k \in \mathbb{N} : (f_1(k), \ldots, f_n(k)) \in U\}$ is infinite. □

Theorem 5.23 *Let K be a real-closed field of cardinality $\mathfrak{c}$. Then there is a free ultrafilter $\mathcal{U}$ on $\mathbb{N}$ such that there is an embedding of K^0 into $c_0/\mathcal{U}$.*

Proof Let g_0 and S be as specified in Lemma 5.22.

Take B to be a transcendence base for K over $\mathbb{R}$ with $B \subset K^{0+}$, so that $|B| \le \mathfrak{c}$ and there is an injection $\pi : B \to S \subset c_0$.

For $m, n \in \mathbb{N}$, define

$$\begin{aligned}\mathcal{E}_{m,n} = \{(p_1, \ldots, p_m; b_1, \ldots, b_n) &: p_1, \ldots, p_m \in \mathbb{R}[X_1, \ldots, X_n] ,\\ b_1, \ldots, b_n \in B, &\quad \text{and} \ \ p_i(b_1, \ldots, b_n) > 0 \quad (i = 1, \ldots, m)\} ,\end{aligned}$$

and define $\mathcal{E} = \bigcup\{\mathcal{E}_{m,n} : m, n \in \mathbb{N}\}$. The field K is uniquely determined (up to isomorphism) by B and $\mathcal{E}$.

Take $p = (p_1, \ldots, p_m) \in (\mathbb{R}[X_1, \ldots, X_n])^m$ such that

$$(p; b) = (p_1, \ldots, p_m; b_1, \ldots, b_n) \in \mathcal{E}_{m,n}$$

for some $b_1, \ldots, b_n \in B$, and set

$$U_p = \{x \in \mathbb{R}^n : p_i(x) > 0 \quad (i = 1, \ldots, m)\} ,$$

so that U_p is an open set in $\mathbb{R}^n$. We *claim* that $0 \in \overline{U}_p$. For fix $\varepsilon > 0$, set $L = \mathcal{R}(\mathbb{R}(b_1, \ldots, b_n))$, and let $\mathcal{V}$ be a free ultrafilter on $\mathbb{N}$. By Propositions 2.26 and 2.27, L is an α_1-subfield of K, and, by Theorem 4.29, $\mathbb{R}^{\mathbb{N}}/\mathcal{V}$ is a real-closed η_1-field, and so, by Theorem 2.30, there is an embedding ι of L into $\mathbb{R}^{\mathbb{N}}/\mathcal{V}$. Choose $f_1, \ldots, f_n \in \mathbb{R}^{\mathbb{N}}$ such that

$$\iota(b_j) = [f_j]_{\mathcal{V}} \quad (j = 1, \ldots, n) .$$

Since $b_1, \ldots, b_n \in K^0$, we have $\iota(b_1), \ldots, \iota(b_n) \in (\mathbb{R}^{\mathbb{N}}/\mathcal{V})^0$, and so we may suppose that

$$|f_j(k)| < \varepsilon \quad (j = 1, \ldots, n, \ k \in \mathbb{N}) .$$

We have

$$[p_i(f_1,\dots,f_n)]_{\mathcal{V}} > 0 \quad \text{in} \quad \mathbb{R}^{\mathbb{N}}/\mathcal{V}$$

for $i = 1,\dots,m$, and so there exists $k \in \mathbb{N}$ with

$$p_i(f_1(k),\dots,f_n(k)) > 0 \quad (i = 1,\dots,m).$$

Set $x = (f_1(k),\dots,f_n(k)) \in \mathbb{R}^n$. Then

$$x \in U \cap B_n(0;\varepsilon),$$

and the claim follows.

Now take $(p;b) = (p_1,\dots,p_m;b_1,\dots,b_n) \in \mathcal{E}_{m,n}$, and set

$$F_{(p;b)} = \{k \in \mathbb{N} : (\pi(b_1)(k),\dots,\pi(b_n)(k)) \in U_p\}.$$

The intersection of two sets so obtained has the same form: to see this, take $(p;b) \in \mathcal{E}_{m_1,n_1}$ and $(q;c) \in \mathcal{E}_{m_2,n_2}$, regard the polynomials p_i and q_j as elements of $\mathcal{R}[X_1,\dots,X_{m_1+m_2}]$ by neglecting the first m_1 and last m_2 variables, respectively, and set

$$F = \{k \in \mathbb{N} : (\pi(b)(k), \pi(c)(k)) \in (U_p \times \mathbb{R}^{m_2}) \cap (\mathbb{R}^{m_1} \times U_q)\},$$

so that $F = F_{(p;b)} \cap F_{(q;c)}$.

By the key property of S in Lemma 5.22, each set $F_{(p;b)}$ is infinite, and so the family of such sets is contained in a free ultrafilter, say $\mathcal{U}$, on $\mathbb{N}$. The map $\pi : B \to S$ induces an embedding of K^0 into $c_0/\mathcal{U}$. □

The next lemma will require some elementary results from algebraic geometry, which we first summarize.

Let I be an ideal in $\mathbb{C}[X_1,\dots,X_n]$. Then

$$\mathcal{V}(I) = \{z = (z_1,\dots,z_n) \in \mathbb{C}^n : p(z) = 0 \quad (p \in I)\}$$

is the *variety* of I. Let E be a subset of $\mathbb{C}^n$. Then

$$\mathcal{I}(E) = \{p \in \mathbb{C}[X_1,\dots,X_n] : p(z) = 0 \quad (z \in E)\}$$

is the *ideal* of E. A variety V is *reducible* if there are varieties V_1 and V_2 of some ideals in $\mathbb{C}[X_1,\dots,X_n]$ such that $V_1, V_2 \neq V$ and $V = V_1 \cup V_2$; a variety which is not reducible is *irreducible*. By the Hilbert basis theorem, $\mathbb{C}[X_1,\dots,X_n]$ is Noetherian; it follows that every variety is a finite union of irreducible varieties. By the Hilbert Nullstellensatz, for each ideal I in $\mathbb{C}[X_1,\dots,X_n]$,

$$\mathcal{I}(\mathcal{V}(I)) = \{p \in \mathbb{C}[X_1,\dots,X_n] : p^k \in I \ \text{for some} \ k \in \mathbb{N}\},$$

and so $\mathcal{V}(I)$ is irreducible if and only if I is a prime ideal in $\mathbb{C}[X_1,\dots,X_n]$, and there is a bijective correspondence between irreducible varieties and prime ideals in $\mathbb{C}[X_1,\dots,X_n]$.

Let V be an irreducible variety in $\mathbb{C}^n$, and let U be an open set in $\mathbb{C}^n$ such that $U \cap V \neq \emptyset$. It is standard that, if $p \in \mathbb{C}[X_1,\dots,X_n]$ and $p \mid (U \cap V) = 0$, then $p \in \mathcal{I}(V)$: this is the *identity theorem for irreducible varieties*.

Theorem 5.24 *Let A be a non-unital, complex integral domain. Then there is a cardinal κ and a free ultrafilter $\mathcal{U}$ on κ such that there is an embedding of A into $(\mathbb{C}^\kappa/\mathcal{U})^0$.*

Proof Let J be the collection of all sequences of the form

$$s = (\delta; a_1, \dots, a_n), \tag{5.3}$$

where $\delta > 0$ and $(a_1, \dots, a_n)$ is an ordered set of distinct elements of $A \setminus \{0\}$.

Fix an element s in J of the form (5.3); we shall associate with s a point $z \in \mathbb{C}^n$. Indeed, define

$$P_s = \{p \in \mathbb{C}[X_1, \dots, X_n] : p(a_1, \dots, a_n) = 0\} .$$

Then P_s is an ideal in $\mathbb{C}[X_1, \dots, X_n]$, and P_s is a prime ideal because A is an integral domain. Set $V_s = \mathcal{V}(P_s)$. Since A is non-unital, necessarily $0 \in V_s$. Take W to be the union of the coordinate axes in $\mathbb{C}^n$. It cannot be that there is a neighbourhood U of 0 in $\mathbb{C}^n$ such that $V_s \cap U \subset W \cap U$, for, if there were such a neighbourhood, it would follow from the identity theorem for irreducible varieties that the polynomial $X_1 \cdots X_n$ belonged to P_s, and this is not the case because $a_1 \cdots a_n \neq 0$ in A. Thus there exists a point $z = (z_1, \dots, z_n)$ in $(V_s \cap B_n(0; \delta)) \setminus W$, and hence such that

$$p(z_1, \dots, z_n) = 0 \quad (p \in P_s) \quad \text{and} \quad 0 < |z_j| < \delta \quad (j = 1, \dots, n) .$$

It is this point $z \in \mathbb{C}^n$ which corresponds to s.

We now define certain subsets of J. For $\varepsilon > 0$ and a finite subset $\mathbf{b}$ of $A \setminus \{0\}$,

$$\sigma(\varepsilon, \mathbf{b}) = \{s = (\delta; a_1, \dots, a_n) \in J : \mathbf{b} \subset \{a_1, \dots, a_n\}\} .$$

Clearly the family of sets $\sigma(\varepsilon, \mathbf{b})$ has the finite intersection property because

$$\sigma(\varepsilon, \mathbf{b} \cup \mathbf{c}) \subset \sigma(\varepsilon_1, \mathbf{b}) \cap \sigma(\varepsilon_2, \mathbf{c}) ,$$

where $\varepsilon = \min\{\varepsilon_1, \varepsilon_2\}$. Also, given $s \in J$, there are members of the family that do not contain s.

We regard the set J as a cardinal κ, and then the family of sets of the form $\sigma(\varepsilon; \mathbf{b})$ is contained in a free ultrafilter, say $\mathcal{U}$, on κ.

The next step is to define a map $\theta : A \to \mathbb{C}^\kappa$. Take $a \in A$ and $s \in J$ of the form (5.3). Suppose that a appears in the list $(a_1, \dots, a_n)$ as the element a_i. Then we set

$$\theta(a)(s) = z_i ,$$

where $z = (z_1, \dots, z_n)$ corresponds to s. (The element a appears at most once in the list $(a_1, \dots, a_n)$.) If a does not appear in the list, then we set

$$\theta(a)(s) = 0 .$$

Note that, if $a = 0$, then $\theta(a) = 0$ in $\mathbb{C}^\kappa$. The equivalence class in $\mathbb{C}^\kappa/\mathcal{U}$ containing $\theta(a)$ is

$$[\theta(a)]_{\mathcal{U}} = \pi(a) .$$

Take $a \in A$ and $\alpha \in \mathbb{C}$. We *claim* that $\pi(\alpha a) = \alpha\pi(a)$. We may suppose that $a \in A \setminus \{0\}$ and $\alpha \neq 0, 1$, so that a and αa are distinct elements of $A \setminus \{0\}$. Consider the set $\sigma = \sigma(1, \{a, \alpha a\})$. For $s = (\delta; (a_1, \ldots, a_n)) \in \sigma$, there exist distinct integers i, j in $\{1, \ldots, n\}$ with $a_i = a$ and $a_j = \alpha a$. The polynomial $p = \alpha X_i - X_j$ belongs to P_s, and so $\alpha z_i = z_j$, where $(z_1, \ldots, z_n)$ corresponds to s. Thus $\alpha\theta(a) \,|\sigma = \theta(\alpha a)|\,\sigma$, and so $\pi(\alpha a) = \alpha\pi(a)$, as claimed.

Now take $a \in A$. We *claim* that $\pi(a^2) = \pi(a)^2$. We may suppose that $a \neq 0$, and so $a \neq a^2$ because A is a non-unital integral domain, and then the above argument applies, using the polynomial $X_i^2 - X_j$ for suitable i and j.

The next *claim* is that $\pi(a + b) = \pi(a) + \pi(b)$ for $a, b \in A$. We may suppose that a, b, and $a + b$ are distinct elements of $A \setminus \{0\}$, and then the above argument applies, using the polynomial $X_i + X_j - X_k$ for suitable i, j, and k.

Finally, we *claim* that, for $a \in A \setminus \{0\}$, we have $\pi(a) \in (\mathbb{C}^\kappa/\mathcal{U})^0 \setminus \{0\}$. Indeed, take $\varepsilon > 0$. Then, for $s \in \sigma(\varepsilon; \{a\})$, we have $\theta(a)(s) = z_i$ for some $z_i \in \mathbb{C}$ with $0 < |z_i| < \varepsilon$, giving the result.

We conclude that $\pi : A \to (\mathbb{C}^\kappa/\mathcal{U})^0$ is the required embedding. □

We can now prove the universality of the non-unital integral domains $c_0(\mathbb{C})/\mathcal{U}$ by combining the above results. It is important to note that our result is a theorem of ZFC.

Theorem 5.25 *The class of integral domains of the form $c_0(\mathbb{C})/\mathcal{U}$, where $\mathcal{U}$ is a free ultrafilter on $\mathbb{N}$, is universal in the class of all non-unital, complex integral domains of cardinality $\mathfrak{c}$.*

Proof Let A be a non-unital, complex integral domain of cardinality $\mathfrak{c}$. By Theorem 5.24, there is a cardinal κ, a free ultrafilter $\mathcal{V}$ on κ, and an embedding $\pi : A \to (\mathbb{C}^\kappa/\mathcal{V})^0$. Choose a real-closed field $K \subset (\mathbb{R}^\kappa/\mathcal{V})^0$ with $|K| = \mathfrak{c}$ such that $\pi(A) \subset K^0 + \mathrm{i}K^0$. By Theorem 5.23, there is a free ultrafilter $\mathcal{U}$ on $\mathbb{N}$ and an embedding $\iota : K^0 \to c_0/\mathcal{U}$, and ι gives an embedding $\iota : K^0 + \mathrm{i}K^0 \to c_0(\mathbb{C})/\mathcal{U}$. The composition $\iota \circ \pi : A \to c_0(\mathbb{C})/\mathcal{U}$ is an embedding. □

Theorem 5.26 (CH) *(Esterle) Let A be a complex integral domain of cardinality $\mathfrak{c}$.*

(i) *Suppose that A is non-unital. Then A is normable.*

(ii) *Suppose that A is unital. Then A is normable if and only if there is a character on A.*

Proof (i) This follows from Theorems 5.6 and 5.25 and Corollary 4.30.

(ii) Suppose that the unital algebra A has a character φ. By (i), the algebra $\ker\varphi$ is normable, and so $A = \mathbb{C}e \oplus \ker\varphi$ is normable.

Conversely, it is a standard fact of elementary Banach algebra theory that a unital normed algebra has a character. Indeed, suppose that $(A, \|\cdot\|)$ is a normed algebra with completion B. Then B contains a maximal ideal, say M. The ideal M is the kernel of a character, say φ, on B. But then $\varphi \mid A$ is a character on A. □

Let K be a complex field with $|K| = \mathfrak{c}$, but with $K \not\cong \mathbb{C}$ (for example, take $K = \mathbb{C}((X))$). Then K is a unital integral domain of cardinality $\mathfrak{c}$ which is not normable.

It seems almost clear from the above results and from Corollary 2.33 that, with CH, any one algebra $M_P(\mathbb{C})$ is universal for the class of non-unital, complex integral domains of cardinality $\mathfrak{c}$. A little care is required, however, because this result requires that we embed an arbitrary algebra of the class into $M_P(\mathbb{C})$, and not just into $K_P(\mathbb{C})^0$. To prove the result, we require a preliminary lemma about semi-η_1-fields.

Lemma 5.27 *Let K be a real-closed semi-η_1-field, and let H be an η_1-group which is a divisible subgroup of Γ_K. Then there is an embedding $\theta : \mathbf{R} \to K$ such that $v(\theta(a)) \in H \quad (a \in \mathbf{R} \setminus \{0\})$.*

Proof By Theorem 2.36, we may set $\mathbf{R} = \bigcup\{\mathbf{R}^{(\sigma)} : \sigma < \rho_0\}$, where $\mathbf{R}^{(0)} = \mathbb{R}$, $\mathbf{R}^{(\sigma+1)} = \mathcal{R}(\mathbf{R}^{(\sigma)}(a_\sigma))$ for some $a_\sigma \in \mathbf{R} \setminus \mathbf{R}^{(\sigma)}$ such that a_σ defines a countable gap in $\mathbf{R}^{(\sigma)}$, and $\mathbf{R}^{(\sigma)} = \bigcup\{\mathbf{R}^{(\tau)} : \tau < \sigma\}$ for limit ordinals σ. Write Γ_σ for the value group of $\mathbf{R}^{(\sigma)}$. We define θ inductively.

Indeed, assume that θ has been defined on $\mathbf{R}^{(\sigma)}$, and let the gap in $\mathbf{R}^{(\sigma)}$ defined by a_σ be $\langle A, B\rangle$. There are two cases, according to whether $\langle A, B\rangle$ is or is not a pseudo-gap.

Case 1 $\langle A, B\rangle$ is a pseudo-gap.

By Corollary 2.18, we have $\Gamma_{\sigma+1} = \Gamma_\sigma$; by the remark following Definition 1.30, $\langle A, B\rangle$ is a (κ, λ)-gap in $\mathbf{R}^{(\sigma)}$ with both κ and λ infinite, and so $\langle\theta(A), \theta(B)\rangle$ is an $(\aleph_0, \aleph_0)$-pregap in K. Since K is a semi-η_1-field, there exists $x \in K$ with $\theta(A) << x << \theta(B)$. Set $\theta(a_\sigma) = x$, and extend θ to $\mathbf{R}^{(\sigma+1)}$ in the usual way.

Case 2 $\langle A, B\rangle$ is not a pseudo-gap.

We may suppose that

$$A \cup B \subset \mathbf{R}^{(\sigma)+}.$$

By the definition of a pseudo-gap, there exists $c \in \mathbf{R}^{(\sigma)}$ such that, for each $a \in A$ and $b \in B$, v is not constant on $[a-c,\, b-c]$. The gap $\langle A - c,\, B - c\rangle$ in $\mathbf{R}^{(\sigma)}$ then defines a gap $\langle v(B - c),\, v(A - c)\rangle$ in H, and so the pregap $\langle\theta(A - c),\, \theta(B - c)\rangle$ in K defines a pregap, say $\langle C, D\rangle$, in Γ_K; $\langle C, D\rangle$ is actually contained in H. Since $\langle A, B\rangle$ is a countable gap, $\langle C, D\rangle$ is a countable gap in H, and so, since H is an η_1-set, there exists $s \in H$ with $C << s << D$. Choose $x \in K^+$ with $v(x) = s$. Then

$$\theta(A) << x + \theta(c) << \theta(B)$$

in K. Set $\theta(a_\sigma) = x + \theta(c)$, and again extend θ to $\mathbf{R}^{(\sigma+1)}$. Since H is a divisible subgroup of Γ_K, we see that $v(\theta(a)) \in H \quad (a \in \mathbf{R}^{(\sigma+1)} \setminus \{0\})$.

Suppose that $\sigma < \rho_0$ is a limit ordinal. Define $\theta : \mathbf{R}^{(\sigma)} \to K$ in the obvious way to extend all the maps $\theta : \mathbf{R}^{(\tau)} \to K$ for $\tau < \sigma$.

It follows that the required embedding $\theta : \mathbf{R} \to K$ exists. □

Theorem 5.28 *Let Ω be a compact space, and let P be a non-maximal, prime ideal in $C(\Omega, \mathbb{C})$. Then :*

(i) *there is an embedding of $\mathbf{C}^0$ in $M_P(\mathbb{C})$;*

(ii) (CH) *$M_P(\mathbb{C})$ is universal in the class of non-unital, complex integral domains of cardinality $\mathfrak{c}$.*

Proof (i) By Theorem 4.41, there is a non-modular, prime ideal Q in c_0 such that $c_0 \subset Q$ and such that there is an embedding of c_0/Q into M_P, and so there is an embedding $\pi : c_0(\mathbb{C})/Q_{\mathbb{C}} \to M_P(\mathbb{C})$, where

$$Q_{\mathbb{C}} = \{f \in C(\Omega, \mathbb{C}) : |f| \in C(\Omega)\}.$$

As usual, K_Q is the quotient field of $M_Q = c_0/Q$, and Γ_Q is its value group.

By Proposition 4.36, M_Q is an interval in K_Q. Choose $a_0 \in M_Q^+ \setminus \{0\}$, and set $x_0 = v_Q(a_0) \in \Gamma_Q$. Choose $y_0 \in \Gamma_Q$ with $nx_0 < y_0 \quad (n \in \mathbb{N})$, and set

$$G = \{x \in \Gamma_Q : n|x| < y_0 \quad (n \in \mathbb{N})\}.$$

Then G is a divisible subgroup of Γ_Q, G is a β_1-group because $|\Gamma_Q| = \aleph_1$, and G is an η_1-set because K_Q is a semi-η_1-field by Corollary 4.38. By Corollary 1.49, we may identify G with $\mathbf{G}$, and, by Proposition 1.51, there is a copy H of $\mathbf{G}$ in Γ_Q with $H \cap [-x_0, x_0] = \{0\}$.

By Lemma 5.27, there is an embedding $\theta : \mathbf{R} \to K_Q$ such that

$$v_Q(\theta(a)) \in H \quad (a \in \mathbf{R}).$$

In particular, $v_Q(\theta(a)) > x_0$ for $a \in \mathbf{R}^0 \setminus \{0\}$, and so $\theta(\mathbf{R}^0) \subset M_Q$. Thus there is an embedding $\theta : \mathbf{C} \to K_Q(\mathbb{C})$ with $\theta(\mathbf{C}^0) \subset M_Q(\mathbb{C})$.

The map $\pi \circ \theta : \mathbf{C}^0 \to M_P(\mathbb{C})$ is the required embedding.

(ii) Let A be a non-unital, complex integral domain of cardinality $\mathfrak{c}$. By Theorem 5.25, there is a free ultrafilter $\mathcal{U}$ on $\mathbb{N}$ and an embedding

$$\psi : A \to c_0(\mathbb{C})/\mathcal{U} \subset (\mathbb{C}^{\mathbb{N}}/\mathcal{U})^0.$$

By Corollary 4.30, there is an isomorphism $\iota : (\mathbb{C}^{\mathbb{N}}/\mathcal{U})^0 \to \mathbf{C}^0$. Let $\pi \circ \theta$ be as in (i). Then the map $\pi \circ \theta \circ \iota \circ \psi : A \to M_P(\mathbb{C})$ is the required embedding. □

Notes 5.29 The relation between the normability of a real algebra A and its complexification $A_{\mathbb{C}}$ is given in Rickart (1960, 1.3.2). The Gelfand–Mazur theorem for Banach algebras is Bonsall and Duncan (1973, §14) and Rudin (1991, 10.14); the spectral radius formula is Bonsall and Duncan (1973, 5.8) and Rudin (1991, 10.13b).

The first substantial study of Kaplansky's problem after the original paper of Kaplansky (1949) was given by Bade and Curtis (1960), who analysed the structure of an arbitrary homomorphism from an algebra $C(\Omega, \mathbb{C})$ and showed that it is necessarily continuous on a dense subalgebra of $C(\Omega, \mathbb{C})$. The reformulation of the problem as condition (c) of Theorem 5.2 is due to Sinclair (1975); see also Esterle (1978*a*) and Johnson (1976). For an account of this work, see Dales and Woodin (1987, Chapter 1) and Dales (1997).

Constructions of discontinuous homomorphisms from $C(\Omega, \mathbb{C})$ for each infinite compact space Ω were given by Dales (1979) and by Esterle (1978*b*). The key step in Esterle's proof of this result is his proof, in (1978*b*), of Theorem 5.6. The proof that it is relatively consistent with ZFC that all homomorphisms from each algebra $C(\Omega, \mathbb{C})$ are continuous was first given by Solovay in 1976, using a condition of Woodin. Shortly afterwards a different and easier approach to the theorem was developed by Woodin and included in his Berkeley thesis (which has not been published); a full account of this work, intended to be accessible to analysts, is given by Dales and Woodin (1987), and an account for logicians is given by (Jech 1987).

It is not known whether or not all infinite compact spaces Ω are equivalent for Kaplansky's problem; this is so if and only if the existence of a discontinuous homomorphism from $c_0(\mathbb{C})$ into a Banach algebra entails the existence of such a homomorphism from $l^\infty(\mathbb{C})$. This question, and several related questions, are raised in Dales and Woodin (1987, Chapter 3); we are not aware of any progress on these questions.

Theorem 5.3 and Proposition 5.5 are essentially Esterle (1978*a*, 6.4(c) and 7.1), the latter extending Gillman and Jerison (1960, 12.7). Esterle's proof of Theorem 5.6 is completed in Esterle (1978*b*, Théorème 4.1); the algebra C_{ω_1} of that reference is the algebra $\mathbf{C}^\# = \mathfrak{F}_{(1)}(\mathbb{C}, \mathbf{G})^\#$ of our notation.

We proved in Proposition 5.14 that (with GCH) $\widetilde{\mathbf{R}}$ is an example of a real-closed η_1-field K with $|K| = \aleph_2$ and $|\Gamma_K| = \aleph_1$. The existence of such a field K also follows from Chang's two-cardinal theorem (Chang and Kiesler 1990, Theorem 7.2.7).

For a proof of the Tarski–Seidenberg theorem, see Kreisel and Krivine (1967). The results about semi-algebraic sets used in the proof of Theorem 5.23 can be found in Brumfiel (1979) and Benedetti and Risler (1990), for example.

We showed in Corollary 4.30 that, with CH, $\mathbb{R}^\mathbb{N}/\mathcal{U}$ is a β_1-field and hence $\mathbb{R}^\mathbb{N}/\mathcal{U} \cong \mathbf{R}$ for each free ultrafilter $\mathcal{U}$ on $\mathbb{N}$. The situation in the case where CH does not hold is the following. If CH fails, it is a theorem of Dow (1984) that there exist two free ultrafilters $\mathcal{U}$ and $\mathcal{V}$ on $\mathbb{N}$ such that $\mathbb{R}^\mathbb{N}/\mathcal{U}$ and $\mathbb{R}^\mathbb{N}/\mathcal{V}$ are not isomorphic, and so there exists a free ultrafilter $\mathcal{U}$ on $\mathbb{N}$ such that $\mathbb{R}^\mathbb{N}/\mathcal{U}$ is not isomorphic to $\mathbf{R}$, and hence such that $\mathbb{R}^\mathbb{N}/\mathcal{U}$ is not a β_1-field. In the theory ZFC + MA $+\neg$CH (where MA is Martin's Axiom), there is no ultrafilter $\mathcal{U}$ on $\mathbb{N}$ such that $\mathbb{R}^\mathbb{N}/\mathcal{U}$ is isomorphic to $\mathbf{R}$ (Dales and Woodin 1987, Corollary 6.28). Nevertheless it is possible to have CH false and for there to exist an ultrafilter $\mathcal{U}$ with $\mathbb{R}^\mathbb{N}/\mathcal{U} \cong \mathbf{R}$ (Woodin 1993). In fact, in this theorem, $\mathfrak{c} = \aleph_2$; it is not known whether it is possible to have $\mathfrak{c} \geq \aleph_3$ in such an example. This result shows that there are models in which CH is false, but such that there are discontinuous homomorphisms from $C(\Omega, \mathbb{C})$ for each infinite compact space Ω.

Details of the elementary facts about algebraic geometry in $\mathbb{C}^n$ that we stated can be found in many texts, including Kendig (1977) and Zariski and Samuel (1960). The identity theorem for algebraic varieties is Kendig (1977, IV.2.II).

The weaker form of Theorem 5.25 is that the class of integral domains of the form $(\mathbb{C}^{\mathbb{N}}/\mathcal{U})^0$, where $\mathcal{U}$ is a free ultrafilter on $\mathbb{N}$, is universal in the class of all non-unital, complex integral domains of cardinality $\mathfrak{c}$. It can also be shown that the family of algebras $(\mathbb{C}^{\aleph_1}/\mathcal{U})^0$ for $\mathcal{U}$ a uniform ultrafilter on $\aleph_1$ is universal in the class of all non-unital, complex integral domains of cardinality $2^{\aleph_1}$. With CH, each fixed algebra $(\mathbb{C}^{\mathbb{N}}/\mathcal{U})^0$ is universal in the class of non-unital, complex integral domains of cardinality $\mathfrak{c}$. However, it is not true that, even with GCH, all the algebras $(\mathbb{C}^{\aleph_1}/\mathcal{U})^0$ are isomorphic.

Theorem 5.26 was explicitly proved by Esterle (1979, Corollaire 5.3). His proof first uses Theorem 2.13 to embed A in a complex valuation algebra B with quotient field K, and then uses the complex form of Kaplansky's embedding theorem to embed K in $\mathbf{C}$. Thus Esterle's route to 5.26 is more direct than our route through Theorem 5.23.

A preliminary version of some of the results of this chapter was given in Dales (1989).

6
The operational calculus and the field $\widehat{\mathbf{R}}$

There have been two major themes in our work so far. The first, discussed in Chapter 4, was to study the relationships between various classes of real-closed η_1-fields, and in particular of super-real η_1-fields. A diagram showing the relations that are almost immediately obvious was drawn on page 104. The second theme, from Chapter 5, was to wonder for which super-real fields K_P is the algebra $K_P^{\#}$ normable (in the theory ZFC + GCH); the question was resolved for all fields save those with $|K_P| = \aleph_2$ and $|\Gamma_P| = \aleph_1$, and it was left open whether there are super-real fields satisfying this combination of cardinalities.

The first natural candidate for such a missing field is the η_1-field $\widehat{\mathbf{R}}$ (see Corollary 5.12). The main result of the present chapter is that $\widehat{\mathbf{R}}$ is *not* a super-real field. A related result is that $\widehat{\mathbf{R}}$ is not isomorphic to a super-real field K_P as an ordered group for any exponential prime ideal and, in particular, for any prime z-ideal P.

The second natural candidate for such a missing field is the η_1-field $\widetilde{\mathbf{R}}$ (see Proposition 5.14); it is much more difficult to place this field in the hierarchy of classes, and this will be left to Chapter 9.

A proof that $\widehat{\mathbf{R}}$ is not a super-real field, and, more generally, a proof that distinguishes two classes of the scheme of Chapter 4 will follow from the discovery of an *invariant*; this is a property that is possessed by all members of the smaller class, which is preserved under isomorphisms (of fields, of totally order groups, or of ordered sets, as appropriate), but which some member of the larger class does not possess. The first invariant of ordered fields that we shall consider in this chapter is that of having an exponentiation on a convex subgroup of the field; this invariant is preserved under isomorphisms of ordered fields. The second invariant that we shall consider here applies to totally ordered groups: we consider, for a totally ordered group $(G, +, \leq)$, whether or not $w(G) = |\Gamma_G|$.

The proofs that we shall give in this chapter depend on the construction of an *operational calculus* for super-real fields. This operational calculus will also be used later in the book, and we develop it here in the generality that will be required eventually.

Let K be an ordered field, and set

$$\mathbb{I}_K = \{a \in K : 0 \leq a \leq 1\},$$

so that $\mathbb{I}_K$ is the closed unit interval of K. Take $a \in \mathbb{I}_K$. In the case where $p \in \mathbb{R}[X]$, the element $p(a)$ is defined in K. Further, if $r = p/q \in \mathbb{R}(X)$ is a rational function, then $r(a)$ is defined in K whenever $q(a) \neq 0$. An operational calculus is an extension of this notion; we seek to define $F(a)$ for a general function $F \in C(\mathbb{I})$ in such a way that we generalize the above definitions of $p(a)$ and $r(a)$, and so that the map

$$F \mapsto F(a), \quad C(\mathbb{I}) \to K,$$

is a unital homomorphism. In general, we cannot obtain such a strong operational calculus, even in the case where K is a super-real field, and so we must restrict to subalgebras of $C(\mathbb{I})$. Thus we have the following definition. The *coordinate functional* Z is the function defined by the rule

$$Z : t \mapsto t, \quad \mathbb{I} \to \mathbb{I}.$$

Definition 6.1 *Let* $\mathfrak{A}$ *be a unital subalgebra of* $C(\mathbb{I})$ *containing* Z*, and let* K *be an ordered field. An* $\mathfrak{A}$-operational calculus *for* K *is a family* $\{\Theta_a : a \in \mathbb{I}_K\}$ *of unital homomorphisms from* $\mathfrak{A}$ *into* $K^{\#}$ *such that*

$$\Theta_a(Z) = a \quad (a \in \mathbb{I}_K).$$

We shall develop an operational calculus for super-real fields. Thus, let Ω be a compact space, and let P be a non-maximal, prime ideal in $C(\Omega)$. As in Chapter 4, we set $A_P = C(\Omega)/P$, and we take K_P to be the super-real field which is the quotient of A_P, so that K_P is a real-closed field. We define

$$\mathbb{I}_{(P)} = \{a \in K_P : 0 \leq a \leq 1\}.$$

The field K_P and its subsets, including the subinterval $\mathbb{I}_{(P)}$, are considered as topological spaces with respect to the order topology. Thus we have acquired the notion of a continuous function from a subset of K_P into a topological space. Recall from Definition 3.13 that a function $H : \mathbb{I}_{(P)} \to K_P$ is *uniformly continuous* if, for each $\varepsilon > 0$ in K_P, there exists $\delta > 0$ in K_P such that $|H(a) - H(b)| < \varepsilon$ whenever $a, b \in \mathbb{I}_{(P)}$ with $|a - b| < \delta$. By Theorem 4.21(ii), $A_P^+ \setminus \{0\}$ is coinitial in $K_P^+ \setminus \{0\}$, and so it suffices to suppose that $\varepsilon \in A_P^+ \setminus \{0\}$. Note that there are continuous functions $H : \mathbb{I}_{(P)} \to K_P$ which are not uniformly continuous.

In the case of a general prime ideal P, we shall define an operational calculus for an algebra of Lipschitz functions on $\mathbb{I}$.

Definition 6.2 *Let* $\alpha \in (0, 1]$. *A function* $F : \mathbb{I} \to \mathbb{R}$ *belongs to* $\mathrm{Lip}_\alpha \mathbb{I}$ *if the set*

$$\left\{ \frac{|F(s) - F(t)|}{|s - t|^\alpha} : s, t \in \mathbb{I}, s \neq t \right\}$$

is bounded. A function F *on* $\mathbb{I}$ *is* Lipschitz *if* $F \in \mathrm{Lip}_\alpha \mathbb{I}$ *for some* $\alpha \in (0, 1]$.

We denote the set of these Lipschitz functions by $\mathfrak{L}(\mathbb{I})$, so that

$$\mathfrak{L}(\mathbb{I}) = \bigcup \{\mathrm{Lip}_\alpha \mathbb{I} : 0 < \alpha \leq 1\}.$$

For each $\alpha \in (0,1]$, the space $\mathrm{Lip}_\alpha \mathbb{I}$ is a Banach algebra with respect to the norm $\|\cdot\|_\alpha$, where

$$\|F\|_\alpha = |F|_{\mathbb{I}} + \sup\left\{\frac{|F(s)-F(t)|}{|s-t|^\alpha} : s,t \in \mathbb{I}, s \neq t\right\} \quad (F \in \mathrm{Lip}_\alpha \mathbb{I}).$$

Clearly $\mathfrak{L}(\mathbb{I})$ is a unital subalgebra of $C(\mathbb{I})$ containing the coordinate functional Z. However $\mathfrak{L}(\mathbb{I}) \neq C(\mathbb{I})$ because the function $t \mapsto L(t/2)$, where L was defined in equation (4.9), belongs to $C(\mathbb{I})$, but not to $\mathfrak{L}(\mathbb{I})$.

As a preliminary to our main result, we define an $\mathfrak{L}(\mathbb{I})$-operational calculus for elements $a \in A_P \cap \mathbb{I}_{(P)}$.

Lemma 6.3 *Let Ω be a compact space, let P be a prime ideal in $C(\Omega)$, and let $F \in \mathfrak{L}(\mathbb{I})$. Suppose that $f, g \in C(\Omega, \mathbb{I})$ with $f - g \in P$. Then $F \circ f - F \circ g \in P$.*

Proof There exist $n \in \mathbb{N}$ and $M > 0$ such that

$$|F(s) - F(t)| \leq M\,|s-t|^{1/n} \quad (s,t \in \mathbb{I}).$$

We have

$$|(F\circ f)(x) - (F \circ g)(x)|^n \leq M^n\,|f(x) - g(x)| \quad (x \in \Omega),$$

and so $|F \circ f - F \circ g|^n \leq M^n\,|f-g|$ in $C(\Omega)$. It follows from Proposition 4.7(ii) that $F \circ f - F \circ g \in P$. □

Let $a \in A_P \cap \mathbb{I}_{(P)}$ and $F \in \mathfrak{L}(\mathbb{I})$. Choose $f \in C(\Omega, \mathbb{I})$ with $\pi_P(f) = a$; the above lemma shows that the element $\pi_P(F \circ f)$ of A_P is independent of the choice of f, and so $\pi_P(F \circ f)$ is a natural definition of the element $F_{(P)}(a)$ in A_P. The following result is clear.

Proposition 6.4 *Let P be a prime ideal in $C(\Omega)$, and let $a \in A_P \cap \mathbb{I}_{(P)}$. Then the map*

$$\Theta_a : F \to F_{(P)}(a), \quad \mathfrak{L}(\mathbb{I}) \to K_P^\#$$

is a unital homomorphism with $\Theta_a(Z) = a$. □

We must generalize the above construction to apply to all elements $a \in \mathbb{I}_P$. We should also like to generalize the definition so that it applies to each function $F \in C(\mathbb{I})$, rather than just to Lipschitz functions; however, for this to succeed, we must restrict consideration to the case where P is z-ideal in $C(\Omega)$.

Let Ω be a compact space, let P be a non-maximal, prime ideal in $C(\Omega)$, let $F \in C(\mathbb{I})$, and let $a \in \mathbb{I}_{(P)}$. The recipe for the definition of $F_{(P)}(a)$ is as follows. There exist $f \in C(\Omega)^+$ and $g \in C(\Omega)^+ \setminus P$ with $\pi_P(f) = a\pi_P(g)$ in K_P and $f(x) \leq g(x) \quad (x \in \Omega)$. Define

$$h(x) = F\left(\frac{f(x)}{g(x)}\right) \quad (x \in \Omega \setminus \mathbf{Z}(g)),$$

so that $h \in C^b(\Omega \setminus \mathbf{Z}(g))$. The function gh, defined to be 0 on $\mathbf{Z}(g)$, belongs to $C(\Omega)$. Define

$$F_{(P)}(a) = \pi_P(gh)/\pi_P(g) \in K_P^\# . \tag{6.1}$$

In general, this definition of $F_{(P)}(a)$ depends on the choice of f and g; we shall establish cases in which $F_{(P)}(a)$ is well defined. Our results on $F_{(P)}(a)$ will be easy consequences of the following calculation.

Take $a_1, a_2 \in \mathbb{I}_{(P)}$, and choose $f_1, f_2 \in C(\Omega)^+$ and $g_1, g_2 \in C(\Omega)^+ \setminus P$ with $\pi_P(f_1) = a_1\pi_P(g_1)$ and $\pi_P(f_2) = a_2\pi_P(g_2)$ in K_P and such that

$$f_1(x) \leq g_1(x), \quad f_2(x) \leq g_2(x) \quad (x \in \Omega).$$

For $F \in C(\mathbb{I})$, set $h_1 = F \circ (f_1/g_1)$ on $\Omega \setminus \mathbf{Z}(g_1)$ and $h_2 = F \circ (f_2/g_2)$ on $\Omega \setminus \mathbf{Z}(g_2)$, and set $k_1 = g_1h_1$ and $k_2 = g_2h_2$, so that $k_1, k_2 \in C(\Omega)$. Now suppose that $F \in \mathfrak{L}(\mathbb{I})$. Then there exist $n \in \mathbb{N}$ and $M > 0$ such that

$$|F(s) - F(t)|^n \leq M^n |s - t| \quad (s, t \in \mathbb{I}). \tag{6.2}$$

We then have

$$|h_1(x) - h_2(x)|^n \leq M^n \left|\frac{f_1(x)}{g_1(x)} - \frac{f_2(x)}{g_2(x)}\right| \quad (x \in \Omega \setminus (\mathbf{Z}(g_1) \cup \mathbf{Z}(g_2))),$$

and so

$$|k_1g_2 - k_2g_1|^n \leq M^n g_1^{n-1} g_2^{n-1} |f_1g_2 - f_2g_1|. \tag{6.3}$$

Theorem 6.5 *Let P be a prime ideal in $C(\Omega)$, and let $F \in \mathfrak{L}(\mathbb{I})$.*

(i) *For each $a \in \mathbb{I}_{(P)}$, $F_{(P)}(a)$ is well defined in $K_P^{\#}$.*

(ii) *The map $F_{(P)} : \mathbb{I}_{(P)} \to K_P^{\#}$ is uniformly continuous.*

(iii) *Suppose that F is increasing on $\mathbb{I}$. Then $F_{(P)}$ is increasing on $\mathbb{I}_{(P)}$.*

(iv) *Suppose that there exist $m \in \mathbb{N}$ and $\delta > 0$ such that*

$$|F(s) - F(t)| \geq \delta |s - t|^m \quad (s, t \in \mathbb{I}).$$

Then the map $a \mapsto F_{(P)}(a)$, $\mathbb{I}_{(P)} \to K_P^{\#}$, is an injection.

Proof (i) We apply the above calculation in the case where $a_1 = a_2$. We then have $f_1g_2 - f_2g_1 \in P$, and so, by (6.3), $k_1g_2 - k_2g_1 \in P$. Hence

$$\pi_P(k_1)\pi_P(g_2) = \pi_P(k_2)\pi_P(g_1),$$

and so $F_{(P)}(a_1) = F_{(P)}(a_2)$ in $K_P^{\#}$. Thus $F_{(P)}(a)$ is well defined.

(ii) Let $\varepsilon \in A_P^{\#+} \setminus \{0\}$, say $\varepsilon = \pi_P(h_0)$, where $h_0 \in C(\Omega)^+ \setminus P$, and set $\delta = (\varepsilon/M)^n$, where n and M are as in (6.2). Suppose that $|a_1 - a_2| < \delta$ and that we have the above representations of a_1 and a_2. Then we may suppose that $|f_1g_2 - f_2g_1| < h_0^n g_1g_2/M^n$ in $C(\Omega)$, and so, by (6.3),

$$|k_1g_2 - k_2g_1| \leq g_1g_2h_0.$$

Thus $\left|F_{(P)}(a_1) - F_{(P)}(a_2)\right| < \pi_P(h_0) = \varepsilon$, and so the function $F_{(P)}$ is uniformly continuous on $\mathbb{I}_{(P)}$.

(iii) Let $a_1, a_2 \in \mathbb{I}_{(P)}$ with $a_1 \leq a_2$, and take the above representations of a_1 and a_2. Then

$$f_1(x)/g_1(x) \leq f_2(x)/g_2(x) \quad (x \in \Omega \setminus (\mathbf{Z}(g_1) \cup \mathbf{Z}(g_2))),$$

and so

$$h_1(x) \le h_2(x) \quad (x \in \Omega \setminus (\mathbf{Z}(g_1) \cup \mathbf{Z}(g_2)))$$

because F is increasing. It follows that $g_1g_2h_1 \le g_1g_2h_2$ in $C(\Omega)$, and so $F_{(P)}(a_1) \le F_{(P)}(a_2)$ in K_P.

(iv) Again, take $a_1, a_2 \in \mathbb{I}_{(P)}$ with the above representations. By hypothesis,

$$\delta \left| \frac{f_1(x)}{g_1(x)} - \frac{f_2(x)}{g_2(x)} \right|^m \le |h_1(x) - h_2(x)| \quad (x \in \Omega \setminus (\mathbf{Z}(g_1) \cup \mathbf{Z}(g_2))),$$

and so

$$\delta\, |f_1g_2 - f_2g_1|^m \le |k_1g_2 - k_2g_1|\, g_1^{m-1}g_2^{m-1}$$

in $C(\Omega)$. It follows that, if $F_{(P)}(a_1) = F_{(P)}(a_2)$, so that $k_1g_2 - k_2g_1 \in P$, then $f_1g_2 - f_2g_1 \in P$, and hence $a_1 = a_2$. □

For $a \in \mathbb{I}_{(P)}$, we define

$$\Theta_a : F \mapsto F_{(P)}(a), \quad \mathfrak{L}(\mathbb{I}) \to K_P^{\#}.$$

The following result is now clear.

Theorem 6.6 *Let P be a prime ideal in $C(\Omega)$. Then the family*

$$\{\Theta_a : a \in \mathbb{I}_{(P)}\}$$

is an $\mathfrak{L}(\mathbb{I})$-operational calculus for K_P. □

We are able to define a stronger operational calculus in the special case where P is a z-ideal.

Theorem 6.7 *Let P be a prime z-ideal in $C(\Omega)$, and let $F \in C(\mathbb{I})$.*

(i) *For each $a \in \mathbb{I}_{(P)}$, $F_{(P)}(a)$ is well defined in $K_P^{\#}$.*

(ii) *The map $F_{(P)} : \mathbb{I}_{(P)} \to K_P^{\#}$ is uniformly continuous.*

(iii) *Suppose that F is increasing on $\mathbb{I}$. Then $F_{(P)}$ is increasing on $\mathbb{I}_{(P)}$.*

(iv) *Suppose that F is strictly increasing on $\mathbb{I}$. Then the map $a \mapsto F_{(P)}(a)$, $\mathbb{I}_{(P)} \to K_P^{\#}$, is an injection.*

Proof (i) Let $a_1, a_2 \in \mathbb{I}_{(P)}$, and take $f_1, f_2, g_1, g_2, h_1, h_2$, and k_1, k_2 as above. Suppose that $a_1 = a_2$. Then there exists $Z \in \mathbf{Z}[P]$ with

$$f_1(x)g_2(x) = f_2(x)g_1(x) \quad (x \in Z),$$

and this implies that

$$F\left(\frac{f_1(x)}{g_1(x)}\right) = F\left(\frac{f_2(x)}{g_2(x)}\right) \quad (x \in Z \setminus (\mathbf{Z}(g_1) \cup \mathbf{Z}(g_2)))$$

and hence that

$$k_1(x)g_2(x) = k_2(x)g_1(x) \quad (x \in Z \setminus (\mathbf{Z}(g_1) \cup \mathbf{Z}(g_2))).$$

This latter equation also holds for $x \in \mathbf{Z}(g_1) \cup \mathbf{Z}(g_2)$ because both sides are zero in this case. Thus $\mathbf{Z}(k_1 g_2 - k_2 g_1) \in \mathbf{Z}[P]$. Since P is a z-ideal, we have $k_1 g_2 - k_2 g_1 \in P$, and so $F_{(P)}(a_1) = F_{(P)}(a_2)$ in $K_P^{\#}$.

(ii) Take $\varepsilon \in A_P^{\#+} \setminus \{0\}$, say $\varepsilon = \pi_P(h_0)$, where $h_0 \in C(\Omega)^+ \setminus P$. We *claim* that there exists $k_0 \in C(\Omega)^+$ such that $\mathbf{Z}(k_0) = \mathbf{Z}(h_0)$ and such that, for each $x \in \Omega \setminus \mathbf{Z}(h_0)$, we have

$$|F(s) - F(t)| < h_0(x)$$

whenever $s, t \in \mathbb{I}$ with $|s - t| < k_0(x)$. For choose $\delta_n \in (0, 1/n)$ such that $|F(s) - F(t)| < 1/n$ whenever $s, t \in \mathbb{I}$ with $|s - t| \leq \delta_n$. Define

$$\begin{cases} k_0(x) = \delta_n & (x \in \Omega \text{ with } h_0(x) = 1/n)\,, \\ k_0(x) = 0 & (x \in \mathbf{Z}(h_0))\,. \end{cases}$$

Then k_0 has an extension to a function in $C(\Omega)$ such that

$$\delta_{n+1} \leq k_0(x) \leq \delta_n \quad \text{when} \quad 1/(n+1) \leq h_0(x) \leq 1/n\,,$$

and this k_0 is the required function. Note that $k_0 \notin P$ because P is a z-ideal and $\mathbf{Z}(h_0) = \mathbf{Z}(k_0)$.

Define $\delta = \pi_P(k_0) \in A_P^+ \setminus \{0\}$, and take $a_1, a_2 \in \mathbb{I}_{(P)}$ with $|a_1 - a_2| \leq \delta$. Then

$$\left| \frac{f_1(x)}{g_1(x)} - \frac{f_2(x)}{g_2(x)} \right| \leq k_0(x) \quad (x \in \Omega \setminus (\mathbf{Z}(g_1) \cup \mathbf{Z}(g_2)))\,.$$

Set $X = \Omega \setminus (\mathbf{Z}(g_1) \cup \mathbf{Z}(g_2) \cup \mathbf{Z}(h_0))$. Then it follows from the definition of k_0 that

$$|h_1(x) - h_2(x)| < h_0(x) \quad (x \in X)\,,$$

and so

$$|k_1(x) g_2(x) - k_2(x) g_1(x)| \leq h_0(x) g_1(x) g_2(x) \quad (x \in X)\,. \tag{6.4}$$

Let $x \in \mathbf{Z}(h_0) \setminus (\mathbf{Z}(g_1) \cup \mathbf{Z}(g_2))$. Then $f_1(x)/g_1(x) = f_2(x)/g_2(x)$, and hence $h_1(x) = h_2(x)$, whence (6.4) holds for x. The inequality (6.4) also holds for $x \in \mathbf{Z}(g_1) \cup \mathbf{Z}(g_2)$, and hence for all $x \in \Omega$.

It follows that $\left|F_{(P)}(a_1) - F_{(P)}(a_2)\right| \leq \varepsilon$, and so $F_{(P)}$ is uniformly continuous.

(iii) This is a minor modification of the proof of (i).

(iv) Take $a_1, a_2 \in \mathbb{I}_{(P)}$ with $F_{(P)}(a_1) = F_{(P)}(a_2)$. Since F is strictly increasing on $\mathbb{I}$,

$$\mathbf{Z}(k_1 g_2 - k_2 g_1) \subset \mathbf{Z}(f_1 g_2 - f_2 g_1) \cup \mathbf{Z}(g_1) \cup \mathbf{Z}(g_2)\,.$$

But $k_1 g_2 - k_2 g_1 \in P$ and $g_1 g_2 \notin P$, and so $\mathbf{Z}(f_1 g_2 - f_2 g_1) \in \mathbf{Z}[P]$. Since P is a z-ideal, $f_1 g_2 - f_2 g_1 \in P$, and so $a_1 = a_2$. □

Theorem 6.8 *Let P be a prime z-ideal in $C(\Omega)$. Then the family*

$$\{\Theta_a : a \in \mathbb{I}_{(P)}\}$$

is a $C(\mathbb{I})$-operational calculus for K_P. □

The difficulty of defining a $C(\mathbb{I})$-operational calculus for an arbitrary prime ideal P in $C(\Omega)$ is exemplified by the following example.

Let P be a prime ideal which is not an exponential prime; such a prime was constructed in Chapter 4. Take $f \in P^+$ with $f < 1$, but with $L \circ f \notin P$, where the function L was defined in (4.9). Then $\pi_P(0) = \pi_P(f)$, but $\pi_P(L \circ 0) \neq \pi_P(L \circ f)$ in A_P, and so $L_{(P)}(0)$ is not well defined in A_P.

In Chapter 10 (but not in the present chapter), we shall require a several-variable version of the $\mathfrak{L}(\mathbb{I})$- and $C(\mathbb{I})$-operational calculi.

For $j = 1, \ldots, m$, the *coordinate functionals* Z_j are the functions defined by the rule

$$Z_j : (t_1, \ldots, t_m) \mapsto t_j, \quad \mathbb{I}^m \to \mathbb{I}.$$

Definition 6.9 *Let* $\mathfrak{A}$ *be a unital subalgebra of* $C(\mathbb{I}^m)$ *containing the functionals* $Z_1, \ldots, Z_m$*, and let* K *be an ordered field. An* $\mathfrak{A}$-operational calculus *for* K *is a family* $\{\Theta_a : a \in \mathbb{I}_K^m\}$ *of unital homomorphisms from* $\mathfrak{A}$ *into* $K^\#$ *such that* $\Theta_a(Z_j) = a_j \quad (a = (a_1, \ldots, a_m) \in \mathbb{I}_K^m)$.

A function F defined on $\mathbb{I}^m$ is *Lipschitz* if there exists $\alpha \in (0, 1)$ such that the set

$$\left\{ \frac{|F(s) - F(t)|}{d(s,t)^\alpha} : s, t \in \mathbb{I}^m, \ s \neq t \right\}$$

is bounded, where

$$d\left((s_j), (t_j)\right) = \sum_{j=1}^{m} |s_j - t_j| \quad ((s_j), (t_j) \in \mathbb{I}^m).$$

The set of these functions is denoted by $\mathfrak{L}(\mathbb{I}^m)$. Clearly $\mathfrak{L}(\mathbb{I}^m)$ is a unital subalgebra of $C(\mathbb{I}^m)$ containing $Z_1, \ldots, Z_m$.

Now let $F \in C(\mathbb{I}^m)$ and take $a_1, \ldots, a_m \in \mathbb{I}_{(P)}^m$. Then there exist functions $f_1, \ldots, f_m \in C(\Omega)$ and $g_1, \ldots, g_m \in C(\Omega) \setminus P$ with $\pi_P(f_j) = a_j \pi_P(g_j)$ in K_P and $f_j(x) \leq g_j(x)$ $(x \in \Omega)$ for $j = 1, \ldots, m$. Define $g = g_1 \cdots g_m$ and

$$h(x) = F\left(\frac{f_1(x)}{g_1(x)}, \ldots, \frac{f_m(x)}{g_m(x)}\right) \quad (x \in \Omega \setminus \mathbf{Z}(g)),$$

so that $h \in C^b(\Omega \setminus \mathbf{Z}(g))$. The function gh, defined to be 0 on $\mathbf{Z}(g)$, belongs to $C(\Omega)$. Define

$$F_{(P)}(a_1, \ldots, a_m) = \pi_P(gh)/\pi_P(g) \in K_P^\#. \tag{6.5}$$

In the following result and for the remainder of this book, the space $\mathbb{I}_{(P)}^m$ has the product topology from the topology on $\mathbb{I}_{(P)}$.

Theorem 6.10 *Let* P *be a prime ideal in* $C(\Omega)$*, and let* $F \in \mathfrak{L}(\mathbb{I}^m)$.

(i) *For each* $a_1, \ldots, a_m \in \mathbb{I}_{(P)}$, $F_{(P)}(a_1, \ldots, a_m)$ *is well defined in* $K_P^\#$.

(ii) *The map* $F_{(P)} : \mathbb{I}_{(P)}^m \to K_P^\#$ *is uniformly continuous.*

Proof Since $F \in \mathfrak{L}(\mathbb{I}^m)$, there exist $n \in \mathbb{N}$ and $M > 0$ such that

$$|F(s_1, \ldots, s_m) - F(t_1, \ldots, t_m)|^n \leq M^n \sum_{j=1}^{m} |s_j - t_j|$$

for $s_1, \ldots, s_m, t_1, \ldots, t_m \in \mathbb{I}$. Using this, the results follow by essentially the same calculations as those in Theorem 6.5, (i) and (ii), respectively. □

For $a = (a_1, \ldots, a_m) \in \mathbb{I}^m_{(P)}$, we define

$$\Theta_a : F \mapsto F_{(P)}(a_1, \ldots, a_m), \;\; \mathfrak{L}(\mathbb{I}^m) \to K_P \,.$$

Theorem 6.11 *Let P be a prime ideal in $C(\Omega)$. Then the family*

$$\{\Theta_a : a \in \mathbb{I}^m_{(P)}\}$$

is an $\mathfrak{L}(\mathbb{I}^m)$-operational calculus for K. □

In a similar way, we obtain the following result by a minor modification of the proofs of Theorem 6.7, (i) and (ii), respectively.

Theorem 6.12 *Let P be a prime z-ideal in $C(\Omega)$, and let $F \in C(\mathbb{I}^m)$.*

(i) *For each $a_1, \ldots, a_m \in \mathbb{I}_{(P)}$, $F_{(P)}(a_1, \ldots, a_m)$ is well defined in $K^{\#}_P$.*

(ii) *The map $F_{(P)} : \mathbb{I}^m_{(P)} \to K^{\#}_P$ is uniformly continuous.* □

Theorem 6.13 *Let P be a prime z-ideal in $C(\Omega)$. Then the family*

$$\{\Theta_a : a \in \mathbb{I}^m_{(P)}\}$$

is a $C(\mathbb{I}^m)$-operational calculus for K_P. □

We shall first use the above-constructed operational calculus to define an exponentiation on a large part of an arbitrary super-real field.

Let P be a prime ideal in $C(\Omega)$, and let $F \in \mathfrak{L}(\mathbb{I})$. Then we have shown above how F acts on $\mathbb{I}_{(P)}$ by composition; $F_{(P)}(a)$ is defined in $K^{\#}_P$ for each $a \in \mathbb{I}_{(P)}$. By replacing the interval $\mathbb{I}$ by $[-n, n]$ for suitably large $n \in \mathbb{N}$, we may also define $F_{(P)}(a)$ for $a \in K_P$ with $|a| \leq n$ whenever $F \in \mathfrak{L}([-n, n])$. We set

$$\mathfrak{L}_{\text{loc}}(\mathbb{R}) = \{F \in C(\mathbb{R}) : F \mid [-n, n] \in \mathfrak{L}([-n, n]) \quad (n \in \mathbb{N})\} \,,$$

so that $F_{(P)}(a)$ is defined whenever $F \in \mathfrak{L}_{\text{loc}}(\mathbb{R})$ and $a \in K^{\#}_P$. Similarly we set

$$\mathfrak{L}_{\text{loc}}(\mathbb{R}^+) = \{F \in C(\mathbb{R}^+) : F \mid [0, n] \in \mathfrak{L}([0, n]) \quad (n \in \mathbb{N})\} \,,$$

so that $F_{(P)}(a)$ is defined in $K^{\#}_P$ whenever $F \in \mathfrak{L}_{\text{loc}}(\mathbb{R}^+)$ and $a \in K^{\#+}_P$.

In particular, taking $F(t) = \mathrm{e}^t \;\; (t \in \mathbb{R})$, we define $\exp a \in K^{\#}_P$ for each $a \in K^{\#}_P$. Our idea is to extend this definition of $\exp a$ to the whole of K_P, or at least to a large part of it, to obtain an *exponentiation* on K_P. The property of having an exponentiation is a feature of super-real fields, but not necessarily of more general fields.

We now give our formal definition of exponentiation.

Definition 6.14 *Let K be an ordered field. A* convex subgroup *of K is a subgroup I of $(K,+)$ containing* 1 *which is an absolutely convex set. A* convex subsemigroup *of K^+ is a subset of K^+ of the form I^+, where I is a convex subgroup of K.*

Clearly, if I is a convex subgroup of K, then $I \supset K^{\#}$.

Definition 6.15 *Let K be an ordered field, and let I be a convex subgroup of K. Then an* exponentiation on I *is a map*

$$\exp : I \to K^+ \setminus \{0\}$$

such that:

(i) $\exp(a+b) = (\exp a)(\exp b) \quad (a, b \in I)$;

(ii) $\exp 0 = 1$ *and* $\exp 1 = \mathrm{e}1$;

(iii) $\exp a < \exp b$ *whenever* $a, b \in I$ *with* $a < b$;

(iv) *the range of* exp *is* $K^+ \setminus \{0\}$.

It follows from (i)–(iii) that $\exp(\alpha 1) = \mathrm{e}^{\alpha} 1 \quad (\alpha \in \mathbb{Q})$, that

$$\exp(K^{\#}) \subset K^{\#+} \setminus \{0\},$$

and that $\exp(I^+ \setminus K^{\#}) \subset K^+ \setminus K^{\#}$. The main effect of clause (iv) is to ensure that $I \neq K^{\#}$.

The map from $K^+ \setminus \{0\}$ onto I which is the inverse of exp is called log, so we have a function

$$\log : K^+ \setminus \{0\} \to I$$

such that:

(i) $\log(ab) = \log a + \log b \quad (a, b \in K^+ \setminus \{0\})$;

(ii) $\log 1 = 0$ and $\log(\mathrm{e}1) = 1$;

(iii) $\log a < \log b$ whenever $a, b \in K^+ \setminus \{0\}$ with $a < b$.

Let J be a convex subsemigroup of K^+. An *exponentiation on J* is a bijection

$$\exp : J \to [1, \infty)$$

satisfying (i) - (iii) of Definition 6.15. In this situation, set $I = \{a \in K : |a| \in J\}$, so that I is a convex subgroup of K, and extend exp to I by setting

$$\exp(-a) = (\exp a)^{-1} \quad (a \in J).$$

Then exp is an exponential on I.

Note that, if there is an exponentiation on a convex subgroup of an ordered field K and if K is isomorphic as an ordered field to L, then there is an exponentiation on a convex subgroup of L.

We shall give some examples of exponentiations shortly. First we note that exponentiation extends to the Cauchy completion $\widetilde{K}$ of an ordered field K. Let I be a convex subgroup of K, and recall that

$$\widetilde{I} = \{x \in \widetilde{K} : x \leq |a| \text{ for some } a \in I\}$$

is the corresponding convex subgroup of $\widetilde{K}$, as in Chapter 3.

Proposition 6.16 *Let K be an ordered field, and let $\exp : I \to K^+ \setminus \{0\}$ be an exponentiation on a convex subgroup I of K.*

(i) *For each $a \in I$, $\exp \mid ((-\infty, a] \cap I)$ is uniformly continuous.*

(ii) *For each $a \in K^+ \setminus \{0\}$, $\log \mid [a, \infty)$ is uniformly continuous.*

(iii) *There is an exponentiation on $\widetilde{I}$ extending the given exponentiation.*

Proof (i) Take $\varepsilon \in K^+ \setminus \{0\}$. Then there exists $\delta \in I^+ \setminus \{0\}$ with

$$\exp \delta = 1 + \varepsilon(\exp a)^{-1}.$$

For $x, y \in (-\infty, a] \cap I$ with $x \le y < x + \delta$, we have

$$\exp x \le \exp y \le (\exp x)(\exp \delta) < \exp x + \varepsilon,$$

and so $\exp \mid ((-\infty, a] \cap I)$ is uniformly continuous.

(ii) Take $\varepsilon \in I^+ \setminus \{0\}$, and set $\delta = a(\exp \varepsilon - 1)$, so that $\delta > 0$. For $u, v \in [a, \infty)$ with $a \le u \le v < u + \delta$, set $x = \log u$ and $y = \log v$. Assume that $y \ge x + \varepsilon$. Then

$$v - u \ge (\exp x)(\exp \varepsilon) - \exp x \ge \delta,$$

a contradiction. Hence $x \le y < x + \varepsilon$, and this implies that $\log \mid [a, \infty)$ is uniformly continuous.

(iii) By (i) and Theorem 3.14, the function $\exp : I \to K$ has a uniformly continuous extension to $\widetilde{\exp} : \widetilde{I} \to \widetilde{K}^+ \setminus \{0\}$, and it is immediate that $\widetilde{\exp}$ satisfies (i)–(iii) of Definition 6.15. In a similar way, the map $\log$ has an extension to a function $\widetilde{\log} : \widetilde{K}^+ \setminus \{0\} \to \widetilde{I}$, and $\widetilde{\log}$ is the inverse of $\widetilde{\exp}$, and so clause 6.15(iv) is satisfied. □

We now prove that super-real fields have an exponentiation that arises in a natural way from the pointwise definition of exponentiation on algebras of continuous functions.

Let P be a prime ideal in an algebra $C(\Omega)$, so that $K_P^\#$ is a convex subgroup of K_P, and set

$$F(t) = \mathrm{e}^t \quad (t \in \mathbb{R}).$$

Then $F \in \mathfrak{L}_{\mathrm{loc}}(\mathbb{R})$, and so

$$F_{(P)}(a) = \exp a$$

belongs to $K_P^\#$ for each $a \in K_P^\#$. Clearly the map

$$\exp : a \mapsto \exp a, \quad K_P^\# \to K_P^{\#+} \setminus \{0\},$$

satisfies (i) and (ii) of Definition 6.15, and, by Theorem 6.5, (iii) and (iv), it also satisfies clause (iii). Our aim is now to extend the domain of $\exp$ so that clause (iv) also holds. We now again consider the specific function $E : \mathbb{R}^+ \to [0, 1)$ that was defined in equation (4.8), so that $E(t) = \exp(-1/t)$ for $t > 0$. Certainly $E \in \mathfrak{L}_{\mathrm{loc}}(\mathbb{R}^+)$, and hence $E_{(P)}(a) \in K_P^\#$ whenever $a \in K_P^{\#+}$. We set

$$J_E = \{a \in K_P^{\#+} : E_{(P)}(a) = 0\}.$$

Since $E(t) > 0$ if $t > 0$, we have $J_E \subset K_P^0$. It is clear that $J_E = \{0\}$ if and only if P is an exponential prime in the sense of Definition 4.45; in particular, $J_E = \{0\}$ whenever P is a z-ideal.

Lemma 6.17 (i) *Suppose that* $a \in J_E$ *and* $b \in K_P$ *with* $0 \leq b \leq a$. *Then* $b \in J_E$.

(ii) *Suppose that* $a \in J_E$. *Then* $2a \in J_E$.

Proof (i) This follows because E is increasing on $\mathbb{R}^+$.

(ii) This follows because $E(2t)^2 = E(t) \quad (t \in \mathbb{R}^+)$. □

Now set

$$I_E = \{a \in K_P^+ \setminus \{0\} : a^{-1} \notin J_E\} \cup \{0\} = \{b^{-1} : b \in K_P^+ \setminus J_E\} \cup \{0\}.$$

Clearly $K_P^{\#+} \subset I_E$, and $I_E = K_P^+$ if and only if P is an exponential prime.

Lemma 6.18 *The set* I_E *is a convex subsemigroup of* K_P^+.

Proof Take $a \in I_E$ and $b \in K_P$ with $0 < b \leq a$. Then $b^{-1} \geq a^{-1}$, and so $b^{-1} \notin J_E$, for otherwise, by Lemma 6.17(i), $a^{-1} \in J_E$. Thus $b \in I_E$.

Take $a \in I_E \setminus \{0\}$. Then $a^{-1} \notin J_E$, and so $(2a)^{-1} = a^{-1}/2 \notin J_E$ by Lemma 6.17(ii). Thus $2a \in I_E$.

Certainly $1 \in I_E$. □

We extend exp to I_E by the formula:

$$\exp a = (E_{(P)}(a^{-1}))^{-1} \quad (a \in I_E \setminus K_P^{\#+}).$$

Let $a \in I_E \setminus K_P^{\#+}$. Then $a^{-1} \in K_P^0 \setminus J_E$, so that $E_{(P)}(a^{-1})$ is defined and $E_{(P)}(a^{-1}) \neq 0$.

Lemma 6.19 *The map* $\exp : I_E \to K_P$ *is an exponentiation on* I_E.

Proof We first verify that, if $a, b \in I_E$, then

$$\exp(a + b) = (\exp a)(\exp b). \tag{6.6}$$

The equation certainly holds if $a, b \in K_P^{\#+}$.

Suppose that $a, b \in I_E \setminus K_P^{\#+}$ (so that $a + b \in I_E \setminus K_P^{\#+}$). We must verify in this case that

$$E_{(P)}((a + b)^{-1}) = E_{(P)}(a^{-1})E_{(P)}(b^{-1}). \tag{6.7}$$

Take $f_1, f_2, g_1, g_2 \in C(\Omega)^+$ with $\pi_P(f_1) = a^{-1}\pi_P(g_1)$, $\pi_P(f_2) = b^{-1}\pi_P(g_2)$, and $f_1(x) \leq g_1(x)$, $f_2(x) \leq g_2(x) \quad (x \in \Omega)$. Set $X = \mathbf{Z}(f_1) \cup \mathbf{Z}(f_2)$, and set

$$h(x) = \frac{f_1(x)f_2(x)}{f_1(x)g_2(x) + f_2(x)g_1(x)} \quad (x \in \Omega \setminus X),$$

so that $h \in C^b(\Omega \setminus X)$. Define

$$k(x) = \begin{cases} (g_1g_2h)(x) & (x \in \Omega \setminus X), \\ 0 & (x \in X). \end{cases}$$

Since

$$k(x) \leq \max\{g_1(x)f_2(x), g_2(x)f_1(x)\} \quad (x \in \Omega \setminus X),$$

we see that $k \in C(\Omega)$. Set $c = \pi_P(k)(\pi_P(g_1g_2))^{-1} \in K_P^+$. Since

$$f_1f_2g_1g_2 = k(f_1g_2 + f_2g_1)$$

in $C(\Omega)$, we have $(a^{-1} + b^{-1})c = a^{-1}b^{-1}$ in $K_P^\#$, and so $(a + b)c = 1$ and $c = (a + b)^{-1}$. For each $x \in \Omega$, we have

$$g_1(x)g_2(x)\exp\left(\frac{-1}{h(x)}\right) = g_1(x)\exp\left(\frac{-g_1(x)}{f_1(x)}\right) g_2(x)\exp\left(\frac{-g_2(x)}{f_2(x)}\right),$$

and so $\pi_P(g_1g_2)E_{(P)}(c) = \pi_P(g_1)E_{(P)}(a^{-1})\pi_P(g_2)E_{(P)}(b^{-1})$, from which (6.7) follows.

Now suppose that $a \in I_E \setminus K_P^{\#+}$ and $b \in K_P^{\#+}$ (so that $a + b \in I_E \setminus K_P^{\#+}$). We must verify in this case that

$$E_{(P)}(a^{-1}) = E_{(P)}((a + b)^{-1})\exp b\,. \tag{6.8}$$

Take $f_1, f_2, g_1, g_2 \in C(\Omega)^+$ with $\pi_P(f_1) = a^{-1}\pi_P(g_1)$, $\pi_P(f_2) = b\pi_P(g_2)$, and $f_1(x) \leq g_1(x)$, $f_2(x) \leq g_2(x)$ $(x \in \Omega)$. Set $X = \mathbf{Z}(g_1) \cup \mathbf{Z}(g_2)$, and set

$$h(x) = \frac{f_1(x)g_2(x)}{f_1(x)f_2(x) + g_1(x)g_2(x)} \quad (x \in \Omega \setminus X),$$

so that $h \in C^b(\Omega \setminus X)$. Again the function g_1g_2h has a continuous extension to $k \in C(\Omega)$ with $k(x) = 0$ $(x \in X)$. Set $c = \pi_P(k)(\pi_P(g_1g_2))^{-1} \in K_P^+$. Since we have

$$k(f_1f_2 + g_1g_2) = f_1g_2g_1g_2$$

in $C(\Omega)$, necessarily $c(a^{-1}b + 1) = a^{-1}$ in $K_P^\#$, and so $c = (a + b)^{-1}$. Again, for each $x \in \Omega$, we have

$$g_1(x)\exp\left(\frac{-g_1(x)}{f_1(x)}\right) g_2^2(x) = g_1(x)g_2(x)\exp\left(\frac{-1}{h(x)}\right) g_2(x)\exp\left(\frac{f_2(x)}{g_2(x)}\right),$$

and so $\pi_P(g_1)E_{(P)}(a^{-1})\pi_P(g_2^2) = \pi_P(g_1g_2)E_{(P)}(c)\pi_P(g_2)\exp b$, from which (6.8) follows.

We have verified (6.6) in each case.

Clearly $\exp 0 = 1$ and $\exp 1 = e1$, and $\exp a < \exp b$ whenever $a, b \in I_E$ with $a < b$.

We finally verify that the range of exp on I_E is $[1, \infty)$. Take $c \in K_P$ with $c \geq 1$. If $c \in K_P^\#$, set $a = H_{(P)}(c - 1)$, where

$$H(t) = \log(1 + t) \quad (t \in \mathbb{R}^+);$$

the element a is well defined because $H \in \mathfrak{L}_{\mathrm{loc}}(\mathbb{R}^+)$, and it is clear that $\exp a = c$. Now suppose that $c \in K_P^+ \setminus K_P^\#$, and set $b = c^{-1} \in K_P^{0+} \setminus \{0\}$; we require

$a \in K_P^0 \setminus J_E$ with $E_{(P)}(a) = b$. Take $f, g \in C(\Omega)^+$ with $\pi_P(f) = b\pi_P(g)$ and $f(x) \leq g(x)/2 \quad (x \in \Omega)$; this is possible because b is an infinitesimal. Let $L : [0, 1) \to \mathbb{R}^+$ be as defined in (4.9), so that $(G \circ L)(t) = t$ for $t \in [0, 1)$, and set

$$h(x) = L\left(\frac{f(x)}{g(x)}\right) \quad (x \in \Omega \setminus \mathbf{Z}(g)).$$

Then $h \in C^b(\Omega \setminus \mathbf{Z}(g))$ because L is bounded on $[0, 1/2]$, and so $gh \in C(\Omega)$; set

$$a = \pi_P(gh)/\pi_P(g) \in K_P^{\#}.$$

(We cannot say that $a = L_{(P)}(b)$ in $K_P^{\#}$ because $L \notin \mathfrak{L}(\mathbb{I})$.) Then

$$\pi_P(g)E_{(P)}(a) = \pi_P(g \cdot (E \circ L \circ (f/g))) = \pi_P(f),$$

and so $E_{(P)}(a) = b$, as required. □

We have now established the following result.

Theorem 6.20 *Let Ω be a compact space, and let P be a prime ideal in $C(\Omega)$. Then there is an exponentiation on a convex subgroup of K_P. In the case where P is an exponential prime, there is an exponentiation on K_P itself.* □

Corollary 6.21 *(Alling) Let K be a hyper-real field. Then there is an exponentiation on K.* □

In an entirely similar way we see that there is a function

$$\exp_{(2)}$$

on the convex subgroup I_E of K_P; this is a function, *exponentiation to base* 2, which satisfies the same definition as exp, save that $\exp_{(2)}1 = 2$. In particular, in the case where P is an exponential prime, we have a map

$$a \mapsto 2^a = \exp_{(2)}(a), \quad K_P \to K_P. \tag{6.9}$$

We immediately apply the above theorem to show that the field $\widehat{\mathbf{R}}$ is not a super-real field.

Theorem 6.22 *It is not the case that there is an exponentiation on a convex subgroup of* $\widehat{\mathbf{R}}$.

Proof To obtain a contradiction, assume that I is a convex subgroup of $\widehat{\mathbf{R}}$ and

$$\exp : I \to \widehat{\mathbf{R}}$$

is an exponentiation on I. Then there exists $f_0 \in I^+ \setminus \widehat{\mathbf{R}}^{\#}$, and $v(f_0) < 0$.

The interval $(v(f_0), 0)$ in $\mathbf{G}$ is an η_1-set, and so, by Theorem 1.13, there is an isotonic map $\iota : \mathbf{Q} \to (v(f_0), 0)$. We identify $\mathbf{Q}$ with $\iota(\mathbf{Q})$, and then we may regard $\mathfrak{F}(\mathbb{R}, \mathbf{Q})$ as a subgroup of $\widehat{\mathbf{R}}$. For each $f \in \mathfrak{F}(\mathbb{R}, \mathbf{Q})^+ \setminus \{0\}$, we have $v(f) > v(f_0)$, and so $f < f_0$ in $\widehat{\mathbf{R}}$, whence $f \in I^+$. Thus we have $\mathfrak{F}(\mathbb{R}, \mathbf{Q})^+ \subset I^+$ and $\mathfrak{F}(\mathbb{R}, \mathbf{Q})^+ \cap \widehat{\mathbf{R}}^{\#} = \{0\}$.

Define a map

$$\psi : f \mapsto -v(\exp f), \quad \mathfrak{F}(\mathbb{R}, \mathbf{Q})^+ \to \mathbf{G}.$$

We *claim* that ψ is isotonic. For take $f, g \in \mathfrak{F}(\mathbb{R}, \mathbf{Q})^+$ with $f < g$, say $g = f + h$, where $h \in I \setminus \widehat{\mathbf{R}}^{\#}$. Since exp $g =$ (exp f)(exp h), we have $\psi(g) = \psi(f) + \psi(h)$. But $\exp h \in \widehat{\mathbf{R}}^+ \setminus \widehat{\mathbf{R}}^{\#}$, so that $v(\exp h) < 0$ in $\mathbf{G}$, and hence $\psi(h) > 0$ in $\mathbf{G}$. Thus $\psi(f) < \psi(g)$, as required.

The existence of such an isotonic map ψ is a contradiction of Proposition 1.41, and so there is no exponentiation on a strong interval of K. □

In the case where CH holds, the recourse to Proposition 1.41 in the above proof can be avoided. The argument shows that the map ψ is an injection from $\mathfrak{F}(\mathbb{R}, \mathbf{Q})^+$ into $\mathbf{G}$. However, by Proposition 1.9(v), $\mathbf{Q}$ contains a well-ordered subset of cardinality $\aleph_1$, and so

$$\left|\mathfrak{F}(\mathbb{R}, \mathbf{Q})^+\right| \geq 2^{\aleph_1},$$

whereas $|\mathbf{G}| = 2^{\aleph_0}$ by Proposition 1.44(i). With CH, $2^{\aleph_1} > 2^{\aleph_0}$, and so we have obtained a contradiction.

Theorem 6.23 *The real-closed η_1-field $\widehat{\mathbf{R}}$ is not isomorphic as a field to any super-real field.*

Proof By Theorem 6.20, there is an exponentiation on a convex subgroup of K for each super-real field K. By Theorem 6.22, there is no exponentiation on a convex subgroup of $\widehat{\mathbf{R}}$. □

Thus, in the terminology of Chapter 4, the classes **real-closed η_1-fields** and **super-real η_1** are distinct. In fact, the argument given shows that $\widehat{\mathbf{R}}$ is not isomorphic as a ring to any super-real field. This distinction will be recorded in our next version of the main diagram, Fig. 9.2. on page 248.

In contrast to Theorem 6.21, we have the following result about $\mathbf{R}$ and $\widetilde{\mathbf{R}}$.

Theorem 6.24 *There is an exponentiation on $\mathbf{R}$ and on $\widetilde{\mathbf{R}}$.*

Proof By Theorem 2.23, $\mathbf{R}$ is a real-closed β_1-η_1-field, and so the groups $(\mathbf{R}, +)$ and $(\mathbf{R}^+ \setminus \{0\}, \cdot)$ are both divisible, totally ordered β_1-η_1-groups. By the isomorphism theorem, Corollary 1.49, there is an isotonic group morphism

$$\exp : (\mathbf{R}, +) \to (\mathbf{R}^+ \setminus \{0\}, \cdot),$$

and we may suppose that $\exp 1 = \mathrm{e}1$. The map exp is an exponentiation. By Proposition 6.16(iii), this exponentiation has an extension to an exponentiation on $\widetilde{\mathbf{R}}$. □

The field $\mathbf{R} = \mathfrak{F}_{(1)}(\mathbb{R}, \mathbf{G})$ has cardinality $\aleph_1$ (with CH), rather than $\aleph_2$, but it is still of interest to enquire whether or not $\mathbf{R}$ is a super-real field. In fact, this question is independent of ZFC. First, we have noted in Corollary 4.30 that, with CH, $\mathbf{R}$ is a hyper-real field. We now consider the situation when CH is false; in fact, we work in the theory ZFC + MA + $\neg$CH, where MA is Martin's Axiom.

Proposition 6.25 (MA + ¬CH) *Let* K_P *be a super-real field. Then* M_P *contains a strictly decreasing sequence of length* ω_2.

Proof By Theorem 4.41, there is a prime ideal Q in c_0 such that c_0/Q embeds in M_P. Let $\mathcal{U}$ be a free ultrafilter on $\mathbb{N}$ associated with Q, so that there is an epimorphism from $c_0/\mathcal{U}$ onto c_0/Q. It is a standard consequence of MA that $\operatorname{coi}((c_0/\mathcal{U})^+ \setminus \{0\}) = \mathfrak{c}$. It follows that, with $\neg CH$, the algebra c_0/Q contains a strictly decreasing sequence of length ω_2, and so M_P also contains such a sequence. □

Corollary 6.26 (MA + ¬CH) *None of the real-closed* η_1*-fields* $\mathbf{R}$, $\widetilde{\mathbf{R}}$, *or* $\widehat{\mathbf{R}}$ *is order-isomorphic to a super-real field.*

Proof By Theorem 2.23(v), the field $\widehat{\mathbf{R}}$ is an α_2-set, and so each subfield is an α_2-set. By the proposition, no super-real field is an α_2-set. □

We now consider a second invariant of totally ordered fields K, namely whether or not $w(K) = |\Gamma_K|$. Certainly we always have $w(K) \geq |\Gamma_K|$.

Lemma 6.27 *Let* K *be an ordered field with infinite elements, and suppose that there is an exponentiation on a convex subgroup of* K. *Then there is no element* $a \in K^+$ *such that* $\{a^n : n \in \mathbb{N}\}$ *is cofinal in* K^+.

Proof Let $\exp : I \to K^+ \setminus \{0\}$ be an exponentiation on the convex subgroup I of G. To obtain a contradiction, assume that $\{a^n : n \in \mathbb{N}\}$ is cofinal in K^+; necessarily a is infinite. There exists $b \in I^+$ with $\exp b = a$. For each $n \in \mathbb{N}$, we have $ab > nb$, and so

$$\exp(ab) > \exp(nb) = (\exp b)^n = a^n \quad (n \in \mathbb{N}),$$

a contradiction of the fact that $\{a^n : n \in \mathbb{N}\}$ is cofinal in K^+. □

Theorem 6.28 *Let* K *be an ordered field with* Γ_K *infinite, and suppose that there is an exponentiation on* K. *Then* $w(K) = |\Gamma_K|$. *In the case where it is also true that* $\delta(K) = \aleph_1$ *and* Γ_K *is a* β_1*-set,* $(K^+, \leq)$ *contains an order-dense subset which is a* β_1*-set.*

Proof Set $\delta = \delta(K)$, so that $\delta \leq |\Gamma_K|$, and let $(\varepsilon_\sigma : \sigma < \delta)$ be a coinitial sequence in $K^+ \setminus \{0\}$. We may suppose that $(\varepsilon_\sigma) \subset K^0$, so that $(\varepsilon_\sigma^{1/n})$ is an increasing sequence for each $\sigma < \delta$.

For $\sigma < \delta$, set $a \sim_\sigma b$ for $a, b \in K^+$ if there exists $n \in \mathbb{N}$ such that

$$|b - a|^n < \varepsilon_\sigma \, .$$

Clearly, $\sim_\sigma$ is an equivalence relation on K^+, and the equivalence classes are intervals. Form a set S_σ by choosing one representative of each equivalence class, and set $S = \bigcup \{S_\sigma : \sigma < \delta\}$.

Take $a, b \in K^+$ with $0 \leq a < b$. The sequence $((b - a)^n)$ is not coinitial in $K^+ \setminus \{0\}$, for otherwise $((b-a)^{-n})$ would be cofinal in K^+, and this is not possible by Lemma 6.27. Thus there exists $\sigma < \delta$ such that $\varepsilon_\sigma < (b-a)^n \quad (n \in \mathbb{N})$. Set $c = (b-a)/2$, and let I be the equivalence class with respect to $\sim_\sigma$ containing c. Then $I \subset (a,b)$, and so $(a,b) \cap S_\sigma \neq \emptyset$. Thus S is order-dense in K^+.

For $\sigma < \delta$, define a map

$$\psi : a \mapsto -v\left(\exp\left(\frac{a}{\varepsilon_\sigma}\right)\right), \quad S_\sigma \to \Gamma_K .$$

For $a, b \in S_\sigma$ with $a > b$,

$$\exp\left(\frac{a}{\varepsilon_\sigma}\right)\exp\left(\frac{b-a}{\varepsilon_\sigma}\right) = \exp\left(\frac{b}{\varepsilon_\sigma}\right).$$

But $b - a > \varepsilon_\sigma^{1/2}$, and so $(b-a)/\varepsilon_\sigma \in K^+ \setminus K^\#$, whence

$$\exp((b-a)/\varepsilon_\sigma) \in K^+ \setminus K^\# .$$

This shows that $\psi(a) < \psi(b)$, and so ψ is isotonic. It follows that $|S_\sigma| \leq |\Gamma_K|$. We now have

$$|S| \leq \delta \cdot |\Gamma_K| \leq |\Gamma_K|^2 = |\Gamma_K|$$

because Γ_K is infinite. Thus $w(K) \leq |S| \leq |\Gamma_K|$. Since $w(K) \geq |\Gamma_K|$, we have $w(K) = |\Gamma_K|$.

Now suppose that Γ_K is a β_1-set. Then each S_σ is a β_1-set. If, further, $\delta = \aleph_1$, then S is a β_1-set which is order-dense in K^+. □

Theorem 6.29 *Let P be a non-maximal, exponential prime ideal in an algebra $C(\Omega)$. Then $w(K_P) = |\Gamma_P|$.*

Proof Since P is an exponential prime, there is an exponentiation on all of K_P by Theorem 6.20, and so, by the theorem, $w(K_P) = |\Gamma_P|$. □

In particular, $w(K_P) = |\Gamma_P|$ whenever P is a non-maximal, prime z-ideal.

The next result is related to Theorem 6.23.

Theorem 6.30 *The totally ordered group $(\widehat{\mathbf{R}}, +)$ is not isomorphic as an ordered group to $(K_P, +)$ for any exponential prime ideal P in an algebra $C(\Omega)$.*

Proof Assume towards a contradiction that P is an exponential prime ideal in an algebra $C(\Omega)$ with $(K_P, +) \simeq (\widehat{\mathbf{R}}, +)$. Necessarily P is non-maximal. The isomorphism preserves the base numbers and the value sets of K_P and $\widehat{\mathbf{R}}$. Thus $\delta(K_P) = \delta(\widehat{\mathbf{R}}) = \aleph_1$ by Theorem 2.23(iv), and Γ_P is a β_1-set. By Theorem 6.28, K_P contains an order-dense β_1-set, and hence $\widehat{\mathbf{R}}$ contains an order-dense β_1-set. But this is a contradiction of Theorem 2.23(viii). □

In particular, the field $\widehat{\mathbf{R}}$ is a real-closed η_1-field such that $(\widehat{\mathbf{R}}, +)$ is not isomorphic to $(K, +)$ for any hyper-real field K.

The third example given in Chapter 5 of a real-closed field K with $|K| = \aleph_2$ and $|\Gamma_K| = \aleph_1$ was the field $\widetilde{\mathbf{R}}$ of Definition 3.19. We ask if $\widetilde{\mathbf{R}}$ is a super-real field in the theory ZFC + CH? By Theorem 6.24, there is an exponentiation on $\widetilde{\mathbf{R}}$, and so we cannot use the argument of Theorem 6.23 to exclude the possibility that $\widetilde{\mathbf{R}}$ be a super-real field. It is an altogether difficult challenge to determine whether or not $\widetilde{\mathbf{R}}$ is a super-real field. This problem was our motivation for the study that we report on in Chapters 8 and 9.

Notes 6.31 We believe that the main results of this chapter are new. Theorem 6.23 answers Open Problem (8.9)(a) of Antonovskij *et al.* (1981).

Let $\alpha \in (0,1]$. Then $(\mathrm{Lip}_\alpha \mathbb{I}, \|\cdot\|_\alpha)$ is a Banach function algebra on $\mathbb{I}$ with character space $\mathbb{I}$; see Dales (1997, §3.3).

The axioms for an exponentiation in Definition 6.15 are independent (save that the condition '$\exp 0 = 1$' follows from 6.15(i)). The notion of an *exponentially closed field* was introduced by Alling (1962*b*): a field K is exponentially closed if there is an exponentiation on K, in the sense of Definition 6.15. Alling proved that a hyper-real field is exponentially closed (this is our Corollary 6.21), and essentially gave our Theorem 6.24.

A study of 'exponential fields' is contained in Kuhlmann (1996). This work is continued in Kuhlmann and Kuhlmann (1994), where the notion of a strong exponential is introduced. A *strong exponential* on an ordered field K is an exponential exp on K which satisfies additional axioms, suggested by the properties of the usual exponential function on $\mathbb{R}$, relating the growth of exp with the growth of polynomials. For example, it is required that,

$$\exp a > \sum_{k=0}^{n} \frac{a^k}{k!} \quad (n \in \mathbb{N}).$$

The exponentiation that we constructed on a super-real field K_P in the case where P is an exponential prime is a strong exponential in this sense.

There has been a systematic study of the structure $\langle \mathbb{R}, +, \cdot, \exp \rangle$, and this study has recently yielded some striking results. In particular, Wilkie (1996) has shown that the first-order theory $T_{\exp}$ of this structure is model-complete, in the sense that every formula is equivalent, modulo $T_{\exp}$, to an existential formula. Further, van den Dries, Macintyre, and Marker (1994) have developed a generalization of formal power series algebras to produce fields with exponentiation (satisfying $T_{\exp}$) that also admit local definitions of all analytic functions.

For a discussion of Martin's Axiom MA, see Dales and Woodin (1987, Chapter 5) and Fremlin (1984), for example.

7

Examples

So far, we have sought real-closed η_1-fields K in particular classes of the classification scheme of Chapter 4 with the property that (with GCH),

$$|K| = \aleph_2 \quad \text{and} \quad |\Gamma_K| = \aleph_1 .$$

(We explained in Chapter 5 that it is known for each ordered field K *not* satisfying this condition whether or not $K^{\#}$ is normable.) Examples of fields with these properties were given in Chapter 5; the main examples were the fields $\widehat{\mathbf{R}}$ and $\widetilde{\mathbf{R}}$, which do belong to the class **real-closed η_1-fields**. However, by Theorem 6.23, $\widehat{\mathbf{R}}$ is not in the class **super-real η_1**, and we have not yet resolved which class $\widetilde{\mathbf{R}}$ belongs to. We now continue our search for super-real η_1-fields K, and for fields in smaller classes, that do satisfy the above condition. The main example of the present chapter is given in Theorem 7.2: we shall prove (as a theorem of ZFC + GCH) that there is indeed a super-real η_1-field K with the above property. In fact the field K has the form K_P for a prime z-ideal in $C(\Omega)$, so that K belongs to the class **z-ideal η_1**.

We do not know if the example given in Theorem 7.2 necessarily belongs to the smaller class **hyper-real**. However we can exhibit a hyper-real field K in our class if we are allowed the combinatorial principle 'diamond' ($\diamondsuit$); $\diamondsuit$ is known to be consistent with ZFC.

Finally we shall discuss whether there are examples of fields K satisfying the above condition in the yet smaller class **ultrapowers**. We shall state without proof in Theorem 7.18 a 'relative consistency' result: under the assumption that it is consistent with ZFC that a certain large cardinal exists, it is also consistent with ZFC that there is an ultrapower in the class. Thus it cannot be proved in ZFC that there is not an ultrapower K with $|K| = \aleph_2$ and $|\Gamma_K| = \aleph_1$.

In a sense, the examples of this chapter (except for the unproved Theorem 7.18) are redundant because examples with stronger properties will be given later. However, those examples, whilst incorporating some of the ideas of the present examples, require a deeper background and much longer proofs; the results of the present chapter provide a gentler version of those later constructions.

Throughout this chapter, the ordinal ω_1 is regarded as a topological space with respect to the order topology.

We shall require the following well-known result. Recall that a filter on a set of cardinality $\aleph_1$ is uniform if it contains no countable sets.

Proposition 7.1 *Let $\mathcal{F}$ be a uniform filter on a set S of cardinality $\aleph_1$. Then $|\mathbb{R}^S/\mathcal{F}| = 2^{\aleph_1}$.*

Proof Let $\{s_\alpha : \alpha < \omega_1\}$ be an enumeration of S, and set $I = \{0,1\}^S$, so that $|I| = 2^{\aleph_1}$. We shall define a map $H : I \to \mathbb{R}^S$ such that

$$[H(a)]_{\mathcal{F}} \neq [H(b)]_{\mathcal{F}} \quad \text{in} \quad \mathbb{R}^S/\mathcal{F}$$

whenever $a, b \in I$ with $a \neq b$, thus proving that $|\mathbb{R}^S/\mathcal{F}| \geq |I|$. Since certainly $|\mathbb{R}^S/\mathcal{F}| \leq |\mathbb{R}^S| = 2^{\aleph_1}$, this will imply the result.

For each $\alpha < \omega_1$, set $a \sim_\alpha b$ for $a, b \in I$ if

$$a(s_\beta) = b(s_\beta) \quad (\beta < \alpha)\,.$$

Then $\sim_\alpha$ is an equivalence relation on I; the equivalence class containing $a \in I$ is $[a]_\alpha$. Clearly

$$|\{[a]_\alpha : a \in I\}| = |\{0,1\}^\alpha| \leq 2^{\aleph_0}\,,$$

and so there is an injective map

$$\pi_\alpha : \{[a]_\alpha : a \in I\} \to \mathbb{R}\,.$$

For $a \in I$ and $\alpha < \omega_1$, define $H(a)(s_\alpha) = \pi_\alpha([a]_\alpha)$, so that $H : I \to \mathbb{R}^S$ is a map. Take $a, b \in I$ with $a \neq b$. Then there exists $\alpha < \omega_1$ such that $a(s_\alpha) \neq b(s_\alpha)$. Thus, for each $\beta > \alpha$, we have $[a]_\beta \neq [b]_\beta$, and so $H(a)(s_\beta) \neq H(b)(s_\beta)$. Hence

$$\{s \in S : H(a)(s) = H(b)(s)\} \subset \{s_\beta : \beta \leq \alpha\}\,,$$

a countable set. Since $\mathcal{F}$ is uniform, $[H(a)]_{\mathcal{F}} \neq [H(b)]_{\mathcal{F}}$ in $\mathbb{R}^S/\mathcal{F}$. □

The proofs of Theorems 7.2 and 7.14, below, will incorporate several lemmas; in these lemmas, we maintain the same notation. Note that Theorem 7.2 is proved in the theory ZFC + GCH, without any further axioms.

Theorem 7.2 (GCH) *There is a compact space Ω and a non-maximal, prime z-ideal P in $C(\Omega)$ such that K_P is a real-closed η_1-field, such that $|K_P| = \aleph_2$, and such that $w(K_P) = |\Gamma_P| = \aleph_1$.*

We begin by fixing a set S of cardinality $\aleph_1$, and we fix a decomposition $S = \bigcup\{S_\alpha : \alpha < \omega_1\}$ of S, where the subsets S_α of S satisfy the following properties:

$$S_\alpha \text{ is countable and infinite} \qquad (\alpha < \omega_1)\,;$$
$$S_{\alpha+1} \setminus S_\alpha \neq \emptyset \qquad (\alpha < \omega_1)\,;$$
$$S_\alpha \subset S_\beta \qquad (\alpha \leq \beta < \omega_1)\,;$$
$$S_\alpha = \bigcup\{S_\beta : \beta < \alpha\} \text{ for each limit ordinal } \alpha < \omega_1\,.$$

Clearly each countable subset of S is contained in S_α for some $\alpha < \omega_1$. The set S is a topological space with respect to the discrete topology.

Set

$$X = S^{\mathbb{N}}, \quad X_\alpha = S_\alpha^{\mathbb{N}} \quad (\alpha < \omega_1). \tag{7.1}$$

Then X is a topological space with respect to the product topology; X is a metrizable, and hence completely regular, space. (The space X is not discrete – but it is rather close to being discrete!) Clearly

$$X_\alpha \subset X_\beta \quad (\alpha \le \beta < \omega_1).$$

Since each sequence in S is contained in S_α for some $\alpha < \omega_1$, we have

$$X = \bigcup\{X_\alpha : \alpha < \omega_1\}.$$

Clearly $|X| = 2^{\aleph_1}$, and so, with GCH, $|X| = \aleph_2$.

Let $S^{<\mathbb{N}}$ and $S_\alpha^{<\mathbb{N}}$ be the families of all finite sequences of elements of S and S_α, respectively. Then

$$S^{<\mathbb{N}} = \bigcup\{S_\alpha^{<\mathbb{N}} : \alpha < \omega_1\}$$

and, for each limit ordinal $\alpha < \omega_1$,

$$S_\alpha^{<\mathbb{N}} = \bigcup\{S_\beta^{<\mathbb{N}} : \beta < \alpha\}.$$

Also we clearly have $\left|S^{<\mathbb{N}}\right| = \aleph_1$ and $\left|S_\alpha^{<\mathbb{N}}\right| = \aleph_0 \ (\alpha < \omega_1)$. Suppose that

$$\mathbf{s} = (s_1, \dots, s_n) \in S^{<\mathbb{N}}.$$

Then the *length* of $\mathbf{s}$, $l(\mathbf{s})$, is n. For each $\mathbf{s} \in S^n$, set

$$U_{\mathbf{s}} = \{x \in X : x \restriction n = \mathbf{s}\}.$$

Then $U_{\mathbf{s}}$ is an open and closed subset of X, and $\{U_{\mathbf{s}} : \mathbf{s} \in S^{<\mathbb{N}}\}$ is a basis for the topology of X. Since $\left|S^{<\mathbb{N}}\right| = \aleph_1$, the space X has a dense subset of cardinality $\aleph_1$.

Take $\alpha < \omega_1$. We note that X_α is a closed subset of X with $\operatorname{int} X_\alpha = \emptyset$, and that, since $S_\alpha^{<\mathbb{N}}$ is countable, X_α is second countable, and hence separable. Further $U_{\mathbf{s}} \cap X_\alpha = \emptyset$ for each $\mathbf{s} \in S^{<\mathbb{N}} \setminus S_\alpha^{<\mathbb{N}}$, and so

$$\{\mathbf{s} \in S^{<\mathbb{N}} : U_{\mathbf{s}} \cap X_\alpha \ne \emptyset\}$$

is countable because it is contained in $S_\alpha^{<\mathbb{N}}$.

Let $\mathcal{F}$ be a z-filter on X. Then $\mathcal{F}$ is separable (see Definition 4.11) if and only if there exists $\alpha < \omega_1$ such that $X_\alpha \in \mathcal{F}$.

We can now begin the sequence of lemmas that will lead to the proof of Theorem 7.2.

Lemma 7.3 *Let U be a dense, open subset of X, and set*

$$C = \{\alpha < \omega_1 : U \cap X_\alpha \text{ is dense in } X_\alpha\}.$$

Then C is closed and unbounded in ω_1.

Proof Suppose that $\alpha < \omega_1$ and that $\alpha = \sup \beta_n$, where

$$\{\beta_n : n \in \mathbb{N}\} \subset C.$$

Take $\mathbf{s} \in S_\alpha^{<\mathbb{N}}$. Then $\mathbf{s} \in S_{\beta_n}^{<\mathbb{N}}$ for some $n \in \mathbb{N}$, and so $U_\mathbf{s} \cap U \cap X_{\beta_n} \neq \emptyset$. Thus

$$U_\mathbf{s} \cap U \cap X_\alpha \neq \emptyset,$$

and so $U \cap X_\alpha$ is dense in X_α. This shows that C is closed in ω_1.

Now fix $\alpha_0 < \omega_1$, and take $\beta < \omega_1$. For each $\mathbf{s} \in S_\beta^{<\mathbb{N}}$, there exists $\mathbf{t} \in S^{<\mathbb{N}}$ with $U_\mathbf{t} \subset U_\mathbf{s} \cap U$, and there exists $\gamma_\mathbf{s} < \omega_1$ with $\mathbf{t} \in S_{\gamma_\mathbf{s}}^{<\mathbb{N}}$. Since $S_\beta^{<\mathbb{N}}$ is countable, there exists $\gamma < \omega_1$ such that $\gamma_\mathbf{s} < \gamma$ for each $\mathbf{s} \in S_\beta^{<\mathbb{N}}$. The ordinal γ has the property that, for each $\mathbf{s} \in S_\beta^{<\mathbb{N}}$, there exists $\mathbf{t} \in S_\gamma^{<\mathbb{N}}$ with $U_\mathbf{t} \subset U_\mathbf{s} \cap U$; we take $f(\beta)$ to be the minimum ordinal with this property. In this way, we have defined a map $f : \omega_1 \to \omega_1$.

There is a limit ordinal $\alpha < \omega_1$ with $\alpha > \alpha_0$ such that $f(\beta) < \alpha$ for each $\beta < \alpha$. We *claim* that $U \cap X_\alpha$ is dense in X_α. For take $\mathbf{s} \in S^{<\mathbb{N}}$ with $U_\mathbf{s} \cap X_\alpha \neq \emptyset$. Then $\mathbf{s} \in S_\alpha^{<\mathbb{N}}$, and so $\mathbf{s} \in S_\beta^{<\mathbb{N}}$ for some $\beta < \alpha$. There exists

$$\mathbf{t} \in S_{f(\beta)}^{<\mathbb{N}} \subset S_\alpha^{<\mathbb{N}}$$

with $U_\mathbf{t} \subset U_\mathbf{s} \cap U$, and so $U \cap X_\alpha$ is dense in X_α. Since $\alpha > \alpha_0$, C is unbounded in ω_1. □

We shall consider certain z-filters on X. First we require some further notation.

Let $x \in X$. Then R_x is the range of x: if $x = (x_n)$, then

$$R_x = \{x_n : n \in \mathbb{N}\} \subset S.$$

Now let Z be a closed subset of X. Then

$$R_Z = \bigcup \{R_x : x \in Z\},$$

the union of the ranges of the elements of Z, so that R_Z is a non-empty subset of S. Finally let $\mathcal{F}$ be a z-filter on X. Then

$$R_\mathcal{F} = \{R_Z : Z \in \mathcal{F}\}.$$

Lemma 7.4 *Let $\mathcal{F}$ be a z-filter on X. Then:*

(i) *$R_\mathcal{F}$ is a filter on S;*

(ii) *in the case where $\mathcal{F}$ is non-separable, $R_\mathcal{F}$ is a uniform filter on S.*

Proof (i) Let T be a subset of S with $T \supset R_Z$ for some $Z \in \mathcal{F}$, and set $W = T^{<\mathbb{N}}$. Then W is closed in X and $W \supset Z$, so that $W \in \mathcal{F}$. Thus $T = R_W \in R_\mathcal{F}$. Now take $Z_1, Z_2 \in \mathcal{F}$, and set $Z = Z_1 \cap Z_2$. Then $Z \in \mathcal{F}$ and

$$R_{Z_1} \cap R_{Z_2} \supset R_Z \in R_\mathcal{F}.$$

Thus $R_\mathcal{F}$ is a filter.

(ii) Assume towards a contradiction that there exists $Z \in \mathcal{F}$ with $|R_Z| < |S|$. Then R_Z is countable, and so there exists $\alpha < \omega_1$ such that $R_Z \subset S_\alpha$. Thus $Z \subset R_Z^\mathbb{N} \subset S_\alpha^\mathbb{N} = X_\alpha$, and so $X_\alpha \in \mathcal{F}$, a contradiction of the fact that $\mathcal{F}$ is non-separable. Hence $|R_Z| = |S|$ for each $Z \in \mathcal{F}$, and this proves that $R_\mathcal{F}$ is uniform. □

Lemma 7.5 (GCH) *Let $\mathcal{F}$ be a z-filter on X. Then:*

(i) $|C^b(X)/\mathcal{F}| \leq \aleph_2$;

(ii) $|C^b(X)/\mathcal{F}| = \aleph_2$ *if and only if $\mathcal{F}$ is non-separable.*

Proof We set $A = C^b(X)/\mathcal{F}$.

(i) Since X has a dense subset of cardinality $\aleph_1$, certainly

$$|C(X)| \leq 2^{\aleph_1} = \aleph_2 ,$$

and so $|A| \leq \aleph_2$.

(ii) Suppose that $\mathcal{F}$ is separable, say $X_\alpha \in \mathcal{F}$, where $\alpha < \omega_1$. Then we have $A \cong C^b(X_\alpha)/\mathcal{F}$, and so $|A| \leq |C^b(X_\alpha)| = \aleph_1 < \aleph_2$.

Conversely, suppose that $\mathcal{F}$ is non-separable.

By Lemma 7.4(ii), $R_\mathcal{F}$ is a uniform filter on S, and so, by Proposition 7.1, $|\mathbb{R}^S/R_\mathcal{F}| = \aleph_2$ (with GCH). Thus there is a sequence $(f_\zeta : \zeta < \omega_2)$ of functions in $\mathbb{R}^S$ such that $[f_{\zeta_1}] \neq [f_{\zeta_2}]$ whenever $\zeta_1 \neq \zeta_2$. Since $|\mathbb{R}| = |\mathbb{N}^\mathbb{N}|$, we may regard each f_ζ as a map from S into $\mathbb{N}^\mathbb{N}$.

For each $\zeta < \omega_2$, define $F_\zeta : X \to \mathbb{N}^{\mathbb{N}\times\mathbb{N}}$ by setting

$$(F_\zeta(x))(i,j) = (f_\zeta(x_i))(j) \quad (x = (x_n) \in X,\ i,j \in \mathbb{N}) .$$

Each map F_ζ is continuous. Since $\mathbb{N}^{\mathbb{N}\times\mathbb{N}}$ is homeomorphic to $\mathbb{N}^\mathbb{N}$, and since $\mathbb{N}^\mathbb{N}$ is homeomorphic to the set of irrationals in $(0,1)$ with the relative topology from $(0,1)$, we may regard each F_ζ as a continuous map from X into $\mathbb{I}$; we do this, and so $F_\zeta \in C(X,\mathbb{I})$ $(\zeta < \omega_2)$.

Suppose that $[F_{\zeta_1}] = [F_{\zeta_2}]$ in A. Then there exists $Z \in \mathcal{F}$ with $F_{\zeta_1} \mid Z = F_{\zeta_2} \mid Z$. We *claim* that this implies that

$$f_{\zeta_1} \mid R_Z = f_{\zeta_2} \mid R_Z .$$

For take $t \in R_Z$, say there exists $x = (x_n) \in Z$ and $i \in \mathbb{N}$ with $x_i = t$. Since $F_{\zeta_1}(x) = F_{\zeta_2}(x)$, we have

$$(f_{\zeta_1}(x_i))(j) = (f_{\zeta_2}(x_i))(j) \quad (i,j \in \mathbb{N}) ,$$

and so $f_{\zeta_1}(t) = f_{\zeta_2}(t)$, giving the claim. But now $[f_{\zeta_1}] = [f_{\zeta_2}]$ in $\mathbb{R}^S/R_\mathcal{F}$, and so $\zeta_1 = \zeta_2$. It follows that the map

$$\zeta \mapsto [F_\zeta], \quad \omega_2 \to A ,$$

is an injection, and so $|A| \geq \aleph_2$. By (i), we have $|A| \leq \aleph_2$, and so $|A| = \aleph_2$ in the case where $\mathcal{F}$ is non-separable. □

We now define a particular subset Q of $C^b(X)$.

Definition 7.6 *The set Q consists of the functions $h \in C^b(X)$ such that $h \mid U$ is locally constant for some dense, open subset U of X.*

Let $\mathcal{F}$ be a z-filter on X. Then

$$Q/\mathcal{F} = \{[h] \in C^b(X)/\mathcal{F} : h \in Q\} .$$

Lemma 7.7 *Let $\mathcal{F}$ be a prime z-filter on X. Then $Q/\mathcal{F}$ is order-dense in $C^b(X)/\mathcal{F}$.*

Proof Set $P = C^b(X) \cap \mathbf{Z}^{-1}[\mathcal{F}]$, a prime ideal in $C^b(X) = C(\beta X)$.

Take $a, b \in C^b(X)/\mathcal{F} = C(\beta X)/P$ with $a < b$, and take $f, g \in C^b(X)$ with $\pi_P(f) = a$ and $\pi_P(g) = b$. Set

$$Z = \{x \in X : f(x) \geq g(x)\} \quad \text{and} \quad W = \{x \in X : f(x) < g(x)\}.$$

Then $Z \notin \mathcal{F}$. Since $\overline{W} \cup Z = X \in \mathcal{F}$ and $\mathcal{F}$ is a prime z-filter, $\overline{W} \in \mathcal{F}$. There exists $h \in C^b(X)$ with

$$h(x) > 0 \quad (x \in \operatorname{int} Z) \quad \text{and} \quad h(x) = 0 \quad (x \in \overline{W}),$$

and $\pi_P((f \vee g) + h) = \pi_P(g) = b$, and so, by replacing g by $(f \vee g) + h$, we may suppose that W is a dense, open set in X.

A basic open set $U_{\mathbf{s}}$ is *separating* for f and g if there exists $x_{\mathbf{s}} \in U_{\mathbf{s}}$ such that

$$f(y) < \tfrac{1}{2}(f(x_{\mathbf{s}}) + g(x_{\mathbf{s}})) < g(y) \quad (y \in U_{\mathbf{s}}).$$

Let $\mathcal{V}$ be a maximal collection of pairwise disjoint, separating sets for f and g in X, and set

$$V = \bigcup\{U_{\mathbf{s}} : U_{\mathbf{s}} \in \mathcal{V}\}.$$

Then V is an open subset of W. Clearly, each point of W has a neighbourhood which is a separating set for f and g, and so V is dense in W, and hence in X.

Define $h : V \to \mathbb{R}$ by setting

$$h(y) = \tfrac{1}{2}(f(x_{\mathbf{s}}) + g(x_{\mathbf{s}})) \quad (y \in U_{\mathbf{s}})$$

for each $U_{\mathbf{s}} \in \mathcal{V}$. Then h is well defined and locally constant on V, and

$$f(y) < h(y) < g(y) \quad (y \in V).$$

We now set

$$h(x) = \tfrac{1}{2}(f(x) + g(x)) \quad (x \in X \setminus V),$$

and we *claim* that h is continuous on X. To show this, it suffices to show that $h(x^i) \to h(x)$ whenever $x = (x_n) \in X \setminus V$, $(x^i) \subset V$ and $x^i \to x$ as $i \to \infty$. For each $i \in \mathbb{N}$, take $\mathbf{s}^i \in S^{<\mathbb{N}}$ such that $x^i \in U_{\mathbf{s}^i} \in \mathcal{V}$.

Assume first that $(l(\mathbf{s}^i))$ is bounded. By passing to a subsequence of (x^i), we may suppose that there exists $j \in \mathbb{N}$ such that $l(\mathbf{s}^i) = j \ (i \in \mathbb{N})$. Since S is discrete, there exists $k \in \mathbb{N}$ such that $\mathbf{s}^i = (x_1, \dots, x_j) \ (i \geq k)$, and then $x \in U_{\mathbf{s}^k} \subset V$, a contradiction. Thus $(l(\mathbf{s}^i))$ is unbounded, and indeed we may suppose that $(l(\mathbf{s}^i))$ is strictly increasing. Since $x_{\mathbf{s}^i} \in U_{\mathbf{s}^i}$, it is now true that $x_{\mathbf{s}^i} \to x$ as $i \to \infty$, and so

$$h(x^i) = \tfrac{1}{2}(f(x_{\mathbf{s}^i}) + g(x_{\mathbf{s}^i})) \to \tfrac{1}{2}(f(x) + g(x)) = h(x)$$

as $i \to \infty$. This establishes the claim that h is continuous on X.

It is now clear that $h \in Q$ and that $f(x) \leq h(x) \leq g(x) \quad (x \in X)$. Set $c = [h]$. Then $c \in Q/\mathcal{F}$ and $a \leq c \leq b$. This shows that $Q/\mathcal{F}$ is order-dense in $C^b(X)/\mathcal{F}$. □

We introduce the following notation. Let T be a subset of $S^{<\mathbb{N}}$, and take $\alpha < \omega_1$. Then

$$V_T = \bigcup\{U_{\mathbf{s}} : \mathbf{s} \in T\},$$
$$V_T[\alpha] = \bigcup\{U_{\mathbf{s}} : \mathbf{s} \in T \cap S_\alpha^{<\mathbb{N}}\}.$$

Thus: each $V_T[\alpha]$ is an open subset of V_T with $V_T[\alpha] \cap X_\alpha = V_T \cap X_\alpha$; for $\alpha \leq \beta < \omega_1$, we have $V_T[\alpha] \subset V_T[\beta]$; and $V_T = \bigcup\{V_T[\alpha] : \alpha < \omega_1\}$.

Definition 7.8 *Let $\mathcal{F}$ be a z-filter on X. Then:*

(i) *$\mathcal{F}$ is* indecomposable *if, for each $T \subset S^{<\mathbb{N}}$ such that V_T is dense in X, there exists $\alpha < \omega_1$ such that $V_T[\alpha] \in \mathcal{F}$;*

(ii) *$\mathcal{F}$ is* thick *if $\overline{V} \in \mathcal{F}$ for each open subset V of X such that $V \cap X_\alpha$ is dense in X_α for some $\alpha < \omega_1$.*

Suppose that $\mathcal{F}$ is a thick z-filter on X, and take $T \subset S^{<\mathbb{N}}$ such that V_T is dense in X. By Lemma 7.3, there exists $\alpha < \omega_1$ such that $V_T \cap X_\alpha$ is dense in X_α, and hence such that $V_T[\alpha] \cap X_\alpha$ is dense in X_α. Since $\mathcal{F}$ is thick, we have $\overline{V_T[\alpha]} \in \mathcal{F}$, and so $\mathcal{F}$ is indecomposable. Thus each thick z-filter on X is indecomposable.

Lemma 7.9 (CH) *Let $\mathcal{F}$ be an indecomposable z-filter on X. Then*

$$|Q/\mathcal{F}| \leq \aleph_1 .$$

Proof Assume towards a contradiction that $(h_\zeta : \zeta < \omega_2)$ is a sequence of length ω_2 in Q such that $[h_{\zeta_1}] \neq [h_{\zeta_2}]$ in $Q/\mathcal{F}$ whenever $\zeta_1 \neq \zeta_2$.

For each $\zeta < \omega_2$, there is a dense, open subset W_ζ of X such that $h_\zeta \mid W_\zeta$ is locally constant. Take

$$T_\zeta = \{\mathbf{s} \in S^{<\mathbb{N}} : h_\zeta \mid U_{\mathbf{s}} \text{ is constant}\}.$$

Then $W_\zeta \subset V_{T_\zeta}$, and we may suppose that $W_\zeta = V_{T_\zeta}$; write $W_\zeta[\alpha]$ for $V_{T_\zeta}[\alpha]$. Since $\mathcal{F}$ is indecomposable, there exists $\alpha_\zeta < \omega_2$ such that

$$\overline{W_\zeta[\alpha_\zeta]} \in \mathcal{F}.$$

We have a map $\zeta \mapsto \alpha_\zeta$, $\omega_2 \to \omega_1$. By passing to a subsequence of the initial sequence $(h_\zeta : \zeta < \omega_2)$, we may suppose that there exists $\alpha < \omega_1$ such that $\alpha = \alpha_\zeta$ for all ζ. We then have

$$\overline{W_\zeta[\alpha]} \in \mathcal{F} \quad (\zeta < \omega_2).$$

Since $|C(X_\alpha)| = \aleph_1$, we may also suppose that there exists $h \in C(X_\alpha)$ with

$$h_\zeta \mid X_\alpha = h \quad (\zeta < \omega_2).$$

Since $S_\alpha^{<\mathbb{N}}$ is countable, there are only $\aleph_1$ possibilities for $T_\zeta \cap S_\alpha^{<\mathbb{N}}$, and so we may further suppose that $T_\zeta \cap S_\alpha^{<\mathbb{N}}$ is independent of ζ, and hence that $W_\zeta[\alpha]$ is independent of ζ, say

$$W_\zeta[\alpha] = W \quad (\zeta < \omega_2),$$

and then $\overline{W} \in \mathcal{F}$.

We now make the *key claim* of the proof of Theorem 7.2: if $\zeta_1, \zeta_2 < \omega_1$, then

$$h_{\zeta_1} \mid \overline{W} = h_{\zeta_2} \mid \overline{W}. \tag{7.2}$$

To establish this, it suffices to show that $h_{\zeta_1}(x) = h_{\zeta_2}(x)$ for each $x \in W$. Take $x \in W$, say $x \in U_{\mathbf{s}}$, where

$$\mathbf{s} \in T_{\zeta_1} \cap S_\alpha^{<\mathbb{N}} = T_{\zeta_2} \cap S_\alpha^{<\mathbb{N}}.$$

Since h_{ζ_1} and h_{ζ_2} are constant on $U_{\mathbf{s}}$ and $U_{\mathbf{s}} \cap X_\alpha \neq \emptyset$, there exists $y \in U_{\mathbf{s}} \cap X_\alpha$ such that $h_{\zeta_1}(x) = h_{\zeta_1}(y)$ and $h_{\zeta_2}(x) = h_{\zeta_2}(y)$. But certainly $h_{\zeta_1}(y) = h_{\zeta_2}(y) = h(y)$, and so $h_{\zeta_1}(x) = h_{\zeta_2}(x)$, as required.

Since $\overline{W} \in \mathcal{F}$, it follows from (7.2) that $[h_{\zeta_1}] = [h_{\zeta_2}]$ in $Q/\mathcal{F}$, a contradiction of the choice of (h_ζ). Thus $|Q/\mathcal{F}| \leq \aleph_1$, giving the result. □

Lemma 7.10 (GCH) *Let $\mathcal{F}$ be an indecomposable, non-separable, prime z-filter on X, and let K be the quotient field of $C^b(X)/\mathcal{F}$. Then K is a super-real η_1-field with $|K| = \aleph_2$ and $w(K) = |\Gamma_K| = \aleph_1$.*

Proof Since $\mathcal{F}$ is a prime z-filter, $\mathbf{Z}^{-1}[\mathcal{F}]$ is a prime ideal in $C(X)$, and so $C^b(X)/\mathcal{F}$ is an integral domain. Since $\mathcal{F}$ is non-separable, $|C^b(X)/\mathcal{F}| = \aleph_2$ by Lemma 7.5(ii), and so $|K| = \aleph_2$. By Lemma 7.7, $Q/\mathcal{F}$ is order-dense in $C^b(X)/\mathcal{F}$. Since $\mathcal{F}$ is indecomposable, $|Q/\mathcal{F}| \leq \aleph_1$ by Lemma 7.9, and so $w(K) \leq \aleph_1$. Since $|\Gamma_K| \geq \aleph_1$, it follows that $w(K) = |\Gamma_K| = \aleph_1$.

It follows from Corollary 1.21 that $\delta(K) = \aleph_1$, and so, by Theorem 4.28(ii), K is an η_1-field. □

It is now clear that, to conclude the proof of Theorem 7.2, we must exhibit an indecomposable, non-separable, prime z-filter on X. In fact, we shall now construct a thick, non-separable, prime z-filter on X.

Proof of Theorem 7.2 Let $\mathcal{G}$ be the collection of open subsets V of X such that $V \cap X_\alpha$ is dense in X_α for some $\alpha < \omega_1$.

We first *claim* that $V \neq \emptyset$ for each subset V of X of the form

$$V = V_1 \cap \cdots \cap V_n,$$

where $V_1, \ldots, V_n \in \mathcal{G}$. For take $\alpha_1, \ldots, \alpha_n < \omega_1$ such that $V_j \cap X_{\alpha_j}$ is dense in X_{α_j} for $j = 1, \ldots, n$. We may suppose that $\alpha_1 \leq \cdots \leq \alpha_n$. Take $\mathbf{s}^1 \in S_{\alpha_1}^{<\mathbb{N}}$ such that $U_{\mathbf{s}^1} \subset V_1$. Given $\mathbf{s}^j \in S_{\alpha_j}^{<\mathbb{N}}$, there exists $\mathbf{s}^{j+1} \in S_{\alpha_{j+1}}^{<\mathbb{N}}$ such that

$$U_{\mathbf{s}^{j+1}} \subset V_{j+1} \cap U_{\mathbf{s}^j};$$

in this way, we successively construct the elements $\mathbf{s}^1, \ldots, \mathbf{s}^n \in S^{<\mathbb{N}}$. Now set $U = U_{\mathbf{s}^n}$. Then

$$U \subset (V_n \cap U_{\mathbf{s}^{n-1}}) \subset (V_n \cap V_{n-1} \cap U_{\mathbf{s}^{n-2}}) \subset \cdots \subset (V_n \cap \cdots \cap V_1),$$

establishing the claim.

Let $\mathcal{F}_0$ be the z-filter on X generated by $\{\overline{V} : V \in \mathcal{G}\}$. Clearly $\mathcal{F}_0$ is a thick z-filter. For each $Z \in \mathcal{F}_0$, we have $\operatorname{int} Z \neq \emptyset$, and so it follows that $X_\alpha \notin \mathcal{F}$ $(\alpha < \omega_1)$; the filter $\mathcal{F}_0$ is non-separable.

Now take $\mathcal{F}$ to be a z-filter on X which is maximal in the family of all non-separable z-filters on X which contain $\mathcal{F}_0$. Clearly there is such a maximal z-filter $\mathcal{F}$, and $\mathcal{F}$ is thick and non-separable.

For each $Z \in \mathbf{Z}(X) \setminus \mathcal{F}$, there exists $Z' \in \mathcal{F}$ such that $Z \cap Z' \subset X_\alpha$ for some $\alpha < \omega_1$. For assume that, for each $Z' \in \mathcal{F}$, we have $Z \cap Z' \not\subset X_\alpha$ for each $\alpha < \omega_1$. Then $\mathcal{F} \cup \{Z\}$ generates a non-separable z-filter strictly containing $\mathcal{F}$, a contradiction of the maximality of $\mathcal{F}$.

We now prove that $\mathcal{F}$ is a prime z-filter. For take $Z_1, Z_2 \in \mathbf{Z}(X) \setminus \mathcal{F}$. Then there exist $Z_1', Z_2' \in \mathcal{F}$ and $\alpha_1, \alpha_2 < \omega_1$ with $Z_1 \cap Z_1' \subset X_{\alpha_1}$ and $Z_2 \cap Z_2' \subset X_{\alpha_2}$. Set $\alpha = \max\{\alpha_1, \alpha_2\}$. Then $Z_1' \cap Z_2' \in \mathcal{F}$ and

$$(Z_1 \cap Z_2) \cap (Z_1' \cap Z_2') \subset X_\alpha\,,$$

and so $Z_1 \cap Z_2 \notin \mathcal{F}$. Thus $\mathcal{F}$ is a prime z-filter.

Set $P = C^b(X) \cap \mathbf{Z}^{-1}[\mathcal{F}]$, and set $\Omega = \beta X$. Then P is a non-maximal, prime z-ideal in $C(\Omega)$, and, by Lemma 7.10, K_P is a super-real η_1-field with $|K_P| = \aleph_2$ and $w(K_P) = |\Gamma_P| = \aleph_1$. □

The above theorem has been proved in the theory ZFC + GCH. Now suppose that we are working in the theory ZFC itself. The construction gives a non-maximal, prime z-ideal P in $C(\Omega)$ such that K_P is a real-closed field with

$$|K_P| = 2^{\aleph_1} \quad \text{and} \quad w(K_P) = |\Gamma_P| = 2^{\aleph_0}\,.$$

Thus we have found an example of a prime ideal P in an algebra $C(\Omega)$ for which Question 2 of Chapter 4 is open.

The example given in Theorem 7.2 lies in class **z-ideal η_1** of the scheme in Chapter 4. We do not know whether or not this example is in the class **hyper-real**; it is not obviously a hyper-real field because the z-filter $\mathcal{F}$ constructed at the end of the proof is not a z-ultrafilter. In fact, a z-ultrafilter on X cannot be both thick and non-separable. However, the proof of Theorem 7.2 only required a prime z-filter that was both indecomposable and non-separable. (See Lemma 7.10.) We now seek to construct a z-ultrafilter on X with these two properties. To do this, we shall call in aid the combinatorial principle called diamond and denoted by $\diamondsuit$.

Definition 7.11 *The principle $\diamondsuit$ asserts that there is a sequence $(\sigma_\alpha : \alpha < \omega_1)$ of ordinals such that:*

(i) $\sigma_\alpha \le \alpha \quad (\alpha < \omega_1)$;

(ii) *for each subset A of ω_1 and each closed, unbounded subset C of ω_1, there exists $\alpha \in C$ such that $A \cap \alpha = \sigma_\alpha$.*

The principle $\diamondsuit$ implies the Continuum Hypothesis, but it is strictly stronger. It was introduced by Jensen, who proved that it holds in Gödel's constructible universe, and hence that it is relatively consistent with ZFC.

We continue to investigate the topological space X which was specified in (7.1).

Lemma 7.12 ($\diamondsuit$) *There is a sequence $(W_\alpha : \alpha < \omega_1)$ of open subsets of X such that:*

(i) *$W_\alpha \cap X_\alpha$ is dense in X_α for each $\alpha < \omega_1$;*

(ii) *for each subset T of $S^{<\mathbb{N}}$ such that V_T is dense in X, there exists $\alpha < \omega_1$ such that $W_\alpha = V_T[\alpha]$.*

Proof Fix a bijection $\pi : \omega_1 \to S^{<\mathbb{N}}$, and set

$$C_1 = \{\alpha < \omega_1 : \pi([0,\alpha)) = S_\alpha^{<\mathbb{N}}\}.$$

Since $S_\alpha^{<\mathbb{N}} = \bigcup\{S_\beta^{<\mathbb{N}} : \beta < \alpha\}$ for each limit ordinal $\alpha < \omega_1$, C_1 is closed in ω_1. Now take $\alpha < \omega_1$, and choose sequences (α_n) and (β_n) in ω_1 with

$$\alpha < \alpha_0 < \beta_0 < \alpha_1 < \beta_1 < \cdots$$

and such that $\pi([0,\alpha_n)) \subset S_{\beta_n}^{<\mathbb{N}} \subset \pi([0,\alpha_{n+1}))$ for $n \in \mathbb{N}$: this is possible because $\pi([0,\alpha_n))$ and $S_{\beta_n}^{<\mathbb{N}}$ are countable. Set $\gamma = \sup \alpha_n = \sup \beta_n$. Then

$$\pi([0,\gamma)) = \bigcup \pi([0,\alpha_n)) \subset \bigcup S_{\beta_n}^{<\mathbb{N}} \subset \bigcup \pi([0,\alpha_{n+1})) = \pi([0,\gamma))$$

and $\bigcup S_{\beta_n}^{<\mathbb{N}} = S_\gamma$, and so $S_\gamma = \pi([0,\gamma))$. Thus $\gamma \in C_1$ and $\gamma > \alpha$. This shows that C_1 is an unbounded subset of ω_1.

Let $(\sigma_\alpha : \alpha < \omega_1)$ be the sequence whose existence is asserted by $\diamondsuit$. For $\alpha < \omega_1$, set

$$T_\alpha = \{\pi(\beta) : \beta \in \sigma_\alpha\},$$

and set $W_\alpha = V_{T_\alpha}$ if $V_{T_\alpha} \cap X_\alpha$ is dense in X_α and $W_\alpha = X$ otherwise. (Here we are using the notation given before Definition 7.8.) Then each W_α is an open set in X, and $W_\alpha \cap X_\alpha$ is dense in X_α for each $\alpha < \omega_1$.

Take T to be a subset of $S^{<\mathbb{N}}$ such that V_T is dense in X, and set:

$$\begin{aligned} A &= \{\beta < \omega_1 : \pi(\beta) \in T\}; \\ C_2 &= \{\beta < \omega_1 : V_T \cap X_\beta \text{ is dense in } X_\beta\}; \\ C &= C_1 \cap C_2. \end{aligned}$$

By Lemma 7.3, C_2 is closed and unbounded in ω_1, and so C also has this property. There exists $\alpha \in C$ with $A \cap \alpha = \sigma_\alpha$. Take $\beta < \omega_1$, and set $\mathbf{s} = \pi(\beta) \in S^{<\mathbb{N}}$. Since $\alpha \in C_1$, we have $\pi([0,\alpha)) = S_\alpha^{<\mathbb{N}}$, and so $\mathbf{s} \in T \cap S_\alpha^{<\mathbb{N}}$ if and only if $\beta \in A \cap \alpha$. Thus

$$T \cap S_\alpha^{<\mathbb{N}} = T_\alpha,$$

and hence $V_{T_\alpha} = V_T[\alpha]$. Since $\alpha \in C_2$, the set $V_T[\alpha] \cap X_\alpha$ is dense in X_α, and so $V_{T_\alpha} \cap X_\alpha$ is dense in X_α, whence $W_\alpha = V_{T_\alpha}$. Finally, this shows that $W_\alpha = V_T[\alpha]$.

This completes the proof of the lemma. □

It is clear that the family $\{V_{T_\alpha} : \alpha < \omega_1\}$ has the finite intersection property, and so $\{W_\alpha : \alpha < \omega_1\}$ has this property. Let $\mathcal{F}_1$ be the z-filter generated by $\{\overline{W}_\alpha : \alpha < \omega_1\}$. Then $\mathcal{F}_1$ is indecomposable.

Lemma 7.13 ($\diamondsuit$) *There is an indecomposable, non-separable, z-ultrafilter $\mathcal{U}$ on X with $\mathcal{U} \supset \mathcal{F}_1$.*

Proof We shall construct by induction on β a sequence $(Z_\beta : \beta < \omega_1)$ of open and closed subsets of X such that:

(i) for each $\beta < \omega_1$, $Z_\beta \cap X_\beta = \emptyset$;

(ii) for each $n \in \mathbb{N}$ and $\beta_1, \dots, \beta_n$ with $\beta_1 < \beta_2 < \cdots < \beta_n < \omega_1$,

$$\{s_1 : \mathbf{s} \in S^{<\mathbb{N}}_{\beta_n+1} \text{ and } U_{\mathbf{s}} \subset Z_{\beta_1} \cap \cdots \cap Z_{\beta_n} \cap W_{\beta_1} \cap \cdots \cap W_{\beta_n}\}$$

is infinite. (Here s_1 is the first term of the sequence $\mathbf{s}$.)

Assume that $\alpha < \omega_1$ and that $(Z_\beta : \beta < \alpha)$ has been constructed to satisfy condition (i) for each $\beta < \alpha$ and condition (ii) for each $\beta_1 < \beta_2 < \cdots < \beta_n < \alpha$. Let $(\tau_k : k \in \mathbb{N})$ be an enumeration of all the finite subsets of α. We shall define

$$(\mathbf{s}^k : k \in \mathbb{N}) \subset S^{<\mathbb{N}}_{\alpha+1} \setminus S^{<\mathbb{N}}_\alpha$$

by the following recipe. Suppose that $\mathbf{s}^1, \dots, \mathbf{s}^k$ have been defined, and that $\tau_{k+1} = \{\beta_1, \dots, \beta_m\}$, where $\beta_1 < \beta_2 < \cdots < \beta_m < \alpha$. Then there exists $\mathbf{s} \in S^{<\mathbb{N}}_{\beta_m+1}$ with

$$U_{\mathbf{s}} \subset Z_{\beta_1} \cap \cdots \cap Z_{\beta_m} \cap W_{\beta_1} \cap \cdots \cap W_{\beta_m}$$

and with $s_1 \notin \{s_1^1, \dots, s_1^k\}$. Since $\beta_m < \alpha$, we have $\mathbf{s} \in S^{<\mathbb{N}}_\alpha$. Since $W_\alpha \cap X_\alpha$ is dense in X_α, there exists $\mathbf{t} \in S^{<\mathbb{N}}_\alpha$ with $\mathbf{t}$ extending $\mathbf{s}$ so that $U_{\mathbf{t}} \subset W_\alpha$. Finally, since $S_{\alpha+1} \setminus S_\alpha \neq \emptyset$, there exists $t \in S_{\alpha+1} \setminus S_\alpha$: set $\mathbf{s}^{k+1} = \mathbf{t}^\frown t$, so that

$$\mathbf{s}^{k+1} \in S^{<\mathbb{N}}_{\alpha+1} \setminus S^{<\mathbb{N}}_\alpha .$$

Thus we have an inductive definition of $(\mathbf{s}^k)$.

Now define $Z_\alpha = \bigcup\{U_{\mathbf{s}^k} : k \in \mathbb{N}\}$, so that Z_α is an open set in X. If $i \neq j$, then $s_1^i \neq s_1^j$, and this easily implies that Z_α is closed in X.

We *claim* that Z_α is a set that continues the inductive construction of the sequence (Z_β).

Indeed, since $(\mathbf{s}^k) \subset S^{<\mathbb{N}} \setminus S^{<\mathbb{N}}_\alpha$, we have $Z_\alpha \cap X_\alpha = \emptyset$, and so (i) holds for each $\beta \leq \alpha$.

Now take $\beta_1 < \beta_2 < \cdots < \beta_n < \beta_{n+1} = \alpha$. For each $k \in \mathbb{N}$ with $\{\beta_1, \dots, \beta_{n+1}\} \subset \tau_k$, we have

$$U_{\mathbf{s}^k} \subset Z_{\beta_1} \cap \cdots \cap Z_{\beta_n} \cap W_{\beta_1} \cap \cdots \cap W_{\beta_n} \cap Z_\alpha \cap W_\alpha .$$

But there are infinitely many values of k such that $\{\beta_1, \dots, \beta_{n+1}\} \subset \tau_k$, and for any two such values the corresponding first terms of the sequence $\mathbf{s}^k$ are distinct, and so

$$\{s_1 : \mathbf{s} \in S^{<\mathbb{N}} \text{ and } U_{\mathbf{s}} \subset Z_{\beta_1} \cap \cdots \cap Z_{\beta_{n+1}} \cap W_{\beta_1} \cap \cdots \cap W_{\beta_{n+1}}\}$$

is infinite. Thus (ii) holds in this case.

The inductive construction of $(Z_\beta : \beta < \omega_1)$ continues.

We now let $\mathcal{U}$ be any z-ultrafilter on X which contains the family

$$\{Z_{\beta_1} \cap \cdots \cap Z_{\beta_n} \cap \overline{W}_{\beta_1} \cap \cdots \cap \overline{W}_{\beta_n} : \beta_1, \dots, \beta_n < \omega_1\} .$$

For each $\alpha < \omega_1$, we have $Z_\alpha \in \mathcal{U}$ and $Z_\alpha \cap X_\alpha = \emptyset$, and so $X_\alpha \notin \mathcal{U}$. Thus $\mathcal{U}$ is non-separable. For each $T \subset S^{<\mathbb{N}}$ such that V_T is dense in X, there exists $\alpha < \omega_1$ such that $W_\alpha = V_T[\alpha]$, and $\overline{W}_\alpha \in \mathcal{U}$. Thus $\mathcal{U}$ is indecomposable. □

As we said in the introduction to this chapter, examples with stronger properties will be given later; the existence of a topological space X with the properties given in the following theorem will be established in Theorem 9.26 without the use of $\diamondsuit$.

Theorem 7.14 (GCH + $\diamondsuit$) *There is a maximal ideal M in the algebra $C(X)$ such that the hyper-real field $K = C(X)/M$ has the following properties:*

$$|K| = \aleph_2 \quad \text{and} \quad w(K) = |\Gamma_K| = \aleph_1 .$$

Proof Take $\mathcal{U}$ to be an indecomposable, non-separable, z-ultrafilter on X whose existence is asserted in Lemma 7.13, and set $K = C(X)/M$, where $M = \mathbf{Z}^{-1}[\mathcal{U}]$. Then K is a hyper-real field, and, by Lemma 7.10, K is as required. □

We now turn to examine the case of ultrapowers (see Chapter 4). We shall calculate the invariants $|K|$, $w(K)$, and $|\Gamma_K|$ in the case where $K = \mathbb{R}^\kappa/\mathcal{U}$. A second proof of the following result will be noted on page 208.

Theorem 7.15 *Let κ be a cardinal, let $\mathcal{U}$ be a free ultrafilter on κ, and set $K = \mathbb{R}^\kappa/\mathcal{U}$. Then*

$$|\Gamma_K| = w(K) = |\mathbb{Q}^\kappa/\mathcal{U}| .$$

Proof Certainly $w(K) \geq |\Gamma_K|$.

We shall show that $w(K) \leq |\mathbb{Q}^\kappa/\mathcal{U}|$ by showing that $\mathbb{Q}^\kappa/\mathcal{U}$ is order-dense in K. Indeed, take $[f], [g] \in K$ with $[f] < [g]$, and let

$$S = \{\sigma < \kappa : f(\sigma) < g(\sigma)\} .$$

Then $S \in \mathcal{U}$. For each $\sigma \in S$, choose $h(\sigma) \in \mathbb{Q} \cap (f(\sigma), g(\sigma))$. Then $h \in \mathbb{Q}^\kappa/\mathcal{U}$ and $[f] < [h] < [g]$ in K, so that $\mathbb{Q}^\kappa/\mathcal{U}$ is order-dense in K.

We finally show that $|\mathbb{Q}^\kappa/\mathcal{U}| \leq |\Gamma_K|$ by giving an injection

$$\psi : \mathbb{Q}^\kappa/\mathcal{U} \to \Gamma_K .$$

Let $\iota : \mathbb{Q} \to \mathbb{N}$ be an injection, and let f_0 be a fixed element of $\mathbb{R}^\kappa$ (with $f_0(\sigma) \geq 1$ $(\sigma < \kappa)$) and $[f_0]$ infinitely large in K. For $f \in \mathbb{R}^\kappa$, set

$$\theta(f) : \sigma \mapsto f_0(\sigma)^{\iota(f(\sigma))}, \quad \kappa \to \mathbb{R} ,$$

and set $\psi([f]) = v_K([\theta(f)])$. Then $\psi([f])$ is well defined on Γ_K. Take $[f], [g] \in K$ with $[f] \neq [g]$, and set

$$S = \{\sigma < \kappa : \iota(f(\sigma)) > \iota(g(\sigma))\}$$

and

$$T = \{\sigma < \kappa : \iota(f(\sigma)) < \iota(g(\sigma))\} .$$

Then $S \cup T \in \mathcal{U}$ because ι is an injection, and so either $S \in \mathcal{U}$ or $T \in \mathcal{U}$, say $S \in \mathcal{U}$. For each $\sigma \in S$,

$$\theta(f)(\sigma) = f_0(\sigma)^{\iota(f(\sigma)) - \iota(g(\sigma))}\theta(g)(\sigma) \geq f_0(\sigma)\theta(g)(\sigma) ,$$

and so $[\theta(f)] \geq n[\theta(g)]$ for each $n \in \mathbb{N}$. Thus $\psi(f) < \psi(g)$, and so ψ is an injection, as required. □

The following result is deeper than the above theorems; its proof is not given here. It is an example of a consistency result whose statement involves the assumption that the theory consisting of ZFC together with a certain 'large cardinal' axiom is consistent; the large cardinal axiom is the assertion that an 'almost huge' cardinal exists. It is known that the conclusion of Theorem 7.16 implies the consistency of *some* large cardinal axiom, but an exact match is not known. For all practical purposes, the conclusion of the following theorem is that it cannot be proved in the theory ZFC that there is not a uniform ultrafilter $\mathcal{U}$ on ω_1 such that $\left|\mathbb{Q}^{\omega_1}/\mathcal{U}\right| = \aleph_1$.

Theorem 7.16 *Assume that it is consistent with* ZFC *that there exists an almost huge cardinal. Then it is also consistent with* ZFC *that there exists a uniform ultrafilter $\mathcal{U}$ on ω_1 such that $\left|\mathbb{Q}^{\omega_1}/\mathcal{U}\right| = \aleph_1$.* □

The route to Theorem 7.16 passes through the following result. In this theorem, $X = S^{\mathbb{N}}$ is the topological space constructed at the beginning of the proof of Theorem 7.2.

Theorem 7.17 *Assume that it is consistent with* ZFC *that there exists an almost huge cardinal. Then it is also consistent with* ZFC + CH *that there is a set $Z \subset X$ such that:*

(i) *Z is nowhere meagre;*

(ii) *every subset Y of Z differs from an open set by a meagre set.* □

Condition (i) states that $Z \cap U$ is non-meagre (or second category) for each non-empty, open subset U of X. Condition (ii) states that, for each subset Y of Z, there is an open subset U of X such that $(Y \setminus U) \cup (U \setminus Y)$ is meagre (or first category).

Let us assume the existence of the subset Z of X specified in Theorem 7.17, and sketch the argument that there exists a uniform ultrafilter $\mathcal{U}$ on Z such that $\left|\mathbb{Q}^Z/\mathcal{U}\right| = \aleph_1$. By condition (i) of Theorem 7.17, $|Z| = \aleph_1$, and so this is sufficient for the proof of Theorem 7.16.

We use the notation from the proof of Theorem 7.2. Thus $\mathcal{G}$ is the collection of open subsets V of X such that $V \cap X_\alpha$ is dense in X_α for some $\alpha < \omega_1$. We proved that $\mathcal{G}$ has the finite intersection property. Now let

$$\mathcal{G}_1 = \{V_1 \cap \cdots \cap V_n : V_1, \ldots, V_n \in \mathcal{G}\},$$

$$\mathcal{F} = \{Y \subset Z : (Y \setminus V) \cup (V \setminus Y) \text{ is meagre for some } V \in \mathcal{G}_1\}.$$

Then $\mathcal{F}$ is a uniform filter on Z. Let $\mathcal{U}$ be an ultrafilter extending $\mathcal{F}$. The key point is that, by condition (ii) of Theorem 7.17, for each function $f : Z \to \mathbb{Q}$, there exists $Z' \subset Z$ such that $f \mid Z'$ is continuous and $Z \setminus Z'$ is meagre. Given this the argument of Lemma 7.9 can be adapted to show that $\left|\mathbb{Q}^Z/\mathcal{U}\right| = \aleph_1$.

By combining Theorems 7.15 and 7.16 and Proposition 7.1, we finally obtain the following result.

Theorem 7.18 *Assume that it is consistent with* ZFC *that there exists an almost huge cardinal. Then it is also consistent with* ZFC *that there exists a uniform ultrafilter* $\mathcal{U}$ *on* ω_1 *such that*

$$|\Gamma_K| = w(K) = \aleph_1 \text{ and } |K| = 2^{\aleph_1} ,$$

where K *is the ultrapower* $\mathbb{R}^{\omega_1}/\mathcal{U}$. □

Notes 7.19 Again, we believe that the results of this chapter after Proposition 7.1 are new.

Proposition 7.1 is related to Comfort and Negrepontis (1974, Theorem 12.21). The homeomorphism between $\mathbb{N}^{\mathbb{N}}$ and the irrationals in $(0,1)$ is given in detail in Jech (1978, p.36).

As we noted in the Preface, 'relative consistency' results are discussed in Dales and Woodin (1987, Chapter 4).

We make some remarks about the history of Theorem 7.16. The consistency of the existence of an ultrafilter $\mathcal{U}$ as specified in the theorem was first established by Laver (1982) from the relative consistency of a strong form of the Axiom of Determinacy. The first such consistency result to follow from the relative consistency of a large cardinal axiom is given by Foreman *et al.* (1988, Theorem 9), although that theorem requires a slightly larger cardinal than the 'almost huge' one of Theorem 7.16; subsequently Shelah has weakened the hypothesis on the large cardinal required. Theorem 7.17 is an unpublished result of W.H. Woodin.

8

Non-standard structures for super-real fields and the gap theorem

Let $\mathcal{U}$ be a free ultrafilter on $\mathbb{N}$, and let $\mathbb{R}^{\mathbb{N}}/\mathcal{U}$ be the corresponding ultrapower. As we remarked in Chapter 4, $\mathbb{R}^{\mathbb{N}}/\mathcal{U}$ is of the form K_P for a certain valuation prime z-ideal $P = J_{\mathfrak{p}}$ in $C(\beta\mathbb{N})$. From the point of view of model theory, $\mathbb{R}^{\mathbb{N}}/\mathcal{U}$ has a special importance: it is the underlying field of a non-standard model of analysis. Thus $\mathbb{R}^{\mathbb{N}}/\mathcal{U}$ is a real-closed η_1-field of cardinality $\mathfrak{c}$, but, further than this, much of the structure of $\mathbb{R}$ can be 'transferred' to $\mathbb{R}^{\mathbb{N}}/\mathcal{U}$; $\mathbb{R}^{\mathbb{N}}/\mathcal{U}$ with this additional structure is what is termed a 'non-standard model of analysis'. (See Chapter 10 for a brief resume of model theory.) We shall show that a (in general, weaker) version of this additional structure exists for an arbitrary super-real field K_P. This structure will allow us to obtain additional information about super-real fields, and in particular about the structure of their gaps. As our main theorem of this chapter, we shall prove a *gap theorem* (Theorem 8.61): whenever $\langle A, B\rangle$ is an $(\aleph_0, \lambda)$-gap in a super-real semi-η_1-field K_P with λ infinite, then necessarily $\lambda = \operatorname{cof} K_P^0$. Hyper-real fields are necessarily η_1-fields, and so the gap theorem applies to all hyper-real fields. The gap theorem is not true for an arbitrary real-closed η_1-field, and so we obtain (in ZFC) a strong form of counter-example to the suggestion that every real-closed η_1-field is a hyper-real field: Theorem 8.63 exhibits a real-closed η_1-field which is not even order-isomorphic to any super-real field.

In Chapters 9 and 10, we shall build on the results of the present chapter, developing a more sophisticated form of the additional structure, and proving further theorems. These additional theorems will allow us to separate some of the classes of the scheme of Chapter 4 that we cannot separate in any other way. Thus, independent of their application to the proof of the gap theorem, most of the preliminary results of this chapter will be required in Chapters 9 and 10.

The additional structure that we impose on a super-real field K_Q is rather straightforward to define when Q is a prime z-ideal in $C(\Omega)$. (By applying our construction in the case where the field is an ultrapower, we shall obtain the usual structure for non-standard analysis.) However the definition of the structures in K_P in the case where P is an arbitrary prime ideal in an algebra $C(\Omega)$ is considerably more subtle; care is required because, in the general case, $C(\Omega)$ may contain rather few functions as compared with the space $\mathbb{R}^{\Omega}$. The structures on K_P will be defined by transferring structures from a field K_Q, where Q is a prime z-ideal in $C(\Omega)$ below P.

We shall begin with a general definition, and then describe our additional structure on K_Q in the case where Q is a prime z-ideal.

Let I be an ideal in $C(\Omega)$. Recall from Definition 4.19(ii) that the hull of I is $h(I) = \bigcap\{\mathbf{Z}(f) : f \in I\}$.

Definition 8.1 *Let Ω be a compact space, and let I be an ideal in $C(\Omega)$. A family $\{Z_j : j \in \mathbb{N}\}$ of sets in $\mathbf{Z}(\Omega)$, whose union is denoted by Z_∞, is an* I-partition with limit set contained in Z_0 *if:*

(i) $Z_0 \cap Z_\infty = \emptyset$;

(ii) $Z_0 \cup Z_\infty \in \mathbf{Z}[I]$, *and* $Z_0 \in \mathbf{Z}(\Omega) \setminus \mathbf{Z}[I]$;

(iii) $Z_j \cap \overline{(Z_\infty \setminus Z_j)} = \emptyset \quad (j \in \mathbb{N})$;

(iv) $h(I) \subset Z_0$.

Note that, if $\{Z_j : j \in \mathbb{N}\}$ is such a partition, then

$$\overline{Z}_\infty \subset Z_0 \cup Z_\infty .$$

Also $Z_j \notin \mathbf{Z}[I]$ for $j \in \mathbb{N}$. The 'limit set' implicitly referred to in the above definition is $Z_0 \cap \overline{Z}_\infty$; in fact, this set will not occur explicitly in our presentation. In the case where P is a prime ideal in $C(\Omega)$, we have $h(P) = \{x_P\}$, and so $x_P \in Z_0$. If Ω is metrizable and P is a non-maximal, prime ideal in $C(\Omega)$, we need only consider the relatively straightforward situation where $Z_0 = \{x_P\}$; in this case, Z_∞ is just the union of a pairwise disjoint sequence of closed sets which accumulates at $\{x_P\}$, the only further requirement being that $\{x_P\} \cup Z_\infty \in \mathbf{Z}[P]$. However, such a simplification is not possible in general. For example, let I be the ideal $J_{\mathfrak{p}}$ in $C(\beta\mathbb{N})$, where $\mathfrak{p} \in \beta\mathbb{N} \setminus \mathbb{N}$. Set $Z_j = \{j\}$ for $j \in \mathbb{N}$, and set $Z_0 = \beta\mathbb{N} \setminus \mathbb{N}$. Then $\{Z_j : j \in \mathbb{N}\}$ is a $J_{\mathfrak{p}}$-partition with limit set contained in Z_0, but we cannot take the limit set to be the singleton $\{\mathfrak{p}\}$.

Let $\{Z_j : j \in \mathbb{N}\}$ be a sequence in $\mathbf{Z}(\Omega)$ with union Z_∞ satisfying (iii), above. First note that, if f is a function on Z_∞ such that $f \mid Z_j$ is continuous for each j and $|f|_{Z_j} \to 0$ as $j \to \infty$, then, by Tietze's extension theorem, f is the restriction to Z_∞ of an element of $C(\Omega)$.

Now suppose that I is an ideal in $C(\Omega)$, that $\{Z_j : j \in \mathbb{N}\}$ is an I-partition with limit set contained in Z_0, and that σ is a subset of $\mathbb{N}$. Take $f_1 \in C(\Omega)$ with $\mathbf{Z}(f_1) = Z_0 \cup Z_\infty$, and take $f_2 \in C(\Omega)$ with

$$f_2(Z_j) = \{0\} \quad (j \in \sigma \cup \{0\}) \quad \text{and} \quad f_2(Z_j) = \{1/j\} \quad (j \in \mathbb{N} \setminus \sigma) .$$

Then

$$Z_0 \cup \bigcup\{Z_j : j \in \sigma\} = \mathbf{Z}(f_1^2 + f_2^2) \in \mathbf{Z}(\Omega) . \tag{8.1}$$

Thus $\{Z_j : j \in \sigma\}$ is an I-partition with limit set contained in Z_0 provided only that

$$Z_0 \cup \bigcup\{Z_j : j \in \sigma\} \in \mathbf{Z}[I] .$$

Our first task is to show that a non-maximal, prime z-ideal Q does indeed have a Q-partition.

Theorem 8.2 *Let Ω be a compact space, and let Q be a non-maximal, prime z-ideal in $C(\Omega)$. Then there is a Q-partition.*

Proof Take

$$g \in M_{x_Q}^{+} \setminus Q\,,$$

so that $\pi_Q(g) \in M_Q^{+} \setminus \{0\}$; we may suppose that $0 \leq g(x) \leq 1 \quad (x \in \Omega)$. We construct a certain partition of an element of $\mathbf{Z}[Q]$. Indeed, set $Z_0 = \mathbf{Z}(g)$ and, for each $j \in \mathbb{N}$, set

$$X_j = \left\{x \in \Omega : g(x) \in \left[\frac{1}{j+1}, \frac{1}{j}\right]\right\}, \tag{8.2}$$

so that $X_j \in \mathbf{Z}(\Omega)$. Finally, set

$$\Omega_1 = Z_0 \cup \bigcup\{X_{2j} : j \in \mathbb{N}\}, \quad \Omega_2 = Z_0 \cup \bigcup\{X_{2j+1} : j \in \mathbb{N}\}\,.$$

It is clear that $\Omega_1, \Omega_2 \in \mathbf{Z}(\Omega)$ and that $\Omega_1 \cup \Omega_2 = \Omega$. By Proposition 4.10(i), $\mathbf{Z}[Q]$ is a prime z-filter, and so either $\Omega_1 \in \mathbf{Z}[Q]$ or $\Omega_2 \in \mathbf{Z}[Q]$. In the former case, we set $Z_j = X_{2j} \quad (j \in \mathbb{N})$, and, in the latter case, we set $Z_j = X_{2j-1} \quad (j \in \mathbb{N})$.

Clearly $\{Z_j : j \in \mathbb{N}\}$ is a Q-partition. □

The above partition is a *standard partition* for Q corresponding to g.

It is easy to see that a prime ideal P which is not a z-ideal has a P-partition if and only if $\mathbf{Z}^{-1}[\mathbf{Z}\,[\mathbf{P}]\,]$, the minimum prime z-ideal containing P, is not equal to the unique maximal ideal M_{x_P} containing P, and so there are many examples of prime ideals P without a P-partition.

The following straightforward remark is very useful.

Proposition 8.3 *Let Q be a prime z-ideal in $C(\Omega)$, and let $\{Z_j : j \in \mathbb{N}\}$ be a Q-partition with union Z_∞.*

(i) *Suppose that $f \in C(\Omega)$ with $\mathbf{Z}(f) \supset Z_\infty$. Then $f \in Q$.*

(ii) *Suppose that $f, g \in C(\Omega)$ with $f(x) \geq g(x)$ $(x \in Z_\infty)$. Then*

$$\pi_Q(f) \geq \pi_Q(g)\,.$$

Proof The partition $\{Z_j : j \in \mathbb{N}\}$ has limit set contained in Z_0, say.

(i) As above, there exists $g \in C(\Omega)^+ \setminus Q$ with $\mathbf{Z}(g) = Z_0$. We see that $\mathbf{Z}(fg) \supset Z_0 \cup Z_\infty$, and so $\mathbf{Z}(fg) \in \mathbf{Z}[Q]$ because $Z_0 \cup Z_\infty \in \mathbf{Z}[Q]$. Since Q is a z-ideal, we have $fg \in Q$, and so $f \in Q$.

(ii) Set

$$h = (f \vee g) - f \in C(\Omega)\,.$$

Then $h(x) = 0 \quad (x \in Z_\infty)$, and so $h \in Q$ by (i). It is now immediate that $\pi_Q(f) = \pi_Q(f \vee g) \geq \pi_Q(g)$. □

Definition 8.4 *Let I be an ideal in $C(\Omega)$. An I-partition $\{Z_j : j \in \mathbb{N}\}$* refines *an I-partition $\{W_k : k \in \mathbb{N}\}$ if, for each $j \in \mathbb{N}$, there exists $k \in \mathbb{N}$ with $W_k \supset Z_j$.*

For example, let $I = J_{\mathfrak{p}}$ in $C(\beta\mathbb{N})$, as above, and let U be a neighbourhood of $\mathfrak{p}$ in $\beta\mathbb{N}$, so that $U \cap \mathbb{N}$ belongs to the ultrafilter associated with $\mathfrak{p}$, say $U \cap \mathbb{N} = \{n_k : k \in \mathbb{N}\}$. Then $\{\{n_k\} : k \in \mathbb{N}\}$ is a $J_{\mathfrak{p}}$-partition which refines the $J_{\mathfrak{p}}$-partition $\{\{j\} : j \in \mathbb{N}\}$.

The next lemma shows that the family of Q-partitions is a directed set with respect to refinement whenever Q is a prime z-ideal in $C(\Omega)$.

Lemma 8.5 *Let Q be a prime z-ideal in $C(\Omega)$. Suppose that $\{Z_j : j \in \mathbb{N}\}$ and $\{W_k : k \in \mathbb{N}\}$ are Q-partitions with limit sets contained in Z_0 and W_0, respectively. Then $\{Z_j \cap W_k : j, k \in \mathbb{N}\}$ is a Q-partition with limit set contained in*

$$(Z_0 \cap (W_0 \cup W_\infty)) \cup (W_0 \cap (Z_0 \cup Z_\infty)),$$

and the partition $\{Z_j \cap W_k : j, k \in \mathbb{N}\}$ refines both of the given partitions.

Proof Set $Z = Z_0 \cup Z_\infty$ and $W = W_0 \cup W_\infty$, so that $Z, W \in \mathbf{Z}[Q]$.

Now define $U_{j,k} = Z_j \cap W_k$ $(j, k \in \mathbb{N})$ and $U_\infty = \bigcup\{U_{j,k} : j, k \in \mathbb{N}\}$. The family $\{U_{j,k} : j, k \in \mathbb{N}\}$ is contained in $\mathbf{Z}(\Omega) \setminus \mathbf{Z}[Q]$ because Q is an ideal. We clearly have

$$U_{j,k} \cap \overline{(U_\infty \setminus U_{j,k})} = \emptyset \quad (j, k \in \mathbb{N}).$$

Set $U_0 = (Z_0 \cap W) \cup (W_0 \cap Z)$, so that $U_0 \in \mathbf{Z}(\Omega)$ and $U_0 \cap U_\infty = \emptyset$. Since $Z_0, W_0 \notin \mathbf{Z}[Q]$ and Q is a prime ideal, necessarily $Z_0 \cup W_0 \notin \mathbf{Z}[Q]$. But $U_0 \subset Z_0 \cup W_0$ and $Z_0 \cup W_0 \in \mathbf{Z}(\Omega)$, and so $U_0 \notin \mathbf{Z}[Q]$. We have

$$U_0 \cup U_\infty = (Z_0 \cap W) \cup (W_0 \cap Z) \cup (Z_\infty \cap W_\infty) = Z \cap W \in \mathbf{Z}[Q],$$

and so the result follows. □

Lemma 8.6 *Let Q be a prime z-ideal, and suppose that $\{Z_j : j \in \mathbb{N}\}$ is a Q-partition with limit set contained in Z_0. Then*

$$\left\{\sigma \subset \mathbb{N} : Z_0 \cup \bigcup\{Z_j : j \in \sigma\} \in \mathbf{Z}[Q]\right\}$$

is a free ultrafilter on $\mathbb{N}$.

Proof Let $\mathcal{U}_Q$ be the family of subsets of $\mathbb{N}$ defined in the statement of the lemma. Since $Z_0 \notin \mathbf{Z}[Q]$, necessarily $\emptyset \notin \mathcal{U}_Q$. Suppose that $\sigma, \tau \in \mathcal{U}_Q$. Then $\sigma \cap \tau \in \mathcal{U}_Q$ because $\mathbf{Z}[Q]$ is a z-filter. Suppose that $\sigma \in \mathcal{U}_Q$ and that $\sigma \subset \tau \subset \mathbb{N}$. Then $\tau \in \mathcal{U}_Q$ since it follows from (8.1) that

$$Z_0 \cup \bigcup\{Z_j : j \in \tau\} \in \mathbf{Z}(\Omega).$$

Suppose that $\sigma, \tau \subset \mathbb{N}$ with $\sigma \cup \tau = \mathbb{N}$. Then either $\sigma \in \mathcal{U}_Q$ or $\tau \in \mathcal{U}_Q$ because $\mathbf{Z}[Q]$ is a prime z-filter. This shows that $\mathcal{U}_Q$ is an ultrafilter on $\mathbb{N}$.

Finally assume that $\{j\} \in \mathcal{U}_Q$ for some $j \in \mathbb{N}$. Then $Z_0 \cup Z_j \in \mathbf{Z}[Q]$, and so either $Z_0 \in \mathbf{Z}[Q]$ or $Z_j \in \mathbf{Z}[Q]$, a contradiction. Thus $\mathcal{U}_Q$ is a free ultrafilter on $\mathbb{N}$. □

Let Q be a non-maximal, prime z-ideal in $C(\Omega)$. Then the following three important statements follow from Theorem 8.2 and the above two lemmas.

(1) There is a Q-partition.

(2) Suppose that functions f and g are each defined on the unions of some Q-partitions. Then we may suppose that f and g are defined on the union of the *same* Q-partition.

(3) Suppose that $\{Z_j : j \in \mathbb{N}\}$ is a Q-partition, and that, for each $j \in \mathbb{N}$, $\mathcal{P}_j$ is a property that may or may not hold. Then we may suppose *either* that $\mathcal{P}_j$ holds for all $j \in \mathbb{N}$ *or* that the negation of $\mathcal{P}_j$ holds for all $j \in \mathbb{N}$.

We shall frequently use the above three facts in the following calculations.

Our next step is to define 'bigger' versions of $\mathbb{Q}$ and $\mathbb{R}$ inside the field K_Q.

Definition 8.7 *Let Q be a non-maximal, prime z-ideal in an algebra $C(\Omega)$. For a subset S of $\mathbb{R}$, $B_Q(S)$ is the set of functions f defined on the union Z_∞ of some Q-partition $\{Z_j : j \in \mathbb{N}\}$ and taking values in S with the property that $f \mid Z_j$ is constant for each $j \in \mathbb{N}$.*

Let $f, g \in B_Q(S)$. Then

$$f \sim g$$

if there is a Q-partition $\{Z_j : j \in \mathbb{N}\}$ with union Z_∞ such that f and g are defined on Z_∞ and $f \mid Z_\infty = g \mid Z_\infty$.

It follows from Proposition 8.3(ii) that, if $f, g \in C(\Omega)$ and $f \sim g$ in $B_Q(\mathbb{R})$, then $\pi_Q(f) = \pi_Q(g)$ in A_Q.

We now fix a non-maximal, prime z-ideal Q in $C(\Omega)$. At some key points we shall require that K_Q be a semi-η_1-field, but most of the technical preliminaries do not need this.

It is clear that the relation $\sim$ of Definition 8.7 is an equivalence relation on the set $B_Q(\mathbb{R})$; we denote the equivalence class containing f by $[f]$. Now take $f, g \in B_Q(\mathbb{R})$. Then, by passing to other elements of the equivalence classes, we may suppose that f and g are both defined on the union of the same Q-partition. We define $f + g$, αf, and fg pointwise on this union, and then define

$$[f] + [g] = [f + g]\,, \quad \alpha[f] = [\alpha f]\,, \quad [f][g] = [fg]\,,$$

(where $\alpha \in \mathbb{R}$). We also set

$$[f] \leq [g] \quad \text{in} \quad (B_Q(\mathbb{R})/\sim)$$

if there is a Q-partition with union Z_∞ such that

$$f(x) \leq g(x) \quad (x \in Z_\infty)\,.$$

Clearly the various operations are well defined.

It is easily checked that $(B_Q(\mathbb{R})/\sim)$ is a real-closed ordered field; the calculations are essentially the same as those which established that $\mathbb{R}^\kappa/\mathcal{U}$ is such a field for each cardinal κ and each ultrafilter $\mathcal{U}$ on κ. For example, take $f \in B_Q(\mathbb{R})$, and take a Q-partition $\{Z_j : j \in \mathbb{N}\}$ such that

$$f(Z_j) = \{r_j\} \quad (j \in \mathbb{N})\,.$$

We may suppose either that $r_j = 0$ for all $j \in \mathbb{N}$ or that $r_j \neq 0$ for all $j \in \mathbb{N}$. In the former case, $f \sim 0$ and so $[f] = 0$ in $(B_Q(\mathbb{R})/\sim)$. In the latter case, take $g \in B_Q(\mathbb{R})$ such that

$$g(Z_j) = \{1/r_j\} \quad (j \in \mathbb{N}).$$

Then $[f]\,[g] = 1$ in $(B_Q(\mathbb{R})/\sim)$. In this latter case, we may also suppose either that $r_j > 0$ for all $j \in \mathbb{N}$ (i.e. $[f] > 0$) or that $r_j < 0$ for all $j \in \mathbb{N}$ (i.e. $[f] < 0$); this verifies that the order on $(B_Q(\mathbb{R})/\sim)$ is a total order.

For example, take $Q = J_{\mathfrak{p}}$ in $C(\beta\mathbb{N})$, where $\mathfrak{p} \in \beta\mathbb{N}\setminus\mathbb{N}$. Then $f \sim g$ in $B_Q(\mathbb{R})$ if and only if $f|U = g|U$ for some U in the ultrafilter $\mathcal{U}_{\mathfrak{p}}$, and so the field $B_Q(\mathbb{R})$ is just the ultrapower $\mathbb{R}^{\mathbb{N}}/\mathcal{U}_{\mathfrak{p}}$.

We now define a map

$$\pi : (B_Q(\mathbb{R})/\sim) \to K_Q .$$

Indeed, take $f \in B_Q(\mathbb{R})$. If $[f] = 0$, then $\pi([f]) = 0$. If $[f] > 0$, there are a Q-partition $\{Z_j : j \in \mathbb{N}\}$ with limit set contained in Z_0 and a sequence (r_j) in $\mathbb{R}^+ \setminus \{0\}$ such that $f(Z_j) = \{r_j\}$ $(j \in \mathbb{N})$. Choose sequences (α_j) and (β_j) in $\mathbb{R}^+ \setminus \{0\}$ such that $\beta_j = \alpha_j r_j$ $(j \in \mathbb{N})$ and such that $\alpha_j \to 0$ and $\beta_j \to 0$ as $j \to \infty$. There exist $g, h \in C(\Omega)^+ \setminus Q$ with

$$g(x) = \beta_j, \;\; h(x) = \alpha_j \quad (x \in Z_j). \tag{8.3}$$

We have

$$g(x) = f(x)h(x) \quad (x \in Z_0 \cup Z_\infty).$$

Define

$$\pi([f]) = \pi_Q(g)(\pi_Q(h))^{-1} \in K_Q .$$

If $[f] < 0$, define $\pi([f]) = -\pi(-[f])$.

We *claim* that $\pi([f])$ is well defined in K_Q. For let

$$\pi_Q(g_1)(\pi_Q(h_1))^{-1} \quad \text{and} \quad \pi_Q(g_2)(\pi_Q(h_2))^{-1}$$

be two possible interpretations of $\pi([f])$. Then there is a Q-partition

$$\{Z_j : j \in \mathbb{N}\}$$

with limit set contained in Z_0 such that $h_1 g_2 = h_2 g_1$ on $Z_0 \cup Z_\infty$. Hence $h_1 g_2 - h_2 g_1 \in Q$ by Proposition 8.3(i), and so

$$\pi_Q(g_1)(\pi_Q(h_1))^{-1} = \pi_Q(g_2)(\pi_Q(h_2))^{-1},$$

as required.

It is now routine to verify that $\pi : (B_Q(\mathbb{R})/\sim) \to K_Q$ is an isotonic monomorphism.

Definition 8.8 *For a subset S of $\mathbb{R}$, define*

$$S_Q = \pi(B_Q(S)/\sim).$$

Thus, for example, $\mathbb{Q}_Q$ is a rational subfield of $\mathbb{R}_Q$. Three remarks are in order.

First, the map π may or may not be a surjection of $(B_Q(\mathbb{R})/\sim)$ onto K_Q. For example, suppose that $Q = J_\mathfrak{p}$ in $C(\beta\mathbb{N})$. Then $\mathbb{R}_Q = \mathbb{R}^\mathbb{N}/\mathcal{U}_\mathfrak{p} = K_Q$. On the other hand, suppose that $\Omega = \beta\mathbb{R}$ and that Q is a non-maximal, prime z-ideal in $C(\beta\mathbb{R})$. Let $\mathcal{F}$ be the prime z-filter on $\mathbb{R}$ corresponding to Q. It is not difficult to see that $\mathbb{R}_Q = K_Q$ if and only if $\mathcal{F}$ contains a discrete subset of $\mathbb{R}$. Now each z-filter on $\mathbb{R}$ which is maximal among the z-filters which do not contain a discrete set is necessarily a prime z-filter, and the ideal, say Q, corresponding to such a z-filter is a non-maximal, prime z-ideal in $C(\mathbb{R})$ for which $\mathbb{R}_Q \neq K_Q$.

The second remark is that $\mathbb{R}_Q \cap K_Q^{\#}$ is not necessarily contained in A_Q. For example, suppose that $\mathfrak{p} \in \beta\mathbb{N} \setminus \mathbb{N}$, and that $Q = c \cap J_\mathfrak{p}$, a non-maximal, prime z-ideal in c. Then

$$\mathbb{R}_Q \cap K_Q^{\#} = K_Q^{\#} \cong C(\beta\mathbb{N})/J_\mathfrak{p} \quad \text{and} \quad \mathbb{R}_Q \cap K_Q^{\#} \not\subset c/(c \cap J_\mathfrak{p}) = A_Q\,.$$

Third, we note that, in our setting, we are essentially viewing K_Q as the *real line of a non-standard model* of (some fragment of) analysis. We shall develop this point of view, and discuss models, in Chapter 10.

A key result in our theory is that $\mathbb{Q}_Q$ is always order-dense in K_Q, even in the case where $\mathbb{R}_Q \subsetneq K_Q$. The proof is a refinement of the proof in Theorem 8.2 that there is a Q-partition.

Theorem 8.9 *Let Ω be a compact space, and let Q be a non-maximal, prime z-ideal in $C(\Omega)$. Then $\mathbb{Q}_Q$ and $K_Q \setminus \mathbb{Q}_Q$ are order-dense in the ordered field K_Q.*

Proof Since $\mathbb{Q}_Q^+ \setminus \{0\}$ is a subgroup of $(K_Q^+ \setminus \{0\}, \cdot)$, it follows from Proposition 4.26 that it is sufficient to show that $\mathbb{Q}_Q \cap [c, d] \neq \emptyset$ whenever $c < d$ in $M_Q^+ \setminus \{0\}$.

Thus take $c, d \in M_Q$ with $0 < c < d$, and take $f, g \in M_{xQ}^+ \setminus Q$ with $\pi_Q(f) = c$, with $\pi_p(g) = d$, and with

$$0 \le f(x) \le g(x) \le 1 \quad (x \in \Omega)\,.$$

Set $h = g - f$, so that $h \in M_{xQ}^+ \setminus Q$, and let $\{Z_k : k \in \mathbb{Z}^+\}$ be a standard partition corresponding to h, say

$$Z_k = X_{2k} = \left\{x \in \Omega : h(x) \in \left[\frac{1}{2k+1}, \frac{1}{2k}\right]\right\}\,.$$

For each $k, j \in \mathbb{N}$, set

$$Y_{k,j} = \left\{x \in X_{2k} : f(x) \in \left[\frac{j-1}{3k}, \frac{j}{3k}\right]\right\}$$

(so that $Y_{k,j} = \emptyset$ if $j \ge 3k+1$), and set

$$Y_1 = Z_0 \cup \bigcup\{Y_{k,2j} : k, j \in \mathbb{N}\}, \quad Y_2 = Z_0 \cup \bigcup\{Y_{k,2j-1} : k, j \in \mathbb{N}\}\,.$$

Then $Y_1, Y_2 \in \mathbf{Z}(\Omega)$ and $Y_1 \cup Y_2 = \Omega_1 \in \mathbf{Z}[Q]$. Again, either $Y_1 \in \mathbf{Z}[Q]$ or $Y_2 \in \mathbf{Z}[Q]$, say $Y_1 \in \mathbf{Z}[Q]$. Define f on Y_1 by setting

$$F(x) = \frac{2j}{3k} \qquad (x \in Y_{k,2j})\,. \tag{8.4}$$

Take $x \in Y_{k,2j}$. Certainly $f(x) \leq F(x)$. Also

$$\begin{aligned} g(x) &= f(x) + h(x) \\ &\geq \frac{2j-1}{3k} + \frac{1}{2k+1} \\ &\geq \frac{2j}{3k} = F(x)\,, \end{aligned}$$

so that $F(x) \in [f(x), g(x)]$. Thus $[f] \leq [F] \leq [g]$ in $(B_Q(\mathbb{R})/\sim)$, and hence

$$c = \pi([f]) \leq \pi([F]) \leq \pi([g]) = d\,.$$

But $\pi([F]) \in \mathbb{Q}_Q$, and so the result follows.

It follows immediately that $K_Q \setminus \mathbb{Q}_Q$ is order-dense in K_Q. □

Let S be a subset of $\mathbb{R}$. Then we have defined S_Q as a subset of K_Q. We now wish to define a subset $S_{(Q)}$ of K_Q that is different from S_Q in an important conceptual manner: $S_{(Q)}$ is the subset of K_Q constructed in an *analogous* way to that in which S is constructed as a subset of $\mathbb{R}$. More generally we shall many times construct an entity denoted by $S_{(Q)}$ (or $S^{(Q)}$, if notationally this is more convenient) which is related to K_Q in an analogous way to that in which an entity S is related to $\mathbb{R}$. We begin with the basic version of this construction.

Definition 8.10 *Let S be a subset of $\mathbb{Q}$. Then $S_{(Q)} = S_Q$. Also*

$$[a,b]_{(Q)} = \{x \in K_Q : a \leq x \leq b\}$$

for $a, b \in K_Q$ with $a \leq b$.

In particular, we have defined $\mathbb{N}_{(Q)} = \mathbb{N}_Q$, $\mathbb{Z}^+_{(Q)} = \mathbb{Z}^+_Q$, and $\mathbb{Z}_{(Q)} = \mathbb{Z}_Q$. However, $[0,1]_{(Q)} \neq [0,1]_Q$, for example. Our convention would lead us to define $\mathbb{R}_{(Q)} = K_Q$ (so that $\mathbb{R}_{(Q)}$ is not necessarily equal to $\mathbb{R}_Q$), but we prefer to maintain the notation K_Q. Of course, for $S \subset \mathbb{Q}$, we shall identify S itself with the elements of $S_{(Q)}$ of the form $\pi([f])$, where f is constant and S-valued on a Q-partition.

The zero and identity of K_Q are still denoted by 0 and 1, respectively; clearly,

$$\mathbb{Z}^+_{(Q)} = \mathbb{N}_{(Q)} \cup \{0\}\,.$$

Also, as in Chapter 6, $\mathbb{I}_{(Q)} = [0,1]_{(Q)}$.

The set $\mathbb{N}_{(Q)}$ is the *set of natural numbers* in the non-standard model corresponding to K_Q; it is a 'non-standard model of number theory' (see Chapter 10). The set $\mathbb{N}$ is an initial segment of $\mathbb{N}_{(Q)}$; this means that, if $a \in \mathbb{N}$ and $b \in \mathbb{N}_{(Q)}$ with $b \leq a$, then $b \in \mathbb{N}$. The set $\mathbb{N}_{(Q)} \setminus \mathbb{N}$ is the collection of *infinite natural numbers*. Clearly, if $a, b \in \mathbb{N}_{(Q)}$, then $a + b \in \mathbb{N}_{(Q)}$ and $ab \in \mathbb{N}_{(Q)}$, so that $\mathbb{N}_{(Q)}$ 'acts like' $\mathbb{N}$. However, trivially, the totally ordered set $\mathbb{N}_{(Q)}$ is not well-ordered: for example, there is no minimum infinite natural number.

An order-dense subset of K_Q contains an element in each interval $[a, a+1]_{(Q)}$ for $a \in \mathbb{N}_{(Q)}$, and so $|\mathbb{N}_{(Q)}| \leq w(K_Q)$. Each element of $\mathbb{Q}^+_{(Q)} \setminus \{0\}$ has the form

a/b, where $a, b \in \mathbb{N}_{(Q)}$, and so $|\mathbb{N}_{(Q)}| = |\mathbb{Q}_{(Q)}|$. It follows from Theorems 6.29 and 8.9 that

$$|\mathbb{N}_{(Q)}| = w(K_Q) = |\Gamma_Q|, \tag{8.5}$$

where, as in Chapter 4, Γ_Q is the value group of the ordered field K_Q.

For each $a \in \mathbb{N}_{(Q)} \setminus \mathbb{N}$, the element a^{-1} belongs to $\mathbb{Q}^+_{(Q)} \setminus \{0\}$, and it is an infinitesimal element of K_Q. For each $b \in K^+_Q \setminus \{0\}$, there exists $c \in \mathbb{Q}_{(Q)}$ with $0 < c < b$ by Theorem 8.9, and there exists $a \in \mathbb{N}_{(Q)} \setminus \mathbb{N}$ with $0 < a^{-1} < c$. Thus the map

$$a \mapsto a^{-1}, \quad \mathbb{N}_{(Q)} \setminus \mathbb{N} \to K^+_Q \setminus \{0\}$$

is an anti-isotonic map with coinitial range, and so

$$\operatorname{cof} \mathbb{N}_{(Q)} = \operatorname{cof} K_Q = \delta(K_Q).$$

Similarly, we see that

$$\operatorname{coi}(\mathbb{N}_{(Q)} \setminus \mathbb{N}) = \operatorname{cof} K^0_Q. \tag{8.6}$$

The following typical short calculation shows that, for each $a \in \mathbb{N}_{(Q)}$, there exists $f \in C(\Omega)^+ \setminus Q$ and a Q-partition $\{Z_j : j \in \mathbb{N}\}$ such that

$$\frac{1}{f(x)} \in \mathbb{N} \quad (x \in \mathbb{Z}_\infty) \text{ and } \pi_Q(f) = 1/a.$$

Indeed, take $f \in C(\Omega)^+ \setminus Q$ with $\pi_Q(f) = 1/a$. Then there exist $g, h \in C(\Omega) \setminus Q$, a Q-partition $\{W_j : j \in \mathbb{N}\}$, and a sequence (n_j) in $\mathbb{N}$ such that $\pi_Q(g) = a\pi_Q(h)$, and such that

$$g(x) = n_j h(x), \quad h(x) \neq 0 \quad (x \in W_j).$$

Since $\pi_Q(h) = \pi_Q(f)\pi_Q(g)$, there exists $Z \in \mathbf{Z}[Q]$ such that

$$h(x) = f(x)g(x) \quad (x \in Z).$$

Set $Z_j = Z \cap W_j \ (j \in \mathbb{N})$. Then $\{Z_j : j \in \mathbb{N}\}$ is a Q-partition, and

$$f(x) = \frac{1}{n_j} \quad (x \in Z_j),$$

giving the required conclusion.

A general principle shows that definitions and theorems of number theory, which are statements about $\mathbb{N}$ or $\mathbb{Z}$, are also definitions and theorems of non-standard number theory holding about $\mathbb{N}_{(Q)}$ or $\mathbb{Z}_{(Q)}$, when suitably interpreted. We exemplify this by showing in detail how two facts about $\mathbb{Z}$ are translated into facts about $\mathbb{Z}_{(Q)}$. The proofs are by what we shall call *standard pointwise arguments.*

First, we define

$$a^b \in \mathbb{Q}_{(Q)}$$

for $a \in \mathbb{N}_{(Q)}$ and $b \in \mathbb{Z}_{(Q)}$. Indeed, choose elements $f, g \in B_Q(\mathbb{Q})$, a Q-partition $\{Z_j : j \in \mathbb{N}\}$, and sequences (m_j) in $\mathbb{N}$ and (n_j) in $\mathbb{Z}$ such that $\pi([f]) = a$,

$\pi([g]) = b$, and, for each $j \in \mathbb{N}$, $f(Z_j) = \{m_j\}$ and $g(Z_j) = \{n_j\}$. Define h in the obvious way by requiring that

$$h(Z_j) = \{m_j^{n_j}\} \quad (j \in \mathbb{N}).$$

Then $h \in B_Q(\mathbb{Q})$, and we can set

$$a^b = \pi([h]).$$

It is clear that a^b is well defined in $\mathbb{Q}_{(Q)}$, and that the usual rules involving a^b apply. For example, for each $n \in \mathbb{Z}_{(Q)}$, the element 2^n of $\mathbb{Q}_{(Q)}$ coincides with the element $\exp_{(2)}(n)$ of K_Q defined in (6.9). We note that $\{2^{-a} : a \in \mathbb{N}_{(Q)}\}$ is also coinitial in $K_Q^+ \setminus \{0\}$.

More generally, functions $f : \mathbb{Q} \times \mathbb{Q} \to \mathbb{Q}$ and $g : \mathbb{N} \times \mathbb{N} \to \mathbb{N}$ induce functions

$$f_{(Q)} : \mathbb{Q}_{(Q)} \times \mathbb{Q}_{(Q)} \to \mathbb{Q}_{(Q)} \quad \text{and} \quad g_{(Q)} : \mathbb{N}_{(Q)} \times \mathbb{N}_{(Q)} \to \mathbb{N}_{(Q)},$$

etc., by the obvious pointwise definitions; these are the *induced functions*.

Second, we note that, for each $a \in \mathbb{Z}_{(Q)}$, the least element of $\mathbb{Z}_{(Q)}$ strictly greater than a is $a + 1$. For suppose that $f, g \in B_Q(\mathbb{Q})$ with $\pi([f]) = a$ and $\pi([g]) > a$. Then there are a Q-partition $\{Z_j : j \in \mathbb{N}\}$ and sequences (m_j) and (n_j) in $\mathbb{Z}$ such that, for each $j \in \mathbb{N}$, $f(Z_j) = \{m_j\}$, $g(Z_j) = \{n_j\}$, and $n_j > m_j$. But then $n_j \geq m_j + 1$ $(j \in \mathbb{N})$, and so $\pi([g]) \geq a + 1$, as required. Similarly the greatest element of $\mathbb{Z}_{(Q)}$ strictly less than a is $a - 1$.

We now form in $\mathbb{Q}_{(Q)}$ the analogue $\mathbb{D}_{(Q)}$ of the dyadic rationals $\mathbb{D}$ in $\mathbb{I}$. It will be seen later that it is more convenient to work with $\mathbb{D}_{(Q)}$ than with $\mathbb{Q}_{(Q)}$ in many of our calculations. We first recall the definition of $\mathbb{D}$.

For $k \in \mathbb{Z}^+$, set

$$\mathbb{D}_k = \{r \in \mathbb{Q} \cap \mathbb{I} : 2^k r \in \mathbb{Z}^+\} = \{0, 1/2^k, 2/2^k, \ldots, (2^k - 1)/2^k, 1\},$$

and set

$$\mathbb{D} = \bigcup \{\mathbb{D}_k : k \in \mathbb{Z}^+\}.$$

Then $\mathbb{D}$ is the set of *dyadic rationals* in $\mathbb{I}$. The set $\mathbb{D}_{k+1}$ is formed from $\mathbb{D}_k$ by adding to $\mathbb{D}_k$ the mid-point of every complementary interval of $\mathbb{D}_k$ in $\mathbb{I}$. Note that, for each $r \in \mathbb{D}$, there is a minimum element $k \in \mathbb{Z}^+$ such that $r \in \mathbb{D}_k$.

Definition 8.11 *For $a \in \mathbb{Z}^+_{(Q)}$, set*

$$\mathbb{D}_a^{(Q)} = \{r \in \mathbb{Q}_{(Q)} \cap \mathbb{I}_{(Q)} : 2^a r \in \mathbb{Z}^+_{(Q)}\},$$

and set

$$\mathbb{D}_{(Q)} = \bigcup \{\mathbb{D}_a^{(Q)} : a \in \mathbb{Z}^+_{(Q)}\}.$$

Then $\mathbb{D}_{(Q)}$ is the set of dyadic rationals *in $\mathbb{I}_{(Q)}$.*

Clearly our new definition of $\mathbb{D}_{(Q)}$ coincides with that given by Definition 8.10. The set $\mathbb{D}_{a+1}$ is formed from $\mathbb{D}_a$ by adding to $\mathbb{D}_a$ the mid-point of every complementary interval of $\mathbb{D}_a$ in $\mathbb{I}_{(Q)}$. The following lemma shows that $\mathbb{D}_{(Q)}$ acts as a subset of $\mathbb{I}_{(Q)}$ in essentially the same way as $\mathbb{D}$ acts as a subset of $\mathbb{I}$.

Lemma 8.12 (i) *For $a, b \in \mathbb{Z}^+_{(Q)}$ with $a \leq b$, $\mathbb{D}^{(Q)}_a \subset \mathbb{D}^{(Q)}_b$.*

(ii) *$\mathbb{D}_{(Q)}$ and $(\mathbb{Q}_{(Q)} \cap \mathbb{I}_{(Q)}) \setminus \mathbb{D}_{(Q)}$ are order-dense in $\mathbb{I}_{(Q)}$.*

(iii) *Let $a \in \mathbb{Z}^+_{(Q)}$, and take $r, s \in \mathbb{D}^{(Q)}_a$ with $r < s$. Then $2^{-a} \leq s - r$. Further, $\mathbb{D}^{(Q)}_a \cap (r, s) = \emptyset$ if and only if $s - r = 2^{-a}$.*

(iv) *Let $a \in \mathbb{Z}^+_{(Q)}$, and take $r, s \in \mathbb{Q}_{(Q)} \cap \mathbb{I}_{(Q)}$ with $r < s$. Suppose that $\mathbb{D}^{(Q)}_a \cap (r, s) = \emptyset$. Then $\mathbb{D}^{(Q)}_{a+1} \cap (r, s)$ contains at most one element.*

(v) *Take $r, s \in \mathbb{Q}_{(Q)} \cap \mathbb{I}_{(Q)}$ with $0 < r < s < 1$. Then there exists $a \in \mathbb{Z}^+_{(Q)}$ such that $\mathbb{D}^{(Q)}_a \cap (r, s) = \emptyset$ and $\mathbb{D}^{(Q)}_{a+1} \cap (r, s) \neq \emptyset$.*

(vi) *Let $a \in \mathbb{Z}^+_{(Q)}$ and $r \in \mathbb{Q}_{(Q)} \cap \mathbb{I}_{(Q)}$. Then $\mathbb{D}^{(Q)}_a \cap [0, r]$ has a maximum element, say r', and $\mathbb{D}^{(Q)}_a \cap [r, 1]$ has a minimum element, say r''. Further, if $r \notin \mathbb{D}^{(Q)}_a$, then $r'' - r' = 2^{-a}$ and $\mathbb{D}^{(Q)}_a \cap (r', r'') = \emptyset$.*

Proof All these results are proved by standard pointwise arguments. For example, (ii) follows easily from Theorem 8.9. As a further sample, we prove part of (vi).

Take $f, g \in B_Q(\mathbb{Q})$, a Q-partition $\{Z_j : j \in \mathbb{N}\}$, and sequences (n_j) in $\mathbb{N}$ and (r_j) in $\mathbb{Q} \cap \mathbb{I}$ such that $\pi([f]) = a$, $\pi([g]) = r$, and, for each $j \in \mathbb{N}$, $f(Z_j) = \{n_j\}$ and $g(Z_j) = \{r_j\}$. Set

$$r'_j = \max(\mathbb{D}_{n_j} \cap [0, r_j]) \quad (j \in \mathbb{N}),$$

define $h \in B_Q(\mathbb{Q})$ by the condition that $h(Z_j) = \{r'_j\}$ $(j \in \mathbb{N})$, and finally set $r' = \pi([h])$. Then it is easily checked that r' is the maximum element of the set $\mathbb{D}^{(Q)}_a \cap [0, r]$. □

We now come to the standard functions ε_r whose non-standard analogues will be of the greatest importance and will drive most of our later calculations. For each $r \in \mathbb{I} \setminus \mathbb{D}$, let $\varepsilon_r : \mathbb{N} \to \{0, 1\}$ be the function that gives the coefficients of the binary expansion of r, so that

$$r = \sum_{n=1}^{\infty} \frac{\varepsilon_r(n)}{2^n}.$$

Since $r \notin \mathbb{D}$, the sequence ε_r is unambiguously defined, and the sets

$$\{n \in \mathbb{N} : \varepsilon_r(n) = 0\} \quad \text{and} \quad \{n \in \mathbb{N} : \varepsilon_r(n) = 1\}$$

are both infinite. We recall that, for each $k \in \mathbb{N}$,

$$\varepsilon_r \restriction k : \{1, \ldots, k\} \to \{0, 1\}$$

is a restriction of ε_r. The successive complementary open intervals of $\mathbb{D}_k$ in $\mathbb{I}$ are labelled by the 2^k elements of $\{0, 1\}^k$, ordered lexicographically, and $\varepsilon_r \restriction k$ is constant and equal to the corresponding label on each of these intervals.

The following result is immediate.

Proposition 8.13 (i) *Let $r \in \mathbb{I} \setminus \mathbb{D}$ and $k \in \mathbb{N}$. Then*

$$\sum_{n=1}^{k} \frac{\varepsilon_r(n)}{2^n} \quad \text{and} \quad \sum_{n=1}^{k} \frac{\varepsilon_r(n)}{2^n} + \frac{1}{2^k}$$

are, respectively, the maximum element of $\mathbb{D}_k \cap [0, r]$ and the minimum element of $\mathbb{D}_k \cap [r, 1]$.

(ii) *Let $r, s \in \mathbb{I} \setminus \mathbb{D}$ with $r < s$, and let $k \in \mathbb{N}$. Then $\varepsilon_r \upharpoonright k = \varepsilon_s \upharpoonright k$ if and only if $\mathbb{D}_k \cap (r, s) = \emptyset$, and*

$$\varepsilon_r(n) = \varepsilon_s(n) \quad (n > k) \tag{8.7}$$

if and only if $s - r \in \mathbb{D}_k$. □

Now let $r \in \mathbb{D} \setminus \{0, 1\}$. Then r has a unique expansion in the form

$$r = \sum_{n=1}^{n_0} \frac{\varepsilon_r(n)}{2^n},$$

where $\varepsilon_r(1), \ldots, \varepsilon_r(n_0) \in \{0, 1\}$ and $\varepsilon_r(n_0) = 1$. The two 'binary expansions' of r are ε_r^+ and ε_r^-, where:

$$\varepsilon_r^+(n) = \begin{cases} \varepsilon_r(n) & (n = 1, \ldots, n_0), \\ 0 & (n \geq n_0 + 1); \end{cases}$$

$$\varepsilon_r^-(n) = \begin{cases} \varepsilon_r(n) & (n = 1, \ldots, n_0 - 1), \\ 0 & (n = n_0), \\ 1 & (n \geq n_0 + 1). \end{cases}$$

We have

$$r = \sum_{n=1}^{\infty} \frac{\varepsilon_r^+(n)}{2^n} = \sum_{n=1}^{\infty} \frac{\varepsilon_r^-(n)}{2^n}.$$

It is very important for all that follows to define a 'Q-version' $\varepsilon_r^{(Q)}$ of the functions ε_r. We consider first the case where $r \in (\mathbb{Q}_{(Q)} \cap \mathbb{I}_{(Q)}) \setminus \mathbb{D}_{(Q)}$. We shall define

$$\varepsilon_r^{(Q)} : \mathbb{N}_{(Q)} \to \{0, 1\}$$

by again working with representatives of elements of $B_Q(\mathbb{Q})$. To do this we take $a \in \mathbb{N}_{(Q)}$, and choose functions $f, g \in B_Q(\mathbb{Q})$, a Q-partition $\{Z_j : j \in \mathbb{N}\}$, and sequences (n_j) in $\mathbb{N}$ and (r_j) in $\mathbb{Q} \cap \mathbb{I}$ such that $\pi([f]) = a$, $\pi([g]) = r$, and, for each $j \in \mathbb{N}$, $f(Z_j) = \{n_j\}$ and $g(Z_j) = \{r_j\}$. We may suppose that, for each $j \in \mathbb{N}$, we have $r_j \in (\mathbb{Q} \cap \mathbb{I}) \setminus \mathbb{D}$, and then we may suppose either that $\varepsilon_{r_j}(n_j) = 0$ for each $j \in \mathbb{N}$ or that $\varepsilon_{r_j}(n_j) = 1$ for each $j \in \mathbb{N}$: in the former case, set $\varepsilon_r^{(Q)}(a) = 0$, and, in the latter case, set $\varepsilon_r^{(Q)}(a) = 1$. The element $\varepsilon_r^{(Q)}(a)$ is well defined.

For example, take $Q = J_p$ in $C(\beta\mathbb{N})$, and let $r \in (\mathbb{Q}_{(Q)} \cap \mathbb{I}_{(Q)}) \setminus \mathbb{D}_{(Q)}$, say $r = [f]_{\mathcal{U}}$, where $f : \mathbb{N} \to (\mathbb{Q} \cap \mathbb{I}) \setminus \mathbb{D}$. Take $a \in \mathbb{N}_{(Q)}$, say $a = [g]_{\mathcal{U}}$, where $g : \mathbb{N} \to \mathbb{N}$, and define $h : \mathbb{N} \to \{0,1\}$ by $h(n) = \varepsilon_{f(n)}(g(n))$. Then

$$\varepsilon_r^{(Q)}(a) = [h]_{\mathcal{U}} = \begin{cases} 1 & \text{if } h^{-1}(1) \in \mathcal{U}, \\ 0 & \text{if } h^{-1}(0) \in \mathcal{U}. \end{cases}$$

Note that we have so far not defined $\varepsilon_r^{(Q)}$ for all $r \in \mathbb{I}_{(Q)} \setminus \mathbb{D}_{(Q)}$; the reason for this is of course that such an r is not necessarily of the form $\pi([g])$ for an element g of $B_Q(\mathbb{R})$.

We slightly extend the notation for the restriction of a function. As in the preface, we write $\operatorname{dom} F$ and $\operatorname{ran} F$ for the domain and range, respectively, of a function F. Let F be a function with domain an initial segment of $\mathbb{N}_{(Q)}$ (for example, $\operatorname{dom} F = [1,a] \cap \mathbb{N}_{(Q)}$ for some $a \in \mathbb{N}_{(Q)}$), and take $b \in \operatorname{dom} F$. Then

$$(F \restriction b)(x) = F(x) \quad (x \in [1,b] \cap \mathbb{N}_{(Q)}),$$

so that $F \restriction b$ is formed by restricting the domain of F to another initial segment of $\mathbb{N}_{(Q)}$.

The next lemma follows from the 'pointwise' version in Proposition 8.13(ii).

Lemma 8.14 *Let $a \in \mathbb{N}_{(Q)}$, and take $r, s \in (\mathbb{Q}_{(Q)} \cap \mathbb{I}_{(Q)}) \setminus \mathbb{D}_{(Q)}$ with $r < s$. Then $\varepsilon_r^{(Q)} \restriction a = \varepsilon_s^{(Q)} \restriction a$ if and only if $\mathbb{D}_a^{(Q)} \cap (r,s) = \emptyset$, and $\varepsilon_r^{(Q)}(b) = \varepsilon_s^{(Q)}(b)$ $(b > a)$ if and only if $s - r \in \mathbb{D}_a^{(Q)}$.* □

By using Lemma 8.14, we can now define $\varepsilon_r^{(Q)}$ for $r \in \mathbb{I}_{(Q)} \setminus \mathbb{D}_{(Q)}$. Indeed, take $r \in \mathbb{I}_{(Q)} \setminus \mathbb{D}_{(Q)}$, and let $a \in \mathbb{N}_{(Q)}$. By Theorem 8.9, $\mathbb{Q}_{(Q)}$ is order-dense in K_Q, and so, using Lemma 8.12(vi), we see that there exists $s \in (\mathbb{I}_{(Q)} \cap \mathbb{Q}_{(Q)}) \setminus \mathbb{D}_{(Q)}$ with $s > r$ and $\mathbb{D}_a^{(Q)} \cap (r,s) = \emptyset$. Define

$$\varepsilon_r^{(Q)}(a) = \varepsilon_s^{(Q)}(a).$$

By Lemma 8.14, $\varepsilon_r^{(Q)}(a)$ is well defined, and so we have a function

$$\varepsilon_r^{(Q)} : \mathbb{N}_{(Q)} \to \{0,1\}.$$

Lemma 8.15 *Let $r, s \in \mathbb{I}_{(Q)} \setminus \mathbb{D}_{(Q)}$ with $r < s$.*

(i) *For each $a \in \mathbb{Z}^+_{(Q)}$, $\mathbb{D}_a^{(Q)} \cap (r,s) = \emptyset$ and $\mathbb{D}_{a+1}^{(Q)} \cap (r,s) \neq \emptyset$ if and only if $\varepsilon_r^{(Q)} \restriction a = \varepsilon_s^{(Q)} \restriction a$, $\varepsilon_r^{(Q)}(a+1) = 0$, and $\varepsilon_s^{(Q)}(a+1) = 1$.*

(ii) *There exists $a \in \mathbb{Z}^+_{(Q)}$ such that $\varepsilon_r^{(Q)} \restriction a = \varepsilon_s^{(Q)} \restriction a$, $\varepsilon_r^{(Q)}(a+1) = 0$, and $\varepsilon_s^{(Q)}(a+1) = 1$.*

Proof (i) This follows from Lemma 8.14 and the definition of $\varepsilon_r^{(Q)}$.

(ii) Take $d \in \mathbb{N}_{(Q)}$ with $2^{-d} < s - r$, and hence such that $(r,s) \cap \mathbb{D}_d^{(Q)} \neq \emptyset$. We have $\varepsilon_r^{(Q)} \restriction d \neq \varepsilon_s^{(Q)} \restriction d$, and so, by Lemma 8.12(v) and (i) and Lemma 8.14, there exists $a \leq d$ with the required properties. □

The next lemma gives a more direct way of computing the value of $\varepsilon_r^{(Q)}$ for $r \in \mathbb{I}_{(Q)} \setminus \mathbb{D}_{(Q)}$.

Lemma 8.16 *Let $r \in \mathbb{I}_{(Q)} \setminus \mathbb{D}_{(Q)}$, and let $a \in \mathbb{N}_{(Q)}$. Set*

$$s_1 = \max(\mathbb{D}_a^{(Q)} \cap [0,r])\,, \quad s_2 = \max(\mathbb{D}_{a+1}^{(Q)} \cap [0,r])\,,$$
$$t_1 = \min(\mathbb{D}_a^{(Q)} \cap [r,1])\,, \quad t_2 = \min(\mathbb{D}_{a+1}^{(Q)} \cap [r,1])\,.$$

Then:

(i) $s_1 \le s_2 < r < t_2 \le t_1$;

(ii) *either*

$$s_1 = s_2, \quad t_1 \neq t_2 \quad \text{and} \quad \varepsilon_r^{(Q)}(a+1) = 0$$

or

$$s_1 \neq s_2, \quad t_1 = t_2, \quad \text{and} \quad \varepsilon_r^{(Q)}(a+1) = 1\,.$$

Proof (i) This is immediate because $\mathbb{D}_a^{(Q)} \subset \mathbb{D}_{a+1}^{(Q)}$.

(ii) By Lemma 8.12(vi), $t_1 - s_1 = 2^{-a}$ and $t_2 - s_2 = 2^{-(a+1)}$. Since $r \notin \mathbb{D}_{(Q)}$, we have $\mathbb{D}_a^{(Q)} \cap (s_1, t_1) = \emptyset$, and so, by Lemma 8.12(iv), the set $\mathbb{D}_{a+1}^{(Q)} \cap (s_1, t_1)$ contains at most one element. Suppose that $t_1 = t_2$. Then $s_1 < s_2$, and there exists $r' \in (s_1, s_2) \setminus \mathbb{D}_{(Q)}$; since $(r', r) \cap \mathbb{D}_a^{(Q)} = \emptyset$ and $(r', r) \cap \mathbb{D}_{a+1}^{(Q)} \neq \emptyset$, we have $\varepsilon_r^{(Q)}(a+1) = 1$ by Lemma 8.15(i). Similarly, $\varepsilon_r^{(Q)}(a+1) = 0$ if $s_1 = s_2$. □

Lemma 8.17 *Let $r \in \mathbb{I}_{(Q)} \setminus \mathbb{D}_{(Q)}$. Suppose that $a_1, \dots, a_m \in \mathbb{N}_{(Q)}$ are such that $a_1 < \cdots < a_m$ and $\varepsilon_r^{(Q)}(a_i) = 1$ $(i = 1, \dots, m)$. Then*

$$2^{-a_1} + \cdots + 2^{-a_m} < r\,.$$

Proof The proof is by induction on m; we give just the inductive step because the initial step is similar, but easier. Thus, assume that $q < r$, where

$$q = 2^{-a_1} + \cdots + 2^{-a_m}\,.$$

It follows from Lemma 8.12 that, for each $b \in \mathbb{N}_{(Q)}$ with $a_m \le b$, we have $q \in \mathbb{D}_b^{(Q)}$, $q + 2^{-b} \in \mathbb{D}_b^{(Q)}$, and $\mathbb{D}_b^{(Q)} \cap (q, q + 2^{-b}) = \emptyset$. Assume towards a contradiction that $r \le q + 2^{-a_{m+1}}$, and take $s \in \mathbb{I}_{(Q)} \setminus \mathbb{D}_{(Q)}$ with

$$s \in (q + 2^{-a_{m+1}}, q + 2 \cdot 2^{-a_{m+1}})\,.$$

Set $b = a_{m+1} - 1$. We have $\mathbb{D}_b^{(Q)} \cap (r, s) = \emptyset$ and $\mathbb{D}_{b+1}^{(Q)} \cap (r, s) \neq \emptyset$, and so, by Lemma 8.15(i), $\varepsilon_r^{(Q)}(a_{m+1}) = 0$, a contradiction. Thus

$$r > q + 2^{-a_{m+1}} = 2^{-a_1} + \cdots + 2^{-a_m} + 2^{-a_{m+1}},$$

and the induction continues. □

We have still not defined the two 'binary expansions' $\varepsilon_r^{+(Q)}$ and $\varepsilon_r^{-(Q)}$ for $r \in \mathbb{D}_{(Q)}$; we shall do this when these expansions are needed, in Definition 10.6.

Our next task is to build an analogue of the set of finite binary sequences. For each $k \in \mathbb{N}$, $\{0,1\}^k$ is the set of binary sequences of length k, and we set

$$\{0,1\}^{<\mathbb{N}} = \bigcup \{\{0,1\}^k : k \in \mathbb{N}\}$$

for the set of binary sequences of finite length.

Definition 8.18 *Let $a \in \mathbb{N}_{(Q)}$. Then*

$$\{0,1\}^a_{(Q)} = \{\varepsilon_r^{(Q)} \restriction a : r \in \mathbb{I}_{(Q)} \setminus \mathbb{D}_{(Q)}\}\,.$$

Also

$$\{0,1\}^{<\mathbb{N}}_{(Q)} = \bigcup \left\{\{0,1\}^a_{(Q)} : a \in \mathbb{N}_{(Q)}\right\}.$$

Thus $\{0,1\}^a_{(Q)}$ is a subset of the set of binary sequences 'of length a', and $\{0,1\}^{<\mathbb{N}}_{(Q)}$ is a subset of the set of binary sequences 'of length a for some $a \in \mathbb{N}_{(Q)}$'.

We caution that $\{0,1\}^a_{(Q)}$ is a restricted class of functions from the initial segment $[1,a] \cap \mathbb{N}_{(Q)}$ into $\{0,1\}$. For example, for many functions

$$f : [1,a] \cap \mathbb{N}_{(Q)} \to \{0,1\}\,,$$

it is not true that $\{b \leq a : f(b) = 1\}$ has either a minimum or maximum element. However, for each $f \in \{0,1\}^a_{(Q)}$, the corresponding set has both a minimum and a maximum element. More generally, take $f, g \in \{0,1\}^a_{(Q)}$ with $f \neq g$. It follows from Lemma 8.15(ii) that

$$\{b \in [1,a] \cap \mathbb{N}_{(Q)} : f(b) \neq g(b)\}$$

has a minimum, which we denote by $\delta(f,g)$, and also a maximum. The purpose of the introduction of the special functions $\varepsilon_r^{(Q)}$, as above, was exactly to enable us to distinguish the subset $\{0,1\}^a_{(Q)}$, with its special properties.

Although $\{0,1\}^a_{(Q)}$ is a restricted class of functions, it is closed under many operations. For example, suppose that $f, g \in \{0,1\}^a_{(Q)}$. Then $f \vee g$, $f \wedge g$, and $1 - f$ also belong to $\{0,1\}^a_{(Q)}$. Further, suppose that h is defined by setting $h(b) = 1$ if $f(b) = g(b)$ and $h(b) = 0$ if $f(b) \neq g(b)$. Then $h \in \{0,1\}^a_{(Q)}$.

The *lexicographic order* $\preceq$ on $\{0,1\}^a_{(Q)}$ is defined in the usual way. Indeed, let $f, g \in \{0,1\}^a_{(Q)}$ with $f \neq g$. Then

$$f \prec g \quad \text{if} \quad f(\delta(f,g)) = 0 \quad \text{and} \quad g(\delta(f,g)) = 1\,;$$

we set $f \preceq g$ if $f \prec g$ or $f = g$.

Take $r, s \in \mathbb{I}_{(Q)} \setminus \mathbb{D}_{(Q)}$ with $r \leq s$. Then clearly we have

$$\varepsilon_r^{(Q)} \restriction a \preceq \varepsilon_s^{(Q)} \restriction a \quad (a \in \mathbb{N}_{(Q)})\,,$$

and

$$\varepsilon_r^{(Q)} \restriction a \prec \varepsilon_s^{(Q)} \restriction a \;\text{ if }\; \mathbb{D}_a^{(Q)} \cap (r,s) \neq \emptyset\,.$$

Our situation continues to be that Q is a fixed, non-maximal, prime z-ideal in $C(\Omega)$, that K_Q is the quotient field of A_Q, and that $\mathbb{Q}_{(Q)}$ is a specific rational subfield of K_Q such that $\mathbb{Q}_{(Q)}$ is order-dense in K_Q. We now come to a key lemma; it is the point at which we require that K_Q be a semi-η_1-field.

Lemma 8.19 *Let Q be such that K_Q is a semi-η_1-field, and let (f_k) be a sequence in $\{0,1\}^{<\mathbb{N}}_{(Q)}$. Suppose that there is a sequence (a_k) in $\mathbb{N}_{(Q)}$ and an element a in $\mathbb{N}_{(Q)}$ such that, for each $k \in \mathbb{N}$:*

(i) $a_k < a_{k+1} < a$;

(ii) $\operatorname{dom} f_k = [1, a_k] \cap \mathbb{N}_{(Q)}$;

(iii) $f_{k+1} \restriction a_k = f_k$.

Then there exists $f \in \{0,1\}^a_{(Q)}$ such that $f \restriction a_k = f_k$ for each $k \in \mathbb{N}$.

Proof Take $k \in \mathbb{N}$, and choose $r_k \in \mathbb{I}_{(Q)} \setminus \mathbb{D}_{(Q)}$ such that $\varepsilon^{(Q)}_{r_k} \restriction a_k = f_k$. Set

$$u_k = \max\left(\mathbb{D}^{(Q)}_{a_k} \cap [0, r_k]\right), \quad v_k = \min\left(\mathbb{D}^{(Q)}_{a_k} \cap [r_k, 1]\right),$$

so that $v_k = u_k + 2^{-a_k}$. By (iii), we have

$$\varepsilon^{(Q)}_{r_{k+1}} \restriction a_k = \varepsilon^{(Q)}_{r_k} \restriction a_k .$$

Consider the case where $r_k < r_{k+1}$. By Lemma 8.15(i), $\mathbb{D}^{(Q)}_{a_k} \cap (r_k, r_{k+1}) = \emptyset$, and so, by Lemma 8.12(vi), $r_{k+1} \in (u_k, v_k)$. This also holds in the case where $r_{k+1} < r_k$, and so

$$[u_{k+1}, v_{k+1}] \subset [u_k, v_k] .$$

Since $v_k - u_k = 2^{-a_k}$, we have $2^{-a} \leq v_k - u_k$.

We *claim* that there exist $u, v \in \mathbb{I}_{(Q)}$ such that

$$u_k < u < v < v_k \quad (k \in \mathbb{N}) .$$

If (u_k) eventually takes the constant value u_0, take $u = u_0 + 2^{-(a+2)}$ and set $v = u_0 + 2^{-(a+1)}$. If (v_k) is eventually constant, we obtain u and v similarly. If neither (u_k) nor (v_k) is eventually constant, then (u_k) has a cofinal, strictly increasing subsequence and (v_k) has a coinitial, strictly decreasing subsequence; since K_Q is a semi-η_1-field, there exists $u_0 \in \mathbb{I}_{(Q)}$ with $u_k < u_0 < v_k \quad (k \in \mathbb{N})$. We may suppose that $u_0 + 2^{-(a+1)} < v_k \ (k \in \mathbb{N})$, and we take $u = u_0$ and $v = u_0 + 2^{-(a+1)}$ to satisfy the condition. The claim is established.

Finally, since $\mathbb{I}_{(Q)} \setminus \mathbb{D}_{(Q)}$ is order-dense in $\mathbb{I}_{(Q)}$, there exists $r \in \mathbb{I}_{(Q)} \setminus \mathbb{D}_{(Q)}$ with $u < r < v$. We have $\mathbb{D}^{(Q)}_{a_k} \cap (u_k, v_k) = \emptyset \ (k \in \mathbb{N})$ and $r, r_k \in (u_k, v_k)$, and so, by Lemma 8.15(i) again,

$$\varepsilon^{(Q)}_{r} \restriction a_k = \varepsilon^{(Q)}_{r_k} \restriction a_k \quad (k \in \mathbb{N}) .$$

Set $f = \varepsilon^{(Q)}_r \restriction a$. Then f has the required properties. □

We shall continue to build our additional structures by forming the analogues of $\mathbb{N}^{<\mathbb{N}}$ (in this chapter) and of $\mathbb{N}^{\mathbb{N}}$ (in Chapter 9). The idea is to identify a function with its graph, and to identify subsets of $\mathbb{N} \times \mathbb{N}$ with subsets of $\mathbb{N}$ by a 'coding map' that takes $\mathbb{N} \times \mathbb{N}$ onto $\mathbb{N}$. The coding map that we choose has to have certain modest additional properties, but its exact definition is not important.

To define the coding map, we first specify a total order $\leq$ on $\mathbb{N} \times \mathbb{N}$ such that $(\mathbb{N} \times \mathbb{N}, \leq)$ and $(\mathbb{N}, \leq)$ are order-isomorphic. Indeed, let $\preceq$ be the lexicographic order on $\mathbb{N} \times \mathbb{N} \times \mathbb{N}$, and set $(m_1, n_1) \leq (m_2, n_2)$ in $\mathbb{N} \times \mathbb{N}$ if

$$(m_1 \vee n_1, m_1, n_1) \preceq (m_2 \vee n_2, m_2, n_2).$$

Note that $(m, n_1) < (m, n_2)$ in $\mathbb{N} \times \mathbb{N}$ if and only if $n_1 < n_2$ in $\mathbb{N}$. Let

$$\Phi : (\mathbb{N} \times \mathbb{N}, \leq) \to (\mathbb{N}, \leq) \quad \text{and} \quad \rho = \Phi^{-1} : (\mathbb{N}, \leq) \to (\mathbb{N} \times \mathbb{N}, \leq)$$

be the corresponding isotonic maps. For example $\Phi(1, 4) = 7$. It will be technically important for us later to note that there are bounds on the size of $\Phi(m, n)$ and $\rho(k)$. Indeed, it is easily checked that

$$m \vee n \leq \Phi(m, n) \leq (m \vee n)^2 \leq m^2 n^2 \qquad (m, n \in \mathbb{N}) \tag{8.8}$$

and that

$$\rho(k) \in [1, k] \times [1, k] \qquad (k \in \mathbb{N}). \tag{8.9}$$

We now define the induced functions $\Phi_{(Q)}$ and $\rho_{(Q)}$ in the usual pointwise fashion, obtaining bijections

$$\Phi_{(Q)} : \mathbb{N}_{(Q)} \times \mathbb{N}_{(Q)} \to \mathbb{N}_{(Q)} \quad \text{and} \quad \rho_{(Q)} : \mathbb{N}_{(Q)} \to \mathbb{N}_{(Q)} \times \mathbb{N}_{(Q)}$$

which are inverses of each other. The bounds we have noted above give us the following bounds for $\Phi_{(Q)}$ and $\rho_{(Q)}$:

$$a \vee b \leq \Phi_{(Q)}(a, b) \leq (a \vee b)^2 \leq a^2 b^2 \quad (a, b \in \mathbb{N}_{(Q)}); \tag{8.10}$$

$$\rho_{(Q)}(c) \in [1, c] \times [1, c] \quad (c \in \mathbb{N}_{(Q)}). \tag{8.11}$$

First let g be a function with $\operatorname{dom} g \subset \mathbb{N}$ and $\operatorname{ran} g \subset \{0, 1\}$, and set

$$R_g = \{\rho(n) : n \in \operatorname{dom} g, \quad g(n) = 1\}, \tag{8.12}$$

so that R_g is just a subset of $\mathbb{N} \times \mathbb{N}$, and hence may or may not be a partial function from a subset of $\mathbb{N}$ into $\mathbb{N}$.

Now let g be a function with $\operatorname{dom} g \subset \mathbb{N}_{(Q)}$ and $\operatorname{ran} g \subset \{0, 1\}$, and set

$$R_g^{(Q)} = \{\rho_{(Q)}(a) : a \in \operatorname{dom} g, \quad g(a) = 1\}, \tag{8.13}$$

so that $R_g^{(Q)}$ is a subset of $\mathbb{N}_{(Q)} \times \mathbb{N}_{(Q)}$. We are going to use the sets $R_g^{(Q)}$ in a definition of the analogue $\mathbb{N}_{(Q)}^{<\mathbb{N}}$ of the family $\mathbb{N}^{<\mathbb{N}}$ of all functions from a finite initial segment of $\mathbb{N}$ into $\mathbb{N}$.

Definition 8.20 *Let $a \in \mathbb{N}_{(Q)}$. Then $\mathbb{N}_{(Q)}^a$ is the collection of all functions F with $\operatorname{dom} F = [1, a] \cap \mathbb{N}_{(Q)}$ and $\operatorname{ran} F \subset \mathbb{N}_{(Q)}$ such that $\operatorname{Gr} F = R_g^{(Q)}$ for some $g \in \{0, 1\}_{(Q)}^{<\mathbb{N}}$. Set*

$$\mathbb{N}_{(Q)}^{<\mathbb{N}} = \bigcup \{\mathbb{N}_{(Q)}^a : a \in \mathbb{N}_{(Q)}\}.$$

There is a connection between $\{0,1\}^{<\mathbb{N}}_{(Q)}$ and $\mathbb{N}^{<\mathbb{N}}_{(Q)}$: it can easily be checked by pointwise arguments that, for each $a \in \mathbb{N}_{(Q)}$,

$$\{F+1 : F \in \{0,1\}^a_{(Q)}\} = \{G \in \mathbb{N}^a_{(Q)} : \operatorname{ran} G \subset \{1,2\}\}.$$

We now give a sequence of lemmas (Lemmas 8.21–8.24) about $\mathbb{N}^{<\mathbb{N}}_{(Q)}$; all can be proved by standard pointwise arguments.

Lemma 8.21 *Let $g \in \{0,1\}^a_{(Q)}$, where $a \in \mathbb{N}_{(Q)}$. Then*

$$\{c \in [1,a] \cap \mathbb{N}_{(Q)} : R^{(Q)}_{g\restriction c} \text{ is a function in } \mathbb{N}^{<\mathbb{N}}_{(Q)}\}$$

is empty or has a maximum element. □

A function $F : [1,b] \cap \mathbb{N}_{(Q)} \to \mathbb{N}_{(Q)}$ is *increasing* if $F(a_1) \le F(a_2)$ whenever $a_1 \le a_2$ in $[1,b] \cap \mathbb{N}_{(Q)}$.

Lemma 8.22 *Let $F \in \mathbb{N}^a_{(Q)}$, where $a \in \mathbb{N}_{(Q)}$. Then*

$$\{b \in [1,a] \cap \mathbb{N}_{(Q)} : F \restriction b \text{ is increasing}\}$$

has a maximum element. □

Lemma 8.23 *Let $F \in \mathbb{N}^{<\mathbb{N}}_{(Q)}$, and take $c \in \mathbb{N}_{(Q)}$. Then:*

(i) $(\operatorname{ran} F) \cap [1,c] \cap \mathbb{N}_{(Q)}$ *is either empty or has a maximum element;*

(ii) $F^{-1}(\{c\})$ *is either empty or has a minimum element.* □

Let $F \in \mathbb{N}^{<\mathbb{N}}_{(Q)}$. Then it follows from the definition of $\mathbb{N}^{<\mathbb{N}}_{(Q)}$ that F is bounded, and so, by Lemma 8.23(i), $\operatorname{ran} F$ has a maximum element.

The following two lemmas are stronger than are needed for the gap theorem of this chapter, but they will be needed in Chapter 9.

Lemma 8.24 (i) *Let $a \in \mathbb{N}_{(Q)}$, and let $H : [1,a] \cap \mathbb{N}_{(Q)} \to \mathbb{N}_{(Q)}$ be a constant function. Then $H \in \mathbb{N}^a_{(Q)}$.*

(ii) *Let $a, b \in \mathbb{N}_{(Q)}$ with $a < b$, and let $F \in \mathbb{N}^b_{(Q)}$. Then $F \restriction a \in \mathbb{N}^a_{(Q)}$.*

(iii) *Let $a \in \mathbb{N}_{(Q)}$, and let $F, G, H \in \mathbb{N}^a_{(Q)}$. Then the following functions also belong to $\mathbb{N}^a_{(Q)}$: $F+G$, FG, $F \vee G$, $F \wedge G$, and $(F-G) \vee H$.*

(iv) *Let $a, b \in \mathbb{N}_{(Q)}$ with $a < b$, and let $F \in \mathbb{N}^a_{(Q)}$ and $G \in \mathbb{N}^b_{(Q)}$. Define*

$$H : [1,b] \cap \mathbb{N}_{(Q)} \to \mathbb{N}_{(Q)}$$

by setting

$$H(c) = F(c) \quad (c \le a), \quad H(c) = G(c) \quad (a < c \le b).$$

Then $H \in \mathbb{N}^b_{(Q)}$.

(v) *Let $a, b \in \mathbb{N}_{(Q)}$, and let $F \in \mathbb{N}^a_{(Q)}$ and $G \in \mathbb{N}^b_{(Q)}$ with $\operatorname{ran} G \subset [1,a]$. Then $F \circ G \in \mathbb{N}^b_{(Q)}$.* □

We next prove the analogue of Lemma 8.19 for the set $\mathbb{N}_{(Q)}^{<\mathbb{N}}$. The method is to use our coding device to reduce the result to the earlier one.

Lemma 8.25 *Let Q be such that K_Q is a semi-η_1-field, and let (F_k) be a sequence in $\mathbb{N}_{(Q)}^{<\mathbb{N}}$. Suppose that there is a sequence (b_k) in $\mathbb{N}_{(Q)}$ and an element d in $\mathbb{N}_{(Q)}$ such that, for each $k \in \mathbb{N}$:*

(i) $b_k < b_{k+1} < d$;

(ii) $\operatorname{dom} F_k = [1, b_k] \cap \mathbb{N}_{(Q)}$;

(iii) $\operatorname{ran} F_k \subset [1, d] \cap \mathbb{N}_{(Q)}$;

(iv) $F_{k+1} \restriction b_k = F_k$.

Then there exists $F \in \mathbb{N}_{(Q)}^d$ such that $F \restriction b_k = F_k$ for each $k \in \mathbb{N}$.

Proof We first suppose additionally that, for each $k \in \mathbb{N}$,

$$F_k(x) \le x \quad (x \in [1, b_k] \cap \mathbb{N}_{(Q)}) . \tag{8.14}$$

Set $a = d^2$ and, for each $k \in \mathbb{N}$, take $a_k \in \mathbb{N}_{(Q)}$ and $g_k \in \{0,1\}_{(Q)}^{a_k}$ such that $\operatorname{Gr} F_k = R_{g_k}^{(Q)}$; by (8.14) and (8.10), we may suppose that $a_k \le b_k^2$. By the assumptions on the sequence (F_k), we have, for each $k \in \mathbb{N}$:

$$a_k < a_{k+1} < a ; \quad \operatorname{dom} g_k = [1, a_k] \cap \mathbb{N}_{(Q)} ; \quad g_{k+1} \restriction a_k = g_k .$$

Thus, by Lemma 8.19, there exists $g \in \{0,1\}_{(Q)}^a$ such that $g \restriction a_k = g_k$ for each $k \in \mathbb{N}$.

By Lemma 8.21, the set

$$\{c \in [1, a] \cap \mathbb{N}_{(Q)} : R_{g \restriction c}^{(Q)} \text{ is a function}\}$$

has a maximum element, say c_0. But we have $R_{g\restriction a_k}^{(Q)} = R_{g_k}^{(Q)} = \operatorname{Gr} F_k$, and so $a_k < c_0 \le a$ $(k \in \mathbb{N})$. Take F so that $R_{g\restriction c_0}^{(Q)} = \operatorname{Gr} F$ and $F \in \mathbb{N}_{(Q)}^{<\mathbb{N}}$. We have $\operatorname{dom} F \supset \operatorname{dom} F_k$ $(k \in \mathbb{N})$. By Lemma 8.24(iv), we may extend F so that $\operatorname{dom} F = [1, d] \cap \mathbb{N}_{(Q)}$ and $F \in \mathbb{N}_{(Q)}^d$. Since $g \restriction a_k = g_k$, we have $F \restriction b_k = F_k$ for each $k \in \mathbb{N}$, and so F has the required properties.

For the general case, set $c = 2d$ and $c_k = d + b_k$ $(k \in \mathbb{N})$. For each $k \in \mathbb{N}$, define the function $\overline{F}_k : [1, c_k] \cap \mathbb{N}_{(Q)} \to \mathbb{N}_{(Q)}$ by setting

$$\begin{cases} \overline{F}_k(a) = 1 & (a \in [1, d] \cap \mathbb{N}_{(Q)}) , \\ \overline{F}_k(a) = F_k(a - d) & (a \in [d+1, d+b_k] \cap \mathbb{N}_{(Q)}) . \end{cases}$$

Then, by Lemma 8.24, $\overline{F}_k \in \mathbb{N}_{(Q)}^{c_k}$ $(k \in \mathbb{N})$. For each $k \in \mathbb{N}$, we have

$$\overline{F}_{k+1} \restriction c_k = \overline{F}_k \quad \text{and} \quad \overline{F}_k(a) \le a \ (a \in [1, c_k]) ,$$

and so the first part of the proof shows that there exists a function $\overline{F} \in \mathbb{N}_{(Q)}^c$ with $\overline{F} \restriction c_k = \overline{F}_k$ $(k \in \mathbb{N})$. Define

$$F : a \longmapsto \overline{F}(d + a) , \quad [1, d] \cap \mathbb{N}_{(Q)} \to \mathbb{N}_{(Q)} .$$

Then $F \in \mathbb{N}_{(Q)}^d$, and F has the required properties. □

So far we have dealt only with super-real fields of the form K_Q, where Q is a non-maximal, prime z-ideal in $C(\Omega)$. We now turn to the more difficult, general case, in which the prime ideal in question is not necessarily a z-ideal. Thus now fix Ω to be a compact space and P to be an arbitrary non-maximal, prime ideal in $C(\Omega)$. As before,

$$\pi_P : C(\Omega) \to C(\Omega)/P = A_P$$

is the quotient map. By Proposition 4.23, there is a maximum z-ideal Q below P in $C(\Omega)$, and Q is a prime ideal. We shall define the structures

$$\mathbb{Q}_{(P)}, \quad \mathbb{N}_{(P)}, \quad \mathbb{D}_a^{(P)}, \quad \mathbb{D}_{(P)}, \quad \{0,1\}^{<\mathbb{N}}_{(P)}, \quad \mathbb{N}^{<\mathbb{N}}_{(P)},$$

and the map $r \mapsto \varepsilon_r^{(P)}$ from $(\mathbb{Q}_{(P)} \cap \mathbb{I}_{(P)}) \setminus \mathbb{D}_{(P)}$ (and eventually from $\mathbb{I}_{(P)} \setminus \mathbb{D}_{(P)}$) into $\{0,1\}^{\mathbb{N}_{(P)}}$, all of which are associated with the prime ideal P, by using Q-partitions, rather than P-partitions. We are working with a specific prime z-ideal Q below P, but, in fact, the structures that we are defining are independent of the choice of the ideal Q satisfying these conditions; for this, see the remarks at the end of this chapter.

Why do we follow an apparently circuitous route to define the structures associated with P, and not work directly with P-partitions? The fundamental difficulty is that there may not be a P-partition. However, even if there is a P-partition, it may be that two functions which agree on the union of a P-partition do not necessarily define the same elements of A_P (as they do in the case of z-ideals by Proposition 8.3(ii)), and this defeats the pointwise arguments that we have used many times for z-ideals. However, if $f, g \in C(\Omega)$ with $f(x) = g(x)$ for x in the union of a Q-partition, then $f - g \in Q$ by Proposition 8.3, and so $\pi_P(f) = \pi_P(g)$. The transfer method (using Q-partitions) is a convenient device that allows us to again use pointwise arguments in some cases. There are some remaining subtle points, however, and a reason for care will be pointed out after the definition of $\mathbb{Q}_{(P)}$.

In the case of a prime z-ideal Q, we defined two sets $\mathbb{Q}_Q$ and $\mathbb{Q}_{(Q)}$ that were actually equal; now we shall define $\mathbb{Q}_{(P)}$ as the analogue in K_P of $\mathbb{Q}$, but we have no definition of $\mathbb{Q}_P$.

Recall that when working with a prime z-ideal Q, we simultaneously defined the rational field $\mathbb{Q}_{(Q)}$ and picked out the subset $\mathbb{N}_{(Q)}$. Our approach now is to first define $\mathbb{N}_{(P)}$, and then to define the rational subfield $\mathbb{Q}_{(P)}$ of K_P in terms of $\mathbb{N}_{(P)}$. In fact, for reasons that we shall explain shortly, we start by defining a set $\mathbb{M}_{(P)}$.

Let P and Q be arbitrary prime ideals in $C(\Omega)$ with Q below P. Then the natural mapping

$$\pi_{PQ} : f + Q \mapsto f + P, \quad A_Q \to A_P$$

is an order-preserving epimorphism, but of course there is no obvious extension of π_{PQ} to a map from K_Q into K_P. Also, there is no obvious embedding of K_P in K_Q.

Definition 8.26 *Let Ω be a compact space, let P be a non-maximal, prime ideal in $C(\Omega)$, and let Q be the maximal z-ideal in $C(\Omega)$ below P. Then the set $\mathbb{M}_{(P)}$ consists of the elements $a \in K_P^+$ such that there exist functions $f, g \in C(\Omega)^+ \setminus P$, a Q-partition $\{Z_j : j \in \mathbb{N}\}$, and a sequence (n_j) in $\mathbb{N}$ with:*

(i) $f(x) = n_j g(x) \qquad (x \in Z_j,\ j \in \mathbb{N})$;

(ii) $\pi_P(f) = a\pi_P(g)$.

Note that, by refining the Q-partition if necessary, we may suppose that $f(x) > 0$ and $g(x) > 0$ for $x \in Z_\infty$, and we shall do this henceforth.

We first observe that $\mathbb{M}_{(P)}$ contains a copy of $\mathbb{N}$ and at least some infinitely large elements of K_P; indeed, $\mathbb{M}_{(P)}$ is cofinal in K_P^+. Clearly, if $a, b \in \mathbb{M}_{(P)}$, then $a + b \in \mathbb{M}_{(P)}$ and $ab \in \mathbb{M}_{(P)}$. Further $a - b \in \mathbb{M}_{(P)}$ in the case where $a \geq b + 1$. For each $a \in \mathbb{M}_{(P)}$, the least element of $\mathbb{M}_{(P)}$ strictly greater than a is $a + 1$.

The first lemma will lead to the fact that $\mathbb{M}_{(P)}$ embeds in $\mathbb{N}_{(Q)}$. (The complete pattern of inclusions that we are establishing will be illustrated diagrammatically in Fig. 8.1. on page 183.)

Lemma 8.27 *Let $f_1, g_1, f_2, g_2 \in C(\Omega)^+ \setminus P$, and let $\{Z_j : j \in \mathbb{N}\}$ be a Q-partition. Suppose that $f_1g_2 - f_2g_1 \in P$ and that $\{n_j^{(1)} : j \in \mathbb{N}\}$ and $\{n_j^{(2)} : j \in \mathbb{N}\}$ are subsets of $\mathbb{N}$ such that*

$$f_1(x) = n_j^{(1)} g_1(x), \quad f_2(x) = n_j^{(2)} g_2(x) \qquad (x \in Z_j,\ j \in \mathbb{N}). \tag{8.15}$$

Then $\{Z_j : n_j^{(1)} = n_j^{(2)}\}$ is a Q-partition.

Proof To obtain a contradiction, assume that $\{Z_j : n_j^{(1)} = n_j^{(2)}\}$ is not a Q-partition. By the remarks following Lemma 8.6, we may suppose that

$$\{Z_j : n_j^{(1)} > n_j^{(2)}\} = \{Z_j : n_j^{(1)} \geq n_j^{(2)} + 1\}$$

is a Q-partition, say it is $\{W_k : k \in \mathbb{N}\}$. By (8.15),

$$(f_1g_2 - f_2g_1)(x) \geq (g_1g_2)(x) \qquad (x \in W_\infty),$$

where $W_\infty = \bigcup W_k$. By Proposition 8.3(ii), $\pi_Q(f_1g_2 - f_2g_1) \geq \pi_Q(g_1g_2)$, and so

$$\pi_P(f_1g_2 - f_2g_1) \geq \pi_P(g_1g_2) \geq 0.$$

But $\pi_P(f_1g_2 - f_2g_1) = 0$, and so $g_1g_2 \in P$, a contradiction because P is a prime ideal. Thus the result holds. □

Now take $a \in \mathbb{M}_{(P)}$ to be as defined in Definition 8.26 in terms of f and g, and define $\iota(a)$ to be the element of K_Q such that $\pi_Q(f) = \iota(a)\pi_Q(g)$. Lemma 8.27 shows that $\iota(a)$ is a well-defined element of $\mathbb{N}_{(Q)}$. The map $\iota : \mathbb{M}_{(P)} \to \mathbb{N}_{(Q)}$ is an isotonic injection, and $\iota(a + b) = \iota(a) + \iota(b)$ and $\iota(ab) = \iota(a)\iota(b)$ for $a, b \in \mathbb{M}_{(P)}$. We now regard $\mathbb{M}_{(P)}$ as a subset of $\mathbb{N}_{(Q)}$.

Note that, although $\mathbb{M}_{(P)}$ is closed under addition and multiplication, it may not be closed under exponentiation (to base 2): if $a \in \mathbb{M}_{(P)}$, then 2^a is calculated in $\mathbb{N}_{(Q)}$, but there is no reason in general why it should be that $2^a \in \mathbb{M}_{(P)}$. However, if P is an exponential prime (as in Definition 4.45), then $\mathbb{M}_{(P)}$ is closed under exponentiation. For let $a \in \mathbb{M}_{(P)}$ be as specified in Definition 8.26 in terms of f and g (with $f \leq 1/2$). Define $F, G \in C(\Omega)$ by setting

$$G(x) = 2^{-1/g(x)}, \quad F(x) = 2^{(f(x)-1)/g(x)} \qquad (x \in \Omega \setminus \mathbf{Z}(g)),$$

and $F(x) = G(x) = 0$ $(x \in \mathbf{Z}(g))$. We have $F(x) = 2^{n_j} G(x)$ $(x \in Z_j)$, and so there exists $b \in \mathbb{M}_{(P)}$ with $\pi_P(F) = b\pi_P(G)$. Since $\iota(b) = 2^a$, we have $2^a \in \mathbb{M}_{(P)}$.

Definition 8.28 *The rational subfield of K_P generated by $\mathbb{M}_{(P)}$ is $\mathbb{Q}_{(P)}$.*

Thus we may regard $\mathbb{Q}_{(P)}$ as a rational subfield of $\mathbb{Q}_{(Q)} \subset K_Q$. Let $a \in K_P$. For future reference, we note explicitly that there exist functions $f \in C(\Omega)$ and $g \in C(\Omega) \setminus P$ such that $a = \pi_P(f)/\pi_P(g)$ in $\mathbb{Q}_{(P)}$ and $a = \pi_Q(f)/\pi_Q(g)$ in $\mathbb{Q}_{(Q)}$ under the identification of $\mathbb{Q}_{(P)}$ as a subset of $\mathbb{Q}_{(Q)}$ that we are making. We caution that the above representation of a as $\pi_Q(f)/\pi_Q(g)$ is quite special: in the case where $Q \neq P$, there exists $h \in P \setminus Q$, and then $a = \pi_P(f+h)/\pi_P(g)$ in $\mathbb{Q}_{(P)}$, but $a \neq \pi_Q(f+h)/\pi_Q(g)$ in $\mathbb{Q}_{(Q)}$. This point will be discussed further in Chapter 10. We also caution that, whilst $\mathbb{Q} = \mathbb{R} \cap \mathbb{Q}_{(Q)}$ (where we are identifying $\mathbb{R}$ with its canonical image in K_Q), it can happen that $\mathbb{Q} \subsetneq \mathbb{R} \cap \mathbb{Q}_{(P)}$ (now identifying $\mathbb{R}$ with its canonical image in K_P); the embedding of $\mathbb{Q}_{(P)}$ in $\mathbb{Q}_{(Q)}$ does not necessarily respect the interpretations of $\mathbb{R}$ in K_P and K_Q.

We now explain a feature of the definition of $\mathbb{Q}_{(P)}$ that requires care. Suppose that $f, g \in C(\Omega)^+ \setminus P$, that $\{Z_j : j \in \mathbb{N}\}$ is a Q-partition, and that

$$f(x) = r_j g(x) \qquad (x \in Z_j,\ j \in \mathbb{N}).$$

for a sequence (r_j) in $\mathbb{Q}$. Then f and g define an element $a \in \mathbb{Q}_{(Q)}$. But, in general, $a \notin \mathbb{Q}_{(P)}$. The difficulty is the following: it may be that, for each $h \in C(\Omega)$ such that $|h|_{Z_j} \to 0$ as $j \to \infty$, necessarily $h \in P$. This places severe constraints on the functions that we can use to build new elements of $\mathbb{M}_{(P)}$.

Nevertheless, the next two results show, first, that there are sufficiently many functions in $C(\Omega) \setminus P$ to justify the pointwise arguments that we shall need in order to establish the appropriate properties of $\mathbb{M}_{(P)}$ and $\mathbb{Q}_{(P)}$, and, second, that $\mathbb{Q}_{(P)}$ is still a sufficiently large set for it to be order-dense in K_P.

Theorem 8.29 *Let $m \in \mathbb{N}$. Suppose that:*

(i) *$F : \mathbb{N}^m \to \mathbb{N}$ is a function such that, for some constant C,*

$$F(n_1, \dots, n_m) \leq C n_1 \cdots n_m \qquad (n_1, \dots, n_m \in \mathbb{N});$$

(ii) *$\{Z_j : j \in \mathbb{N}\}$ is a Q-partition with union Z_∞;*

(iii) *$f_1, \dots, f_m, g_1, \dots, g_m \in C(\Omega) \setminus P$ and $n_{1,j}, \dots, n_{m,j} \in \mathbb{N}$ for $j \in \mathbb{N}$, and, for $i = 1, \dots, m$,*

$$f_i(x) = n_{i,j} g_i(x) \qquad (x \in Z_j,\ j \in \mathbb{N}).$$

Then there exist f and g in $C(\Omega)^+ \setminus P$ such that

$$f(x) = F(n_{1,j}, \dots, n_{m,j}) g(x) \qquad (x \in Z_j,\ j \in \mathbb{N}). \tag{8.16}$$

Proof Set $g = g_1^2 \cdots g_n^2$, so that $g \in C(\Omega)^+ \setminus P$, and define f on Z_∞ by formula (8.16). We *claim* that f has a continuous extension to $\overline{Z}_\infty$.

Indeed, take $y \in \overline{Z}_\infty \setminus Z_\infty$ and (y_ν) in Z_∞ with $y_\nu \to y$, say $y_\nu \in Z_{j_\nu}$ for some $j_\nu \in \mathbb{N}$. To define $f(y)$, first suppose that $g(y) \neq 0$. Then, for $i = 1, \dots, m$, we have $g_i(y_\nu) \neq 0$ for ν sufficiently large, and so

$$\lim_\nu n_{i,j_\nu} = \lim_\nu \frac{f_i(y_\nu)}{g_i(y_\nu)} = \frac{f_i(y)}{g_i(y)};$$

the net (n_{i,j_ν}) is eventually constant, with value n_i, say. We set

$$f(y) = F(n_1, \dots, n_m)g(y) = \lim_\nu f(y_\nu).$$

Second, suppose that $g(y) = 0$, say $g_{i_0}(y) = 0$, where $i_0 \in \{1, \dots, m\}$, and define $f(y) = 0$. Set

$$M = \max\{|g_i|_\Omega, |f_i|_\Omega : i = 1, \dots, m\}.$$

Then, for each ν, we have

$$\begin{aligned} 0 \le f(y_\nu) &= F(n_{1,j_\nu}, \dots, n_{m,j_\nu})g(y_\nu) \\ &\le Cn_{1,j_\nu} \cdots n_{m,j_\nu} g_1(y_\nu)^2 \cdots g_m(y_\nu)^2 \qquad \text{by (i)} \\ &= Cf_1(y_\nu) \cdots f_m(y_\nu) g_1(y_\nu) \cdots g_m(y_\nu) \\ &\le CM^{2m-1} g_{i_0}(y_\nu), \end{aligned}$$

and so $\lim_\nu f(y_\nu) = 0$. Thus, in each case,

$$\lim_\nu f(y_\nu) = f(y),$$

and the claim that f has a continuous extension to $\overline{Z}_\infty$ follows.

For $x \in Z_\infty$, we have $0 \le g(x) \le f(x)$, and so

$$0 \le g(x) \le f(x) \quad (x \in \overline{Z}_\infty).$$

There is a continuous extension of f to Ω such that $0 \le g \le f$ in $C(\Omega)$; since $g \notin P$, we have $f \notin P$, and so $f \in C(\Omega)^+ \setminus P$. □

The following corollary is the key to our constructions in the field K_P. It shows that $\mathbb{M}_{(P)}$ is a 'rather rich' subset of $\mathbb{N}_{(Q)}$ in that the ranges of certain functions are contained in $\mathbb{M}_{(P)}$. This will allow us to transfer our non-standard structures from K_Q to K_P. The result is called the *transfer principle*; it will be used many times.

Corollary 8.30 *Let $F : \mathbb{N}^m \to \mathbb{N}$ be a function such that, for some constant C,*

$$F(n_1, \dots, n_m) \le Cn_1 \cdots n_m \qquad (n_1, \dots, n_m \in \mathbb{N}),$$

and let $F_{(Q)} : \mathbb{N}^m_{(Q)} \to \mathbb{N}_{(Q)}$ be the induced function. Then

$$F_{(Q)}(\mathbb{M}^m_{(P)}) \subset \mathbb{M}_{(P)}.$$

Proof This follows immediately from the theorem. □

The proof of the following lemma is rather similar to the proof of Theorem 8.29 on $\mathbb{M}_{(P)}$; it is a more complicated version of the proof of Theorem 8.9.

Lemma 8.31 *The rational field $\mathbb{Q}_{(P)}$ is order-dense in K_P.*

Proof By Lemma 4.26, it suffices to show that, for each $f, g \in C(\Omega) \setminus P$ with

$$0 < \pi_P(f) < \pi_P(g) \leq 1,$$

there exists $a \in \mathbb{Q}_{(P)}$ with $\pi_P(f) \leq a \leq \pi_P(g)$.

Set $h = g - f \in C(\Omega)^+ \setminus P \subset C(\Omega)^+ \setminus Q$, and form the standard Q-partition corresponding to h, say the partition is $\{Z_j : j \in \mathbb{N}\}$ with union Z_∞, where

$$Z_j = \left\{x \in \Omega : h(x) \in \left[\frac{1}{2j+1}, \frac{1}{2j}\right]\right\} \qquad (j \in \mathbb{N}).$$

Set $Z_0 = \mathbf{Z}(h)$.

For each $j \in \mathbb{N}$, each $k \in \{0, 1, 2\}$, and each $m \in \mathbb{Z}^+$, define

$$W_{j,k,m} = \left\{x \in Z_j : f(x) \leq \frac{3m+k}{2j+1} \leq g(x)\right\}$$

and

$$W_{j,k} = \bigcup \{W_{j,k,m} : m \in \mathbb{Z}^+\}.$$

Each $W_{j,k,m}$ belongs to $\mathbf{Z}(\Omega)$, and $W_{j,k,m} = \emptyset$ for $m \geq j+1$ because $g \leq 1$, and so $W_{j,k} \in \mathbf{Z}(\Omega)$. For each $j \in \mathbb{N}$, we have $W_{j,0} \cup W_{j,1} \cup W_{j,2} = Z_j$. Set

$$W_{\infty,k} = \bigcup \{W_{j,k} : j \in \mathbb{N}\} \quad (k = 0, 1, 2).$$

Then

$$(Z_0 \cup W_{\infty,0}) \cup (Z_0 \cup W_{\infty,1}) \cup (Z_0 \cup W_{\infty,2}) \in \mathbf{Z}[Q],$$

and so one of the three sets on the left belongs to $\mathbf{Z}[Q]$ by Proposition 4.10(ii), say $Z_0 \cup W_{\infty,0} \in \mathbf{Z}[Q]$. (The other cases are treated similarly.) Thus

$$\{W_{j,0,m} : j \in \mathbb{N},\ m \in \mathbb{Z}^+\} \setminus \{\emptyset\}$$

is a Q-partition with limit set contained in Z_0.

By re-indexing the above partition, we see that there are a Q-partition, say $\{U_i : i \in \mathbb{N}\}$ with union U_∞, sequences (m_i) and (n_i) in $\mathbb{N}$, and $F \in C(\Omega) \setminus P$ such that, for each $i \in \mathbb{N}$:

(i) $n_i F(x) = 1 \quad (x \in U_i)$;

(ii) $f(x) \leq 3m_i/n_i \leq g(x) \quad (x \in U_i)$;

(iii) $g(x) - f(x) \leq 2/n_i \quad (x \in U_i)$.

We now define G on U_∞ by

$$G(x) = 3m_i F^2(x) \qquad (x \in U_i). \tag{8.17}$$

We again *claim* that G has a continuous extension to $\overline{U}_\infty$.

Indeed, take $y \in \overline{U}_\infty \setminus U_\infty$ and (y_ν) in U_∞ with $y_\nu \to y$, say $y_\nu \in U_{j_\nu}$ for some $j_\nu \in \mathbb{N}$. To define $G(y)$, first suppose that $F(y) \neq 0$, so that $F(y_\nu) \neq 0$ for sufficiently large ν and

$$\lim_\nu n_{j_\nu} = \lim_\nu \frac{1}{F(y_\nu)} = \frac{1}{F(y)}.$$

Thus (n_{j_ν}) eventually takes the constant value $n = 1/F(y)$. By (iii), $g(y) - f(y) \leq 2/n$, and, by (ii), for sufficiently large ν,

$$f(y_\nu) \leq \frac{3m_{j_\nu}}{n} \leq g(y_\nu).$$

But $\lim_\nu f(y_\nu) = f(y)$ and $\lim_\nu g(y_\nu) = g(y)$, and so (m_{j_ν}) is eventually constant because there is only one number of the form $3m/n$ in the interval $[f(y), g(y)]$, say $m_{j_\nu} = m$ eventually. Set $G(y) = 3m/n^2$. Then

$$\lim_\nu G(y_\nu) = \lim_\nu 3m_{j_\nu} F^2(y_\nu) = G(y).$$

Second, suppose that $F(y) = 0$. Then $\lim_\nu F(y_\nu) = 0$, and so

$$\lim_\nu n_{j_\nu} F^2(y_\nu) = \lim_\nu F(y_\nu) = 0.$$

But

$$0 \leq G(y_\nu) = 3m_{j_\nu} F^2(y_\nu) \leq n_{j_\nu} F^2(y_\nu),$$

and so $\lim_\nu G(y_\nu) = 0$. Set $G(y) = 0$. Thus, in each case, $\lim_\nu G(y_\nu) = G(y)$, and the claim that G has a continuous extension to $\overline{U}_\infty$ follows.

We have $0 \leq F^2(x) \leq G(x)$ $(x \in U_\infty)$, and so G has a continuous extension to $G \in C(\Omega)^+ \setminus P$.

By Definition 8.26, there exist elements $b, c \in \mathbb{M}_{(P)}$ with $\pi_P(G) = b\pi_P(F^2)$ and $1 = c\pi_P(F)$. Set $a = bc^{-1}$, so that $a = \pi(G)/(\pi_P(F))^{-1} \in \mathbb{Q}_{(P)}$. By (i), (ii), and (8.17), we have

$$F^2(x)f(x) \leq G(x)F(x) \leq F^2(x)g(x) \qquad (x \in U_\infty),$$

and so

$$\pi_P(F^2)\pi_P(f) \leq \pi_P(G)\pi_P(F) \leq \pi_P(F^2)\pi_P(g)$$

in A_P, whence $\pi_P(f) \leq a \leq \pi_P(g)$ in K_P, as required. □

Define $h : \mathbb{N} \times \mathbb{N} \to \mathbb{N}$ by

$$h(m,n) = \begin{cases} 1 & \text{if } 2^m > n, \\ 2^m & \text{if } 2^m \leq n, \end{cases}$$

and let $h_{(Q)} : \mathbb{N}_{(Q)} \times \mathbb{N}_{(Q)} \to \mathbb{N}_{(Q)}$ be the induced map. Since

$$h(m,n) \leq n \leq mn \quad (m, n \in \mathbb{N}),$$

it follows from the transfer principle, Corollary 8.30, that we have a map

$$h_{(Q)} : \mathbb{M}_{(P)} \times \mathbb{M}_{(P)} \to \mathbb{M}_{(P)}.$$

Thus, for each $a \in \mathbb{M}_{(P)}$, $2^a \in \mathbb{M}_{(P)}$ if and only $2^a \leq b$ for some $b \in \mathbb{M}_{(P)}$.

We now come to the definition of the set $\mathbb{N}_{(P)}$.

Definition 8.32 *Let P be a non-maximal, prime ideal in $C(\Omega)$. Then*

$$\mathbb{N}_{(P)} = \{a \in \mathbb{M}_{(P)} : 2^a \in \mathbb{M}_{(P)}\},$$

where 2^a is computed in $\mathbb{N}_{(Q)}$. Also

$$\mathbb{Z}^+_{(P)} = \mathbb{N}_{(P)} \cup \{0\} \quad \textit{and} \quad \mathbb{Z}_{(P)} = \{a \in K_P : |a| \in \mathbb{Z}^+_{(P)}\}.$$

The relationships between the various sets that we have defined are shown in Fig. 8.1.

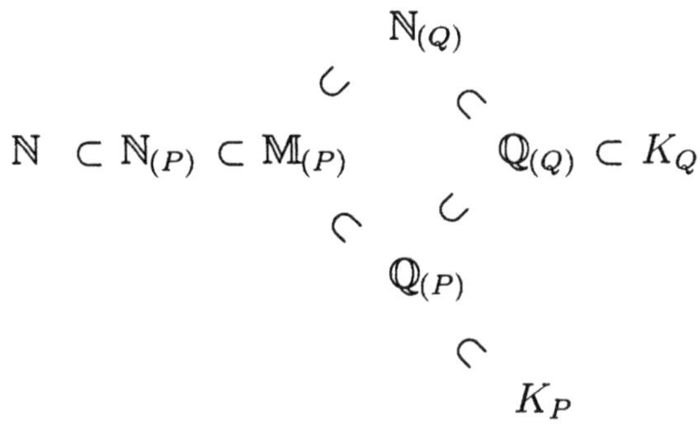

Fig. 8.1. The relationships between some sets

Lemma 8.33 (i) *The set* $\mathbb{N}_{(P)}$ *is an initial segment of* $\mathbb{M}_{(P)}$.

(ii) *Suppose that* $a, b \in \mathbb{N}_{(P)}$. *Then* $a + b \in \mathbb{N}_{(P)}$.

(iii) *For each* $x \in K_P^+$, *there exists a unique element* $a \in \mathbb{M}_{(P)} \cup \{0\}$ *with* $a \leq x < a + 1$.

(iv) *For each* $x \in K_P^+$ *such that* $x \geq 1$, *there exists a unique element* $a \in \mathbb{Z}_{(P)}^+$ *with* $2^a \leq x < 2^{a+1}$.

(v) $\operatorname{cof} \mathbb{N}_{(P)} = \operatorname{cof} \mathbb{M}_{(P)} = \delta(K_P)$.

(vi) *For each* $a \in \mathbb{N}_{(P)}$, $\{b \in \mathbb{N}_{(P)} : b^2 \leq a\}$ *has a maximum element.*

(vii) $\mathbb{N}_{(P)} = \mathbb{M}_{(P)}$ *if and only if* P *is an exponential prime.*

Proof (i) This follows because the function $a \mapsto 2^a$ is increasing.

(ii) For $a, b \in \mathbb{N}_{(P)}$, we have $2^a, 2^b \in \mathbb{M}_{(P)}$, and so $2^{a+b} = 2^a \cdot 2^b \in \mathbb{M}_{(P)}$. Hence $a + b \in \mathbb{N}_{(P)}$.

(iii) By Lemma 8.31, it suffices to suppose that $x \in \mathbb{Q}_{(P)}^+ \setminus \{0\}$, say $x = c/d$, where $c, d \in \mathbb{M}_{(P)}$. Define

$$h : (m, n) \mapsto \max\{k \in \mathbb{N} : k \leq 1 + m/n\}, \quad \mathbb{N}^2 \to \mathbb{N},$$

and let $h_{(Q)} : \mathbb{N}_{(Q)}^2 \to \mathbb{N}_{(Q)}$ be the induced map. We have

$$h(m, n) \leq 2m \quad (m, n \in \mathbb{N}),$$

and so, by Corollary 8.30, $h_{(Q)}(\mathbb{M}_{(P)} \times \mathbb{M}_{(P)}) \subset \mathbb{M}_{(P)}$. Set $a = h_{(Q)}(c, d) - 1$. Then $a \in \mathbb{M}_{(P)} \cup \{0\}$, and $a \leq x < a + 1$. Clearly a is uniquely specified by this condition.

(iv) By (iii), it suffices to suppose that $x \in \mathbb{M}_{(P)}$. Define

$$h : n \mapsto \max\{k \in \mathbb{N} : 2^k \leq 2n\}, \quad \mathbb{N} \to \mathbb{N},$$

so that $h(n) \leq n \quad (n \in \mathbb{N})$, and let $h_{(Q)} : \mathbb{N}_{(Q)} \to \mathbb{N}_{(Q)}$ be the induced map. By Corollary 8.30, $h_{(Q)}(\mathbb{M}_{(P)}) \subset \mathbb{M}_{(P)}$. Define $a = h_{(Q)}(x) - 1$. Then $a \in \mathbb{M}_{(P)} \cup \{0\}$; clearly $2^a \leq x < 2^{a+1}$ and so $a \in \mathbb{Z}^+_{(P)}$.

(v) This follows from (iv).

(vi) Define $h : n \mapsto \max\{k \in \mathbb{N} : k^2 \leq n\}$, $\mathbb{N} \to \mathbb{N}$, and again let $h_{(Q)}$ be the induced map. Then $h_{(Q)}(a) \in \mathbb{M}_{(P)}$ and $h_{(Q)}(a) \leq a$, so that $h_{(Q)}(a) \in \mathbb{N}_{(P)}$ is the required maximum.

(vii) Suppose that P is an exponential prime. We have remarked that $\mathbb{M}_{(P)}$ is closed under exponentiation to base 2, and so $\mathbb{M}_{(P)} = \mathbb{N}_{(P)}$. Conversely, suppose that $\mathbb{M}_{(P)} = \mathbb{N}_{(P)}$. As in Lemma 6.19, the function $\exp_{(2)}$ is defined on a convex subgroup I_E of $(K_P, +)$. Since $\mathbb{N}_{(P)} \subset I_E$ and $\mathbb{M}_{(P)}$ is cofinal in K^+_P, necessarily $I_E = K_P$, and so P is an exponential prime. □

It follows from Lemmas 8.31 and 8.33(i) that the analogue of equation (8.6) holds:

$$\operatorname{coi}(\mathbb{N}_{(P)} \setminus \mathbb{N}) = \operatorname{cof} K^0_P . \tag{8.18}$$

The following calculation of cardinalities gives a significant result that will be used in Chapter 9. It indicates a potential clear difference between $\mathbb{N}_{(P)}$ and $\mathbb{M}_{(P)}$. Of course, in the case where P is an exponential prime, we have $\mathbb{N}_{(P)} = \mathbb{M}_{(P)}$, and, by Theorem 6.29, $w(K_P) = |\Gamma_P|$.

Theorem 8.34 *Let P be a non-maximal, prime ideal in $C(\Omega)$. Then:*

(i) $|\mathbb{M}_{(P)}| = w(K_P)$;

(ii) $|\mathbb{N}_{(P)}| = |\Gamma_P|$.

Proof (i) By Lemma 8.31, $\mathbb{Q}_{(P)}$ is order-dense in K_P, and so

$$w(K_P) \leq |\mathbb{Q}_{(P)}| = |\mathbb{M}_{(P)}| .$$

Let S be an order-dense subset of K_P with $|S| = w(K_P)$. For each $a \in \mathbb{M}_{(P)}$, there exists $s \in S$ with $a \leq s \leq a+1$, and so $w(K_P) \geq |\mathbb{M}_{(P)}|$, giving the result.

(ii) Suppose that $a, b \in \mathbb{N}_{(P)}$ with $\mathbb{N} + a << b$. Then $2^a, 2^b \in \mathbb{M}_{(P)} \subset K_P$, and $v_P(2^a) > v_P(2^b)$ because $2^b \geq 2^n \cdot 2^a \quad (n \in \mathbb{N})$. Hence $|\mathbb{N}_{(P)}| \leq |\Gamma_P|$. It follows from Lemma 8.33(iv) that, for each $x \in K^+_P$ with $x \geq 1$ there exists $a \in \mathbb{Z}^+_{(P)}$ with $v(2^a) = v(x)$, and so the map

$$a \mapsto v(2^a), \quad \mathbb{Z}^+_{(P)} \to \Gamma_P ,$$

has range Γ^+_P. Thus $|\Gamma_P| \leq |\mathbb{N}_{(P)}|$. We have shown that $|\Gamma_P| = |\mathbb{N}_P|$. □

Note that there is no reason to suppose that the set $\mathbb{N}_{(P)}$ is multiplicatively closed in the case where $\mathbb{N}_{(P)} \neq \mathbb{M}_{(P)}$. This is a disappointing feature of a set which purports to be an analogue of $\mathbb{N}$, but nevertheless we shall contend that $\mathbb{N}_{(P)}$, rather than $\mathbb{M}_{(P)}$, is the correct analogue. We shall give an example for which $\mathbb{N}_{(P)}$ is not multiplicatively closed shortly, but first we show how the theory simplifies in the special cases where Q is strongly convex prime or, in particular, a valuation prime (see Chapter 4).

Theorem 8.35 *In the case where Q is a z-ideal which is a strongly convex prime of $C(\Omega)$, $\mathbb{M}_{(P)}$ is an initial segment of $\mathbb{N}_{(Q)}$.*

Proof By the definition of a strongly convex prime, M_Q is an interval in K_Q^0, and certainly $\pi_Q(P)$ is an interval in M_Q, and so $M_Q^+ \setminus \pi_Q(P)$ is an interval in K_Q^0. Set

$$J = \{a \in \mathbb{N}_{(Q)} : 1/a \in M_Q^+ \setminus \pi_Q(P)\}.$$

Then J is an initial segment of $\mathbb{N}_{(Q)} \setminus \mathbb{N}$.

First, we *claim* that $J \subset \mathbb{M}_{(P)}$. For take $a \in J$, and choose a function $f \in C(\Omega)^+ \setminus Q$, a Q-partition $\{Z_j : j \in \mathbb{N}\}$, and a sequence (n_j) in $\mathbb{N}$ such that $1/f(x) = n_j \ (x \in Z_j)$ and $\pi_Q(f) = 1/a$. Since $1/a \in M_Q^+ \setminus \pi_Q(P)$, necessarily $f \in C(\Omega)^+ \setminus P$. We have

$$1 = n_j f(x) \quad (x \in Z_j,\ j \in \mathbb{N})$$

and $\pi_P(1) = a\pi_P(f)$, and so $a \in \mathbb{M}_{(P)}$ by the definition of $\mathbb{M}_{(P)}$, giving the first claim.

Second, we *claim* that J is cofinal in $\mathbb{M}_{(P)}$. For take $c \in \mathbb{M}_{(P)} \setminus K_Q^{\#}$. Then $1/c \in K_Q^+ \setminus \pi_Q(P)$. But clearly $\{1/a : a \in J\}$ is coinitial in the set $M_Q^+ \setminus \pi_Q(P)$, and $M_Q^+ \setminus \pi_Q(P)$ is coinitial in $K_Q^+ \setminus \pi_Q(P)$, and so there exists $a \in J$ with $1/a < 1/c$. We have $a > c$, giving the second claim.

It follows that $\mathbb{M}_{(P)} \setminus \mathbb{N} = J$, and so, since J is an initial segment of $\mathbb{N}_{(Q)} \setminus \mathbb{N}$, $\mathbb{M}_{(P)}$ is an initial segment of $\mathbb{N}_{(Q)}$. □

Note that, in the above proof, we used the fact that Q is strongly convex only to show that J is an initial segment of $\mathbb{N}_{(Q)} \setminus \mathbb{N}$; for every prime z-ideal Q, J is a cofinal subset of $\mathbb{M}_{(P)}$.

Proposition 8.36 *Let Q be a valuation prime z-ideal in $C(\Omega)$.*

(i) *A subset J of $\mathbb{N}_{(Q)}$ has the form $\mathbb{M}_{(P)}$ for some prime ideal P in $C(\Omega)$ above Q if and only if J is a multiplicatively closed initial segment of $\mathbb{N}_{(Q)}$.*

(ii) *A subset I of $\mathbb{N}_{(Q)}$ has the form $\mathbb{N}_{(P)}$ for some prime ideal P in $C(\Omega)$ above Q if and only if I is an additively closed initial segment of $\mathbb{N}_{(Q)}$.*

Proof (i) A set $\mathbb{M}_{(P)}$ is always multiplicatively closed; by Theorem 8.35, $\mathbb{M}_{(P)}$ is an initial segment of $\mathbb{N}_{(Q)}$ in the case where Q is a valuation prime.

Now take J to be a multiplicatively closed initial segment of $\mathbb{N}_{(Q)}$. Define

$$L = \{c \in K_Q^0 : |ac| < 1 \quad (a \in J)\}.$$

Then L is an ideal in $K_Q^{\#} = A_Q$. Take $c_1, c_2 \in K_Q^0 \setminus L$, so that there exist $a_1, a_2 \in J$ with $|a_1c_1| \geq 1$ and $|a_2c_2| \geq 1$. Since J is multiplicatively closed, $a_1a_2 \in J$, and we have

$$|(a_1a_2)(c_1c_2)| \geq 1.$$

Thus $c_1c_2 \notin L$, and so L is a prime ideal. Set $P = \pi_Q^{-1}(L)$. Then P is a prime ideal in $C(\Omega)$ above Q.

Take $a \in J$. Then $1/a \notin L$, and so $a \in \mathbb{M}_{(P)}$ by our characterization of $\mathbb{M}_{(P)}$. Now assume that $b \in \mathbb{M}_{(P)} \setminus J$. Since $b \notin J$, we have $a < b$ $(a \in J)$, and so $1/b \in L$, a contradiction of the fact that $b \in \mathbb{M}_{(P)}$. Thus $J = \mathbb{M}_{(P)}$.

(ii) This follows from (i) because I is additively closed if and only if the set $\{2^a : a \in I\}$ is multiplicatively closed. □

We now give an example to show that $\mathbb{N}_{(P)}$ need not be multiplicatively closed, even in the case of prime ideals P in $C(\beta\mathbb{N})$. The example will be used in Chapter 9.

Proposition 8.37 *Let $\mathfrak{p} \in \beta\mathbb{N}\setminus\mathbb{N}$, and set $Q = J_\mathfrak{p}$ in $C(\beta\mathbb{N})$. Then there exist:*

(i) *an additively closed initial segment I of $\mathbb{N}_{(Q)}$ such that $\operatorname{cof} I = \aleph_1$ and I is not multiplicatively closed;*

(ii) *a multiplicatively closed initial segment J of $\mathbb{N}_{(Q)}$ such that $\operatorname{cof} J = \aleph_1$ and there exists $c \in J$ with $2^c \notin J$;*

(iii) *a prime ideal P in $C(\beta\mathbb{N})$ above Q such that $\mathbb{N}_{(P)}$ is not multiplicatively closed.*

Proof (i) Fix $d \in \mathbb{N}_{(Q)}\setminus\mathbb{N}$, and let I_0 be an initial segment of $\mathbb{N}_{(Q)}$ with $d \in I_0$, with $2d \notin I_0$, and such that $\operatorname{cof} I_0 = \omega_1$; such an interval I_0 exists because K_Q is an η_1-field and the interval $[d, 2d]$ is infinite.

Define

$$I = \{b \in \mathbb{N}_{(Q)} : b \leq 2^{ad} \text{ for some } a \in I_0\},$$

so that I is an initial segment of $\mathbb{N}_{(Q)}$. We *claim* that I is additively closed. Indeed, take $b_1, b_2 \in I$. Then there exist elements $a_1, a_2 \in I_0$ with $b_1 \leq 2^{a_1d}$ and $b_2 \leq 2^{a_2d}$. Set

$$a = (a_1 \vee a_2) + 1\,;$$

we have $a \in I_0$ because I_0 has no maximum element, and $b_1 + b_2 \leq 2^{ad}$, so that $b_1 + b_2 \in I$.

The segment I is not multiplicatively closed: $2^{d^2} \in I$ because $d \in I_0$, but $2^{2d^2} \notin I$ because $a < 2d$ for each $a \in I_0$.

Clearly we have $\operatorname{cof} I = \aleph_1$ because $\operatorname{cof} I_0 = \aleph_1$.

(ii) Take I as specified in (i), and choose $a_0 \in I$ with $a_0^2 \notin I$. Set

$$J = \{a \in \mathbb{N}_{(Q)} : a \leq 2^b \text{ for some } b \in I\}.$$

Then J has the required properties; the role of c is played by 2^{a_0}.

(iii) By Proposition 8.36(ii), there is a prime ideal P in $C(\beta\mathbb{N})$ above Q such that $I = \mathbb{N}_{(P)}$. □

We remark that we proved in Theorem 6.20 that there exists a specific convex subsemigroup I_E of K_P^+ on which there is an exponentiation. It follows in this case that $\mathbb{N}_{(P)}$ is cofinal in I_E and that $\mathbb{N}_{(P)} = I_E \cap \mathbb{M}_{(P)}$. We shall not use this fact.

We now return to the general case, in which P is an arbitrary non-maximal, prime ideal in $C(\Omega)$ and Q is the maximal z-ideal below P, and we continue to define structures in the super-real field K_P analogous to those in K_Q. It will follow from Lemma 8.39, below, that the following definition of $\varepsilon_r^{(P)}$ is valid.

Definition 8.38 *We set:*

(i) $\mathbb{D}_a^{(P)} = \mathbb{D}_a^{(Q)} \cap \mathbb{Q}_{(P)}$ *for* $a \in \mathbb{Z}^+_{(P)}$*;*

(ii) $\mathbb{D}_{(P)} = \bigcup\{\mathbb{D}_a^{(P)} : a \in \mathbb{Z}^+_{(P)}\}$*;*

(iii) $\varepsilon_r^{(P)} = \varepsilon_r^{(Q)} \mid \mathbb{N}_{(P)}$ *for* $r \in (\mathbb{Q}_{(P)} \cap \mathbb{I}_{(P)}) \setminus \mathbb{D}_{(P)}$*;*

(iv) $\{0,1\}^a_{(P)} = \{\varepsilon_r^{(P)} \restriction a : r \in (\mathbb{Q}_{(P)} \cap \mathbb{I}_{(P)}) \setminus \mathbb{D}_{(P)}\}$ *for* $a \in \mathbb{N}_{(P)}$*;*

(v) $\{0,1\}^{<\mathbb{N}}_{(P)} = \bigcup\{\{0,1\}^a_{(P)} : a \in \mathbb{N}_{(P)}\}$.

The set $\mathbb{D}_{(P)}$ is the set of *dyadic rationals* in K_P

Lemma 8.39 *With the above notation,* $\mathbb{D}_{(P)} = \mathbb{D}_{(Q)} \cap \mathbb{Q}_{(P)}$.

Proof We must show that $\mathbb{D}_{(Q)} \cap \mathbb{Q}_{(P)} \subset \mathbb{D}_{(P)}$. Take $r \in \mathbb{D}_{(Q)} \cap \mathbb{Q}_{(P)}$ with $r \neq 0$. Then there exist $a_1, a_2 \in \mathbb{M}_{(P)}$, and $a, c \in \mathbb{N}_{(Q)}$ with

$$r = \frac{a_1}{a_2} = \frac{c}{2^a}\,.$$

We may suppose that c and 2^a are coprime in $\mathbb{N}_{(Q)}$, and hence that $2^a \leq a_2$. By Lemma 8.33(iv), there exists $b \in \mathbb{N}_{(P)}$ with $a_2 \leq 2^b$. We have $a \leq b$, and so $r \in \mathbb{D}_b^{(P)} \subset \mathbb{D}_{(P)}$. □

There is an aspect of the above notation concerning which care is required. Let $r \in (\mathbb{Q}_{(P)} \cap \mathbb{I}_{(P)}) \setminus \mathbb{D}_{(P)}$, and take $a \in \mathbb{N}_{(P)} \subset \mathbb{N}_{(Q)}$. Then the domain of the function $\varepsilon_r^{(P)} \restriction a$ is $[1,a] \cap \mathbb{N}_{(P)}$, whereas that of $\varepsilon_r^{(Q)} \restriction a$ is $[1,a] \cap \mathbb{N}_{(Q)}$; by Theorem 8.35, these two domains are equal in the special case where Q is a strongly convex prime in $C(\Omega)$, but in general the two sets may be different.

Note that we have not yet defined $\varepsilon_r^{(P)}$ for all $r \in \mathbb{I}_{(P)} \setminus \mathbb{D}_{(P)}$.

As in the case for prime z-ideals, $\{0,1\}^{<\mathbb{N}}_{(P)}$ is a rich collection of functions. For example, if $f_1, f_2 \in \{0,1\}^{<\mathbb{N}}_{(P)}$, then $f_1 \vee f_2$, $f_1 \wedge f_2$, and $1 - f_1$ also belong to $\{0,1\}^{<\mathbb{N}}_{(P)}$.

The following lemma explains in part why we do not define $\mathbb{D}_a^{(P)}$ for elements a of $\mathbb{M}_{(P)} \setminus \mathbb{N}_{(P)}$.

Lemma 8.40 *The set* $\mathbb{D}_{(P)}$ *is order-dense in* $\mathbb{I}_{(P)}$.

Proof By Lemma 8.31, it is sufficient to prove that $\mathbb{D}_{(P)}$ is order-dense in $\mathbb{Q}_{(P)} \cap \mathbb{I}_{(P)}$. Define

$$h : (m_1, m_2, m, n) \mapsto \max\{k \leq m : km_2 \leq 2^n m_1\}\,, \qquad \mathbb{N}^4 \to \mathbb{Z}^+\,, \qquad (8.19)$$

with induced map $h_{(Q)} : \mathbb{N}^4_{(Q)} \to \mathbb{Z}^+_{(Q)}$. We have $h(m_1, m_2, m, n) \leq m$, and so, by the transfer principle, $h_{(Q)}(a_1, a_2, a, b) \in \mathbb{M}_{(P)}$ for $a_1, a_2, a, b \in \mathbb{M}_{(P)}$.

Take $x, y \in \mathbb{Q}_{(P)} \cap \mathbb{I}_{(P)}$ with $x < y$, and choose $a_1, a_2 \in \mathbb{M}_{(P)}$ with $y = a_1/a_2$. It follows from Lemma 8.33(iv) that there exists $a \in \mathbb{N}_{(P)}$ with $2^a > (y-x)^{-1}$. We have $2^{-a} < y - x$, and so, by Lemma 8.12(iii), $\mathbb{D}_a^{(Q)} \cap (x, y) \neq \emptyset$. Set

$$c = h_{(Q)}(a_1, a_2, 2^a, a)\,,$$

so that $c \in \mathbb{M}_{(P)}$ because $a_1, a_2 \in \mathbb{M}_{(P)}$ and $a \in \mathbb{N}_{(P)} \subset \mathbb{M}_{(P)}$, the latter implying that $2^a \in \mathbb{M}_{(P)}$.

We have $c/2^a \in \mathbb{D}_a^{(Q)} \subset \mathbb{D}_{(Q)}$, and $c/2^a \in \mathbb{Q}_{(P)}$, and so $c/2^a \in \mathbb{D}_{(P)}$ by Lemma 8.39. Also $ca_2 \leq 2^a a_1$, and so $c/2^a \leq y$. Finally, from the definition of $h_{(Q)}$, either $c = 2^a$ or $(c+1)a_2 > 2^a a_1$. In each case, $c/2^a > y - 2^{-a} > x$, and so $c/2^a \in [x, y]$. □

Lemma 8.41 *Suppose that $r, s \in (\mathbb{Q}_{(P)} \cap \mathbb{I}_{(P)}) \setminus \mathbb{D}_{(P)}$ with $r < s$, that $b \in \mathbb{M}_{(P)}$, and that $\varepsilon_r^{(Q)} \mid b \neq \varepsilon_s^{(Q)} \mid b$. Then there exists $a \in [0, b] \cap \mathbb{Z}^+_{(P)}$ such that $\varepsilon_r^{(Q)} \mid a = \varepsilon_s^{(Q)} \mid a$, $\varepsilon_r^{(Q)}(a+1) = 0$, and $\varepsilon_s^{(Q)}(a+1) = 1$.*

Proof By Lemma 8.15(ii), there is a unique $a \in [0, b] \cap \mathbb{Z}^+_{(Q)}$ with the required properties; we shall show that, in fact, $a \in \mathbb{Z}^+_{(P)}$.

We define $h : \mathbb{N}^5 \to \mathbb{N}$ as follows. In the case where $m_1, m_2, n_1, n_2 \in \mathbb{N}$, $m_1/n_1,\ m_2/n_2 \in \mathbb{I} \setminus \mathbb{D}$, and the set

$$\{k \in \{1, \dots, m\} : \varepsilon_{m_1/n_1}(k) \neq \varepsilon_{m_2/n_2}(k)\}$$

is non-empty, $h(m_1, m_2, n_1, n_2, m)$ is the minimum element of this set; otherwise, $h(m_1, m_2, n_1, n_2, m) = 1$. We have $h(m_1, m_2, n_1, n_2, m) \leq m$ in both cases, and so the induced map $h_{(Q)}$ is such that $h_{(Q)}(\mathbb{M}^5_{(P)}) \subset \mathbb{M}_{(P)}$ by the transfer principle.

Choose $a_1, a_2, b_1, b_2 \in \mathbb{M}_{(P)}$ with $r = a_1/b_1$ and $s = a_2/b_2$. By Lemma 8.39, $a_1/b_1,\ a_2/b_2 \in \mathbb{I}_{(Q)} \setminus \mathbb{D}_{(Q)}$, and, by hypothesis, $\varepsilon_r^{(Q)} \mid b \neq \varepsilon_s^{(Q)} \mid b$. Hence

$$h_{(Q)}(a_1, a_2, b_1, b_2, b) = a\,,$$

showing that $a \in \mathbb{M}_{(P)}$.

By Lemma 8.15(i), $s - r \leq 2^{-a}$. Choose $c \in \mathbb{N}_{(P)}$ with $2^{-c} < s - r$. Then $2^a < 2^{c+1}$, and so $a < c + 1$. Thus $a \in \mathbb{N}_{(P)}$. □

It follows from Lemma 8.41 that we may define the *lexicographic order* on $\{0,1\}^b_{(P)}$ in the usual way: for $f, g \in \{0,1\}^b_{(P)}$ with $f \neq g$, we say that $f \prec g$ if $f(a) = 0$ and $g(a) = 1$, where a is the minimum element c of $\mathbb{N}_{(P)}$ with $f(c) \neq g(c)$. Let $r, s \in \mathbb{I}_{(P)} \setminus \mathbb{D}_{(P)}$ with $r \leq s$. Then $\varepsilon_r^{(P)} \mid b \preceq \varepsilon_s^{(P)} \mid b$.

Corollary 8.42 *Let $r, s \in (\mathbb{Q}_{(P)} \cap \mathbb{I}_{(P)}) \setminus \mathbb{D}_{(P)}$.*

(i) *Suppose that $a \in \mathbb{N}_{(P)}$ and $\varepsilon_r^{(P)} \mid a = \varepsilon_s^{(P)} \mid a$. Then $\varepsilon_r^{(Q)} \mid a = \varepsilon_s^{(Q)} \mid a$.*

(ii) *Suppose that $\varepsilon_r^{(P)} = \varepsilon_s^{(P)}$ on $\mathbb{N}_{(P)}$. Then $\varepsilon_r^{(Q)} \mid \mathbb{M}_{(P)} = \varepsilon_s^{(Q)} \mid \mathbb{M}_{(P)}$.* □

The above corollary illustrates a further reason why we consider $\mathbb{N}_{(P)}$, rather than $\mathbb{M}_{(P)}$: a function $\varepsilon_r^{(Q)}$ on $\mathbb{M}_{(P)}$ is already defined by its values on $\mathbb{N}_{(P)}$.

Lemma 8.43 *Suppose that $r, s \in \mathbb{I}_{(P)} \cap \mathbb{Q}_{(P)}$ with $r < s$, and $a \in \mathbb{N}_{(P)}$. Then:*

(i) $\max(\mathbb{D}_a^{(Q)} \cap [0, s]_{(Q)})$ *exists and belongs to* $\mathbb{D}_a^{(P)}$;

(ii) $\mathbb{D}_a^{(Q)} \cap [r, s]_{(Q)} = \emptyset$ *if and only if* $\mathbb{D}_a^{(P)} \cap [r, s]_{(P)} = \emptyset$.

Proof (i) The maximum element, say q, of $\mathbb{D}_a^{(Q)} \cap [0, s]_{(Q)}$ exists by Lemma 8.12(vi). Choose $a_1, a_2 \in \mathbb{M}_{(P)}$ with $s = a_1/a_2$, define $h : \mathbb{N}^4 \to \mathbb{N}$ by formula (8.19), and set $c = h_{(Q)}(a_1, a_2, 2^a, a)$; as before, $c \in \mathbb{M}_{(P)}$. We have $q = c/2^a$, and so $q \in \mathbb{D}_a^{(P)}$.

(ii) Suppose that $\mathbb{D}_a^{(Q)} \cap [r, s]_{(Q)} \neq \emptyset$, and take q as (i). Then $q \in [r, s]_{(P)}$, and so $\mathbb{D}_a^{(P)} \cap [r, s]_{(P)} \neq \emptyset$. The result follows. □

The next six lemmas show that the structures that have been defined so far for P have many of the properties that would be expected from considerations of the earlier case of a prime z-ideal. The proofs of these lemmas are similar to the arguments just given. The first such lemma is just a copy of Lemma 8.12, with P replacing Q.

Lemma 8.44 (i) *For $a, b \in \mathbb{Z}^+_{(P)}$ with $a \leq b$, $\mathbb{D}_a^{(P)} \subset \mathbb{D}_b^{(P)}$.*

(ii) *$\mathbb{D}_{(P)}$ and $(\mathbb{Q}_{(P)} \cap \mathbb{I}_{(P)}) \setminus \mathbb{D}_{(P)}$ are order-dense in $\mathbb{I}_{(P)}$.*

(iii) *Let $a \in \mathbb{Z}^+_{(P)}$, and take $r, s \in \mathbb{D}_a^{(P)}$ with $r < s$. Then $2^{-a} \leq s - r$. Further, $\mathbb{D}_a^{(P)} \cap (r, s) = \emptyset$ if and only if $s - r = 2^{-a}$.*

(iv) *Let $a \in \mathbb{Z}^+_{(P)}$, and take $r, s \in \mathbb{Q}_{(P)} \cap \mathbb{I}_{(P)}$ with $r < s$. Suppose that $\mathbb{D}_a^{(P)} \cap (r, s) = \emptyset$. Then $\mathbb{D}_{a+1}^{(P)} \cap (r, s)$ contains at most one element.*

(v) *Take $r, s \in \mathbb{Q}_{(P)} \cap \mathbb{I}_{(P)}$ with $0 < r < s < 1$. Then there exists $a \in \mathbb{N}_{(P)}$ such that $\mathbb{D}_a^{(P)} \cap (r, s) = \emptyset$ and $\mathbb{D}_{a+1}^{(P)} \cap (r, s) \neq \emptyset$.*

(vi) *Let $a \in \mathbb{Z}^+_{(P)}$ and $r \in \mathbb{Q}_{(P)} \cap \mathbb{I}_{(P)}$. Then $\mathbb{D}_a^{(P)} \cap [0, r]$ has a maximum element, say r', and $\mathbb{D}_a^{(P)} \cap [r, 1]$ has a minimum element, say r''. Further, if $r \notin \mathbb{D}_a^{(P)}$, then $r'' - r' = 2^{-a}$ and $\mathbb{D}_a^{(P)} \cap (r', r'') = \emptyset$.*

Proof (i) This is immediate from the definitions and from Lemma 8.12(i).

(ii)–(v) These are proved from the corresponding parts of Lemma 8.12 by introducing certain functions h. As a sample, we prove (v); the other parts are easier.

Take $r, s \in \mathbb{Q}_{(P)} \cap \mathbb{I}_{(P)}$ with $0 < r < s < 1$. By Lemma 8.12(v), there exists $a \in \mathbb{N}_{(Q)}$ with $\mathbb{D}_a^{(Q)} \cap (r, s) = \emptyset$ and $\mathbb{D}_{a+1}^{(Q)} \cap (r, s) \neq \emptyset$. It is clear from 8.12(iii) and 8.12(vi) that $s - r \leq 2^{-a}$.

We define $h : \mathbb{N}^5 \to \mathbb{N}$ as follows. In the case where $m_1, n_1, m_2, n_2, m \in \mathbb{N}$ and

$$\{k \in \{1, \ldots, m\} : \mathbb{D}_k \cap (m_1/n_1, m_2/n_2) = \emptyset\} \neq \emptyset ,$$

$h(m_1, m_2, n_1, n_2, m)$ is the maximum element of this set; otherwise, we set $h(m_1, m_2, n_1, n_2, m) = 1$. The induced map $h_{(Q)}$ is such that $h_{(Q)}(\mathbb{M}^5_{(P)}) \subset \mathbb{M}_{(P)}$.

Choose $a_1, a_2, b_1, b_2 \in \mathbb{M}_{(P)}$ with $r = a_1/b_1$ and $s = a_2/b_2$, and choose $c \in \mathbb{N}_{(P)}$ with $2^{-c} < s - r$, so that $a < c$. Then

$$h_{(Q)}(a_1, a_2, b_1, b_2, c) = a ,$$

showing that $a \in \mathbb{M}_{(P)}$. Since $a < c$, necessarily $a \in \mathbb{N}_{(P)}$.

Since $\mathbb{D}_a^{(P)} \subset \mathbb{D}_a^{(Q)}$, we have $\mathbb{D}_a^{(P)} \cap (r, s) = \emptyset$.

Set $q = \max \left(\mathbb{D}^{(Q)}_{a+1} \cap [0, s]_{(Q)}\right)$. By Lemma 8.43(i), $q \in \mathbb{D}^{(P)}_{a+1}$. If $q = s$, then $r < q - 2^{-(a+1)} < s$ and $q - 2^{-(a+1)} \in \mathbb{D}^{(P)}_{a+1}$, and so $\mathbb{D}^{(P)}_{a+1} \cap (r, s) \neq \emptyset$. If $q < s$, then $r < q < s$, and again $\mathbb{D}^{(P)}_{a+1} \cap (r, s) \neq \emptyset$.

(vi) Set $r' = \max \left(\mathbb{D}_a^{(Q)} \cap [0, r]_{(Q)}\right)$ and $r'' = \min \left(\mathbb{D}_a^{(Q)} \cap [r, 1]_{(Q)}\right)$. By Lemma 8.43(i), $r' \in \mathbb{D}_a^{(P)}$, and $r'' = 2^{-a} + r' \in \mathbb{D}_a^{(P)}$, and so r' and r'' have the required properties. □

Lemma 8.45 *Let $a \in \mathbb{N}_{(P)}$, and take $r, s \in (\mathbb{Q}_{(P)} \cap \mathbb{I}_{(P)}) \setminus \mathbb{D}_{(P)}$ with $r < s$. Then $\varepsilon_r^{(P)} \restriction a = \varepsilon_s^{(P)} \restriction a$ if and only if $\mathbb{D}_a^{(P)} \cap (r, s) = \emptyset$.*

Proof By Lemma 8.41, $\varepsilon_r^{(P)} \restriction a = \varepsilon_s^{(P)} \restriction a$ if and only if $\varepsilon_r^{(Q)} \restriction a = \varepsilon_s^{(Q)} \restriction a$, and, by Lemma 8.44(iii), $\mathbb{D}_a^{(P)} \cap [r, s] = \emptyset$ if and only if $\mathbb{D}_a^{(Q)} \cap [r, s] = \emptyset$. Since $r, s \notin \mathbb{D}_a^{(P)}$, the result follows from Lemma 8.14. □

By using Lemma 8.45 we can now define $\varepsilon_r^{(P)}$ for each $r \in \mathbb{I}_{(P)} \setminus \mathbb{D}_{(P)}$. Indeed, take $r \in \mathbb{I}_{(P)} \setminus \mathbb{D}_{(P)}$. By Lemma 8.31, $\mathbb{Q}_{(P)}$ is order-dense in K_P, and so there exists $s \in (\mathbb{Q}_{(P)} \cap \mathbb{I}_{(P)}) \setminus \mathbb{D}_{(P)}$ with $\mathbb{D}_a^{(P)} \cap (r, s) = \emptyset$. Define

$$\varepsilon_r^{(P)}(a) = \varepsilon_s^{(P)}(a) . \tag{8.20}$$

By Lemma 8.45, $\varepsilon_r^{(P)}$ is well defined, and so we have a function

$$\varepsilon_r^{(P)} : \mathbb{N}_{(P)} \to \{0, 1\} .$$

Clearly, for each $a \in \mathbb{N}_{(P)}$,

$$\{0,1\}^a_{(P)} = \{\varepsilon_r^{(P)} \restriction a : r \in \mathbb{I}_{(P)} \setminus \mathbb{D}_{(P)}\}$$

and so

$$\{0,1\}^{<\mathbb{N}}_{(P)} = \bigcup \left\{ \{\varepsilon_r^{(P)} \restriction a : r \in \mathbb{I}_{(P)} \setminus \mathbb{D}_{(P)}\} : a \in \mathbb{N}_P \right\} . \tag{8.21}$$

For each $r, s \in \mathbb{I}_{(P)} \setminus \mathbb{D}_{(P)}$ with $r \neq s$, there is a minimum element $c \in \mathbb{N}_{(P)}$ with $\varepsilon_r^{(P)}(c) \neq \varepsilon_s^{(P)}(c)$.

To finally pin down the properties of the functions ε_r for $r \in \mathbb{I}_{(P)} \setminus \mathbb{D}_{(P)}$, we record the following, now obvious, lemma; clause (iii) follows from Proposition 8.14 and Corollary 8.42.

Lemma 8.46 *Let $r, s \in \mathbb{I}_{(P)} \setminus \mathbb{D}_{(P)}$ with $r < s$ and let $a \in \mathbb{N}_{(P)}$. Then:*

(i) $\varepsilon_r^{(P)} \upharpoonright a \preceq \varepsilon_s^{(P)} \upharpoonright a$;

(ii) $\varepsilon_r^{(P)} \upharpoonright a = \varepsilon_s^{(P)} \upharpoonright a$ *if and only if* $\mathbb{D}_a^{(P)} \cap (r, s) = \emptyset$, *and in this case* $s - r < 2^{-a}$;

(iii) $\varepsilon_r^{(P)}(b) = \varepsilon_s^{(P)}(b) \quad (b > a)$ *if and only if* $s - r \in \mathbb{D}_a^{(P)}$. □

Lemma 8.47 *Let $r \in \mathbb{I}_{(P)} \setminus \mathbb{D}_{(P)}$, and let $a \in \mathbb{N}_{(P)}$. Set*

$$s_1 = \max(\mathbb{D}_a^{(P)} \cap [0, r])\,, \qquad s_2 = \max(\mathbb{D}_{a+1}^{(P)} \cap [0, r])\,,$$
$$t_1 = \min(\mathbb{D}_a^{(P)} \cap [r, 1])\,, \qquad t_2 = \min(\mathbb{D}_{a+1}^{(P)} \cap [r, 1])\,.$$

Then:

(i) $s_1 \leq s_2 < r < t_2 \leq t_1$;

(ii) *either*

$$s_1 = s_2, \quad t_1 \neq t_2 \quad \textit{and} \quad \varepsilon_r^{(P)}(a+1) = 0$$

or

$$s_1 \neq s_2, \quad t_1 = t_2, \quad \textit{and} \quad \varepsilon_r^{(P)}(a+1) = 1\,.$$

Proof This follows from Lemmas 8.16 and 8.43. □

It follows that $\{a \in \mathbb{N}_{(P)} : \varepsilon_r^{(P)}(a) = 0\}$ and $\{a \in \mathbb{N}_{(P)} : \varepsilon_r^{(P)}(a) = 1\}$ are both cofinal in $\mathbb{N}_{(P)}$ for $r \in \mathbb{I}_{(P)} \setminus \mathbb{D}_{(P)}$.

Lemma 8.48 *Let $r \in \mathbb{D}_{(P)}$. Then there is a minimum $a \in \mathbb{Z}_{(P)}^+$ with $r \in \mathbb{D}_a^{(P)}$.*

Proof We may suppose that $0 < r < 1$. Choose $b \in \mathbb{N}_{(P)}$ with $r \in \mathbb{D}_b^{(P)}$, and take $\delta \in \mathbb{Q}_{(P)}$ with $0 < \delta < 2^{-b}$ and $(r - \delta, r + \delta) \subset (0, 1)$. Then clearly $\mathbb{D}_b^{(P)} \cap (r - \delta, r + \delta) = \{r\}$; by Lemma 8.44(v), there exists $c \in \mathbb{Z}_{(P)}^+$ with $\mathbb{D}_c^{(P)} \cap (r - \delta, r + \delta) = \emptyset$ and $\mathbb{D}_{c+1}^{(P)} \cap (r - \delta, r + \delta) \neq \emptyset$. Thus $c + 1 \leq b$, $\mathbb{D}_{c+1}^{(P)} \cap (r - \delta, r + \delta) = \{r\}$, and $a = c + 1$ has the required properties. □

Lemma 8.49 *Let P be such that K_P is a semi-η_1-field, and let (f_k) be a sequence in $\{0,1\}_{(P)}^{<\mathbb{N}}$. Suppose that there is a sequence (a_k) in $\mathbb{N}_{(P)}$ and an element $a \in \mathbb{N}_{(P)}$ such that, for each $k \in \mathbb{N}$:*

(i) $a_k < a_{k+1} < a$;

(ii) $\operatorname{dom} f_k = [1, a_k] \cap \mathbb{N}_{(P)}$;

(iii) $f_{k+1} \upharpoonright a_k = f_k$.

Then there exists $f \in \{0,1\}_{(P)}^a$ such that $f \upharpoonright a_k = f_k$ for each $k \in \mathbb{N}$.

Proof This is the same as the proof of Lemma 8.19; that result used several properties of the structures associated with Q, and all the analogous results for P have been proved. □

We now recall the pairing functions Φ and ρ defined earlier. The bounds on the sizes of $\Phi(m, n)$ and $\rho(k)$ that were noted in (8.8) and (8.9) lead via the transfer principle to the key facts that $\Phi_{(Q)}(\mathbb{M}_{(P)} \times \mathbb{M}_{(P)}) \subset \mathbb{M}_{(P)}$ and that $\rho_{(Q)}(\mathbb{M}_{(P)}) \subset \mathbb{M}_{(P)} \times \mathbb{M}_{(P)}$. We define

$$\Phi_{(P)} = \Phi_{(Q)} \mid \mathbb{M}_{(P)} \times \mathbb{M}_{(P)}\,, \quad \rho_{(P)} = \rho_{(Q)} \mid \mathbb{M}_{(P)}\,. \tag{8.22}$$

Then the two maps

$$\Phi_{(P)} : \mathbb{M}_{(P)} \times \mathbb{M}_{(P)} \to \mathbb{M}_{(P)}\,, \quad \rho_{(P)} : \mathbb{M}_{(P)} \to \mathbb{M}_{(P)} \times \mathbb{M}_{(P)}$$

are bijections which are inverses of each other.

In an entirely similar way to the definition of $R_g^{(Q)}$ in (8.13), we define

$$R_g^{(P)} = \{\rho_{(P)}(a) : a \in \operatorname{dom} g,\; g(a) = 1\}$$

for a function g with $\operatorname{dom} g \subset \mathbb{M}_{(P)}$ and $\operatorname{ran} g \subset \{0, 1\}$.

We now come to the analogue of Definition 8.20, which defined the collection of functions $\mathbb{N}_{(Q)}^{<\mathbb{N}}$. Since $\mathbb{N}_{(P)}$ may not be multiplicatively closed, to use the direct analogue of Definition 8.20 would produce a meagre collection of functions, and the sum of two functions in this collection would not necessarily be in the collection. For this reason, we modify the previous definition.

Definition 8.50 *Let $a, b \in \mathbb{N}_{(P)}$ be such that $a^2b^2 \in \mathbb{N}_{(P)}$. Then $[1, b]_{(P)}^a$ is the set of functions F with domain $[1, a] \cap \mathbb{N}_{(P)}$ and codomain $[1, b] \cap \mathbb{N}_{(P)}$ such that $\operatorname{Gr} F = R_g^{(P)}$ for some $g = \varepsilon_r^{(P)} \restriction c$ for which $r \in \mathbb{I}_{(P)} \setminus \mathbb{D}_{(P)}$ and $c \in \mathbb{N}_{(P)}$. Further,*

$$\mathbb{N}_{(P)}^{<\mathbb{N}} = \bigcup \left\{ [1, b]_{(P)}^a : a, b \in \mathbb{N}_{(P)} \text{ with } a^2b^2 \in \mathbb{N}_{(P)} \right\}.$$

As usual, we must work through some book-keeping exercises to verify that the structures we have defined have their expected properties.

Lemma 8.51 *Let $a \in \mathbb{N}_{(P)}$, and $r \in (\mathbb{Q}_{(P)} \cap \mathbb{I}_{(P)}) \setminus \mathbb{D}_{(P)}$. Set $f = \varepsilon_r^{(P)} \restriction a$ and $g = \varepsilon_r^{(Q)} \restriction a$. Then:*

(i) $R_f^{(P)} = R_g^{(Q)} \cap (\mathbb{N}_{(P)} \times \mathbb{N}_{(P)})$;

(ii) *$R_f^{(P)}$ is a function if and only if $R_g^{(Q)}$ is a function;*

(iii) *for each $b \in \mathbb{N}_{(P)}$, $R_f^{(P)}$ is a function with domain $[1, b] \cap \mathbb{N}_{(P)}$ if and only if $R_g^{(Q)}$ is a function with domain $[1, b] \cap \mathbb{N}_{(Q)}$;*

(iv) *$R_f^{(P)}$ is increasing if and only if $R_g^{(Q)}$ is increasing.*

Proof (i) This is immediate because $\varepsilon_r^{(P)} = \varepsilon_r^{(Q)} \mid \mathbb{N}_{(P)}$.

(ii) By (i), if $R_g^{(Q)}$ is a function, then $R_f^{(P)}$ is a function.

Define

$$h : (n, n_1, n_2) \mapsto (m, m_1, m_2), \quad \mathbb{N}^3 \to \mathbb{N}^3,$$

as follows. Suppose that $(n, n_1, n_2) \in \mathbb{N}^3$ with $n_1 < n_2$ and $n_1/n_2 \notin \mathbb{D}$, and set

$$\varphi = \varepsilon_{n_1/n_2} \restriction n;$$

if R_φ is not a function, there exist $(m, m_1) \in R_\varphi$ and $(m, m_2) \in R_\varphi$ with $m_1 \neq m_2$, and we define $h(n, n_1, n_2) = (m, m_1, m_2)$. In all other cases, define $h(n, n_1, n_2) = (1, 1, 1)$. Note that $m, m_1, m_2 \leq n$ in each case. In the usual notation, $h_{(Q)} : \mathbb{N}^3_{(Q)} \to \mathbb{N}^3_{(Q)}$ is the induced map.

Now suppose that $R_g^{(Q)}$ is not a function. Choose $a_1, a_2 \in \mathbb{M}_{(P)}$ such that $r = a_1/a_2$, and set

$$(b, b_1, b_2) = h_{(Q)}(a, a_1, a_2) \in \mathbb{N}^3_{(Q)}\,.$$

Then $(b, b_1) \in R_g^{(Q)}$, $(b, b_2) \in R_g^{(Q)}$, and $b_1 \neq b_2$. Also $b, b_1, b_2 \leq a$. By the transfer principle, we have $b, b_1, b_2 \in \mathbb{M}_{(P)}$, and so $b, b_1, b_2 \in \mathbb{N}_{(P)}$ because we have $a \in \mathbb{N}_{(P)}$. By (i), $(b, b_1) \in R_f^{(P)}$ and $(b, b_2) \in R_f^{(P)}$, and so $R_f^{(P)}$ is not a function.

(iii) and (iv). These are similar to (ii); a simple modification of the definition of h suffices. □

There now follow three lemmas that are modifications of Lemmas 8.21, 8.22, and 8.23 respectively; they are slightly changed to take account of the definition that we have given of $\mathbb{N}^{<\mathbb{N}}_{(P)}$. They may be easily deduced from the earlier lemmas and from Lemma 8.51 by again defining suitable maps $h : \mathbb{N}^m \to \mathbb{N}$.

Lemma 8.52 *Let* $g \in \{0,1\}^a_{(P)}$, *where* $a \in \mathbb{N}_{(P)}$. *Then*

$$\{c \in [1, a] \cap \mathbb{N}_{(P)} : R^{(P)}_{g \restriction c} \text{ is a function in } \mathbb{N}^{<\mathbb{N}}_{(P)}\}$$

is empty or has a maximum element. □

Lemma 8.53 *Let* $F \in [1, b]^a_{(P)}$, *where* $a, b \in \mathbb{N}_{(P)}$ *and* $a^2b^2 \in \mathbb{N}_{(P)}$. *Then*

$$\{c \in [1, a] \cap \mathbb{N}_{(P)} : F \restriction b \text{ is increasing}\}$$

has a maximum element. □

Lemma 8.54 *Let* $F \in [1, b]^a_{(P)}$, *where* $a, b \in \mathbb{N}_{(P)}$ *and* $a^2b^2 \in \mathbb{N}_{(P)}$, *and take* $c \in \mathbb{N}_{(P)}$. *Then:*

(i) $(\operatorname{ran} F) \cap [1, c] \cap \mathbb{N}_{(P)}$ *is either empty or has a maximum element;*

(ii) $F^{-1}(\{c\})$ *is either empty or has a minimum element.* □

We are now aiming for an analogue of Lemma 8.25. In fact, we shall prove only a weak analogue, in which the domains of the functions F_k are just the intervals $[1, k]$ in $\mathbb{N}$ (although the more general result is true). We require two preliminary lemmas.

Lemma 8.55 *Let $a \in \mathbb{N}_{(P)}$, and let X be a finite subset of $[1, a] \cap \mathbb{N}_{(P)}$. Then there exists $s \in (\mathbb{Q}_{(P)} \cap \mathbb{I}_{(P)}) \setminus \mathbb{D}_{(P)}$ such that, for each $b \in [1, a] \cap \mathbb{N}_{(P)}$, $\varepsilon_s^{(P)}(b) = 1$ if and only if $b \in X$.*

Proof Let $\{a_1, \ldots, a_m\}$ be an enumeration of X, with $a_1 < a_2 < \cdots < a_m$, and, for each $k \in \{1, \ldots, m\}$, set

$$q_k = 2^{-a_1} + \cdots + 2^{-a_k}.$$

We shall prove the following *claim* by induction on $k \in \mathbb{N}$: for each $a \in \mathbb{N}_{(P)}$ with $a \leq a_k$, for each $s \in \mathbb{Q}_{(P)} \setminus \mathbb{D}_{(P)}$ with

$$s \in (q_k,\, q_k + 2^{-a}),$$

and each $b \in [1, a] \cap \mathbb{N}_{(P)}$, we have $\varepsilon_s^{(P)}(b) = 1$ if and only if $b \in \{a_1, \ldots, a_k\}$. The case where $k = 1$ and the inductive step are similar; we give the latter. Thus we assume that the result holds for k.

Fix $a \in \mathbb{N}_{(P)}$ with $a \geq a_{k+1}$ and $s \in \mathbb{Q}_{(P)} \setminus \mathbb{D}_{(P)}$ with

$$s \in (q_{k+1}, q_{q+1} + 2^{-a}) \subset (q_k, q_k + 2^{-a_k}).$$

First, take $b \in [1, a_k] \cap \mathbb{N}_{(P)}$. Then the result for $\varepsilon_s^{(P)}(b)$ holds by the inductive hypothesis. Next, take $b \in (a_k, a] \cap \mathbb{N}_{(P)}$, and set

$$x_b = \max(\mathbb{D}_b^{(P)} \cap [0, s]).$$

If $b < a_{k+1}$, then $s \in (q_k, q_k + 2^{-b})$. Since $q_k \in \mathbb{D}_{a_k}^{(P)} \subset \mathbb{D}_b^{(P)}$, we have $x_b = q_k$. If $b \geq a_{k+1}$, then $q_{k+1} \in \mathbb{D}_b^{(P)}$, and so $x_b = q_{k+1}$. By Lemma 8.47, $\varepsilon_s^{(P)}(b) = 1$ if and only if $b = a_{k+1}$.

The claim holds for $k + 1$, and so the induction continues to prove the claim for each $k \in \mathbb{N}$.

Clearly the claim we have proved implies the statement of the lemma. □

We next clarify the notation to be used in the following lemma: $[1, b]_{(P)}^k$ is the set defined in Definition 8.50 in the special case where $a = k \in \mathbb{N} \subset \mathbb{N}_{(P)}$, and $([1, b] \cap \mathbb{N}_{(P)})^k$ is the set of all functions with domain $\{1, \ldots, k\}$ and codomain $[1, b] \cap \mathbb{N}_{(P)}$.

Lemma 8.56 *Let $b \in \mathbb{N}_{(P)}$ with $b^2 \in \mathbb{N}_{(P)}$. Then, for each $k \in \mathbb{N}$,*

$$[1, b]_{(P)}^k = ([1, b] \cap \mathbb{N}_{(P)})^k.$$

Proof Take $k \in \mathbb{N}$. Certainly $[1, b]_{(P)}^k \subset ([1, b] \cap \mathbb{N}_{(P)})^k$.

Suppose that $F \in ([1, b] \cap \mathbb{N}_{(P)})^k$, and set

$$X = \{\Phi_{(P)}(j, d) : j \in \{1, \ldots, k\},\ F(j) = d\},$$

a finite subset of $[1, b^2] \cap \mathbb{N}_{(P)}$; by hypothesis, $b^2 \in \mathbb{N}_{(P)}$, and so, by Lemma 8.55, there exists $s \in (\mathbb{Q}_{(P)} \cap \mathbb{I}_{(P)}) \setminus \mathbb{D}_{(P)}$ such that, for each $x \in [1, b] \cap \mathbb{N}_{(P)}$, we have $\varepsilon_s^{(P)}(x) = 1$ if and only if $x \in X$. But, by Definition 8.50, this implies that $F \in [1, b]_{(P)}^k$. □

We now come to the (weak) analogue of Lemma 8.25.

Lemma 8.57 *Let P be such that K_P is a semi-η_1-field, and let $a \in \mathbb{N}_{(P)}$ with $a^2 \in \mathbb{N}_{(P)}$. Suppose that (F_k) is a sequence in $\mathbb{N}_{(P)}^{<\mathbb{N}}$ such that, for each $k \in \mathbb{N}$:*

(i) *F_k is an increasing function;*

(ii) *$F_k \in [1,a]_{(P)}^k$;*

(iii) *$F_{k+1} \restriction k = F_k$.*

Then there exist $d \in \mathbb{N}_{(P)} \setminus \mathbb{N}$ and $F \in [1,a]_{(P)}^d$ such that $F \restriction k = F_k$ for each $k \in \mathbb{N}$.

Proof For $k \in \mathbb{N}$, set $b_k = F_k(k)$ and $a_k = \Phi_{(P)}(k, b_k)$, so that $b_k \leq b_{k+1} \leq a$ and $a_k < a_{k+1} < a^2$. There exists $g_k \in \{0,1\}_{(P)}^{a_k}$ with $R_{g_k}^{(P)} = \mathrm{Gr}\, F_k$. By Lemma 8.49, there exists $g \in \{0,1\}_{(P)}^{a^2}$ with $g \restriction a_k = g_k$ for each $k \in \mathbb{N}$.

By Lemma 8.52, the set

$$\{c \in [1,a^2] \cap \mathbb{N}_{(P)} : R_{g \restriction c}^{(P)} \text{ is a function in } \mathbb{N}_{(P)}^{<\mathbb{N}}\}$$

has a maximum element, say it is c. We have $a_k < c \leq a^2$ ($k \in \mathbb{N}$). Take $F \in \mathbb{N}_{(P)}^{<\mathbb{N}}$ with $R_{g \restriction c}^{(P)} = \mathrm{Gr}\, F$, and take $d \in \mathbb{N}_{(P)}$ with $\mathrm{dom}\, F = [1,d] \cap \mathbb{N}_{(P)}$. Since $g \restriction a_k = g_k$ ($k \in \mathbb{N}$), necessarily $d \in \mathbb{N}_{(P)} \setminus \mathbb{N}$, and $F \restriction k = F_k$ ($k \in \mathbb{N}$), so that $F \in [1,a]_{(P)}^d$, as required. □

Lemma 8.58 *Let $a \in \mathbb{N}_{(P)}$, and let κ and λ be infinite cardinals with $\kappa \neq \lambda$. Assume that there is a (κ, λ)-gap in $(\{0,1\}_{(P)}^a, \preceq)$. Then there is also a (κ, λ)-gap in the initial segment $([1,a] \cap \mathbb{N}_{(P)}, \leq)$.*

Proof Take $\langle A, B \rangle$ to be a (κ, λ)-gap in $(\{0,1\}_{(P)}^a, \preceq)$; we may suppose that A is an initial segment of $\{0,1\}_{(P)}^a$ and that $B = \{0,1\}_{(P)}^a \setminus A$.

For each $f \in A$, define

$$I_f = \{b \in [1,a] \cap \mathbb{N}_{(P)} : \quad \text{there exists } g \in A \text{ with } f \restriction b = g \restriction b \text{ and } f(b+1) < g(b+1)\}.$$

The set I_f is not empty. Define

$$S = \{f \in A : I_f \text{ has a minimum element}\};$$

for $f \in S$, the minimum element of I_f is b_f. Suppose that $f, g \in S$ with $f \prec g$. Then $b_f < b_g$ and $g \restriction b_f = f \restriction b_f$.

We now construct a (κ, λ)-gap $\langle \tilde{A}, \tilde{B} \rangle$ in $[1,a] \cap \mathbb{N}_{(P)}$. There are two cases to consider.

Case 1 Suppose that S is cofinal in A.

$$\tilde{A} = \{b \in [1,a] \cap \mathbb{N}_{(P)} : b < b_f \text{ for some } f \in S\}, \quad \tilde{B} = ([1,a] \cap \mathbb{N}_{(P)}) \setminus \tilde{A}.$$

Then $\tilde{A}$ is an initial segment of $[1,a] \cap \mathbb{N}_{(P)}$, $\langle \tilde{A}, \tilde{B} \rangle$ is a gap, and clearly $\mathrm{cof}\, \tilde{A} = \kappa$.

The function $H : \tilde{A} \to \{0,1\}$ given by $H \upharpoonright b_f = f \upharpoonright b_f$ $(f \in S)$ is well defined. For each $f \in S$, define

$$\overline{f}(b) = \begin{cases} H(b) & (b \in [1, b_f] \cap \mathbb{N}_{(P)}), \\ 1 & (b \in (b_f, a] \cap \mathbb{N}_{(P)}). \end{cases}$$

Then $\overline{f} \in \{0,1\}^a_{(P)}$. Since S is cofinal in A, $\overline{f} >> A$, and so $\overline{f} \in B$. Set

$$\overline{B} = \{g \in B : g \geq \overline{f} \quad \text{for some} \quad f \in S\}.$$

Then $\operatorname{coi} \overline{B} = \operatorname{cof} S = \kappa$. However $\operatorname{coi} B = \lambda$ and $\lambda \neq \kappa$, and so $\overline{B}$ is not coinitial in B. Choose $h_0 \in B \setminus \overline{B}$, so that $h_0 \mid \tilde{A} = H$. For each $h \in B$ with $h \prec h_0$, we have $h \mid \tilde{A} = H = h_0 \mid \tilde{A}$.

Since $\langle A, B \rangle$ is a gap, the set

$$\{b \in \tilde{B} : \text{there exists } \ h \in B \text{ with } \ h \prec h_0, \ \ h \upharpoonright b = h_0 \upharpoonright b, \\ \text{and } \ h(b+1) < h_0(b+1)\}$$

is coinitial in $\tilde{B}$, and this shows that $\operatorname{coi} \tilde{B} = \lambda$.

Case 2 Suppose that S is not cofinal in A.

There exists $f_0 \in A$ such that, for each $f \in A$ with $f_0 \preceq f$, I_f does not have a minimum element. Take $f, g \in A$ with $f_0 \preceq f \prec g$, and take b to be the minimum element of $[1, a] \cap \mathbb{N}_{(P)}$ with $f(b) \neq g(b)$. Then $b - 1 \in I_f$ and $[1, b-1] \cap I_f = [1, b-1] \cap I_g$ is an infinite set. Set

$$A_0 = \{c \in [1, a] \cap \mathbb{N}_{(P)} : b < c \ \text{ for some } \ b \in I_{f_0}\}, \quad B_0 = ([1, a] \cap \mathbb{N}_{(P)}) \setminus A_0 .$$

Then $\langle B_0, A_0 \rangle$ is a gap in $[1, a] \cap \mathbb{N}_{(P)}$, and it follows from the fact that $\operatorname{cof} A = \kappa$ that $\operatorname{coi} A_0 = \kappa$. Once again because $\langle A, B \rangle$ is a gap,

$$\{b \in B_0 : \text{there exists } \ h \in B \ \text{ with } \ h \upharpoonright b = f_0 \upharpoonright b \ \text{ and } \ f_0(b+1) < h(b+1)\}$$

is cofinal in B_0, and so $\operatorname{cof} B_0 = \lambda$.

Finally, define $\tilde{A} = \{a - b : b \in A_0\}$ and $\tilde{B} = \{a - b : b \in B_0\}$. Then $\langle \tilde{A}, \tilde{B} \rangle$ is a (κ, λ)-gap in $[1, a] \cap \mathbb{N}$. □

Lemma 8.59 *Let κ and λ be infinite cardinals with $\kappa \neq \lambda$. Assume that there are a (κ, λ)-gap $\langle A, B \rangle$ in $\mathbb{I}_{(P)}$ and $a \in \mathbb{N}_{(P)}$ such that $s > r + 2^{-a}$ for all $r \in A$ and $s \in B$. Then there is a (κ, λ)-gap in $\mathbb{N}_{(P)}$.*

Proof We may suppose that A is an initial segment of $\mathbb{I}_{(P)}$ and $A \cup B = \mathbb{I}_{(P)}$. Set

$$\tilde{A} = \{\varepsilon_r^{(P)} \upharpoonright a : r \in A \setminus \mathbb{D}_{(P)}\}, \quad \tilde{B} = \{\varepsilon_s^{(P)} \upharpoonright a : s \in B \setminus \mathbb{D}_{(P)}\}.$$

For each $r \in A$ and $s \in B$, we have $s > r + 2^{-a}$. So, given $r_1 \in A$, there exists $r_2 \in A$ with $r_2 > r_1 + 2^{-a}$. By Lemma 8.44(iii), $\mathbb{D}_a^{(P)} \cap (r_1, r_2) \neq \emptyset$, and so, by Lemma 8.45, $\varepsilon_{r_1}^{(P)} \upharpoonright a \prec \varepsilon_{r_2}^{(P)} \upharpoonright a$. Thus $\tilde{A}$ has no maximum element, and hence $\operatorname{cof} \tilde{A} = \kappa$ by Lemma 1.6. Similarly, $\operatorname{coi} \tilde{B} = \lambda$. It follows that $\langle \tilde{A}, \tilde{B} \rangle$ is a (κ, λ)-gap in $\{0,1\}^a_{(P)}$. By Lemma 8.58, there is a (κ, λ)-gap in $\mathbb{N}_{(P)}$. □

Our final gap theorem is proved for a prime ideal P in $C(\Omega)$ such that K_P is a semi-η_1-field, but it is not required that P be a z-ideal. The preliminary lemma is proved for a prime ideal P which is not a z-ideal by using much of the above machinery; in the case where P is a z-ideal, the proof is the same, but it is not necessary to replace the original gap $\langle A, B\rangle$ by the modified gap.

Lemma 8.60 *Let P be a prime ideal in $C(\Omega)$ such that K_P is a semi-η_1-field, and assume that there is an $(\aleph_0, \lambda)$-gap in $\mathbb{N}_{(P)}$ for some infinite cardinal λ. Then $\lambda = \mathrm{coi}(\mathbb{N}_{(P)} \setminus \mathbb{N})$.*

Proof Take $\langle A, B\rangle$ to be an $(\aleph_0, \lambda)$-gap in $\mathbb{N}_{(P)}$; we may suppose that A is an initial segment in $\mathbb{N}_{(P)}$ and that $A \cup B = \mathbb{N}_{(P)}$.

We *claim* that we may suppose that there exists $b \in B$ with $b^2 \in \mathbb{N}_{(P)}$. For suppose that this fails for the original gap $\langle A, B\rangle$, and set

$$A' = \{d \in \mathbb{N}_{(P)} : d^2 \in A\}, \quad B' = \mathbb{N}_{(P)} \setminus A',$$

so that $\langle A', B'\rangle$ is a gap in $\mathbb{N}_{(P)}$. There are again two cases to consider.

Case 1 Suppose that $\{d^2 : d \in A'\}$ is cofinal in A.

Let $b \in B$, and set $c = \max\{d \in \mathbb{N}_{(P)} : d^2 \le b\}$; the element c exists by Lemma 8.33(vi). Then $A << c^2 \le b$, and so $c^2 \in B$, whence $c \in B'$. Thus $\{c^2 : c \in B',\ c^2 \in \mathbb{N}_{(P)}\}$ is coinitial in B, $\langle A', B'\rangle$ is an $(\aleph_0, \lambda)$-gap, and there exists $c \in B'$ with $c^2 \in \mathbb{N}_{(P)}$.

Case 2 Suppose that $\{d^2 : d \in A'\}$ is not cofinal in A.

Fix $c_0 \in A$ with $d^2 < c_0$ $(d \in A')$, and set

$$d_0 = \max\{d \in \mathbb{N}_{(P)} : d^2 \le c_0\},$$

so that $d_0 \in A'$, $d_0 + 1 \in B'$, and $(d_0 + 1)^2 \in B$. Set

$$A'' = \{d - d_0^2 : d \in A,\ d > d_0^2\}, \quad B'' = \mathbb{N}_{(P)} \setminus A''.$$

Then $\langle A'', B''\rangle$ is an $(\aleph_0, \lambda)$-gap in $\mathbb{N}_{(P)}$, being a translation of the original gap $\langle A, B\rangle$. Further $(d_0 + 1)^2 - d_0^2 = 2d_0 + 1 \in B''$ and $(2d_0 + 1)^2 \in \mathbb{N}_{(P)}$.

This establishes the claim.

Fix an increasing, cofinal sequence (a_k) in A. For each $k \in \mathbb{N}$, let F_k be the function with domain $[1, k] \cap \mathbb{N}_{(P)}$ given by

$$F_k(i) = a_i \qquad (i = 1, \ldots, k).$$

Thus, for each $k \in \mathbb{N}$, $F_k \in \mathbb{N}_{(P)}^{<\mathbb{N}}$, F_k is an increasing function, and also $F_{k+1} \restriction k = F_k$. Now take d to be an arbitrary element of B, so that we have $\operatorname{ran} F_k \subset [1, d]$ $(k \in \mathbb{N})$. By Lemma 8.57, there exists $F \in \mathbb{N}_{(P)}^d$ with $F \restriction k = F_k$ for each $k \in \mathbb{N}$. By Lemma 8.53 and the fact that $F \restriction k$ is increasing for each k,

$$\{c \in [1, d] \cap \mathbb{N}_{(P)} : F \restriction c \text{ is increasing}\}$$

has a maximum element, say c_0, with $c_0 >> \mathbb{N}$. By replacing d by c_0, we may suppose that F is increasing on $[1, d] \cap \mathbb{N}_{(P)}$.

We now *claim* that

$$\{F(c) : c \in ([1,d] \cap \mathbb{N}_{(P)}) \setminus \mathbb{N}\}$$

is coinitial in B. To see this, fix $b_1 \in B$. By Lemma 8.54(i),

$$(\operatorname{ran} F) \cap [1, b_1] \cap \mathbb{N}_{(P)}$$

has a maximum element, say b_2. Since $F(k) = a_k < b_2 \quad (k \in \mathbb{N})$, we have $A << b_2$, and so $b_2 \in B$. Take c_2 to be the minimum element of $F^{-1}(\{b_2\})$; c_2 exists by Lemma 8.54(ii). Since F is increasing, we have $c_2 \in ([1,d] \cap \mathbb{N}_{(P)}) \setminus \mathbb{N}$ and $F(c_2) = b_2 \leq b_1$, and so the claim is proved.

Clearly $\operatorname{coi}((\operatorname{dom} F) \setminus \mathbb{N}) = \operatorname{coi} B$. But $(\operatorname{dom} F) \setminus \mathbb{N}$ is coinitial in $\mathbb{N}_{(P)} \setminus \mathbb{N}$, and so $\lambda = \operatorname{coi} B = \operatorname{coi}(\mathbb{N}_{(P)} \setminus \mathbb{N})$. □

At last we have reached the *gap theorem for super-real fields.*

Theorem 8.61 *Let Ω be a compact space, and let P be a non-maximal, prime ideal in $C(\Omega)$ such that K_P is a semi-η_1-field. Let λ be an infinite cardinal such that there is an $(\aleph_0, \lambda)$-gap in K_P. Then $\lambda = \operatorname{cof} K_P^0$.*

Proof Take $\langle A, B\rangle$ to be an $(\aleph_0, \lambda)$-gap in K_P. By scaling and translating if necessary, we may suppose that $\langle A, B\rangle$ is a gap in $\mathbb{I}_{(P)}$. Since K_P is a semi-η_1-field, we have $\lambda > \aleph_0$. By Proposition 3.3, $\langle A, B\rangle$ is not a Cauchy gap: there exists $\varepsilon > 0$ in K_P such that

$$s > r + \varepsilon \quad (r \in A,\ s \in B)\,,$$

and so there exists $a \in \mathbb{N}_{(P)}$ with $s > r + 2^{-a} \quad (r \in A,\ s \in B)$. By Lemma 8.59, there is an $(\aleph_0, \lambda)$-gap in $\mathbb{N}_{(P)}$; by Lemma 8.60, we have $\lambda = \operatorname{coi}(\mathbb{N}_{(P)} \setminus \mathbb{N})$, and so the result follows from (8.18). □

Corollary 8.62 *Let K be a hyper-real field, and let λ be a cardinal such that there is an $(\aleph_0, \lambda)$-gap in K. Then $\lambda = \operatorname{cof} K_P^0$.*

Proof By Theorem 4.29, K is an η_1-field, and so, for each $(\aleph_0, \lambda)$-gap in K, necessarily λ is infinite. □

We pointed out in Chapter 4 that it was suggested in Antonovskij *et al.* (1981) that every real-closed η_1-field is a hyper-real field, that is, that the classes **real-closed η_1** and **hyper-real** of that chapter are not distinct. We proved in Theorem 6.23 that this is not the case, in the sense that the real-closed η_1-field $\widehat{\mathbb{R}}$ is not isomorphic as a field to any super-real field. We now show that there is also a counter-example in the much stronger sense that there is such a field which is not even order-isomorphic to a super-real field.

Theorem 8.63 *There is a real-closed η_1-field which is not order-isomorphic to any super-real field.*

Proof In Proposition 2.25, we exhibited a real-closed η_1-field K (with $|K| = \aleph_2$) such that K contains both $(\aleph_0, \aleph_1)$ and $(\aleph_0, \aleph_2)$ gaps. By Theorem 8.61, no super-real η_1-field can contain gaps of both these types. Thus K is not order-isomorphic to a super-real field. □

There is a final point that we should address in this chapter, namely the relationship between the structures induced by two different prime ideals in the same algebra $C(\Omega)$. This is a natural question, and the conclusions will be needed in the following chapter. The results follow easily from one technical lemma about Q-partitions.

Lemma 8.64 *Let Ω be a compact space, and let Q_1 and Q_2 be non-maximal, prime z-ideals in $C(\Omega)$ with Q_1 below Q_2. Suppose that $\{Z_j : j \in \mathbb{N}\}$ is a Q_2-partition with limit set contained in Z_0 and with union Z_∞. Then:*

(i) *there is a Q_1-partition $\{W_j : j \in \mathbb{N}\}$ such that*

$$W_j \cap (Z_0 \cup Z_\infty) = Z_j \quad (j \in \mathbb{N});$$

(ii) *if $\{W_j' : j \in \mathbb{N}\}$ and $\{W_j'' : j \in \mathbb{N}\}$ are Q_1-partitions with*

$$W_j' \cap (Z_0 \cup Z_\infty) = W_j'' \cap (Z_0 \cup Z_\infty) = Z_j \quad (j \in \mathbb{N}),$$

then $\{W_j' \cap W_j'' : j \in \mathbb{N}\}$ is a Q_1-partition.

Proof (i) Set $Z = Z_0 \cup Z_\infty$. There exists $h \in C(\Omega)^+$ with

$$h(x) = \frac{1}{2j} \quad (x \in Z_j,\, j \in \mathbb{N}), \qquad h(x) = 0 \quad (x \in Z_0).$$

Since $\mathbf{Z}(h) \notin \mathbf{Z}[Q_2]$ and Q_2 is a z-ideal, we have $h \notin Q_2$, and so $h \notin Q_1$. Let $\{W_j : j \in \mathbb{N}\}$ be a standard partition for Q_1 corresponding to h, as in Theorem 8.2. There are two possibilities for W_j: $W_j = h^{-1}(I_j)$, where either $I_j = [1/(2j+1), 1/2j]$ or $I_j = [1/2j, 1/(2j-1)]$. In either case $W_j \cap Z = Z_j$ for $j \in \mathbb{N}$.

(ii) Set

$$U_{j,k} = W_j' \cap W_k'' \quad (j, k \in \mathbb{N}).$$

By Lemma 8.5, $\{U_{j,k} : j, k \in \mathbb{N}\}$ is a Q_1-partition with limit set contained in $(W_0' \cap W'') \cup (W_0'' \cap W')$; say this latter set is V_0. We note that $Z \cap V_0 \subset Z_0$ because $Z \cap W' \subset Z_0$ and $Z \cap W'' \subset Z_0$. Either $\{U_{j,j} : j \in \mathbb{N}\}$ or $\{U_{j,k} : j \neq k\}$ is a Q_1-partition with limit set contained in V_0. Set

$$V_\infty = \bigcup \{U_{j,k} : j \neq k\}.$$

Then $(V_\infty \cup V_0) \cap Z \subset Z_0$, and so it cannot be that $V_\infty \cup V_0 \in \mathbf{Z}[Q_1]$, for this would imply that $Z_0 \in \mathbf{Z}[Q_1] \subset \mathbf{Z}[Q_2]$, which is not the case. Thus $\{U_{j,j} : j \in \mathbb{N}\}$ is a Q_1-partition, and it has the required properties. □

Suppose that P_1 and P_2 are non-maximal, prime ideals in $C(\Omega)$ with P_1 below P_2. We define a partial map

$$\Lambda_{P_1,P_2} : \mathbb{M}_{(P_2)} \to \mathbb{M}_{(P_1)}$$

as follows. The domain of Λ_{P_1,P_2} is the set of elements $a \in \mathbb{M}_{(P_2)}$ such that there exist $f, g \in C(\Omega) \setminus P_2$ with $a = \pi_{P_2}(f)/\pi_{P_2}(g)$ and with $\pi_{P_1}(f)/\pi_{P_1}(g) \in \mathbb{M}_{(P_1)}$, and then $\Lambda_{P_1,P_2}(a) = \pi_{P_1}(f)/\pi_{P_1}(g)$. It is easily verified that $\Lambda_{P_1,P_2}(a)$ is well defined. Clearly the domain of Λ_{P_1,P_2} is an additively and multiplicatively closed subset of $\mathbb{M}_{(P_2)}$ and Λ_{P_1,P_2} preserves addition and multiplication; we do not yet know that $\operatorname{dom}\Lambda_{P_1,P_2}$ is all of $\mathbb{M}_{(P_2)}$.

Lemma 8.65 *Suppose that* $\operatorname{dom}\Lambda_{P_1,P_2} = \mathbb{M}_{(P_2)}$. *Then*

$$\mathbb{M}_{(P_2)} = \{\pi_{P_2}(f)/\pi_{P_2}(g) : f,g \in C(\Omega)\setminus P_2,\ \pi_{P_1}(f)/\pi_{P_1}(g) \in \mathbb{M}_{(P_1)}\}.$$

Proof By hypothesis, $\mathbb{M}_{(P_2)}$ is contained in the set on the right-hand side.

Now take $f,g \in C(\Omega)\setminus P_2$ with $\pi_{P_1}(f)/\pi_{P_1}(g) \in \mathbb{M}_{(P_1)}$. It follows from Lemma 8.33(iii) that there exists $a \in \mathbb{M}_{(P_2)} \cup \{0\}$ with

$$|\pi_{P_2}(f)/\pi_{P_2}(g) - a| \le 1/2.$$

By hypothesis, there exist $f', g' \in C(\Omega)\setminus P_2$ with $a = \pi_{P_2}(f')/\pi_{P_2}(g')$ and with $\pi_{P_1}(f')/\pi_{P_1}(g') \in \mathbb{M}_{(P_1)}$. Thus

$$|\pi_{P_2}(f)/\pi_{P_2}(g) - \pi_{P_2}(f')/\pi_{P_2}(g')| \le 1/2,$$

and so

$$|\pi_{P_1}(f)/\pi_{P_1}(g) - \pi_{P_1}(f')/\pi_{P_1}(g')| \le 1/2.$$

However $\pi_{P_1}(f)/\pi_{P_1}(g),\ \pi_{P_1}(f')/\pi_{P_1}(g') \in \mathbb{M}_{(P_1)}$, and so these two elements of $\mathbb{M}_{(P_1)}$ are equal, i.e. $fg' - f'g \in P_1 \subset P_2$. Thus $\pi_{P_2}(f)/\pi_{P_2}(g) = a \in \mathbb{M}_{(P_2)}$, and so the set on the right-hand side is contained in $\mathbb{M}_{(P_2)}$. □

Lemma 8.66 *Let Ω be a compact space, and let P_1 and P_2 be non-maximal, prime ideals in $C(\Omega)$ with P_1 below P_2. Then* $\operatorname{dom}\Lambda_{P_1,P_2} = \mathbb{M}_{(P_2)}$ *and*

$$\Lambda_{P_1,P_2}(\mathbb{N}_{P_2)}) \subset \mathbb{N}_{(P_1)}.$$

Proof Let Q_1 and Q_2 be the maximum prime z-ideals below P_1 and P_2, respectively, so that $Q_1 \subset Q_2$.

We first note that it is an immediate consequence of the definitions of $\mathbb{M}_{(P_1)}$ and $\mathbb{M}_{(P_2)}$ that $\operatorname{dom}\Lambda_{Q_1,P_1} = \mathbb{M}_{(P_1)}$ and $\operatorname{dom}\Lambda_{Q_2,P_2} = \mathbb{M}_{(P_2)}$.

We second *claim* that $\operatorname{dom}\Lambda_{Q_1,Q_2} = \mathbb{N}_{(Q_2)}$. For take $a \in \mathbb{N}_{(Q_2)}$. Then a is specified by a function $f \in B_{Q_2}(\mathbb{N})$, where f is defined on a Q_2-partition, say $\{Z_j : j \in \mathbb{N}\}$. Let $\{W_j : j \in \mathbb{N}\}$ be a Q_1-partition related to $\{Z_j : j \in \mathbb{N}\}$ as in Lemma 8.64(i), and choose $g \in B_{Q_1}(\mathbb{N})$ so that $\{g(W_j)\} = \{f(Z_j)\}$ $(j \in \mathbb{N})$. The function g defines an element, say $\iota(a)$, of $\mathbb{N}_{(Q_1)}$; by Lemma 8.64(ii), $\iota(a)$ does not depend on the choice of the partition $\{W_j : j \in \mathbb{N}\}$ satisfying the appropriate conditions, and clearly the value of $\iota(a)$ is independent of the choice of f. Thus $a \in \operatorname{dom}\Lambda_{Q_1,Q_2}$ and $\Lambda_{Q_1,Q_2}(a) = \iota(a)$.

By combining the above arguments and again using Lemma 8.64, we see that

$$\operatorname{dom}\Lambda_{Q_1,P_2} = \mathbb{M}_{(P_2)}.$$

Finally, take $a \in \mathbb{M}_{(P_2)} = \operatorname{dom}\Lambda_{Q_1,P_2}$, so that there exist $f,g \in C(\Omega)\setminus P_2$ with $a = \pi_{P_2}(f)/\pi_{P_2}(g)$ and with $\pi_{Q_1}(f)/\pi_{Q_1}(g) \in \mathbb{N}_{(Q_1)}$. Since $\operatorname{dom}\Lambda_{Q_1,P_1} = \mathbb{M}_{(P_1)}$, it follows from Lemma 8.65 applied to the pair $\{P_1, Q_1\}$ that $\pi_{P_1}(f)/\pi_{P_1}(g)$ belongs to $\mathbb{M}_{(P_1)}$. Thus $a \in \operatorname{dom}\Lambda_{P_1,P_2}$, an hence $\operatorname{dom}\Lambda_{P_1,P_2} = \mathbb{M}_{(P_2)}$

Clearly $\Lambda_{P_1,P_2}(\mathbb{N}_{(P_2)}) \subset \mathbb{N}_{(P_1)}$. □

Theorem 8.67 *Let Ω be a compact space, and let P_1 and P_2 be non-maximal, prime ideals in $C(\Omega)$ with P_1 below P_2. Then*

$$\mathbb{M}_{(P_2)} = \{\pi_{P_2}(f)/\pi_{P_2}(g) : f, g \in C(\Omega) \setminus P_2,\ \ \pi_{P_1}(f)/\pi_{P_1}(g) \in \mathbb{M}_{(P_1)}\}.$$

Proof This is now immediate. □

The above theorem shows that, for a non-maximal, prime ideal P and a z-ideal Q which is maximal in the family of z-ideals contained in P, we have

$$\mathbb{M}_{(P)} = \{\pi_P(f)/\pi_P(g) : f, g \in C(\Omega)^+ \setminus P,\ \ \pi_Q(f)/\pi_Q(g) \in \mathbb{N}_{(Q)}\}. \tag{8.23}$$

It may be thought that this is a more natural *definition* of $\mathbb{M}_{(P)}$ than the one we gave in Definition 8.26; however, the definition we chose facilitates some computations. One might be tempted to think that, by analogy with (8.23), the set $\mathbb{Q}_{(P)}$ is equal to

$$\{\pi_P(f)/\pi_P(g) : f, g \in C(\Omega)^+ \setminus P,\ \ \pi_Q(f)/\pi_Q(g) \in \mathbb{Q}_{(Q)}\}.$$

However, if $Q \subsetneq P$, this *fails* to give the subfield of K_P generated by $\mathbb{M}_{(P)}$, which was our definition of $\mathbb{Q}_{(P)}$.

More generally, elements of all the various structures described in the present chapter that are associated with a prime ideal P_2 can be identified with appropriate elements in the corresponding structures associated with P_1, and we shall do this henceforth. We shall use the following result in this area in Chapter 9.

Proposition 8.68 *Let P_1 and P_2 be non-maximal, prime ideals in $C(\Omega)$ with P_1 below P_2, and let $F \in C(\Omega) \setminus P_2$ with $0 \le F \le 1$. Set $r_1 = \pi_{P_1}(F)$ and $r_2 = \pi_{P_2}(F)$.*

(i) *Let $b \in \mathbb{N}_{(P_2)} \subset \mathbb{N}_{(P_1)}$, and suppose that $r_1 \in \mathbb{D}_b^{(P_1)}$. Then $r_2 \in \mathbb{D}_b^{(P_2)}$.*

(ii) *Suppose that $r_2 \in \mathbb{I}_{(P_2)} \setminus \mathbb{D}_{(P_2)}$. Then $\varepsilon_{r_2}^{(P_2)} = \varepsilon_{r_1}^{(P_1)} \mid \mathbb{N}_{(P_2)}$.*

Proof (i) To obtain a contradiction, assume that $r_2 \notin \mathbb{D}_b^{(P_2)}$, and set

$$x = \max(\mathbb{D}_b^{(P_2)} \cap [0, r_2]_{(P_2)}), \quad y = \min(\mathbb{D}_b^{(P_2)} \cap [r_2, 1]_{(P_2)}),$$

so that $x < r_2 < y$, $y - x = 2^{-b}$, and $\mathbb{D}_b^{(P_2)} \cap (x, y)_{(P_2)}) = \emptyset$. Then

$$x, y \in \mathbb{D}_b^{(P_2)} \subset \mathbb{D}_b^{(P_1)}$$

and $y - x = 2^{-b}$ in K_{P_1}, and so, by Lemma 8.44(iii), $\mathbb{D}_b^{(P_1)} \cap (x, y)_{(P_1)} = \emptyset$, a contradiction of the fact that $r_1 \in \mathbb{D}_b^{(P_1)} \cap (x, y)_{(P_1)}$. Thus $r_2 \in \mathbb{D}_b^{(P_2)}$.

(ii) By (i), $r_1 \in \mathbb{I}_{(P_1)} \setminus \mathbb{D}_{(P_1)}$. It suffices to show that $\varepsilon_{r_2}^{(P_2)}(a+1) = \varepsilon_{r_1}^{(P_2)}(a+1)$ for each $a \in \mathbb{N}_{(P_2)}$. Set

$$s_1 = \max(\mathbb{D}_a^{(P_1)} \cap [0, r_1]_{(P_1)}), \quad t_1 = \min(\mathbb{D}_a^{(P_1)} \cap [r_1, 1]_{(P_1)}),$$
$$s_2 = \max(\mathbb{D}_{a+1}^{(P_1)} \cap [0, r_1]_{(P_1)}), \quad t_2 = \min(\mathbb{D}_{a+1}^{(P_1)} \cap [r_1, 1]_{(P_1)}),$$

so that $s_1 \leq s_2 < r_1 < t_2 \leq t_1$ in K_{P_1}. For each $d \in \mathbb{N}_{(P_2)}$, we have

$$\mathbb{D}_d^{(P_2)} = \mathbb{D}_d^{(P_1)} \cap \mathbb{Q}_{(P_2)},$$

and so, by Lemma 8.47(ii), it is sufficient to show that $s_1, s_2, t_1, t_2 \in \mathbb{D}_{(P_2)}$.

We prove that $s_1 \in \mathbb{D}_a^{(P_2)}$. Indeed, set

$$s_1' = \max(\mathbb{D}_a^{(P_2)} \cap [0, r_2]_{(P_2)}), \quad t_1' = \min(\mathbb{D}_a^{(P_2)} \cap [r_2, 1]_{(P_2)}).$$

Then $s_1' \leq s_1 < r_1 < t_1 \leq t_1'$. However, we have $t_1' - s_1' = 2^{-a} = t_1 - s_1$, and so $s_1' = s_1$ and $t_1' = t_1$. In particular $s_1 \in \mathbb{D}_a^{(P_2)}$.

The result follows. □

Notes 8.69 The subject of 'non-standard analysis' has its roots in Leibniz's approach to the foundations of calculus through 'infinitesimals'. This approach was finally given a rigorous foundation by A. Robinson; see Robinson (1966). There are now several elementary texts on non-standard analysis and its applications; see, for example, Stroyan and Luxemburg (1976), Hurd and Loeb (1985), and Cutland (1988).

Our definition of non-standard structures in super-real fields K_Q, where Q is a prime z-ideal, through Q-partitions seems to be new, although it is a rather obvious generalization of the theory for ultrapowers. The theory for general super-real fields K_P, in which our results are proved by using the transfer principle, Corollary 8.30, involves a number of new subtleties that do not arise in the case of ultrapowers.

9

$\widetilde{\mathbf{R}}$ as a hyper-real field

At the end of Chapter 6, we left open the question whether or not $\widetilde{\mathbf{R}}$, the unique real-closed, Cauchy complete η_1-field of weight $\aleph_1$, is a hyper-real field, or even a super-real field. Our main result of the present chapter is that $\widetilde{\mathbf{R}}$ is indeed a hyper-real field (in the theory ZFC + GCH). On the other hand, we shall prove in Theorem 9.6 that $\widetilde{\mathbf{R}}$ is not an ultrapower, and so $\widetilde{\mathbf{R}}$ belongs to the class **hyper-real** of our classification scheme, but not to the class **ultrapower**. In fact, we shall prove in Theorem 9.9 that $\widetilde{\mathbf{R}}$ does not belong to the class **cut prime η_1**.

In Theorem 6.29, we proved that, for a non-maximal, exponential prime ideal P in an algebra $C(\Omega)$, necessarily $w(K_P) = |\Gamma_P|$. In this chapter, we are also seeking an example of a non-maximal, prime ideal P such that K_P is an η_1-field for which $w(K_P) > |\Gamma_P|$. We recall from Theorem 8.34 that $|\mathbb{M}_{(P)}| = w(K_P)$ and $|\mathbb{N}_{(P)}| = |\Gamma_P|$, and so the example we are seeking would show that the case where $|\mathbb{M}_{(P)}| > |\mathbb{N}_{(P)}|$ may arise. Such an example would separate the class **super-real η_1** from the class **exponential η_1**, and hence from the class **z-ideal η_1**, in the scheme in Chapter 4. We do not succeed in constructing an example with $w(K_P) > |\Gamma_P|$ in the theory ZFC + CH, but we do prove that it is relatively consistent with ZFC + GCH that such an example exists. Thus it cannot be proved in the theory ZFC + GCH that the classes **super-real η_1** and **exponential η_1** are equal.

Our method for proving that $\widetilde{\mathbf{R}}$ is a hyper-real field is the obvious one: we shall exhibit a specific completely regular space X (of the form $\mathbb{N} \times Y$ for a certain compact space Y) and a z-ultrafilter $\mathcal{F}_H$ chosen so that the corresponding hyper-real field K has weight $\aleph_1$ and is Cauchy complete, and so is isomorphic to $\widetilde{\mathbf{R}}$. The proof that K is Cauchy complete for this specific ultrafilter uses a characterization of Cauchy completeness which will be given in Theorem 9.3.

The machinery developed for the construction of the above field K is the basis on which all the remaining examples in the final three chapters of this book will be built; the example of a prime ideal P for which $w(K_P) > |\Gamma_P|$, given in the second half of the present chapter, arises from a modification of this machinery, and the examples to be given in Chapters 10 and 11 will be produced by introducing further modifications.

Let P be a non-maximal, prime ideal in an algebra $C(\Omega)$. In Chapter 8, we defined various 'finite' or 'arithmetic' structures associated with P, and in particular we defined $\{0,1\}^{<\mathbb{N}}_{(P)}$, the family of binary sequences 'of length a for some a in $\mathbb{N}_{(P)}$'. We recall that the definitions were transparent for prime z-ideals, but that they were more subtle for arbitrary prime ideals. We shall now define an 'infinite' structure associated with P: this is the set $\{0,1\}^{\mathbb{N}}_{(P)}$. Our characterization of the Cauchy completeness of K_P is in terms of this set. The additional structures that are built in the present chapter will also be of fundamental importance in the final two chapters.

As in the previous chapter, we take Ω to be a compact space and P to be a non-maximal, prime ideal in $C(\Omega)$. We maintain the notation A_P, K_P, π_P, etc.; again, Q is the maximum z-ideal in $C(\Omega)$ below P.

Recall from equation (8.21) that

$$\{0,1\}^{<\mathbb{N}}_{(P)} = \bigcup \left\{ \{\varepsilon_r^{(P)} \restriction a : r \in \mathbb{I}_{(P)} \setminus \mathbb{D}_{(P)}\} : a \in \mathbb{N}_{(P)} \right\}.$$

Recall also from pages 168 and 169 that each $r \in \mathbb{I} \setminus \mathbb{D}$ has a unique binary expansion ε_r, whereas elements $r \in \mathbb{D} \setminus \{0,1\}$ have two binary expansions, ε_r^+ and ε_r^-, each eventually constant.

Definition 9.1 *Let P be a non-maximal, prime ideal in an algebra $C(\Omega)$. The set*

$$\{0,1\}^{\mathbb{N}}_{(P)}$$

consists of the functions $\varepsilon_r^{(P)} : \mathbb{N}_{(P)} \to \{0,1\}$ for some $r \in \mathbb{I}_{(P)} \setminus \mathbb{D}_{(P)}$, together with the functions $f : \mathbb{N}_{(P)} \to \{0,1\}$ such that $f \restriction a = \varepsilon_r^{(P)} \restriction a$ for some $r \in \mathbb{I}_{(P)} \setminus \mathbb{D}_{(P)}$ and $a \in \mathbb{N}_P$, and with $f(b)$ constantly equal to 0 or 1 for all $b \in \mathbb{N}_{(P)}$ with $b > a$.

Clearly, for each $f \in \{0,1\}^{\mathbb{N}}_{(P)}$ and each $a \in \mathbb{N}_{(P)}$, we have

$$f \restriction a \in \{0,1\}^a_{(P)}.$$

The set $\{0,1\}^{\mathbb{N}}_{(P)}$ is the analogue of the set of all binary sequences on $\mathbb{N}$, but it is certainly not equal to the set of all binary sequences on $\mathbb{N}_{(P)}$. For example, the sequence f given by

$$f(a) = 0 \quad (a \in \mathbb{N}), \qquad f(a) = 1 \quad (a \in \mathbb{N}_{(P)} \setminus \mathbb{N})$$

does not belong to $\{0,1\}^{\mathbb{N}}_{(P)}$.

Let $f, g \in \{0,1\}^{\mathbb{N}}_{(P)}$ with $f \neq g$. Then there is a minimum element $a \in \mathbb{N}_{(P)}$ with $f(a) \neq g(a)$; as usual, we set $f \prec g$ if $f(a) = 0$ and $g(a) = 1$. The map

$$r \mapsto \varepsilon_r^{(P)}, \quad \mathbb{I}_{(P)} \setminus \mathbb{D}_{(P)} \to \{\{0,1\}^{\mathbb{N}}, \preceq\},$$

is isotonic.

We noted earlier that the set $\{0,1\}^{<\mathbb{N}}_{(P)}$ is closed under several obvious operations, such as forming $f_1 \wedge f_2$, $f_1 \vee f_2$, and $1 - f_1$ from f_1 and f_2, and one might expect $\{0,1\}^{\mathbb{N}}_{(P)}$ to be similarly closed. In fact, this is not the case; we shall discuss this point further in Chapter 10. Indeed, it is this failure that will drive our definition of 'weak Cauchy completeness' in Chapter 10. We can prove, however, that an 'initial part' of a function in $\{0,1\}^{\mathbb{N}}_{(P)}$ can be changed to be any function from $\{0,1\}^{<\mathbb{N}}_{(P)}$ of the appropriate length.

Lemma 9.2 *Let* $f \in \{0,1\}^{\mathbb{N}}_{(P)}$, *and let* $g \in \{0,1\}^{a}_{(P)}$ *for some* $a \in \mathbb{N}_{(P)}$. *Define* $h : \mathbb{N}_{(P)} \to \{0,1\}$ *by*

$$h(b) = \begin{cases} g(b) & (b \le a)\,, \\ f(b) & (b > a)\,. \end{cases}$$

Then $h \in \{0,1\}^{\mathbb{N}}_{(P)}$.

Proof Choose $s \in \mathbb{I}_{(P)} \setminus \mathbb{D}_{(P)}$ with $g = \varepsilon^{(P)}_s \restriction a$, and set

$$y = \max(\mathbb{D}^{(P)}_a \cap [0,s])\,;$$

the element y exists by Lemma 8.44(vi).

First, suppose that $f = \varepsilon^{(P)}_r$ for some $r \in \mathbb{I}_{(P)} \setminus \mathbb{D}_{(P)}$, and set

$$x = \max(\mathbb{D}^{(P)}_a \cap [0,r])\,.$$

Consider $t = r - x + y$, so that $t \in \mathbb{I}_{(P)} \setminus \mathbb{D}_{(P)}$ and $y < t < s + 2^{-a}$. We suppose that, in fact, $t < s$; the argument in the case where $t \ge s$ is similar. Thus $\mathbb{D}^{(P)}_a \cap (t,s) = \emptyset$, and so $\varepsilon^{(P)}_t \restriction a = \varepsilon^{(P)}_s \restriction a$ by Lemma 8.46(ii). Hence $h \restriction a = \varepsilon^{(P)}_t \restriction a$. For each $b > a$, set

$$x_b = \max(\mathbb{D}^{(P)}_b \cap [0,r])\,, \quad y_b = \min(\mathbb{D}^{(P)}_b \cap [r,1])\,,$$
$$x'_b = \max(\mathbb{D}^{(P)}_b \cap [0,t])\,, \quad y'_b = \min(\mathbb{D}^{(P)}_b \cap [t,1])\,.$$

Since $y - x \in \mathbb{D}^{(P)}_a \subset \mathbb{D}^{(P)}_b$, we have

$$x'_b = x_b - x + y \quad \text{and} \quad y'_b = y_b - x + y\,.$$

So, for example, $x_{b+1} = x_b$ if and only if $x'_{b+1} = x'_b$. It follows from Lemma 8.47 that $\varepsilon^{(P)}_t(b) = \varepsilon^{(P)}_r(b)$. Hence $h(b) = \varepsilon^{(P)}_t(b) \quad (b > a)$. We have shown that $h = \varepsilon^{(P)}_t \in \{0,1\}^{\mathbb{N}}_{(P)}$.

Second, suppose that f is eventually constant. Then h is eventually constant, and so it suffices to show that $h \restriction c \in \{0,1\}^{c}_{(P)}$ for each $c \in \mathbb{N}_{(P)}$. Take $c \in \mathbb{N}_{(P)}$ with $c > a$, and choose $r \in \mathbb{I}_{(P)} \setminus \mathbb{D}_{(P)}$ such that $\varepsilon^{(P)}_r \restriction c = f \restriction c$. Set $\overline{f} = \varepsilon^{(P)}_r$, and define $\overline{h} : \mathbb{N}_{(P)} \to \{0,1\}$ by

$$\overline{h}(b) = \begin{cases} g(b) & (b \le a)\,, \\ \overline{f}(b) & (b > a)\,. \end{cases}$$

Then $\overline{h} \in \{0,1\}^{\mathbb{N}}_{(P)}$ by the first case given above. But clearly $\overline{h} \restriction c = h \restriction c$, and so $h \restriction c \in \{0,1\}^{c}_{(P)}$. □

We discussed the notion of a Cauchy complete field in Chapter 3. We can now prove a key theorem which characterizes when a super-real field K_P is Cauchy complete in terms of the additional structure which we have just defined;

specifically, completeness is characterized by a condition on $\{0,1\}^{\mathbb{N}}_{(P)}$ as a subset of the family of all functions from $\mathbb{N}_{(P)}$ into $\{0,1\}$.

Take $f \in \{0,1\}^{a}_{(P)}$, where $a \in \mathbb{N}_{(P)}$. Then it follows from Lemma 8.46(ii) that

$$\{r \in \mathbb{I}_{(P)} \setminus \mathbb{D}_{(P)} : \varepsilon_r^{(P)} \restriction a = f\} = J_f \cap (\mathbb{I}_{(P)} \setminus \mathbb{D}_{(P)})$$

for some open interval J_f of length 2^{-a} in $\mathbb{I}_{(P)}$. Now start with r_0 in $\mathbb{I}_{(P)} \setminus \mathbb{D}_{(P)}$, and, for $a \in \mathbb{N}_{(P)}$, set $f_a = \varepsilon_{r_0}^{(P)} \restriction a$. Then $\{J_{f_a} : a \in \mathbb{N}_{(P)}\}$ is a nested family of open intervals in $\mathbb{I}_{(P)}$, and the family converges to r_0, in the sense that, for each open interval J in $\mathbb{I}_{(P)}$ with $r_0 \in J$, it is the case that $J_{f_a} \subset J$ for all sufficiently large a. On the other hand, suppose that we start with $r_0 \in \mathbb{D}_{(P)}$. By Lemma 8.48, there is a minimum element $a \in \mathbb{Z}^{+}_{(P)}$ such that $r_0 \in \mathbb{D}_a^{(P)}$, say this minimum is a_0. For each $a \in \mathbb{Z}^{+}_{(P)}$ with $a > a_0$, there is an open interval K_a in $\mathbb{I}_{(P)}$ with $r_0 \in K_a$ such that, for each $r \in K_a$, $\varepsilon_r^{(P)}$ is constant (taking the value 0 or 1) on the interval $[a_0 + 1, a] \cap \mathbb{N}_{(P)}$.

We now come to our characterization of Cauchy completeness of the super-real fields K_P.

Theorem 9.3 *Let P be a non-maximal, prime ideal in $C(\Omega)$ for a compact space Ω. Then K_P is Cauchy complete if and only if*

$$\{0,1\}^{\mathbb{N}}_{(P)} = \{f : \mathbb{N}_{(P)} \to \{0,1\} : \ f \restriction a \in \{0,1\}^{a}_{(P)} \quad (a \in \mathbb{N}_{(P)})\}. \tag{9.1}$$

Proof The set on the right-hand side of (9.1) is denoted by S. Certainly $\{0,1\}^{\mathbb{N}}_{(P)}$ is contained in S. We shall prove that K_P is Cauchy complete if and only if S is contained in $\{0,1\}^{\mathbb{N}}_{(P)}$.

First suppose that K_P is Cauchy complete, and take $f \in S$. For $a \in \mathbb{N}_{(P)}$, let J_a be the open interval of length 2^{-a} in $\mathbb{I}_{(P)}$ such that

$$\{r \in \mathbb{I}_{(P)} \setminus \mathbb{D}_{(P)} : \varepsilon_r^{(P)} \restriction a = f \restriction a\} = J_a \cap (\mathbb{I}_{(P)} \setminus \mathbb{D}_{(P)}),$$

so that $\{J_a : a \in \mathbb{N}_{(P)}\}$ is a nested family of open intervals in $\mathbb{I}_{(P)}$. The sequence $\{2^{-a} : a \in \mathbb{N}_{(P)}\}$ is coinitial in $K_P^{+} \setminus \{0\}$, and so, since K_P is Cauchy complete, there exists $r_0 \in \mathbb{I}_{(P)}$ such that $r_0 \in \overline{J}_a$ for each $a \in \mathbb{N}_{(P)}$; further, the family $\{J_a : a \in \mathbb{N}_{(P)}\}$ converges to r_0.

There are two cases to consider, depending on whether or not $r_0 \in \mathbb{D}_{(P)}$. First, suppose that $r_0 \notin \mathbb{D}_{(P)}$, and fix $a_0 \in \mathbb{N}_{(P)}$. Then there is an open interval J containing r_0 such that

$$\varepsilon_r^{(P)} \restriction a_0 = \varepsilon_{r_0}^{(P)} \restriction a_0 \quad (r \in J \setminus \mathbb{D}_{(P)}).$$

However, $J_a \subset J$ for all sufficiently large a, and so $\varepsilon_{r_0}^{(P)} \restriction a_0 = f \restriction a_0$. Thus $f = \varepsilon_{r_0}^{(P)} \in \{0,1\}^{\mathbb{N}}_{(P)}$. Second, suppose that $r_0 \in \mathbb{D}_{(P)}$, and take a_0 to be the minimum $a \in \mathbb{N}_{(P)}$ with $r_0 \in \mathbb{D}_a^{(P)}$. As above, take open intervals K_a in $\mathbb{I}_{(P)}$ for each $a > a_0$ such that $\varepsilon_r^{(P)}$ is constant on the interval $[a_0 + 1, a] \cap \mathbb{N}_{(P)}$ for each $r \in K_a$. Each interval K_a contains the interval J_b for all sufficiently large

$b \in \mathbb{N}_{(P)}$, and so $f \restriction a = \varepsilon_r^{(P)} \restriction a$ for some $r \in J_b \cap (\mathbb{I}_{(P)} \setminus \mathbb{D}_{(P)}) \subset K_a$. Hence f is constant on $[a_0 + 1, a] \cap \mathbb{N}_{(P)}$. This implies that f is constant on the set

$$\{a \in \mathbb{N}_{(P)} : a \geq a_0 + 1\}.$$

Since $f \in S$, we have $f \restriction a_0 \in \{0,1\}_{(P)}^{a_0}$, and so $f \in \{0,1\}_{(P)}^{\mathbb{N}}$.

For the converse, suppose that $\{0,1\}_{(P)}^{\mathbb{N}} = S$. We take a Cauchy pregap $\langle A, B\rangle$ in K_P; without loss of generality, we may suppose that

$$A \cup B \subset (\mathbb{Q}_{(P)} \cap \mathbb{I}_{(P)}) \setminus \mathbb{D}_{(P)}.$$

Consider $a \in \mathbb{N}_{(P)}$. If there exists $r \in \mathbb{D}_a^{(P)}$ with $A << r << B$, then r interpolates $\langle A, B\rangle$, and so we may suppose that this is not the case. Thus there exist $r_a \in \mathbb{D}_a^{(P)}$, $x_a \in A$, and $y_a \in B$ with

$$r_a < x_a < y_a < r_a + 2^{-a},$$

and r_a is uniquely specified by this condition. Clearly, if $b > a$, then

$$r_b \in [r_a, r_a + 2^{-a}].$$

Further, $\varepsilon_r^{(P)} \restriction a = \varepsilon_s^{(P)} \restriction a$ whenever $r, s \in [r_a, r_a + 2^{-a}] \cap (\mathbb{I}_{(P)} \setminus \mathbb{D}_{(P)})$. We now define $f : \mathbb{N}_{(P)} \to \{0,1\}$ by the requirement that

$$f \restriction a = \varepsilon_{x_a}^{(P)} \restriction a \quad (a \in \mathbb{N}_{(P)}).$$

Then f is well defined, and $f \in S$. By hypothesis, $f \in \{0,1\}_{(P)}^{\mathbb{N}}$. The function f will give an element of K_P which interpolates the pregap $\langle A, B\rangle$.

Again, there are apparently two cases. First, suppose that $f^{-1}(0)$ and $f^{-1}(1)$ are both cofinal in $\mathbb{N}_{(P)}$. Then $f = \varepsilon_r^{(P)}$ for some $r \in \mathbb{I}_{(P)} \setminus \mathbb{D}_{(P)}$. By Lemma 8.46(ii), $r \in (r_a, r_a + 2^{-a})$ for each $a \in \mathbb{N}_{(P)}$, whence $A << r$. Similarly $r << B$, and so r interpolates $\langle A, B\rangle$. Second, suppose that one of $f^{-1}(0)$ and $f^{-1}(1)$, say $f^{-1}(0)$, is bounded in $\mathbb{N}_{(P)}$. Take $a \in \mathbb{N}_{(P)}$ such that $f(b) = 1$ for each $b \in \mathbb{N}_{(P)}$ with $b \geq a$. Then

$$\varepsilon_{x_b}^{(P)} \mid ([a+1, b] \cap \mathbb{N}_{(P)}) = 1 \qquad (b > a),$$

and so $r_b = r_a$ $(b > a)$. But $y_b < r_b + 2^{-b} < x_a$ for b sufficiently large, a contradiction. So in fact this case does not arise.

We have found an interpolating element for $\langle A, B\rangle$ in K_P, and so K_P is Cauchy complete. □

We pause at this stage to examine the additional structures that we have defined in the case of a particular example. In this example, the additional structures are much more transparent than in the general case. Our analysis of the example will allow us to obtain a partial result about $\widetilde{\mathbf{R}}$, namely that $\widetilde{\mathbf{R}}$ is *not* an ultrapower. The example that we considered in Chapter 8 was the prime z-ideal $J_{\mathfrak{p}}$ for a point $\mathfrak{p} \in \beta\mathbb{N} \setminus \mathbb{N}$. The present example is a slight generalization of the previous one.

Thus, let κ be an infinite cardinal, and let $\mathcal{U}$ be a free ultrafilter on κ. In Chapter 4, we described the ultrapower $\mathbb{R}^\kappa/\mathcal{U}$; $\mathbb{R}^\kappa/\mathcal{U}$ has the form K_Q for a certain valuation prime z-ideal Q in $C(\beta\kappa)$, and so

$$\mathbb{R}^\kappa/\mathcal{U} = \mathbb{R}_{(Q)} = \mathbb{R}_Q$$

in the notation of Chapter 8. For each $f \in \mathbb{R}^\kappa$, $[f]_\mathcal{U}$ is the equivalence class containing f.

It is convenient to change our notation somewhat in the case of the above special example. Thus, for a free ultrafilter $\mathcal{U}$ on a cardinal κ, we write

$$K_\mathcal{U},\ \mathbb{N}_{(\mathcal{U})},\ \mathbb{Q}_{(\mathcal{U})},\ \mathbb{D}_{(\mathcal{U})},\ \varepsilon_r^{(\mathcal{U})},\ \{0,1\}^{<\mathbb{N}}_{(\mathcal{U})},\quad \text{and}\quad \{0,1\}^{\mathbb{N}}_{(\mathcal{U})}$$

for the appropriate structures.

We have

$$\begin{aligned}\mathbb{N}_{(\mathcal{U})} &= \{[f]_\mathcal{U} \ :\ f \in \mathbb{N}^\kappa\} = \mathbb{N}^\kappa/\mathcal{U}\,,\\ \mathbb{Q}_{(\mathcal{U})} &= \{[f]_\mathcal{U} \ :\ f \in \mathbb{Q}^\kappa\} = \mathbb{Q}^\kappa/\mathcal{U}\,,\\ \mathbb{D}_{(\mathcal{U})} &= \{[f]_\mathcal{U} \ :\ f \in \mathbb{D}^\kappa\} = \mathbb{D}^\kappa/\mathcal{U}\,.\end{aligned}$$

By (8.5), we have

$$\left|\mathbb{N}_{(\mathcal{U})}\right| = \left|\mathbb{Q}_{(\mathcal{U})}\right| = w(K_\mathcal{U}) = |\Gamma_\mathcal{U}|\,,$$

and this gives a second proof of Theorem 7.15.

Now suppose that F is a function from κ to $\{0,1\}^\mathbb{N}$. For each element $[f]_\mathcal{U} \in \mathbb{N}_{(\mathcal{U})}$, define

$$h : \sigma \mapsto F(\sigma)(f(\sigma)), \qquad \kappa \to \{0,1\}\,,$$

and set

$$F_\mathcal{U}([f]_\mathcal{U}) = [h]_\mathcal{U}\,.$$

Then $F_\mathcal{U}([f]_\mathcal{U})$ is well defined in $\{0,1\}^\kappa/\mathcal{U} = \{0,1\}$, and so from F we have defined a new function $F_\mathcal{U} : \mathbb{N}_{(\mathcal{U})} \to \{0,1\}$. We see that, in this case,

$$\{0,1\}^{\mathbb{N}}_{(\mathcal{U})} = \{F_\mathcal{U} \ :\ F : \kappa \to \{0,1\}^\mathbb{N}\}\,;$$

this is clearly far less complicated than the general definition given in Definition 9.1.

Again, suppose that F is a function from κ to $\{0,1\}^{<\mathbb{N}}$. For each $\sigma \in \kappa$, choose $n_\sigma \in \mathbb{N}$ such that $\operatorname{dom} F(\sigma) = [1, n_\sigma] \cap \mathbb{N}$, and define

$$f : \sigma \mapsto n_\sigma, \quad \kappa \to \mathbb{N}\,.$$

Then $[f]_\mathcal{U} \in \mathbb{N}_{(\mathcal{U})}$, say $a = [f]_\mathcal{U}$. Take $b \in [1,a] \cap \mathbb{N}_{(\mathcal{U})}$, say $b = [g]_\mathcal{U}$. Since $b \le a$ in $\mathbb{N}_{(\mathcal{U})}$, we may suppose that $g(\sigma) \le f(\sigma)$ $(\sigma \in \kappa)$. Define

$$h : \sigma \mapsto F(\sigma)(g(\sigma)), \quad \kappa \to \{0,1\}\,,$$

and set $F_\mathcal{U}(b) = [h]_\mathcal{U}$. Then $F_\mathcal{U}(b)$ is well defined in $\{0,1\}$, and

$$F_\mathcal{U} : [1,a] \cap \mathbb{N}_{(\mathcal{U})} \to \{0,1\}\,.$$

We see that

$$\{0,1\}^{<\mathbb{N}}_{(\mathcal{U})} = \{F_\mathcal{U} \ :\ F : \kappa \to \{0,1\}^{<\mathbb{N}}\}\,.$$

We shall now establish a result that applies in the special example that we are presently considering, but, as we shall see later in this chapter, may fail for a general prime z-ideal Q. Recall from (8.12) that

$$R_g = \{\rho(n) : n \in \mathbb{N},\ \ g(n) = 1\}$$

for $g \in \{0,1\}^{\mathbb{N}}$, so that $R_g \subset \mathbb{N} \times \mathbb{N}$, and, slightly modifying (8.13), that

$$R_g^{(\mathcal{U})} = \{\rho_{(\mathcal{U})}(a) : a \in \mathbb{N}_{(\mathcal{U})},\ \ g(a) = 1\}$$

for $g : \mathbb{N}_{(\mathcal{U})} \to \{0,1\}$, so that $R_g^{(\mathcal{U})} \subset \mathbb{N}_{(\mathcal{U})} \times \mathbb{N}_{(\mathcal{U})}$.

Lemma 9.4 *Let κ be an infinite cardinal, and let $\mathcal{U}$ be a free ultrafilter on κ. Suppose that $h \in \{0,1\}^{\mathbb{N}}_{(\mathcal{U})}$ and that $R_h^{(\mathcal{U})}$ is the graph of a partial function $F : \mathbb{N}_{(\mathcal{U})} \to \mathbb{N}_{(\mathcal{U})}$. Then, for each $a \in \mathbb{N}_{(\mathcal{U})}$, $(\operatorname{dom} F) \cap [1,a]$ is either empty or has a maximum element.*

Proof Choose $H : \kappa \to \{0,1\}^{\mathbb{N}}$ such that $H_{\mathcal{U}} = h$, and define

$$G : \sigma \mapsto R_{H(\sigma)}\,, \quad \kappa \to \mathcal{P}(\mathbb{N} \times \mathbb{N})\,.$$

Then G defines in a natural way a subset $G_{\mathcal{U}}$ of $\mathbb{N}_{(Q)} \times \mathbb{N}_{(Q)}$: indeed, for each $(a,b) \in \mathbb{N}_{(Q)} \times \mathbb{N}_{(Q)}$, we say that $(a,b) \in G_{\mathcal{U}}$ if

$$\{\sigma \in \kappa : (g_a(\sigma), g_b(\sigma)) \in G(\sigma)\} \in \mathcal{U}\,,$$

where $g_a, g_b : \kappa \to \mathbb{N}$ are such that $a = [g_a]_{\mathcal{U}}$ and $b = [g_b]_{\mathcal{U}}$. A moment's chasing round the definitions shows that

$$R_h^{(\mathcal{U})} = G_{\mathcal{U}}\,. \tag{9.2}$$

For example, for $\sigma \in \kappa$, $(g_a(\sigma), g_b(\sigma)) \in R_{H(\sigma)}$ if and only if

$$(g_a(\sigma), g_b(\sigma)) = \rho(n_\sigma)$$

for some $n_\sigma \in \mathbb{N}$ with $H(\sigma)(n_\sigma) = 1$. Set $f : \sigma \mapsto n_\sigma,\ \kappa \to \mathbb{N}$. If $(a,b) \in G_{\mathcal{U}}$, then

$$\{\sigma < \kappa : (g_a(\sigma), g_b(\sigma)) = \rho(f(\sigma))\} \in \mathcal{U}\,.$$

But $[f]_{\mathcal{U}} \in \operatorname{dom} H_{\mathcal{U}}$ with $H_{\mathcal{U}}([f]_{\mathcal{U}}) = 1$, and $\rho_{(Q)}([f]_{\mathcal{U}}) = (a,b)$, showing that $(a,b) \in R_h^{(\mathcal{U})}$. Thus $G_{\mathcal{U}} \subset R_h^{(\mathcal{U})}$. Similarly, $R_h^{(\mathcal{U})} \subset G_{\mathcal{U}}$.

It follows from (9.2) that

$$\{\sigma < \kappa : G(\sigma) \ \text{ is the graph of a partial function from } \mathbb{N} \text{ to } \mathbb{N}\}$$

belongs to $\mathcal{U}$. We may, in fact, suppose that, for each $\sigma < \kappa$, $G(\sigma)$ is the graph of a partial function, also denoted by $G(\sigma) : \mathbb{N} \to \mathbb{N}$.

Fix $a \in \mathbb{N}_{(\mathcal{U})}$, and choose $g : \kappa \to \mathbb{N}$ such that $a = [g]_{\mathcal{U}}$. Define

$$\bar{g} : \sigma \mapsto \max\{(\operatorname{dom} G(\sigma)) \cap [1, g(\sigma)]\}, \quad \kappa \to \mathbb{N}\,,$$

taking $\bar{g}(\sigma) = 0$ in the case where $(\operatorname{dom} G(\sigma)) \cap [1, g(\sigma)] = \emptyset$, and set $b = [\bar{g}]_{\mathcal{U}}$. It is easily checked that $(\operatorname{dom} F) \cap [1,a] = \emptyset$ if and only if $b = 0$, and that, in the case where $(\operatorname{dom} F) \cap [1,a] \neq \emptyset$, the element b is the required maximum element of this set. □

The following proposition will be required only for our special example, but it applies to the general situation.

Proposition 9.5 *Let Q be a prime z-ideal in $C(\Omega)$ with $\delta(K_Q) = \aleph_1$, and let I be an initial segment of $\mathbb{N}_{(Q)}$ with $\operatorname{cof} I = \aleph_1$. Then there exist $F : I \to \mathbb{N}_{(Q)}$ and $h : \mathbb{N}_{(Q)} \to \{0,1\}$ such that*

$$h \restriction a \in \{0,1\}^a_{(Q)} \quad (a \in \mathbb{N}_{(Q)}),$$

$\operatorname{Gr} F = R_h^{(Q)}$, *and $F(I)$ is cofinal in $\mathbb{N}_{(Q)}$.*

Proof By Theorem 4.28(ii), K_Q is an η_1-field, and so Lemma 8.25 applies to K_Q.

Take $(a_\alpha : \alpha < \omega_1)$ to be a strictly increasing, cofinal sequence in I and $(d_\alpha : \alpha < \omega_1)$ to be a cofinal sequence in $\mathbb{N}_{(Q)}$. (We have $\operatorname{cof} \mathbb{N}_{(Q)} = \aleph_1$ because $\operatorname{cof} \mathbb{N}_{(Q)} = \operatorname{cof} K_Q = \aleph_1$.) We shall define by induction on α a sequence $(F_\alpha : \alpha < \omega_1)$ such that the following conditions hold for each $\alpha < \omega_1$:

$(\mathrm{i})_\alpha$ F_α is an increasing function in $\mathbb{N}^{<\mathbb{N}}_{(Q)}$;

$(\mathrm{ii})_\alpha$ $\operatorname{dom} F_\alpha = [1, a_\alpha] \cap \mathbb{N}_{(Q)}$;

$(\mathrm{iii})_\alpha$ $F_\alpha \restriction \beta = F_\beta$ for each $\beta < \alpha$;

$(\mathrm{iv})_\alpha$ $F_\alpha(a_\alpha) > d_\alpha$.

For this construction, first suppose that $\alpha < \omega_1$ and that $(F_\beta : \beta \leq \alpha)$ has been defined so that $(\mathrm{i})_\alpha$–$(\mathrm{iv})_\alpha$ hold. Set

$$F_{\alpha+1}(c) = \begin{cases} F_\alpha(c) & (c \in [1, a_\alpha] \cap \mathbb{N}_{(Q)}), \\ F_\alpha(a_\alpha) + d_{\alpha+1} & (c \in [a_\alpha + 1, a_{\alpha+1}] \cap \mathbb{N}_{(Q)}). \end{cases}$$

Then $(\mathrm{i})_{\alpha+1}$–$(\mathrm{iv})_{\alpha+1}$ hold. By Lemma 8.24(i) and (iv), $F_{\alpha+1} \in \mathbb{N}^{a_{\alpha+1}}_{(Q)}$. Second, suppose that $\alpha < \omega_1$ is a limit ordinal, that $(F_\beta : \beta < \alpha)$ has been defined, and that $(\mathrm{i})_\beta$–$(\mathrm{iv})_\beta$ hold for $\beta < \alpha$. Since $\alpha < \omega_1$, there exists $d \in \mathbb{N}_{(Q)}$ such that

$$d > \max\{a_\beta, F_\beta(a_\beta) : \beta < \alpha\}.$$

By Lemma 8.25, there exists $G \in \mathbb{N}^d_{(Q)}$ such that $G \restriction a_\beta = F_\beta$ for all $\beta < \alpha$. The set

$$\{b \in [1, d] \cap \mathbb{N}_{(Q)} : G \restriction b \text{ is increasing}\}$$

contains a_β for each $\beta < \alpha$, and also, by Lemma 8.22, it contains a maximum element. Thus there exists $b_0 \in [1, d] \cap \mathbb{N}_{(Q)}$ with $a_\beta < b_0 \quad (\beta < \alpha)$ such that $G \restriction b_0$ is increasing. We may suppose that $b_0 + 1 \leq a_\alpha$. Set

$$F_\alpha(c) = \begin{cases} G(c) & (c \in [1, b_0] \cap \mathbb{N}_{(Q)}), \\ G(b_0) + d_\alpha & (c \in [b_0 + 1, a_\alpha] \cap \mathbb{N}_{(Q)}). \end{cases}$$

By Lemma 8.24 again, $F_\alpha \in \mathbb{N}^{a_\alpha}_{(Q)}$. Also, $(\mathrm{i})_\alpha$–$(\mathrm{iv})_\alpha$ hold with this definition of F_α.

This completes the inductive definition of the sequence $(F_\alpha : \alpha < \omega_1)$.

Now define $F : I \to \mathbb{N}_{(Q)}$ by requiring that $F \restriction a_\alpha = F_\alpha \quad (\alpha < \omega_1)$. Then F is well defined, F is an increasing function, $F(I)$ is cofinal in $\mathbb{N}_{(Q)}$, and

$F \restriction a \in \mathbb{N}^a_{(Q)}$ for each $a \in I$. Certainly there is a function $h : \mathbb{N}_{(Q)} \to \{0,1\}$ such that $R^{(Q)}_h = \mathrm{Gr}\, F$.

For each $\alpha < \omega_1$, we have $F_\alpha \in \mathbb{N}^{<\mathbb{N}}_{(Q)}$, and so there exists $h_\alpha \in \{0,1\}^{<\mathbb{N}}_{(Q)}$ such that $R^{(Q)}_{h_\alpha} = \mathrm{Gr}\, F_\alpha$. Since $\rho_{(Q)}(c) \in [1,c] \times [1,c]$ for each $c \in \mathbb{N}_{(Q)}$ by (8.11) and since $(a_\alpha, F_\alpha(a_\alpha)) \in \mathrm{Gr}\, F_\alpha$ with $F_\alpha(a_\alpha) > d_\alpha$, we see that $h_\alpha \in \{0,1\}^{c_\alpha}_{(Q)}$ for some $c_\alpha > d_\alpha$.

We now come to a key point of this proof: we *claim* that

$$h \restriction d_\alpha = h_\alpha \restriction d_\alpha \quad (\alpha < \omega_1)\,.$$

For take $b \in [1, d_\alpha] \cap \mathbb{N}_{(Q)}$. Then $\rho_{(Q)}(b) \in [1, d_\alpha] \times [1, d_\alpha]$. But F is increasing, $F \restriction a_\alpha = F_\alpha$, and $F_\alpha(a_\alpha) > d_\alpha$, and so

$$([1, d_\alpha] \times [1, d_\alpha]) \cap \mathrm{Gr}\, F \subset \mathrm{Gr}\, F_\alpha\,.$$

Thus $h(b) = 1$ if and only if $h_\alpha(b) = 1$, giving the claim.

Since $h_\alpha \in \{0,1\}^{<\mathbb{N}}_{(Q)}$ $(\alpha < \omega_1)$ and $(d_\alpha : \alpha < \omega_1)$ is cofinal in $\mathbb{N}_{(Q)}$, we have $h \restriction a \in \{0,1\}^a_{(Q)}$ for each $a \in \mathbb{N}_{(Q)}$.

This completes the proof of the proposition. □

Theorem 9.6 (i) *Let* κ *be an infinite cardinal, and let* $\mathcal{U}$ *be a free ultrafilter on* κ *such that* $\mathrm{cof}\,(\mathbb{R}^\kappa/\mathcal{U}) = \aleph_1$. *Then* $\mathbb{R}^\kappa/\mathcal{U}$ *is not Cauchy complete.*

(ii) *The ordered field* $\widetilde{\mathbf{R}}$ *is not isomorphic as a totally ordered group to any ultrapower.*

Proof (i) Fix a proper initial segment I of $\mathbb{N}_{(\mathcal{U})}$ with $\mathrm{cof}\, I = \aleph_1$. By Proposition 9.5, there exist $F : I \to \mathbb{N}_{(Q)}$ and $h : \mathbb{N}_{(\mathcal{U})} \to \{0,1\}$ such that

$$h \restriction a \in \{0,1\}^a_{(\mathcal{U})} \quad (a \in \mathbb{N}_{(\mathcal{U})})\,,$$

$\mathrm{Gr}\, F = R^{(\mathcal{U})}_h$, and $F(I)$ is cofinal in $\mathbb{N}_{(\mathcal{U})}$.

To obtain a contradiction, assume that the field $\mathbb{R}^\kappa/\mathcal{U}$ is Cauchy complete. By the characterization of Cauchy completeness given in Theorem 9.3, it follows that $h \in \{0,1\}^{\mathbb{N}}_{(Q)}$. We now use Lemma 9.4 to see that $I \cap [1,a]$ has a maximum element for each $a \in \mathbb{N}_{(Q)}$. Since I is a proper initial segment, there exists $a_0 \in \mathbb{N}_{(Q)} \setminus I$. Clearly $I = I \cap [1, a_0]$ has no maximum element, and this is the required contradiction.

(ii) By Theorem 3.21, $\widetilde{\mathbf{R}}$ is an ordered field which is Cauchy complete and has cofinality $\aleph_1$. By the above theorem, no ultrapower has these two properties. However these two properties are preserved by isomorphisms of totally ordered groups. □

Our next aim is a strengthened version of Theorem 9.6. Recall from Definition 4.35 that a prime ideal P in $C(\Omega)$ is a strongly convex prime if M_P is a convex set in K_P, or, equivalently, if M_P is an ideal in $K^{\#}_P$. For example, every prime ideal in c is strongly convex. By Proposition 4.39, if P_1 and P_2 are primes in $C(\Omega)$ with P_1 below P_2 and if P_1 is strongly convex, then so is P_2. As in Definition

4.42, a prime P is a cut prime if it strictly contains a strongly convex prime. We shall be concerned next with fields in the class **cut prime** η_1; by Theorem 4.43, this class contains the class **ultrapowers**.

Lemma 9.7 *Let* P_1 *and* P_2 *be prime ideals in* $C(\Omega)$ *with* $P_1 \subsetneq P_2$ *and* P_1 *strongly convex. Then:*

(i) $\mathbb{N}_{(P_2)}$ *is an additively closed initial segment of* $\mathbb{N}_{(P_1)}$*;*

(ii) *for each* $c \in \mathbb{N}_{(P_1)} \setminus \mathbb{N}_{(P_2)}$,

$$\{0,1\}^{\mathbb{N}}_{(P_2)} = \{f \mid \mathbb{N}_{(P_2)} : f \in \{0,1\}^{c}_{(P_1)}\}\,.$$

Proof (i) That $\mathbb{N}_{(P_2)}$ is additively closed is Lemma 8.33(ii). To prove that $\mathbb{N}_{(P_2)}$ is an initial segment of $\mathbb{N}_{(P_1)}$, we follow the proof of Lemma 8.35, with P_1 and P_2 for Q and P, respectively. Define J as in that proof; that $J \subset \mathbb{M}_{(P_2)}$ is a consequence of (8.23) and the remainder of the proof is as before.

(ii) Fix $a \in \mathbb{N}_{(P_2)}$ so that the interval $(0, 2^{-a+1})$ of K_{P_j} is contained in M_{P_j} for $j = 1, 2$; there exists such an element a because P_1 and P_2 are strongly convex. By Lemma 9.2, we can throughout work with functions $g \in \{0,1\}^{\mathbb{N}}_{(P_2)}$ such that $g \restriction a = 0$. Take $c \in \mathbb{N}_{(P_1)} \setminus \mathbb{N}_{(P_2)}$ and set

$$S = \{f \mid \mathbb{N}_{(P_2)} : f \in \{0,1\}^{c}_{(P_2)}\}\,.$$

We shall first prove that $\{0,1\}^{\mathbb{N}}_{(P_2)} \subset S$. Indeed, take $g \in \{0,1\}^{\mathbb{N}}_{(P_2)}$ with $g \restriction a = 0$.

First, suppose that $g = \varepsilon^{(P_2)}_{r_2}$ for some $r_2 \in \mathbb{I}_{(P_2)} \setminus \mathbb{D}_{(P_2)}$. Since $g \restriction a = 0$, we have $\mathbb{D}^{(P_2)}_a \cap (0, r_2) = \emptyset$ by Lemma 8.46(ii), and so $r_2 < 2^{-a}$. It follows that $r_2 \in M^+_{P_2} \setminus \{0\}$, say $r_2 = \pi_{P_2}(H)$, where $H \in C(\Omega)^+ \setminus P_2$. Choose $H' \in C(\Omega) \setminus P_1$ with

$$|\pi_{P_1}(H') - \pi_{P_1}(H)| < 2^{-c} \quad \text{and} \quad \pi_{P_1}(H') \in \mathbb{I}_{(P_1)} \setminus \mathbb{D}_{(P_1)}\,;$$

this is possible by Lemma 8.44(ii) and because $(0, 2^{-a}) \subset M_{P_1}$. We see that $H' - H \in P_2$ because $c >> \mathbb{N}_{(P_2)}$, and so $r_2 = \pi_{P_2}(H')$. Set $r_1 = \pi_{P_1}(H')$. By Proposition 8.66(ii), $\varepsilon^{(P_2)}_{r_2} = \varepsilon^{(P_1)}_{r_1} \mid \mathbb{N}_{(P_2)}$. Set $f = \varepsilon^{(P_1)}_{r_1} \restriction c$. Then $f \in \{0,1\}^{c}_{(P_1)}$ and $f \mid \mathbb{N}_{(P_2)} = g$.

Second, suppose that g is eventually constant, say $g(b) = g(d)$ $(b \geq d)$, where $d \in \mathbb{N}_{(P_2)}$. Choose $f' \in \{0,1\}^{c}_{(P_1)}$ with

$$f' \mid ([1, d] \cap \mathbb{N}_{(P_1)}) = g \restriction d\,,$$

and define $f : [1, c] \cap \mathbb{N}_{(P_1)} \to \{0,1\}$ by

$$f(b) = \begin{cases} f'(b) & (b < d)\,, \\ g(d) & (b \geq d)\,. \end{cases}$$

Then $f \in \{0,1\}^{c}_{(P_1)}$ and $f \mid \mathbb{N}_{(P_2)} = g$. Thus $\{0,1\}^{\mathbb{N}}_{(P_2)} \subset S$.

We shall now prove that $S \subset \{0,1\}^{\mathbb{N}}_{(P_2)}$. For this, take $f \in \{0,1\}^{c}_{(P_1)}$ with $f \restriction a = 0$. Then there exists $r_1 \in \mathbb{I}_{(P_1)} \setminus \mathbb{D}_{(P_1)}$ with $\varepsilon^{(P_1)}_{r_1} \restriction c = f$. We have

$0 < r_1 < 2^{-a}$, and so there exists $H \in C(\Omega) \setminus P_1$ with $\pi_{P_1}(H) = r_1$. We may suppose that $f(d) = 1$ for some $d \in \mathbb{N}_{(P_1)}$, for otherwise $f \mid \mathbb{N}_{(P_2)} = 0$, and so certainly $f \mid \mathbb{N}_{(P_2)} \in \{0,1\}^{\mathbb{N}}_{(P_2)}$. By Lemma 8.46(ii), $r_1 \geq 2^{-d} > 0$, and so $H \in C(\Omega)\setminus P_2$. Set $r_2 = \pi_{P_2}(H)$. Again there are two cases to consider.

First, suppose that $r_2 \in \mathbb{I}_{(P_2)} \setminus \mathbb{D}_{(P_2)}$. Then, by Proposition 8.68(ii),

$$\varepsilon^{(P_2)}_{r_2} = \varepsilon^{(P_1)}_{r_1} \mid \mathbb{N}_{(P_2)},$$

and so $f \mid \mathbb{N}_{(P_2)} \in \{0,1\}^{\mathbb{N}}_{(P_2)}$.

Second, suppose that $r_2 \in \mathbb{D}_{(P_2)}$, and choose $b \in \mathbb{N}_{(P_2)}$ with $b \geq a$ and $r_2 \in \mathbb{D}^{(P_2)}_b$. Set

$$x = \max(\mathbb{D}^{(P_1)}_b \cap [0, r_1)), \quad y = \min(\mathbb{D}^{(P_1)}_b \cap (r_1, 1]),$$

so that $0 < x < r_1 < y$ and $y \in (0, 2^{-a+1})$. Choose $H_x, H_y \in C(\Omega)\setminus P_1$ with $\pi_{P_1}(H_x) = x$ and $\pi_{P_1}(H_y) = y$, and set $s_x = \pi_{P_2}(H_x)$ and $s_y = \pi_{P_2}(H_y)$. Then $s_x \leq r_2 \leq s_y$. Since $r_1 \notin \mathbb{D}_{(P_1)}$, we have $y - x = 2^{-b}$, and so $s_y - s_x = 2^{-b}$. By Proposition 8.68(i), we have $s_x, s_y \in \mathbb{D}^{(P_2)}_b$, and so $\mathbb{D}^{(P_2)}_b \cap (s_x, s_y) = \emptyset$, showing that $r_2 \in \{s_x, s_y\}$, say $r_2 = s_x$. Then $r_1 - x < 2^{-d}$ $(d \in \mathbb{N}_{(P_2)})$, and so

$$x = \max(\mathbb{D}^{(P_1)}_d \cap [0, r_1]) \qquad (d \in \mathbb{N}_{(P_2)},\ d \geq b).$$

By Lemma 8.47, $\varepsilon^{(P_1)}_{r_1}(d) = 0$ $(d \in \mathbb{N}_{(P_2)}, d \geq b)$, and so

$$f(d) = 0 \quad (d \in \mathbb{N}_{(P_2)},\ d \geq b),$$

whence f is eventually constant on $\mathbb{N}_{(P_2)}$. Since

$$f \mid ([1,d] \cap \mathbb{N}_{(P_2)}) \in \{0,1\}^d_{(P_2)} \quad (d \in \mathbb{N}_{(P_2)}),$$

it again follows that $f \mid \mathbb{N}_{(P_2)} \in \{0,1\}^{\mathbb{N}}_{(P_2)}$. □

We may interpret Lemma 9.7 by saying that $\{0,1\}^{\mathbb{N}}_{(P_2)}$ is obtained from $\{0,1\}^{<\mathbb{N}}_{(P_1)}$ by 'cutting off' the initial segment $\mathbb{N}_{(P_2)}$ of $\mathbb{N}_{(P_1)}$. This is a further justification for the terminology 'cut prime' of Definition 4.42.

The following lemma is a weaker analogue of Lemma 9.4.

Suppose that P is a non-maximal, prime ideal in $C(\Omega)$, and define

$$J_P = \{a \in \mathbb{N}_{(P)} : a^2 \in \mathbb{N}_{(P)}\}, \tag{9.3}$$

so that $\{a^2 : a \in J_P\}$ is cofinal in $\mathbb{N}_{(P)}$, $\operatorname{cof} J_P = \operatorname{cof} \mathbb{N}_{(P)}$, and the restriction map

$$\Phi_{(P)} : J_P \times J_P \to \mathbb{N}_{(P)}$$

is a bijection. (The map $\Phi_{(P)}$ was defined in (8.22).) Therefore $R^{(P)}_g \subset J_P \times J_P$ whenever $g \in \{0,1\}^{\mathbb{N}}_{(P)}$.

Lemma 9.8 *Let P_1 and P_2 be prime ideals in $C(\Omega)$ with $P_1 \subsetneq P_2$ and P_1 strongly convex. Suppose that $a \in J_{P_2}$ and that $g \in \{0,1\}^{\mathbb{N}}_{(P_2)}$ is such that*

$R_g^{(P_2)} = \operatorname{Gr} F$ *for some function* $F : [1, a] \cap \mathbb{N}_{(P_2)} \to J_{P_2}$. *Then* $\operatorname{ran} F$ *is a bounded set in* J_{P_2}.

Proof Fix $c \in \mathbb{N}_{(P_1)} \setminus \mathbb{N}_{(P_2)}$, and choose $f \in \{0,1\}^c_{(P_1)}$ with $f \mid \mathbb{N}_{(P_2)} = g$; the function f exists by Lemma 9.7(ii).

The set

$$\{d \le c : R_{f \restriction d}^{(P_1)} = \operatorname{Gr} H \text{ for some function } H \text{ with } \operatorname{dom} H \subset \mathbb{N}_{(P_1)}\}$$

contains a cofinal set of elements in $\mathbb{N}_{(P_2)}$. By a routine argument, the set has a maximum element, say d. We have $d \in \mathbb{N}_{(P_1)} \setminus \mathbb{N}_{(P_2)}$ and

$$\operatorname{Gr} F \subset \operatorname{Gr} H = R_{f \restriction d}^{(P_1)}$$

for some function H with $\operatorname{dom} H \subset \mathbb{N}_{(P_1)}$. We see that $F = H \restriction a \in \mathbb{N}^{<\mathbb{N}}_{(P_1)}$ and that $\operatorname{ran}(H \restriction a)$ contains a maximum element, say b, with $b \in J_{P_2}$. The set $\operatorname{ran} F$ is bounded by b in J_{P_2}. □

We can now prove our strengthened version of Theorem 9.6(ii).

Theorem 9.9 *The field* $\widetilde{\mathbf{R}}$ *is not isomorphic as a totally ordered group to any field of the form* K_P *for which* P *is a cut prime in some* $C(\Omega)$.

Proof Assume towards a contradiction that $\widetilde{\mathbf{R}}$ is isomorphic as a totally ordered group to a field K_P, where P is a cut prime. Then K_P is Cauchy complete, and so, by Theorem 9.3,

$$\{0,1\}^{\mathbb{N}}_{(P)} = \{f : \mathbb{N}_{(P)} \to \{0,1\} : \ f \restriction a \in \{0,1\}^a_{(P)} \quad (a \in \mathbb{N}_{(P)})\}. \tag{9.4}$$

Let J_P be defined as in (9.3). Then $\operatorname{cof} J_P = \operatorname{cof} \mathbb{N}_{(P)} = \delta(\widetilde{\mathbf{R}})$ by Lemma 8.33(v), and so $\operatorname{cof} J_P = \aleph_1$.

Fix an element $a \in J_P \setminus \mathbb{N}$. Then there are sequences $(a_\alpha : \alpha < \omega_1)$ and $(b_\alpha : \alpha < \omega_1)$ in J_P with

$$a_\alpha < a_\beta < b_\beta < b_\alpha < a \qquad (\alpha < \beta < \omega_1)$$

defining an $(\aleph_1, \aleph_1)$-gap. Take $(c_\alpha : \alpha < \omega_1)$ to be cofinal in J_P, and set $d_\alpha = c_\alpha^2$, so that $(d_\alpha : \alpha < \omega_1)$ is cofinal in $\mathbb{N}_{(P)}$. It is straightforward to construct inductively functions $g_\alpha \in \{0,1\}^{d_\alpha}_{(P)}$ for $\alpha < \omega_1$ such that $R_{g_\alpha}^{(P)} = F_\alpha$, where

$$\operatorname{dom} F_\alpha = ([1, a_\alpha] \cup [b_\alpha, a]) \cap \mathbb{N}_{(P)} \quad \text{and} \quad F_\alpha(a_\alpha) = c_\alpha$$

for $\alpha < \omega_1$ and where $F_\alpha \subset F_\beta$ and $g_\beta \restriction d_\alpha = g_\alpha$ for $\alpha < \beta < \omega_1$. Set

$$g = \bigcup \{g_\alpha : \alpha < \omega_1\}.$$

By (9.4), $g \in \{0,1\}^{\mathbb{N}}_{(P)}$. We have $R_g^{(P)} = \operatorname{Gr} F$, and $F = \bigcup\{F_\alpha : \alpha < \omega_1\}$ is a function from $[1, a] \cap \mathbb{N}_{(P)}$ whose range contains the cofinal sequence $(c_\alpha : \alpha < \omega_1)$ in J_P. But this contradicts Lemma 9.8, and so the result follows. □

It will be seen that we do not have an analogue in the present context of Theorem 9.6(i). In fact, the following example, whose details we omit, shows that no such analogue exists.

Proposition 9.10 (CH) *There is a compact space* Ω, *and a non-maximal, prime ideal* P *in* $C(\Omega)$ *such that:*

(i) P *strictly contains a* z*-ideal which is a valuation prime;*

(ii) $\operatorname{cof} K_P = \aleph_1$*;*

(iii) K_P *is Cauchy complete.*

Proof By Comfort and Negrepontis (1974, Theorems 10.4 and 13.9), there is a uniform ultrafilter $\mathcal{U}$ on ω_1 such that $K = \mathbb{R}^{\aleph_1}/\mathcal{U}$ is an η_2-field (see Notes 2.37). By Proposition 4.33, K has the form K_L for a valuation prime z-ideal L in some $C(\Omega)$.

Let I be an additively closed initial segment of $\mathbb{N}_{(L)}$ with $\operatorname{cof} I = \aleph_1$. It follows from Lemma 8.36(ii) that $I = \mathbb{N}_{(P)}$ for some prime ideal P in $C(\Omega)$ with $L \subsetneq P$. Then K_P satisfies (i) and (ii), and it can be shown, using Lemma 9.7, that K_P is Cauchy complete, and so satisfies (iii). □

Thus there is a Cauchy complete field in the class **cut prime** $\boldsymbol{\eta_1}$.

We now turn to our main proof that $\widetilde{\mathbf{R}}$ is a hyper-real field. At some points we shall say that we are assuming GCH, but in fact we shall require only CH and the hypothesis that

$$2^{\aleph_1} = \aleph_2 .$$

Our first move is to define a topological space $Y = Y(I)$ associated with a non-empty index set I, taken to be a set of ordinals. Indeed,

$$Y = \{F : \mathbb{N} \times I \to \{0,1\}\} = \{0,1\}^{\mathbb{N}\times I},$$

so that Y is a compact topological space with respect to the product topology. For sets S and T,

$$[S]^{<\omega}$$

is the collection of all the finite subsets of the set S, and

$$\operatorname{Fin}(S,T)$$

is the collection of finite partial functions from S to T. Thus an element of this set is a function whose domain belongs to $[S]^{<\omega}$ and whose codomain is T.

Let A be a subset of $\mathbb{N} \times I$. For $F \in Y$, define

$$\mathcal{R}_A(F) = F \mid A, \tag{9.5}$$

so that $\mathcal{R}_A(F) \in \{0,1\}^A$. For a subset S of Y, we write

$$\mathcal{R}_A[S] \quad \text{for} \quad \{\mathcal{R}_A(F) : F \in S\}.$$

For each $G \in \operatorname{Fin}(\mathbb{N} \times I, \{0,1\})$, define

$$\mathcal{N}_G = \{F \in Y : \mathcal{R}_{\operatorname{dom} G}(F) = G\}. \tag{9.6}$$

Then $\mathcal{N}_G$ is an open and closed subset of Y, and $\mathcal{N}_G \in \mathbf{Z}(Y)$. For each $F \in Y$, the family

$$\{\mathcal{N}_{R_A(f)} : A \in [\mathbb{N} \times I]^{<\omega}\} \tag{9.7}$$

is a base of neighbourhoods of F in Y.

Lemma 9.11 *Let $Z_1, \dots, Z_m \in \mathbf{Z}(Y)$ be such that $\bigcap_{j=1}^m Z_j = \emptyset$. Then there exists A in $[\mathbb{N} \times I]^{<\omega}$ such that $\bigcap_{j=1}^m \mathcal{R}_A[Z_j] = \emptyset$.*

Proof There is an ultrafilter $\mathcal{V}$ on $[\mathbb{N}\times I]^{<\omega}$ such that, for each A in $[\mathbb{N}\times I]^{<\omega}$,

$$\{B \in [\mathbb{N} \times I]^{<\omega} : B \supset A\} \in \mathcal{V}.$$

Assume towards a contradiction that the result fails, so that, for each element $A \in [\mathbb{N} \times I]^{<\omega}$, there exists $f_A \in \bigcap_{j=1}^m \mathcal{R}_A[Z_j]$. Define $H \in Y$ by setting

$$H(k,s) = 1 \ \text{if} \ \{A \in [\mathbb{N} \times I]^{<\omega} : (k,s) \in A, \quad f_A(k,s) = 1\} \in \mathcal{V},$$
$$H(k,s) = 0 \ \text{if} \ \{A \in [\mathbb{N} \times I]^{<\omega} : (k,s) \in A, \quad f_A(k,s) = 1\} \notin \mathcal{V}.$$

Take $(k,s) \in \mathbb{N} \times I$. By the properties of $\mathcal{V}$, $H(k,s) = 0$ if and only if

$$\{A \in [\mathbb{N} \times I]^{<\omega} : (k,s) \in A, \quad f_A(k,s) = 0\} \in \mathcal{V}.$$

There exists $j \in \{1, \dots, m\}$ such that $H \notin Z_j$. Choose $A \in [\mathbb{N} \times I]^{<\omega}$ with $\mathcal{N}_{H|A} \cap Z_j = \emptyset$. We have

$$\{B \in [\mathbb{N} \times I]^{<\omega} : B \supset A, \ f_B \mid A = H \mid A\} \in \mathcal{V},$$

and so there exists $B \in [\mathbb{N}\times I]^{<\omega}$ with $B \supset A$ and $f_B \mid A = H \mid A$. There exists $F \in Z_j$ with $F \mid B = f_B \mid B$, and then $F \in \mathcal{N}_{H|A} \cap Z_j$, a contradiction.

Thus the result holds. □

We now introduce some notation and conventions that will be crucial to the exposition in all the remainder of this book.

Fix a free ultrafilter $\mathcal{U}$ on $\mathbb{N}$. As in the special example discussed on page 208, but now replacing the cardinal κ by the set $\mathbb{N}$, we define the structures

$$\mathbb{N}_{(\mathcal{U})}, \quad \mathbb{Q}_{(\mathcal{U})}, \quad \mathbb{D}_{(\mathcal{U})}, \quad \{0,1\}^{<\mathbb{N}}_{(\mathcal{U})}, \quad \text{and} \quad \{0,1\}^{\mathbb{N}}_{(\mathcal{U})}.$$

We shall define further structures generically denoted by $S_{(\mathcal{U})}$ for a certain set S; in each case there is a canonical bijection from $S^{\mathbb{N}}/\mathcal{U}$ onto $S_{(\mathcal{U})}$, and so we may identify $S_{(\mathcal{U})}$ and $S^{\mathbb{N}}/\mathcal{U}$ as sets. Let $s \in S$. Then the element of $S_{(\mathcal{U})}$ defined by s regarded as a constant function on $\mathbb{N}$ is denoted by $[s]_{\mathcal{U}}$. Note that, if $S = T \cup \{s\}$, where $s \notin T$, then

$$S_{(\mathcal{U})} = T_{(\mathcal{U})} \cup \{[s]_{\mathcal{U}}\}. \tag{9.8}$$

We are to give three definitions of objects depending on $\mathcal{U}$. As a preliminary, we present as a prototype a conceptually straightforward definition. Let S and T be sets, and let $h = (h_k)$ be a sequence of functions from S to T. For $s = (s(k)) \in S^{\mathbb{N}}$, set

$$h(s) = (h_k(s(k))) \in T^{\mathbb{N}},$$

and define

$$h_{\mathcal{U}}([s]_{\mathcal{U}}) = [h(s)]_{\mathcal{U}} \quad ([s]_{\mathcal{U}} \in S_{(\mathcal{U})}). \tag{9.9}$$

It is immediate that $h_{\mathcal{U}} : S^{\mathbb{N}}/\mathcal{U} \to T^{\mathbb{N}}/\mathcal{U}$ is well defined. Using our identification of $S^{\mathbb{N}}/\mathcal{U}$ with $S_{(\mathcal{U})}$, we may regard $h_{\mathcal{U}}$ as a map

$$h_{\mathcal{U}} : S_{(\mathcal{U})} \to T_{(\mathcal{U})}.$$

For the first definition, we apply the prototype in the case where $S = \mathbb{N} \times I$ and $T = \{0,1\}$, and we define $S_{(\mathcal{U})}$ to be $\mathbb{N}_{(\mathcal{U})} \times I_{(\mathcal{U})}$ (and identify $T^{\mathbb{N}}/\mathcal{U}$ with $\{0,1\}$, of course). The set $Y_{(\mathcal{U})}$ is defined by identifying elements of $Y^{\mathbb{N}}/\mathcal{U}$ with certain functions from $\mathbb{N}_{(\mathcal{U})} \times I_{(\mathcal{U})}$ into $\{0,1\}$. Specifically, take $h \in Y^{\mathbb{N}}$, say $h = (h(k))$, where

$$h(k) \in Y = \{0,1\}^{\mathbb{N}\times I} = T^S$$

for each $k \in \mathbb{N}$. An element of $\mathbb{N}_{(\mathcal{U})} \times I_{(\mathcal{U})}$ has the form $([f]_{\mathcal{U}}, [g]_{\mathcal{U}})$, where $f \in \mathbb{N}^{\mathbb{N}}$ and $g \in I^{\mathbb{N}}$. We set

$$h_{\mathcal{U}}([f]_{\mathcal{U}}, [g]_{\mathcal{U}}) = 1 \quad \text{if} \quad \{k \in \mathbb{N} : h(k)(f(k), g(k)) = 1\} \in \mathcal{U},$$
$$h_{\mathcal{U}}([f]_{\mathcal{U}}, [g]_{\mathcal{U}}) = 0 \quad \text{if} \quad \{k \in \mathbb{N} : h(k)(f(k), g(k)) = 0\} \in \mathcal{U}.$$

We see directly from these expressions that

$$h_{\mathcal{U}} : \mathbb{N}_{(\mathcal{U})} \times I_{(\mathcal{U})} \to \{0,1\}$$

is well defined, and then that $h_{\mathcal{U}}$ coincides with the function given by (9.9) in this case. The collection of functions $h_{\mathcal{U}}$ constitutes the set $Y_{(\mathcal{U})}$, so that

$$Y_{(\mathcal{U})} = \{h_{\mathcal{U}} : h \in Y^{\mathbb{N}}\} = \left(\{0,1\}^{\mathbb{N}\times I}\right)^{\mathbb{N}}/\mathcal{U}. \tag{9.10}$$

Next, consider a function $h : \mathbb{N} \to [\mathbb{N} \times I]^{<\omega}$. We define a subset A_h of $\mathbb{N}_{(\mathcal{U})} \times I_{(\mathcal{U})}$ as follows. For $f \in \mathbb{N}^{\mathbb{N}}$ and $g \in I^{\mathbb{N}}$, set

$$A_h = \left\{([f]_{\mathcal{U}}, [g]_{\mathcal{U}}) \in \mathbb{N}_{(\mathcal{U})} \times I_{(\mathcal{U})} : \{k \in \mathbb{N} : (f(k), g(k)) \in h(k)\} \in \mathcal{U}\right\}.$$

Again A_h is well defined; with the usual interpretation, we have

$$[\mathbb{N} \times I]^{<\omega}_{(\mathcal{U})} = \{A_h : \ h : \mathbb{N} \to [\mathbb{N} \times I]^{<\omega}\}. \tag{9.11}$$

Finally, consider a function $h : \mathbb{N} \to \mathrm{Fin}(\mathbb{N} \times I, \{0,1\})$. Then there is a function $\varphi : \mathbb{N} \to [\mathbb{N} \times I]^{<\omega}$ such that

$$\varphi(k) = \operatorname{dom} h(k) \quad (k \in \mathbb{N}),$$

and $h_{\mathcal{U}} : A_\varphi \to \{0,1\}$ is defined as above to give a partial function $h_{\mathcal{U}}$ from $\mathbb{N}_{(\mathcal{U})} \times I_{(\mathcal{U})}$ into $\{0,1\}$; again, we have

$$\mathrm{Fin}(\mathbb{N} \times I, \{0,1\})_{(\mathcal{U})} = \{h_{\mathcal{U}} : \ h : \mathbb{N} \to \mathrm{Fin}(\mathbb{N} \times I, \{0,1\})\}.$$

In summary:

(1) $Y_{(\mathcal{U})}$ is a collection of functions from $\mathbb{N}_{(\mathcal{U})} \times I_{(\mathcal{U})}$ into $\{0,1\}$, and there is a canonical bijection of $Y_{(\mathcal{U})}$ onto $Y^{\mathbb{N}}/\mathcal{U}$;

(2) $[\mathbb{N} \times I]^{<\omega}_{(\mathcal{U})}$ is a collection of subsets of $\mathbb{N}_{(\mathcal{U})} \times I_{(\mathcal{U})}$;

(3) $\mathrm{Fin}(\mathbb{N} \times I, \{0,1\})_{(\mathcal{U})}$ is a collection of partial functions from $\mathbb{N}_{(\mathcal{U})} \times I_{(\mathcal{U})}$ into $\{0,1\}$, the domain of each such function belonging to $[\mathbb{N} \times I]^{<\omega}_{(\mathcal{U})}$.

Objects associated to $\mathcal{U}$ in the above fashion are quite special. For example, $Y_{(\mathcal{U})}$ does not contain all the functions from $\mathbb{N}_{(\mathcal{U})} \times I_{(\mathcal{U})}$ into $\{0,1\}$, just as $\{0,1\}^{\mathbb{N}}_{(\mathcal{U})}$ does not contain all functions from $\mathbb{N}_{(\mathcal{U})}$ into $\{0,1\}$.

Lemma 9.12 (i) *Let $A, B \in [\mathbb{N} \times I]^{<\omega}_{(\mathcal{U})}$. Then $A \cup B$, $A \cap B$, and $A \setminus B$ belong to $[\mathbb{N} \times I]^{<\omega}_{(\mathcal{U})}$.*

(ii) *All finite subsets of $\mathbb{N}_{(\mathcal{U})} \times I_{(\mathcal{U})}$ belong to $[\mathbb{N} \times I]^{<\omega}_{(\mathcal{U})}$.*

(iii) *Suppose that (A_j) is a sequence in $[\mathbb{N} \times I]^{<\omega}_{(\mathcal{U})}$. Then there is an element $B \in [\mathbb{N} \times I]^{<\omega}_{(\mathcal{U})}$ with $\bigcup_{j=1}^{\infty} A_j \subset B$.*

Proof We prove (iii); (i) and (ii) are similar, but easier.

For each $j \in \mathbb{N}$, choose $h_j : \mathbb{N} \to [\mathbb{N} \times I]^{<\omega}$ such that $A_j = A_{h_j}$. Define

$$h : k \mapsto \bigcup_{j=1}^{k} h_j(k), \quad \mathbb{N} \to [\mathbb{N} \times I]^{<\omega},$$

and set $B = A_h$, so that $B \in [\mathbb{N} \times I]^{<\omega}_{(\mathcal{U})}$. We *claim* that, for each $j \in \mathbb{N}$, $A_j \subset B$. Indeed, take $f \in \mathbb{N}^{\mathbb{N}}$ and $g \in I^{\mathbb{N}}$ with $([f]_{\mathcal{U}}, [g]_{\mathcal{U}}) \in A_j$. Then

$$\{k \geq j : (f(k), g(k)) \in h(k)\} \supset \{k \geq j : (f(k), g(k)) \in h_j(k)\} \in \mathcal{U},$$

and so $([f]_{\mathcal{U}}, [g]_{\mathcal{U}}) \in B$. Thus $A_j \subset B$, as required. □

We have a similar lemma for $\mathrm{Fin}(\mathbb{N} \times I, \{0,1\})$. First we require some further notation about partial functions. Let $F, G \in \mathrm{Fin}(\mathbb{N} \times I, \{0,1\})_{(\mathcal{U})}$. Then $F \sim G$ if

$$F \mid (\mathrm{dom}\, F \cap \mathrm{dom}\, G) = G \mid (\mathrm{dom}\, F \cap \mathrm{dom}\, G);$$

if $F \sim G$, then $F \cup G$ is defined on $\mathrm{dom}\, F \cup \mathrm{dom}\, G$ by the requirement that

$$(F \cup G) \mid \mathrm{dom}\, F = F, \quad (F \cup G) \mid \mathrm{dom}\, G = G.$$

The '$\mathcal{U}$-version' of the map $\mathcal{R}_A$ defined in (9.5) is $\mathcal{R}_A^{(\mathcal{U})}$, where

$$\mathcal{R}_A^{(\mathcal{U})}(F) = F \mid A \quad (F \in Y_{(\mathcal{U})})$$

for $A \in [\mathbb{N} \times I]^{<\omega}_{(\mathcal{U})}$; we write $\mathcal{R}_A^{(\mathcal{U})}[S]$ for $\{\mathcal{R}_A^{(\mathcal{U})}(F) : F \in S\}$ in the case where S is a subset of $Y_{(\mathcal{U})}$.

Lemma 9.13 (i) *Suppose that $F \in \mathrm{Fin}(\mathbb{N} \times I, \{0,1\})_{(\mathcal{U})}$ and $A \in [\mathbb{N} \times I]^{<\omega}_{(\mathcal{U})}$. Then*

$$\mathcal{R}_A^{(\mathcal{U})}(F) \in \mathrm{Fin}(\mathbb{N} \times I, \{0,1\})_{(\mathcal{U})}.$$

(ii) *Suppose that $F, G \in \mathrm{Fin}(\mathbb{N} \times I, \{0,1\})_{(\mathcal{U})}$ with $F \sim G$. Then*

$$F \cup G \in \mathrm{Fin}(\mathbb{N} \times I, \{0,1\})_{(\mathcal{U})}.$$

(iii) *Suppose that* $F : \mathbb{N}_{(\mathcal{U})} \times I_{(\mathcal{U})} \to \{0,1\}$ *is a finite partial function. Then*

$$F \in \operatorname{Fin}(\mathbb{N} \times I, \{0,1\})_{(\mathcal{U})}.$$

(iv) *Suppose that* $(F_j : j \in \mathbb{N})$ *is a sequence in* $\operatorname{Fin}(\mathbb{N} \times I, \{0,1\})_{(\mathcal{U})}$ *such that* $F_i \sim F_j \quad (i, j \in \mathbb{N})$. *Then there exists* $F \in \operatorname{Fin}(\mathbb{N} \times I, \{0,1\})_{(\mathcal{U})}$ *such that, for each* $j \in \mathbb{N}$, $\operatorname{dom} F_j \subset \operatorname{dom} F$ *and* $F \sim F_j$.

Proof We just prove (iv). By (ii), we may suppose that

$$\operatorname{dom} F_j \subset \operatorname{dom} F_{j+1} \quad (j \in \mathbb{N}).$$

For each $j \in \mathbb{N}$, choose $h_j : \mathbb{N} \to \operatorname{Fin}(\mathbb{N} \times I, \{0,1\})$ with $F_j = (h_j)_{\mathcal{U}}$. Then

$$\{k \in \mathbb{N} : \operatorname{dom} h_j(k) \subset \operatorname{dom} h_{j+1}(k), \quad h_{j+1}(k) \sim h_j(k)\} \in \mathcal{U}. \tag{9.12}$$

For each $k \in \mathbb{N}$, take j_k to be the minimum element j of the set $\{1, \dots, k\}$ such that either $\operatorname{dom} h_j(k) \not\subset \operatorname{dom} h_{j+1}(k)$ or $h_{j+1}(k) \nsim h_j(k)$, taking $j_k = k$ if no such j exists. Define

$$h : k \mapsto \bigcup_{j=1}^{j_k} h_j(k), \quad \mathbb{N} \to \operatorname{Fin}(\mathbb{N} \times I, \{0,1\}),$$

and set $F = h_{\mathcal{U}}$. For each $j \in \mathbb{N}$, $\{k \in \mathbb{N} : j < j_k\} \in \mathcal{U}$ by (9.12), and so $\operatorname{dom} F_j \subset \operatorname{dom} F$ and $F \sim F_j$. □

We next define a completely regular space X by setting

$$X = \mathbb{N} \times Y = \mathbb{N} \times \{0,1\}^{\mathbb{N} \times I}.$$

For $k \in \mathbb{N}$, the subset $\{k\} \times Y$ of X is denoted by Y_k, so that each Y_k is homeomorphic to Y and $X = \bigcup\{Y_k : k \in \mathbb{N}\}$. For each $\sigma \subset \mathbb{N}$,

$$Y_{[\sigma]} = \bigcup\{Y_k : k \in \sigma\},$$

so that $Y_{[\sigma]} \in \mathbf{Z}(X)$, and $\mathcal{F}_0$, defined to be

$$\mathcal{F}_0 = \{Y_{[\sigma]} : \sigma \in \mathcal{U}\},$$

is a z-filter on X. We shall use many times the trivial fact that, for each sequence (Z_k) contained in $\mathbf{Z}(Y)$, the set

$$\bigcup\{\{k\} \times Z_k : k \in \mathbb{N}\}$$

belongs to $\mathbf{Z}(X)$.

We shall associate to each $Z \in \mathbf{Z}(X)$ a corresponding set $Z_{\mathcal{U}}$ in $Y_{(\mathcal{U})}$. Indeed, for $Z \in \mathbf{Z}(X)$, there exists a sequence (Z_k) in $\mathbf{Z}(Y)$ such that

$$Z \cap Y_k = \{k\} \times Z_k;$$

we say that the sequence (Z_k) is *associated* to Z. Define

$$Z_{\mathcal{U}} = \{h_{\mathcal{U}} : \ h : \mathbb{N} \to Y, \ \{k \in \mathbb{N} : h(k) \in Z_k\} \in \mathcal{U}\}.$$

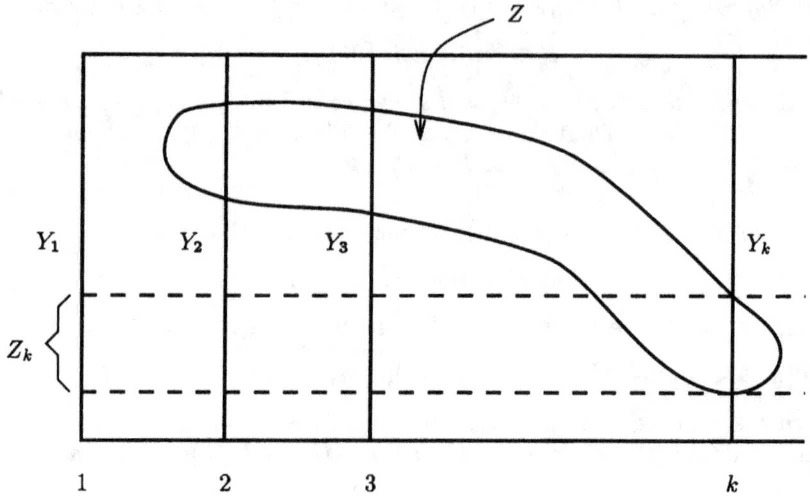

Fig. 9.1. The topological space X

It is clear that $\emptyset_{\mathcal{U}} = \emptyset$ and $Z_{\mathcal{U}} = Y_{(\mathcal{U})}$ for each $Z \in \mathcal{F}_0$. Also, for $Z, W \in \mathbf{Z}(X)$, we have

$$Z_{\mathcal{U}} \cap W_{\mathcal{U}} = (Z \cap W)_{\mathcal{U}} \quad \text{and} \quad Z_{\mathcal{U}} \cup W_{\mathcal{U}} = (Z \cup W)_{\mathcal{U}}\,.$$

Suppose that Z and W have associated sequences (Z_k) and (W_k), respectively, and that $Z_{\mathcal{U}} = W_{\mathcal{U}}$. Then $\{k \in \mathbb{N} : Z_k = W_k\} \in \mathcal{U}$.

We have the now standard definition

$$\mathbf{Z}(Y)_{(\mathcal{U})} = \{Z_{\mathcal{U}} : Z \in \mathbf{Z}(X)\}\,,$$

so that $\mathbf{Z}(Y)_{(\mathcal{U})}$ is a family of subsets of $Y_{(\mathcal{U})}$ containing $\emptyset$ and $Y_{(\mathcal{U})}$ which is closed under finite unions and intersections.

We now come to a key definition.

Definition 9.14 *Let* $H : \mathbb{N}_{(\mathcal{U})} \times I_{(\mathcal{U})} \to \{0, 1\}$ *be such that*

$$H \mid A \in \mathrm{Fin}(\mathbb{N} \times I, \{0,1\})_{(\mathcal{U})}$$

for each $A \in [\mathbb{N} \times I]^{<\omega}_{(\mathcal{U})}$. *Then*

$$\mathcal{F}_H = \{Z \in \mathbf{Z}(X) : H \mid A \in \mathcal{R}^{(\mathcal{U})}_A[Z_{\mathcal{U}}] \text{ for each } A \in [\mathbb{N} \times I]^{<\omega}_{(\mathcal{U})}\}\,.$$

We shall eventually show that $\mathcal{F}_H$ is a z-ultrafilter on X; the main work for this is in the following lemma.

Lemma 9.15 *Let* $\mathcal{F}_H$ *be as above.*

(i) *Suppose that* $Z \in \mathcal{F}_H$ *and* $W \in \mathbf{Z}(X)$ *with* $W \supset Z$. *Then* $W \in \mathcal{F}_H$.

(ii) *Suppose that* $Z_1, \dots, Z_m \in \mathcal{F}_H$. *Then* $\bigcap_{j=1}^m Z_j \neq \emptyset$.

(iii) *For* $Z \in \mathbf{Z}(X)$ *either* $Z \in \mathcal{F}_H$ *or there exists* $W \in \mathcal{F}_H$ *with* $Z \cap W = \emptyset$.

Proof (i) This is immediate from Definition 9.14 because $Z_{\mathcal{U}} \subset W_{\mathcal{U}}$.

(ii) For each $j \in \{1, \ldots, m\}$, take $(Z_{j,k} : k \in \mathbb{N}) \subset \mathbf{Z}(Y)$ to be the associated sequence of Z_j. We shall in fact prove the *claim* that

$$\left\{ k \in \mathbb{N} : \bigcap_{j=1}^{m} (\{k\} \times Z_{j,k}) \neq \emptyset \right\} \in \mathcal{U} .$$

For assume that this claim fails. Then we may suppose that $\bigcap_{j=1}^{m} Z_{j,k} \neq \emptyset$ for each $k \in \mathbb{N}$. By Lemma 9.11, for each $k \in \mathbb{N}$, there exists $A_k \in [\mathbb{N} \times I]^{<\omega}$ such that

$$\bigcap_{j=1}^{m} \mathcal{R}_{A_k}[Z_{j,k}] = \emptyset . \tag{9.13}$$

Define $h : k \mapsto A_k$, $\mathbb{N} \to [\mathbb{N} \times I]^{<\omega}$, so that $A_h \in [\mathbb{N} \times I]^{<\omega}_{(\mathcal{U})}$ and

$$H \mid A_h \in \mathrm{Fin}(\mathbb{N} \times I, \{0,1\})_{(\mathcal{U})} .$$

Choose $f : \mathbb{N} \to \mathrm{Fin}(\mathbb{N} \times I, \{0,1\})$ such that $\mathrm{dom}\, f(k) = A_k \quad (k \in \mathbb{N})$ and $f_{\mathcal{U}} = H \mid A_h$. For $j = 1, \ldots, m$, we have $f_{\mathcal{U}} \in \mathcal{R}^{(\mathcal{U})}_{A_h}[(Z_j)_{\mathcal{U}}]$ because $Z_j \in \mathcal{F}_H$, and so

$$\{k \in \mathbb{N} : f(k) \in \mathcal{R}_{A_k}[Z_{j,k}]\} \in \mathcal{U} .$$

Thus there exists $k \in \mathbb{N}$ such that

$$f(k) \in \mathcal{R}_{A_k}[Z_{j,k}] \quad (j \in \{1, \ldots, m\}) .$$

But this is a contradiction of (9.13), and so the claim holds.

(iii) Take $Z \in \mathbf{Z}(X)$, and suppose that $Z \notin \mathcal{F}_H$. Then there is an element $A \in [\mathbb{N} \times I]^{<\omega}_{(\mathcal{U})}$ such that, for each $G \in Z_{\mathcal{U}}$, we have $G \mid A \neq H \mid A$. Take $h : \mathbb{N} \to \mathrm{Fin}(\mathbb{N} \times I, \{0,1\})$ such that $h_{\mathcal{U}} = H \mid A$.

For $k \in \mathbb{N}$, define $V_k = \mathcal{N}_{h(k)}$ (in the notation of (9.6)), so that V_k is an open set in Y belonging to $\mathbf{Z}(Y)$, and set

$$V = \bigcup \{\{k\} \times V_k : k \in \mathbb{N}\} ,$$

so that $V \in \mathbf{Z}(X)$. For each $G \in V_{\mathcal{U}}$, we have $G \mid A = H \mid A$, and so $V_{\mathcal{U}} \cap Z_{\mathcal{U}} = \emptyset$, whence $\{k \in \mathbb{N} : V_k \cap Z_k = \emptyset\} \in \mathcal{U}$, where (Z_k) is associated to Z. Define

$$W = \bigcup \{\{k\} \times V_k : V_k \cap Z_k = \emptyset\} .$$

Then $W \in \mathbf{Z}(X)$, $W \cap Z = \emptyset$, and $W \in \mathcal{F}_H$. □

Theorem 9.16 *Let $\mathcal{F}_H$ be as above. Then $\mathcal{F}_H$ is a z-ultrafilter on X containing $\mathcal{F}_0$.*

Proof Clearly $\mathcal{F}_H \supset \mathcal{F}_0$.

We *claim* that $Z_1 \cap Z_2 \in \mathcal{F}_H$ whenever $Z_1, Z_2 \in \mathcal{F}_H$. For assume that $Z_1 \cap Z_2 \notin \mathcal{F}_H$. By Lemma 9.15(iii), there exists $W \in \mathcal{F}_H$ with $Z_1 \cap Z_2 \cap W = \emptyset$. But this contradicts Lemma 9.15(ii), and so the claim holds.

It follows from Lemma 9.15(i) and (ii) that $\mathcal{F}_H$ is a z-filter, and from Lemma 9.15(iii) that $\mathcal{F}_H$ is maximal in the family of z-filters on X. □

Let us recall some basic facts from Chapter 4. We write

$$C(X)/\mathcal{F}_H$$

for the quotient algebra $C(X)/\mathbf{Z}^{-1}[\mathcal{F}_H]$; since $\mathcal{F}_H$ is a z-ultrafilter on X, it follows from Proposition 4.6(i) that $\mathbf{Z}^{-1}[\mathcal{F}_H]$ is a maximal ideal in $C(X)$, and so $C(X)/\mathcal{F}_H$ is a hyper-real field.

Definition 9.17 *Denote by Q_H the ideal $C^b(X) \cap \mathbf{Z}^{-1}[\mathcal{F}_H]$ in $C(\beta X)$.*

By Proposition 4.33, Q_H is a valuation prime z-ideal and $C(X)/\mathcal{F}_H \cong K_{Q_H}$. Hence K_{Q_H} belongs to the class **hyper-real** of Chapter 4; by Theorem 4.29, K_{Q_H} is a real-closed η_1-field.

We note that the family $\{Y_k : k \in \mathbb{N}\}$ defined above is a Q_H-partition with limit set contained in $\beta X \setminus X$. Since $\mathcal{F}_H \supset \mathcal{F}_0$, there is a natural embedding of $\mathbb{N}_{(\mathcal{U})}$ into $\mathbb{N}_{(Q_H)}$; we identify $\mathbb{N}_{(\mathcal{U})}$ with its image under this embedding. Our next goal is to show that $\mathbb{N}_{(\mathcal{U})}$ is actually equal to $\mathbb{N}_{(Q_H)}$ (although of course the super-real field K_{Q_H} is very different from the field which is the ultrapower $\mathbb{R}^{\mathbb{N}}/\mathcal{U}$). As usual, we require a preliminary lemma.

Lemma 9.18 *Let $Z \in \mathcal{F}_H$, and suppose that, for each $k \in \mathbb{N}$,*

$$\{W_{1,k}, \dots, W_{m_k,k}\}$$

is a finite partition of $Z \cap Y_k$ in $\mathbf{Z}(Y_k)$. Then there exists $j_k \in \{1, \dots, m_k\}$ for $k \in \mathbb{N}$ such that

$$\bigcup\{W_{j_k,k} : k \in \mathbb{N}\} \in \mathcal{F}_H\,.$$

Proof First note that, by the compactness of each Y_k, for each pairwise disjoint family $\{C_1, \dots, C_m\}$ of closed sets in Y_k, there exists $A \in [\mathbb{N} \times I]^{<\omega}$ such that $\{\mathcal{R}_A[C_1], \dots, \mathcal{R}_A[C_m]\}$ is a pairwise disjoint family in $\{0,1\}^A$.

Choose $\varphi : \mathbb{N} \to [\mathbb{N} \times I]^{<\omega}$ such that, for each $k \in \mathbb{N}$, the family

$$\{\mathcal{R}_{\varphi(k)}[W_{1,k}], \dots, \mathcal{R}_{\varphi(k)}[W_{m_k,k}]\}$$

is pairwise disjoint, and choose $h : \mathbb{N} \to \mathrm{Fin}(\mathbb{N} \times I, \{0,1\})$ such that

$$\operatorname{dom} h(k) = \varphi(k) \quad (k \in \mathbb{N}) \quad \text{and} \quad h_{\mathcal{U}} = H \mid A_\varphi\,,$$

where A_φ is the subset of $\mathbb{N}_{(\mathcal{U})} \times \mathbb{I}_{(\mathcal{U})}$ related to φ. Set

$$S = \{k \in \mathbb{N} : h(k) \in \mathcal{R}_{\varphi(k)}[Z_k]\}\,,$$

where (Z_k) is the sequence associated with Z; since $Z \in \mathcal{F}_H$, we have $S \in \mathcal{U}$.

For each $k \in \mathbb{N}$,

$$\mathcal{R}_{\varphi(k)}[Z_k] = \mathcal{R}_{\varphi(k)}[W_{1,k}] \cup \cdots \cup \mathcal{R}_{\varphi(k)}[W_{m_k,k}]\,,$$

and so, for each $k \in S$, there exists $j_k \in \{1, \dots, m_k\}$ with $h(k) \in R_{\varphi(k)}[W_{j_k,k}]$; if $k \in \mathbb{N} \setminus S$, set $j_k = 1$. We have

$$\{k \in \mathbb{N} : h(k) \in \mathcal{R}_{\varphi(k)}[W_{j_k,k}]\} \in \mathcal{U}\,,$$

and so the result follows. □

Theorem 9.19 *With the above notation,* $\mathbb{N}_{(\mathcal{U})} = \mathbb{N}_{(Q_H)}$.

Proof Since $\{Y_k : k \in \mathbb{N}\}$ is a Q_H-partition, elements of $\mathbb{N}_{(\mathcal{U})}$ correspond to functions $f : X \to \mathbb{N}$ such that $f \mid Y_k$ is constant for each $k \in \mathbb{N}$, whereas elements of $\mathbb{N}_{(Q_H)}$ correspond to functions $f : S_\infty \to \mathbb{N}$, where $\{S_j : j \in \mathbb{N}\}$ is a Q_H-partition of βX with union S_∞. Thus, to obtain the result, it is sufficient to show the following: for each Q_H-partition $\{S_j : j \in \mathbb{N}\}$ of βX, there exists $S \in \mathbf{Z}[Q_H]$ such that, for each $k \in \mathbb{N}$,

$$S \cap S_j \cap Y_k \neq \emptyset \quad \text{for at most one} \quad j \in \mathbb{N}.$$

By refining the Q_H-partition $\{S_j : j \in \mathbb{N}\}$, we may suppose that we have $S_j \subset X \quad (j \in \mathbb{N})$; we take $S_0 \in \mathbf{Z}(\beta X)$ such that the partition has limit set contained in S_0.

Since $S_0 \notin \mathbf{Z}[Q_H]$, we have $S_0 \cap X \notin \mathcal{F}_H$, and so there exists $W \in \mathcal{F}_H$ with $S_0 \cap X \cap W = \emptyset$. Set

$$V_j = S_j \cap W \quad (j \in \mathbb{N}),$$

so that the family $\{V_j : j \in \mathbb{N}\}$ is a Q_H-partition, with union V_∞, say. Since $V_\infty \in \mathbf{Z}(X)$, necessarily $V_\infty \in \mathcal{F}_H$.

For $j, k \in \mathbb{N}$, set

$$W_{j,k} = V_j \cap Y_k .$$

We have $(\overline{V}_\infty \backslash V_\infty) \cap X = \emptyset$, and, for each $k \in \mathbb{N}$, Y_k is compact. Thus $W_{j,k} \neq \emptyset$ for only finitely many values of j, and so $\{W_{j,k} : j \in \mathbb{N}\}$ is a finite partition of $V_\infty \cap Y_k$ in $\mathbf{Z}(Y_k)$. It follows from Lemma 9.18 that there exists $W \in \mathcal{F}_H$ such that, for each $k \in \mathbb{N}$, $W \cap Y_k = W_{j_k,k}$ for some $j_k \in \mathbb{N}$.

Take S to be the closure of W in βX. Then $S \in \mathbf{Z}[Q_H]$, and, for each $k \in \mathbb{N}$, $S \cap S_j \cap Y_k \neq \emptyset$ at most for $j = j_k$. As we explained above, this proves the theorem. □

We now examine the set of functions $\{0,1\}^{\mathbb{N}}_{(Q_H)}$ associated with the prime z-ideal Q_H. We shall first prove that this set contains all sections of H in the following sense.

Definition 9.20 *A function of the form*

$$h : a \mapsto H(a, [f]_{\mathcal{U}}), \quad \mathbb{N}_{(\mathcal{U})} \to \{0,1\}$$

for some $f \in I^{\mathbb{N}}$ *is a* section *of* H.

A generalization of the following lemma will be given in Lemma 9.35.

Lemma 9.21 *Each section of* H *belongs to* $\{0,1\}^{\mathbb{N}}_{(Q_H)}$.

Proof Take a section h of H, say $h(a) = H(a, [f]_{\mathcal{U}}) \quad (a \in \mathbb{N}_{(\mathcal{U})})$, where $f \in I^{\mathbb{N}}$, and define

$$F : (k, z) \mapsto \sum_{j=1}^{\infty} \frac{z(j, f(k))}{2^j}, \quad X = \mathbb{N} \times Y \to \mathbb{I},$$

so that $F \in C^b(X)$.

We may suppose that h is not eventually constant, for otherwise certainly $h \in \{0,1\}^{\mathbb{N}}_{(Q_H)}$ because $h \restriction a \in \{0,1\}^a_{(Q_H)} \quad (a \in \mathbb{N}_{(\mathcal{U})})$.

Our purpose is to define elements $s_a, t_a \in \mathbb{D}_a^{(Q_H)}$ with

$$s_a \leq \pi_{Q_H}(F) \leq t_a \quad \text{and} \quad t_a - s_a = 2^{-a}$$

for each $a \in \mathbb{N}_{(\mathcal{U})}$. To do this, take $a \in \mathbb{N}_{(\mathcal{U})}$ and $g \in \mathbb{N}^{\mathbb{N}}$ with $a = [g]_{\mathcal{U}}$, set

$$A = \{(c, [f]_{\mathcal{U}}) \in \mathbb{N}_{(\mathcal{U})} \times I_{(\mathcal{U})} : c \leq a\},$$

so that $A \in [\mathbb{N} \times I]^{<\omega}_{(\mathcal{U})}$, and choose $G : \mathbb{N} \to \operatorname{Fin}(\mathbb{N} \times I, \{0,1\})$ to be such that $G_{\mathcal{U}} = H \mid A$. We may suppose that

$$\operatorname{dom} G(k) = \{(j, f(k)) : j \in \{1, \ldots, g(k)\}\} \quad (k \in \mathbb{N})$$

since the set of values of k for which this is true belongs to $\mathcal{U}$. Define

$$F_a^-(k, z) = \sum_{j=1}^{g(k)} \frac{z(j, f(k))}{2^j}, \quad F_a^+(k, z) = F_a^-(k, z) + 2^{-g(k)}$$

for $k \in \mathbb{N}$ and $z \in Y$. Then $F_a^-, F_a^+ : X \to \mathbb{I}$ are continuous functions.

For each $k \in \mathbb{N}$, set

$$W_k = \{k\} \times \mathcal{N}_{G(k)} \subset X,$$

and set $W = \bigcup\{W_k : k \in \mathbb{N}\}$. Since $G_{\mathcal{U}} = H \mid A$, we have

$$Z_{\mathcal{U}} = \{\Phi \in Y_{(\mathcal{U})} : \Phi \mid A = H \mid A\},$$

and this shows that $W \in \mathcal{F}_H$ and that $\{W_k : k \in \mathbb{N}\}$ is a Q_H-partition. We note that the following properties of F_a^- and F_a^+ hold for each $k \in \mathbb{N}$:

(i) $F_a^- \mid W_k$ and $F_a^+ \mid W_k$ are both constant, and take values in $\mathbb{D}_{g(k)}$;

(ii) $F_a^-(x) \leq F(x) \leq F_a^+(x) \quad (x \in W_k)$;

(iii) $F_a^+(x) - F_a^-(x) = 2^{-g(k)} \quad (x \in W_k)$.

Set $s_a = \pi_{Q_H}(F_a^-)$ and $t_a = \pi_{Q_H}(F_a^+)$. Then:

$$s_a, t_a \in \mathbb{D}_a^{(Q_H)}; \quad s_a \leq \pi_{Q_H}(F) \leq t_a; \quad t_a - s_a = 2^{-a}.$$

Take $a, b \in \mathbb{N}_{(\mathcal{U})}$ with $a < b$. Then, for each $k \in \mathbb{N}$,

$$F_a^-(x) \leq F_b^-(x) \leq F(x) \leq F_b^+(x) \leq F_a^+(x) \quad (x \in W_k),$$

and so $s_a \leq s_b \leq t_b \leq t_a$.

Set $r = \pi_{Q_H}(F)$. For each $a \in \mathbb{N}_{(Q)}$,

$$F_a^-(x) \leq F(x) \leq F_a^+(x) \quad (x \in W),$$

and so $r \in [s_a, t_a]$. For each $u \in (s_a, t_a)$, we have $\varepsilon_u^{(Q_H)} \restriction a = h \restriction a$; since h is not eventually constant, it follows from Lemma 8.16 that there exists $b \in \mathbb{N}_{(\mathcal{U})}$ with $s_a < s_b < t_b < t_a$. Thus we may conclude that, for each $a \in \mathbb{N}_{(Q_H)}$, we have $r \in (s_a, t_a)$. Since $s_a, t_a \in \mathbb{D}_a^{(Q_H)}$, it follows that $r \in \mathbb{I}_{(Q_H)} \setminus \mathbb{D}_{(Q_H)}$. Hence, for each $a \in \mathbb{N}_{(\mathcal{U})}$, we have $\varepsilon_r^{(Q_H)} \restriction a = h \restriction a$, and so $h = \varepsilon_r^{(Q_H)} \in \{0,1\}^{\mathbb{N}}_{(Q_H)}$, as required. □

The key that we need for the proof that $\widetilde{\mathbf{R}}$ is a hyper-real field is the following theorem that K_{Q_H} is Cauchy complete *provided that* the function H satisfies a certain combinatorial condition. Note that the following theorem does not require any axioms beyond ZFC, but that the verification in Theorem 9.25 that the combinatorial condition can be satisfied does require GCH.

Theorem 9.22 *Let $\mathcal{U}$ be a free ultrafilter on $\mathbb{N}$, and let*

$$H : \mathbb{N}_{(\mathcal{U})} \times I_{(\mathcal{U})} \to \{0,1\}$$

be such that

$$H \mid A \in \mathrm{Fin}(\mathbb{N} \times I, \{0,1\})_{(\mathcal{U})} \quad (A \in [\mathbb{N} \times I]^{<\omega}_{(\mathcal{U})}).$$

Assume that each function $f : \mathbb{N}_{(\mathcal{U})} \to \{0,1\}$ with

$$f \restriction a \in \{0,1\}^a_{(\mathcal{U})} \quad (a \in \mathbb{N})_{(\mathcal{U})}$$

is a section of H. Then the hyper-real field K_{Q_H} is Cauchy complete.

Proof It is immediate from Lemma 9.21 that the condition for the Cauchy completeness of K_{Q_H} given in Theorem 9.3 is satisfied. □

We are within sight of proving that $\widetilde{\mathbf{R}}$ is a hyper-real field; essentially all that remains is to prove that, for a suitable index set I (necessarily with $|I| \geq \aleph_2$), there exists a function $H : \mathbb{N}_{(\mathcal{U})} \times I_{(\mathcal{U})} \to \{0,1\}$ satisfying the combinatorial condition of Theorem 9.22. Almost inevitably, this will be another transfinite induction; the underlying idea is rather simple, but the notation is a problem, for we shall be carrying through the above construction, not just for one fixed index set I, but for a family

$$\{I_\alpha : \alpha < \omega_2\}$$

of index sets, and we have to indicate the further dependence of all our structures on the ordinal α.

Let I and J be index sets (taken to be sets of ordinals) with $I \subset J$. Then $I_{(\mathcal{U})} \subset J_{(\mathcal{U})}$. The next lemma records some obvious relationships between sets associated with I and J; the proof is an immediate verification.

Lemma 9.23 *Let $\mathcal{U}$ be a free ultrafilter on $\mathbb{N}$, and suppose that $I \subset J$. Then:*

(i) $[\mathbb{N} \times I]^{<\omega}_{(\mathcal{U})} \subset [\mathbb{N} \times J]^{<\omega}_{(\mathcal{U})}$;

(ii) $[\mathbb{N} \times I]^{<\omega}_{(\mathcal{U})} = \{A \cap (\mathbb{N}_{(\mathcal{U})} \times I_{(\mathcal{U})}) : A \in [\mathbb{N} \times J]^{<\omega}_{(\mathcal{U})}\}$;

(iii) $\mathrm{Fin}(\mathbb{N} \times I, \{0,1\})_{(\mathcal{U})} \subset \mathrm{Fin}(\mathbb{N} \times J, \{0,1\})_{(\mathcal{U})}$;

(iv) $\mathrm{Fin}(\mathbb{N} \times I, \{0,1\})_{(\mathcal{U})} = \{F \mid (\mathbb{N}_{(\mathcal{U})} \times I_{(\mathcal{U})}) : F \in \mathrm{Fin}(\mathbb{N} \times J, \{0,1\})_{(\mathcal{U})}\}$.

□

We shall now consider a fixed index set I which is the union of a nested family $\{I_\alpha : \alpha < \gamma\}$ of subsets (for some ordinal γ).

Lemma 9.24 *Let $\mathcal{U}$ be a free ultrafilter on* $\mathbb{N}$, *and suppose that* $\operatorname{cof}\gamma > \aleph_0$. *Then:*

(i) $I_{(\mathcal{U})} = \bigcup_{\alpha<\gamma}(I_\alpha)_{(\mathcal{U})}$;

(ii) $[\mathbb{N}\times I]^{<\omega}_{(\mathcal{U})} = \bigcup_{\alpha<\gamma}[\mathbb{N}\times I_\alpha]^{<\omega}_{(\mathcal{U})}$;

(iii) $\mathrm{Fin}(\mathbb{N}\times I,\{0,1\})_{(\mathcal{U})} = \bigcup_{\alpha<\gamma}\mathrm{Fin}(\mathbb{N}\times I_\alpha,\{0,1\})_{(\mathcal{U})}$.

Proof Again, these are immediate verifications. For example, suppose that $A\in[\mathbb{N}\times I]^{<\omega}_{(\mathcal{U})}$, and choose $h:\mathbb{N}\to[\mathbb{N}\times I]^{<\omega}$ such that $A = A_h$. Since $I=\bigcup_{\alpha<\gamma}I_\alpha$ and $\operatorname{cof}\gamma>\aleph_0$, there exists $\alpha<\gamma$ such that

$$h(k)\in[\mathbb{N}\times I_\alpha]^{<\omega}\quad(k\in\mathbb{N}).$$

Thus $A\in[\mathbb{N}\times I_\alpha]^{<\omega}_{(\mathcal{U})}$. □

Note that it follows from Lemmas 9.12 and 9.13 that the above lemma does not hold in the case where $\operatorname{cof}\gamma=\aleph_0$.

Theorem 9.25 (GCH) *Let $\mathcal{U}$ be a free ultrafilter on* $\mathbb{N}$, *and let I be an index set with $|I|=\aleph_2$. Then there exists $H:\mathbb{N}_{(\mathcal{U})}\times I_{(\mathcal{U})}\to\{0,1\}$ such that*

$$H\mid A\in\mathrm{Fin}(\mathbb{N}\times I,\{0,1\})_{(\mathcal{U})}\quad(A\in[\mathbb{N}\times I]^{<\omega}_{(\mathcal{U})})$$

and such that each $f:\mathbb{N}_{(\mathcal{U})}\to\{0,1\}$ with $f\restriction a\in\{0,1\}^a_{(\mathcal{U})}$ $(a\in\mathbb{N}_{(\mathcal{U})})$ is a section of H.

Proof The set of functions $f:\mathbb{N}_{(\mathcal{U})}\to\{0,1\}$ has cardinality $2^{\aleph_1}=\aleph_2$ (with GCH), and so the set of functions f specified in the statement can be indexed as $\{f_s : s\in I\}$. We shall construct $H:\mathbb{N}_{(\mathcal{U})}\times I_{(\mathcal{U})}\to\{0,1\}$ such that:

(1) $H\mid A\in\mathrm{Fin}(\mathbb{N}\times I,\{0,1\})_{(\mathcal{U})}$ for each $A\in[\mathbb{N}\times I]^{<\omega}_{(\mathcal{U})}$;

(2) $H(a,[s]_{\mathcal{U}}) = f_s(a)\quad(a\in\mathbb{N}_{(\mathcal{U})}, s\in I)$.

We begin by fixing a decomposition $I=\bigcup\{I_\alpha:\alpha<\omega_2\}$ of I, where the subsets I_α satisfy the following properties:

$$\begin{aligned}
&I_0=\emptyset;\\
&|I_\alpha|\le\aleph_1\quad(\alpha<\omega_2);\\
&|I_{\alpha+1}\setminus I_\alpha|=1\quad(\alpha<\omega_2);\\
&I_\alpha\subset I_\beta\quad(\alpha\le\beta<\omega_2);\\
&I_\alpha=\bigcup\{I_\beta:\beta<\alpha\}\ \text{ for each limit ordinal }\ \alpha<\omega_2\,.
\end{aligned}$$

We shall inductively construct functions

$$H_\alpha:\mathbb{N}_{(\mathcal{U})}\times(I_\alpha)_{(\mathcal{U})}\to\{0,1\}$$

for $\alpha<\omega_2$ such that:

(i)$_\alpha$ $H_\alpha\mid A\in\mathrm{Fin}(\mathbb{N}\times I_\alpha,\{0,1\})_{(\mathcal{U})}$ for each $A\in[\mathbb{N}\times I_\alpha]^{<\omega}_{(\mathcal{U})}$;

(ii)$_\alpha$ $H_\alpha\mid(\mathbb{N}_{(\mathcal{U})}\times(I_\beta)_{(\mathcal{U})}) = H_\beta\quad(\beta<\alpha)$;

(iii)$_\alpha$ $H_\alpha(a,[s]_{\mathcal{U}}) = f_s(a)\quad(a\in\mathbb{N}_{(\mathcal{U})},\ s\in I_\alpha)$.

There are two cases to consider.

Case 1 We assume that H_α has been constructed to satisfy (i)$_\alpha$–(iii)$_\alpha$, and we construct $H_{\alpha+1}$.

Take $s_\alpha \in I$ so that $I_{\alpha+1} = I_\alpha \cup \{s_\alpha\}$; by (9.8),

$$(I_{\alpha+1})_{(\mathcal{U})} = (I_\alpha)_{(\mathcal{U})} \cup \{[s_\alpha]_\mathcal{U}\}.$$

Define

$$H_{\alpha+1}(a,c) = \begin{cases} H_\alpha(a,c) & (a \in \mathbb{N}_{(\mathcal{U})},\ c \in (I_\alpha)_\mathcal{U}), \\ f_s(a) & (a \in \mathbb{N}_{(\mathcal{U})},\ c = [s_\alpha]_\mathcal{U}), \end{cases}$$

so that $H_{\alpha+1} : \mathbb{N}_{(\mathcal{U})} \times (I_{\alpha+1})_{(\mathcal{U})} \to \{0,1\}$. Clearly $H_{\alpha+1}$ satisfies conditions (ii)$_{\alpha+1}$ and (iii)$_{\alpha+1}$; it remains to prove that $H_{\alpha+1}$ satisfies (i)$_{\alpha+1}$.

To this end, we define

$$h : a \mapsto f_{s_\alpha}(a), \qquad \mathbb{N}_{(\mathcal{U})} \to \{0,1\},$$
$$g : (a, [s_\alpha]_\mathcal{U}) \mapsto f_{s_\alpha}(a), \qquad \mathbb{N}_{(\mathcal{U})} \times \{[s_\alpha]_\mathcal{U}\} \to \{0,1\}.$$

For each $a \in \mathbb{N}_{(\mathcal{U})}$, we have $h \restriction a \in \{0,1\}^a_{(\mathcal{U})}$ by the assumption on f_{s_α}, and so

$$g \mid (([1,a] \cap \mathbb{N}_{(\mathcal{U})}) \times \{[s_\alpha]\}_\mathcal{U}) \in \mathrm{Fin}(\mathbb{N} \times I_{\alpha+1}, \{0,1\})_{(\mathcal{U})},$$

whence

$$g \mid C \in \mathrm{Fin}(\mathbb{N} \times I_{\alpha+1}, \{0,1\})_{(\mathcal{U})} \tag{9.14}$$

for each $C \in [\mathbb{N} \times I_{\alpha+1}]^{<\omega}_{(\mathcal{U})}$. Now take $A \in [\mathbb{N} \times I_{\alpha+1}]^{<\omega}_{(\mathcal{U})}$, and set

$$B = A \cap (\mathbb{N}_{(\mathcal{U})} \times (I_\alpha)_{(\mathcal{U})}), \quad C = A \setminus B.$$

Then $B \in [\mathbb{N} \times I_\alpha]^{<\omega}_{(\mathcal{U})}$ and $C \subset \mathbb{N}_{(\mathcal{U})} \times \{[s_\alpha]_\mathcal{U}\}$, so that $H_{\alpha+1} \mid B = H_\alpha \mid B$ and $H_{\alpha+1} \mid C = g \mid C$. Using (9.14), (i)$_{\alpha+1}$ follows from Lemma 9.13(ii).

Case 2 We suppose that α is a limit ordinal and assume that H_β has been constructed to satisfy (i)$_\beta$–(iii)$_\beta$ for each $\beta < \alpha$, and we construct H_α. There are two subcases, according to whether $\mathrm{cof}\,\alpha = \aleph_1$ or $\mathrm{cof}\,\alpha = \aleph_0$.

Subcase 2.1 $\mathrm{cof}\,\alpha = \aleph_1$.

Define $H_\alpha = \bigcup_{\beta<\alpha} H_\beta$ (again identifying a function with its graph). By Lemma 9.24(i) and (ii), $\mathrm{dom}\, H_\alpha = (I_\alpha)_{(\mathcal{U})}$ and

$$H_\alpha \mid A \in \mathrm{Fin}(\mathbb{N} \times I_\alpha, \{0,1\})_{(\mathcal{U})}$$

for each $A \in [\mathbb{N} \times I_\alpha]^{<\omega}_{(\mathcal{U})}$. Conditions (i)$_\alpha$–(iii)$_\alpha$ are now immediately checked.

Subcase 2.2 $\mathrm{cof}\,\alpha = \aleph_0$, say (α_k) is cofinal in α.

We first fix a sequence $(A_\zeta : \zeta < \omega_1)$ such that:

$A_\zeta \in [\mathbb{N} \times I_\alpha]^{<\omega}_{(\mathcal{U})} \quad (\zeta < \omega_1)$;

$A_\eta \subset A_\zeta \quad (\eta \leq \zeta < \omega_1)$;

for each $A \in [\mathbb{N} \times I_\alpha]^{<\omega}_{(\mathcal{U})}$, there exists $\zeta < \omega_1$ with $A \subset A_\zeta$.

The existence of such a sequence follows from Lemma 9.12 because, with CH,

$$\left|[\mathbb{N} \times I_\alpha]^{<\omega}_{(\mathcal{U})}\right| = |I_\alpha|^{\aleph_0} \leq \aleph_1^{\aleph_0} = \aleph_1.$$

We now construct by induction on ζ a sequence $(F_\zeta : \zeta < \omega_1)$ in

$$\mathrm{Fin}(\mathbb{N} \times I_\alpha, \{0,1\})_{(\mathcal{U})}$$

such that:

$$\begin{aligned} &\mathrm{dom}\, F_\zeta = A_\zeta \quad (\zeta < \omega_1)\,; \\ &F_\zeta \mid A_\eta = F_\eta \quad (\eta \le \zeta < \omega_1)\,; \\ &F_\zeta \mid (A_\zeta \cap \mathrm{dom}\, H_{\alpha_k}) = H_{\alpha_k} \quad (k \in \mathbb{N},\ \zeta < \omega_1)\,. \end{aligned}$$

First, suppose that F_ζ has been defined. For $k \in \mathbb{N}$, set

$$g_k = H_{\alpha_k} \mid (A_{\zeta+1} \cap \mathrm{dom}\, H_{\alpha_k})\,.$$

By Lemma 9.23(ii), $A_{\zeta+1} \cap \mathrm{dom}\, H_{\alpha_k} \in [\mathbb{N} \times I_{\alpha_k}]^{<\omega}_{(\mathcal{U})}$, and so, by Lemma 9.23(iv), $g_k \in \mathrm{Fin}(\mathbb{N} \times I_{\alpha_k}, \{0,1\})_{(\mathcal{U})}$. Finally, by Lemma 9.23(iii),

$$g_k \in \mathrm{Fin}(\mathbb{N} \times I_\alpha, \{0,1\})_{(\mathcal{U})}\,.$$

Now set

$$G_k = F_\zeta \cup g_k \ \text{ on } \ (A_\zeta \cup \mathrm{dom}\, H_{\alpha_k}) \cap A_{\zeta+1}\,;$$

by Lemma 9.13(ii), $G_k \in \mathrm{Fin}(\mathbb{N} \times I_\alpha, \{0,1\})_{(\mathcal{U})}$, noting that $F_\zeta \sim g_k$.

Using the inductive hypothesis on $(H_\beta : \beta < \alpha)$ and the properties of F_ζ, we see that, for each $k \in \mathbb{N}$:

$$\begin{aligned} G_k &\in \mathrm{Fin}(\mathbb{N} \times I_\alpha, \{0,1\})_{(\mathcal{U})}\,; \\ \mathrm{dom}\, G_k &\subset \mathrm{dom}\, G_{k+1} \subset A_{\zeta+1}\,; \\ G_{k+1} &\mid \mathrm{dom}\, G_k = G_k\,. \end{aligned} \tag{9.15}$$

By Lemma 9.13(iv), there exists $G \in \mathrm{Fin}(\mathbb{N} \times I_\alpha, \{0,1\})_{(\mathcal{U})}$ such that, for each $k \in \mathbb{N}$, we have $\mathrm{dom}\, G_k \subset \mathrm{dom}\, G$ and $G \sim G_k$; clearly we may suppose that $\mathrm{dom}\, G \subset A_{\zeta+1}$. Choose $F \in \mathrm{Fin}(\mathbb{N} \times I_\alpha, \{0,1\})_{(\mathcal{U})}$ with $\mathrm{dom}\, F = A_{\zeta+1} \setminus \mathrm{dom}\, G$, and define $F_{\zeta+1} = G \cup F$. Then $F_{\zeta+1}$ satisfies the appropriate conditions.

Second, suppose that ζ is a limit ordinal and that F_η has been defined for each $\eta < \zeta$. Since $\mathrm{cof}\, \zeta = \aleph_0$, there is a cofinal, increasing sequence (η_k) in ζ. For $k \in \mathbb{N}$, set $g_k = H_{\alpha_k} \mid (A_\zeta \cap \mathrm{dom}\, H_{\alpha_k})$ and

$$G_k = F_{\eta_k} \cup g_k \quad \text{on} \quad A_{\eta_k} \cup (A_\zeta \cap \mathrm{dom}\, H_{\alpha_k})\,.$$

Essentially as before, we see that (G_k) satisfies (9.14), and then we construct G from (G_k) by the same recipe, so obtaining the function F_ζ. Again F_ζ satisfies the appropriate conditions.

Now that we have the sequence $(F_\zeta : \zeta < \omega_1)$, we define $H_\alpha = \bigcup_{\zeta<\omega_1} F_\zeta$. We have $\mathrm{dom}\, H_\alpha = \mathbb{N}_{(\mathcal{U})} \times (I_\alpha)_{(\mathcal{U})}$ and

$$H_\alpha \mid (\mathbb{N}_{(\mathcal{U})} \times (I_\beta)_{(\mathcal{U})}) = H_\beta \quad (\beta < \alpha)\,,$$

so that (ii)$_\alpha$ and (iii)$_\alpha$ are satisfied. To see that H_α satisfies (i)$_\alpha$, take a set $A \in [\mathbb{N} \times I_\alpha]^{<\omega}_{(\mathcal{U})}$. Then there exists $\zeta_0 < \omega_1$ with $A \subset A_{\zeta_0}$, and

$$H_\alpha \mid A = F_{\zeta_0} \mid A \in \mathrm{Fin}(\mathbb{N} \times I_\alpha, \{0,1\})_{(\mathcal{U})}\,,$$

as required.

This completes the inductive construction of the sequence $(H_\alpha : \alpha < \omega_2)$. Let

$$H = \bigcup\{H_\alpha : \alpha < \omega_2\}.$$

Then $\operatorname{dom} H = \mathbb{N}_{(\mathcal{U})} \times I_{(\mathcal{U})}$ by Lemma 9.24(i),

$$H \mid A \in \operatorname{Fin}(\mathbb{N} \times I, \{0,1\})_{(\mathcal{U})} \quad (A \in [\mathbb{N} \times I]^{<\omega}_{(\mathcal{U})})$$

by Lemma 9.24(ii), and H satisfies condition (2), above. Finally, for each function $f : \mathbb{N}_{(\mathcal{U})} \to \{0,1\}$ with $f \upharpoonright a \in \{0,1\}^a \quad (a \in \mathbb{N}_{(\mathcal{U})})$, we have

$$H(a, [s]_{\mathcal{U}}) = f(a) \quad (a \in \mathbb{N}_{(\mathcal{U})})$$

for some $s \in I$, so that f is a section of H.

This completes the proof of the theorem. □

By combining Theorems 9.22 and 9.25, we obtain a long-sought goal.

Theorem 9.26 (GCH) *The ordered field* $\widetilde{\mathbf{R}}$ *is a hyper-real field.*

Proof Let H be the function whose existence is asserted in Theorem 9.25, and set $K = K_{Q_H}$. By Theorem 9.16, $\mathcal{F}_H$ is a z-ultrafilter on X, and so K is a hyper-real field; as such, K is a real-closed η_1-field. By Theorem 9.22, K is Cauchy complete.

By Theorem 8.34(i), $w(K_{Q_H}) = |\mathbb{N}_{(Q_H)}|$, and $\mathbb{N}_{(Q_H)} = \mathbb{N}_{(\mathcal{U})}$ by Theorem 9.19. But $|\mathbb{N}_{(\mathcal{U})}| = \mathfrak{c}$, and so, with CH, $w(K_{Q_H}) = \aleph_1$.

By Theorem 3.21(iv), the ordered field $\widetilde{\mathbf{R}}$ is isomorphic to the hyper-real field K_{Q_H}. □

Thus the natural generalization of $\mathbb{R}$ at a higher cardinality – the unique real-closed, Cauchy complete η_1-field of weight $\aleph_1$ – is a hyper-real field. In particular, there is a hyper-real field K with $|K| = \aleph_2$ and $w(K) = |\Gamma_K| = \aleph_1$. The above result subsumes the result of Theorem 7.14.

Note that it is not a theorem of ZFC that $\widetilde{\mathbf{R}}$ is a hyper-real field, for it was proved in Corollary 6.26 that, in the theory MA + $\neg$CH, $\widetilde{\mathbf{R}}$ is not even a super-real field.

Theorem 9.27 (GCH) *There is a hyper-real field that is not isomorphic as a totally ordered group to any super-real field of the form* K_P *for which* P *is a cut prime.*

Proof This is immediate from Theorems 9.9 and 9.26. □

We have obtained the following information about the classes of Chapter 4: **cut prime η_1** is strictly contained in **strongly convex η_1**; **cut prime z-ideal η_1** is strictly contained in **strongly convex z-ideal η_1**; **hyper-real** is not contained in **cut prime z-ideal η_1** . This information will be incorporated in a new version of our main diagram, at the end of the chapter.

In the above construction that led to a proof that $\widetilde{\mathbf{R}}$ is a hyper-real field, we had the freedom to choose the index set I and the function H of Definition 9.14.

(The free ultrafilter $\mathcal{U}$ on $\mathbb{N}$ remains fixed.) By varying I and H, additional examples of hyper-real fields with particular properties can be constructed. In fact we shall introduce a further modest extension of the same general procedure by considering an initial segment J of the structure $\mathbb{N}_{(\mathcal{U})}$; this will lead to a rich collection of super-real fields. It will be by means of this machinery that we shall produce all of the examples that we consider in Chapter 11. The final topic in the present chapter, an attempt to construct an example of a super-real η_1-field K_P with $w(K_P) > |\Gamma_P|$, utilizes such an extension of the machinery. Many of the technical results that are given below will also be required in Chapter 11.

As before, I is an index set, $Y = \{0,1\}^{\mathbb{N}\times I}$ is a compact topological space, $X = \mathbb{N}\times Y$ is a completely regular topological space, $Y_k = \{k\}\times Y$ is a subspace of X for each $k \in \mathbb{N}$, and $\mathcal{U}$ is a free ultrafilter on $\mathbb{N}$. We shall now further extend our terminology by listing some more objects associated with $\mathcal{U}$.

We have already defined the family $\mathbf{Z}(Y)_{(\mathcal{U})}$ of subsets of $Y_{(\mathcal{U})}$, and have noted that $\mathbf{Z}(Y)_{(\mathcal{U})}$ is closed under finite intersections.

Definition 9.28 *A filter on $Y_{(\mathcal{U})}$ contained in $\mathbf{Z}(Y)_{(\mathcal{U})}$ is a $\mathbf{Z}(Y)_{(\mathcal{U})}$-filter.*

We now define $\mathbf{Z}(Y)^{\mathbb{N}}_{(\mathcal{U})}$, a set of functions from $\mathbb{N}_{(\mathcal{U})}$ into $\mathbf{Z}(Y)_{(\mathcal{U})}$; elements of $\mathbf{Z}(Y)^{\mathbb{N}}_{(\mathcal{U})}$ correspond to functions $G : \mathbb{N} \to \mathbf{Z}(X)$. For such a G and $f : \mathbb{N} \to \mathbb{N}$, set

$$W(f,G) = \bigcup\{G(f(k)) \cap Y_k : k \in \mathbb{N}\}, \tag{9.16}$$

so that $W(f,G) \in \mathbf{Z}(X)$, and define

$$G_{\mathcal{U}}([f]_{\mathcal{U}}) = W(f,G)_{\mathcal{U}}.$$

Then $G_{\mathcal{U}}([f]_{\mathcal{U}})$ is well defined in $\mathbf{Z}(Y)_{(\mathcal{U})}$, and $G_{\mathcal{U}} : \mathbb{N}_{(\mathcal{U})} \to \mathbf{Z}(Y)_{(\mathcal{U})}$. We set

$$\mathbf{Z}(Y)^{\mathbb{N}}_{(\mathcal{U})} = \{G_{\mathcal{U}} :\ G : \mathbb{N} \to \mathbf{Z}(X)\}.$$

Further, for each $a \in \mathbb{N}_{(\mathcal{U})}$, we set

$$\mathbf{Z}(Y)^{a}_{(\mathcal{U})} = \{H \restriction a : H \in \mathbf{Z}(Y)^{\mathbb{N}}_{(\mathcal{U})}\}.$$

(On unravelling some identifications, we see that the validity of our definitions requires that there be a canonical injection

$$(\mathbf{Z}(Y)^{\mathbb{N}})^{\mathbb{N}}/\mathcal{U} \to (\mathbf{Z}(Y)_{(\mathcal{U})})^{\mathbb{N}_{(\mathcal{U})}};$$

an attentive reader may wish to check directly that this is the case!)

Elements of $\mathbf{Z}(Y)^{\mathbb{N}}_{(\mathcal{U})}$ and of $\mathbf{Z}(Y)^{a}_{(\mathcal{U})}$ are regarded as sequences of elements of $\mathbf{Z}(Y)$, and they will frequently be written as

$$(W_d : d \in \mathbb{N}_{(\mathcal{U})}) \quad \text{or} \quad (W_d : d \le a).$$

Definition 9.29 *For a z-filter $\mathcal{F}$ on X containing $\mathcal{F}_0$, define*

$$\mathcal{F}_{\mathcal{U}} = \{Z_{\mathcal{U}} : Z \in \mathcal{F}\}.$$

It is easily seen that $\mathcal{F}_{\mathcal{U}}$ is a $\mathbf{Z}(Y)_{(\mathcal{U})}$-filter. To check this, we show, for example, that $\emptyset \notin \mathcal{F}_{(\mathcal{U})}$. For assume that $Z \in \mathcal{F}$ and that $Z_{\mathcal{U}} = \emptyset$, and set

$$\sigma = \{k \in \mathbb{N} : Z_k = \emptyset\}$$

(where (Z_k) is the sequence associated with Z). Then $\sigma \in \mathcal{U}$, $Y_{[\sigma]} \in \mathcal{F}_0 \subset \mathcal{F}$, and $Z \cap Y_{[\sigma]} \in \mathcal{F}$. But $Z \cap Y_{[\sigma]} = \emptyset$, a contradiction. Thus $\emptyset \notin \mathcal{F}_{\mathcal{U}}$.

The map $\mathcal{F} \mapsto \mathcal{F}_{\mathcal{U}}$ is a bijection from the set of z-filters on X containing $\mathcal{F}_0$ onto the set of $\mathbf{Z}(Y)_{(\mathcal{U})}$-filters on $Y_{(\mathcal{U})}$, and the map preserves inclusion, so that z-ultrafilters on X correspond to maximal $\mathbf{Z}(Y)_{(\mathcal{U})}$-filters on $Y_{(\mathcal{U})}$.

We shall analyze z-ultrafilters on X containing $\mathcal{F}_0$. (As in Theorem 9.16, the z-filters $\mathcal{F}_H$ discussed previously have this property, but now we are considering more general z-ultrafilters.) For each such z-ultrafilter $\mathcal{F}$, $Q_{\mathcal{F}} = C^b(X) \cap \mathbf{Z}^{-1}[\mathcal{F}]$ is the valuation prime z-ideal in $C(\beta X)$ corresponding to $\mathcal{F}$. The structures associated to the super-real field $K_{\mathcal{F}}$ which is the quotient of $C(\beta X)/Q_{\mathcal{F}}$ are now denoted by

$$\mathbb{N}_{(\mathcal{F})},\ \mathbb{Q}_{(\mathcal{F})},\ \mathbb{D}_{(\mathcal{F})},\ \{0,1\}^{<\mathbb{N}}_{(\mathcal{F})},\ \{0,1\}^{\mathbb{N}}_{(\mathcal{F})},$$

etc., and the quotient map from $C(\beta X)$ onto $A_{\mathcal{F}} = C(\beta X)/Q_{\mathcal{F}}$ is $\pi_{\mathcal{F}}$, etc. As before, we regard $\mathbb{N}_{(\mathcal{U})}$ as a subset of $\mathbb{N}_{(\mathcal{F})}$; it will follow from the next lemma that $\mathbb{N}_{(\mathcal{U})}$ is cofinal in $\mathbb{N}_{(\mathcal{F})}$. A key point for us in this context is whether or not, given $a \in \mathbb{N}_{(\mathcal{U})}$, $[1,a] \cap \mathbb{N}_{(\mathcal{U})}$ is an initial segment of $\mathbb{N}_{(\mathcal{F})}$; a condition for this will be given in Proposition 9.32.

Let Q be a prime z-ideal in $C(\beta X)$. A Q-partition $\{Z_j : j \in \mathbb{N}\}$ is *well-placed* if, for each $j \in \mathbb{N}$, there exists $k \in \mathbb{N}$ with $Z_j \subset Y_k$, and if, for each $k \in \mathbb{N}$, $\{j \in \mathbb{N} : Z_j \subset Y_k\}$ is finite. Thus a well-placed partition is a collection of finitely many zero sets of the various sets Y_k.

Lemma 9.30 *Let $\mathcal{F}$ be a z-ultrafilter on X containing $\mathcal{F}_0$, and suppose that $\{Z_j : j \in \mathbb{N}\}$ is a $Q_{\mathcal{F}}$-partition. Then there is a well-placed $Q_{\mathcal{F}}$-partition which refines $\{Z_j : j \in \mathbb{N}\}$.*

Proof Since $\{Y_k : k \in \mathbb{N}\}$ is a $Q_{\mathcal{F}}$-partition, we may suppose that, for each $j \in \mathbb{N}$, $Z_j \subset Y_k$ for some $k \in \mathbb{N}$. Take $Z_0 \in \mathbf{Z}(\beta X)$ such that the $Q_{\mathcal{F}}$-partition $\{Z_j : j \in \mathbb{N}\}$ has limit set contained in Z_0. Since $Z_0 \cap X \notin \mathcal{F}$, there exists $Z \in \mathcal{F}$ with $Z \cap Z_0 = \emptyset$. For each $\sigma \subset \mathbb{N}$,

$$\bigcup\{Z \cap Z_j : j \in \sigma\}$$

is closed in X, and so, for each $k \in \mathbb{N}$,

$$\{Z \cap Z_j \cap Y_k : j \in \mathbb{N}\}$$

is finite because Y_k is compact. Set $W_j = Z \cap Z_j \quad (j \in \mathbb{N})$. Then $\{W_j : j \in \mathbb{N}\}$ is a well-placed $Q_{\mathcal{F}}$-partition refining $\{Z_j : j \in \mathbb{N}\}$. □

Lemma 9.31 *Suppose that $a \in \mathbb{N}_{(\mathcal{U})}$ and that $(W_b : b \le a) \in \mathbf{Z}(Y)^a_{(\mathcal{U})}$. Then*

$$\bigcup\{W_b : b \le a\} \in \mathbf{Z}(Y)_{(\mathcal{U})}\,.$$

Proof Fix $f \in \mathbb{N}^{\mathbb{N}}$ with $a = [f]_{\mathcal{U}}$, and fix $g : \mathbb{N} \to \mathbf{Z}(X)$ with

$$g_{\mathcal{U}} \restriction a = (W_b : b \leq a) .$$

For $b \leq a$, take $h_b \in \mathbb{N}^{\mathbb{N}}$ with $b = [h_b]_{\mathcal{U}}$ and $h_b \leq f$. Then

$$W_b = g_{\mathcal{U}}([h_b]_{\mathcal{U}}) = \left(\bigcup\{g(h_b(k)) \cap Y_k : k \in \mathbb{N}\}\right)_{(\mathcal{U})} .$$

Set

$$Z_k = \bigcup\{g(j) \cap Y_k : j \leq f(k)\} \quad (k \in \mathbb{N}) ,$$

and set $Z = \bigcup\{Z_k : k \in \mathbb{N}\}$. Then $Z_k \in \mathbf{Z}(Y_k) \quad (k \in \mathbb{N})$, and so $Z \in \mathbf{Z}(X)$. We have $\bigcup\{W_b : b \leq a\} = Z_{\mathcal{U}} \in \mathbf{Z}(Y)_{(\mathcal{U})}$. □

Proposition 9.32 *Let $\mathcal{F}$ be a z-ultrafilter on X containing $\mathcal{F}_0$, and let $a \in \mathbb{N}_{(\mathcal{U})}$. Then the following conditions are equivalent:*

(a) $[1, a] \cap \mathbb{N}_{(\mathcal{U})}$ *is an initial segment of* $\mathbb{N}_{(\mathcal{F})}$;

(b) *for each sequence* $(W_b : b \leq a)$ *in* $\mathbf{Z}(Y)^a_{(\mathcal{U})}$ *such that*

$$W_b \cap W_c = \emptyset \quad (b \neq c)$$

and such that $\bigcup\{W_b : b \leq a\} \in \mathcal{F}_{\mathcal{U}}$, *there exists* $b \leq a$ *with* $W_b \in \mathcal{F}_{\mathcal{U}}$.

Proof Fix $f \in \mathbb{N}^{\mathbb{N}}$ with $a = [f]_{\mathcal{U}}$, and define $\bar{f}$ on X by

$$\bar{f} \mid Y_k = f(k) \quad (k \in \mathbb{N}) .$$

(a) $\Rightarrow$ (b) Choose $Z \in \mathcal{F}$ with

$$Z_{\mathcal{U}} = \bigcup\{W_b : b \leq a\} ,$$

choose $G : \mathbb{N} \to \mathbf{Z}(X)$ with $G_{\mathcal{U}} \restriction a = (W_b : b \leq a)$, and define

$$Z_{j,k} = G(j) \cap Y_k \quad (j, k \in \mathbb{N}) .$$

Next, define

$$\sigma = \{k \in \mathbb{N} : Z_{i,k} \cap Z_{j,k} = \emptyset \text{ when } i, j \leq f(k) \text{ and } i \neq j\} ,$$

$$\tau = \left\{k \in \mathbb{N} : Z_k = \bigcup\{Z_{j,k} : j \leq f(k)\}\right\} .$$

We have $\sigma, \tau \in \mathcal{U}$, and so $\sigma \cap \tau \in \mathcal{U}$.

We *claim* that

$$\{Z_{j,k} : j \leq f(k), \quad k \in \sigma \cap \tau\}$$

is a $Q_{\mathcal{F}}$-partition. To show this, we need to see that the union, say V, of these sets belongs to $\mathcal{F}$. But

$$V = \bigcup\{G(k) : k \in \sigma \cap \tau\} = Z \cap Y_{[\sigma \cap \tau]} \in \mathcal{F} ,$$

as required.

Define $h : V \to \mathbb{N}$ by setting $h(t) = j$ when $t \in Z_{j,k}$ for some k; $h(t)$ is well defined because the sets $Z_{j,k}$ are pairwise disjoint, and $h \mid V \leq \bar{f} \mid V$.

The element of $\mathbb{N}_{(\mathcal{F})}$ corresponding to h is defined to be b, and $b \leq a$. Since $[1, a] \cap \mathbb{N}_{(\mathcal{U})}$ is an initial segment of $\mathbb{N}_{(\mathcal{F})}$, we have $b \in \mathbb{N}_{(\mathcal{U})}$, say $b = [g]_{\mathcal{U}}$, where $g \in \mathbb{N}^{\mathbb{N}}$ and $g \leq f$. Let

$$W = \bigcup\{Z_{g(k),k} : k \in \sigma \cap \tau\}.$$

Then W is the set of points in V at which two $\mathbb{N}$-valued functions defining the same element b of $\mathbb{N}_{(\mathcal{F})}$ coincide, and so $W \in \mathcal{F}$. But $W = W(g, G)$ as defined in (9.15), and so $W_{\mathcal{U}} = G_{\mathcal{U}}(b) = W_b$. Hence $W_b \in \mathcal{F}_{\mathcal{U}}$, giving (b).

(b) $\Rightarrow$ (a) Take $b_0 \in \mathbb{N}_{(\mathcal{F})}$ with $b_0 \leq a$; we shall show that $b_0 \in \mathbb{N}_{(\mathcal{U})}$.

Choose a $Q_{\mathcal{F}}$-partition $\{Z_j : j \in \mathbb{N}\}$ with union $Z_\infty \in \mathcal{F}$, and a function $h_0 : X \to \mathbb{N}$ with $\pi(h_0) = b_0$; we may suppose that $h_0 \leq \bar{f}$ on X and, by Lemma 9.30, that $\{Z_j : j \in \mathbb{N}\}$ is well-placed.

For $n \in \mathbb{N}$ and $k \in \mathbb{N}$, define

$$W_{n,k} = h_0^{-1}(\{n\}) \cap Y_k .$$

Then each $W_{n,k} \in \mathbf{Z}(Y_k)$ and $W_{n,k} = \emptyset$ if $n > f(k)$, and so

$$\{W_{n,k} : n \in \mathbb{N}, \ \ k \in \mathbb{N}\}$$

is a $Q_{\mathcal{F}}$-partition.

For $n \in \mathbb{N}$, set $V_n = \bigcup\{W_{n,k} : k \in \mathbb{N}\}$, so that $V_n \in \mathbf{Z}(X)$, and define $G : n \mapsto V_n, \quad \mathbb{N} \to \mathbf{Z}(X)$, and $(W_b : b \leq a) = G_{\mathcal{U}} \restriction a$. Since $W_{m,k} \cap W_{n,k} = \emptyset$ whenever $m \neq n$, it follows that $W_b \cap W_c = \emptyset$ whenever $b \neq c$. Since

$$Z_\infty \cap Y_k = \bigcup\{W_{n,k} : n \leq f(k)\} \quad (k \in \mathbb{N}),$$

we have

$$\bigcup\{W_b : b \leq a\} = (Z_\infty)_{\mathcal{U}} \in \mathcal{F}_{\mathcal{U}} .$$

Thus the conditions on the sequence $(W_b : b \leq a)$ from (b) are satisfied.

By (b), there exists $b \leq a$ with $W_b \in \mathcal{F}_{\mathcal{U}}$, say $b = [g]_{\mathcal{U}}$, where $g \in \mathbb{N}^{\mathbb{N}}$, so that $W_b = G_{\mathcal{U}}([g]_{\mathcal{U}}) = W(g, G)_{\mathcal{U}}$, and

$$\begin{aligned} W(g, G) &= \bigcup\{G(g(k)) \cap Y_k : k \in \mathbb{N}\} \\ &= \bigcup\{W_{g(k),k} : k \in \mathbb{N}\} . \end{aligned}$$

Since $W(g, G)_{\mathcal{U}} \in \mathcal{F}_{\mathcal{U}}$, we have $W(g, G) \in \mathcal{F}$.

Finally, for each $k \in \mathbb{N}$, $h_0 \mid (W(g, G) \cap Y_k) = h_0 \mid W_{g(k),k}$, and this function is constantly equal to $g(k)$. Hence $b_0 \in \mathbb{N}_{(\mathcal{U})}$, as required. □

We now bring into consideration an initial segment J of $\mathbb{N}_{(\mathcal{U})}$.

Definition 9.33 *Let $\mathcal{F}$ be a z-ultrafilter on X containing $\mathcal{F}_0$, and let J be an initial segment of $\mathbb{N}_{(\mathcal{U})}$. Then $\mathcal{F}_{\mathcal{U}}$ is J*-complete *if, for each $a \in J$ and each $(W_b : b \leq a)$ in $\mathbf{Z}(Y)^a_{(\mathcal{U})}$ such that $\bigcup\{W_b : b \leq a\} \in \mathcal{F}_{\mathcal{U}}$, there exists $b \leq a$ with $W_b \in \mathcal{F}_{\mathcal{U}}$.*

It follows immediately from Proposition 9.32 that J is necessarily an initial segment of $\mathbb{N}_{(\mathcal{F})}$ in the case where $\mathcal{F}_{\mathcal{U}}$ is J-complete.

Proposition 9.34 *Let $\mathcal{F}$ be a z-ultrafilter on X containing $\mathcal{F}_0$, and let J be an additively closed, proper initial segment of $\mathbb{N}_{(\mathcal{U})}$. Suppose that $\mathcal{F}_{\mathcal{U}}$ is J-complete. Then there is a unique prime ideal P in $C(\beta X)$ such that $Q_{\mathcal{F}} \subsetneq P$ and $\mathbb{N}_{(P)} = J$. The prime ideal P is a cut prime and a valuation prime.*

Proof Since $\mathcal{F}_{\mathcal{U}}$ is J-complete, J is an initial segment of $\mathbb{N}_{(\mathcal{F})}$. Also $Q_{\mathcal{F}}$ is a z-ideal which is a valuation prime, and $Q_{\mathcal{F}} \neq P$ because $J \neq \mathbb{N}_{(\mathcal{U})}$. Thus the result follows from Proposition 8.36(ii). □

We now make a simple observation. Suppose that

$$F_0 : \mathbb{N}_{(\mathcal{U})} \times I_{(\mathcal{U})} \to \{0,1\}$$

is a finite partial function. By Lemma 9.13(iii), $F_0 \in \mathrm{Fin}(\mathbb{N}\times I, \{0,1\})_{(\mathcal{U})}$, and so $F_0 = G_{\mathcal{U}}$ for some $G : \mathbb{N} \to \mathrm{Fin}(\mathbb{N} \times I, \{0,1\})$. For $k \in \mathbb{N}$, set $Z_k = \{k\} \times \mathcal{N}_{G(k)}$, and define

$$Z = \bigcup\{Z_k : k \in \mathbb{N}\} \in \mathbf{Z}(X)\,.$$

Then we observe that

$$\{F \in Y_{(\mathcal{U})} : F \mid \mathrm{dom}\, F_0 = F_0\} = Z_{\mathcal{U}} \in \mathbf{Z}(Y)_{(\mathcal{U})}\,.$$

We now start from a z-ultrafilter $\mathcal{F}$ on X containing $\mathcal{F}_0$; we shall define a function

$$H_{\mathcal{F}} : \mathbb{N}_{(\mathcal{U})} \times I_{(\mathcal{U})} \to \{0,1\}\,.$$

Indeed, take $a \in \mathbb{N}_{(\mathcal{U})}$ and $t \in I_{(\mathcal{U})}$, and set

$$\begin{aligned} W_0 &= \{F \in Y_{(\mathcal{U})} : F(a,t) = 0\}\,, \\ W_1 &= \{F \in Y_{(\mathcal{U})} : F(a,t) = 1\}\,, \end{aligned}$$

so that $W_0 \cap W_1 = \emptyset$ and $W_0 \cup W_1 = Y_{(\mathcal{U})}$. By the above observation, both $W_0, W_1 \in \mathbf{Z}(Y)_{(\mathcal{U})}$. Now $\mathcal{F}_{\mathcal{U}}$ is a maximal $\mathbf{Z}(Y)_{(\mathcal{U})}$-filter, and so exactly one of W_0 and W_1 belongs to $\mathcal{F}_{\mathcal{U}}$; we define

$$H_{\mathcal{F}}(a,t) = \begin{cases} 0 & \text{if} \quad W_0 \in \mathcal{F}_{\mathcal{U}}\,, \\ 1 & \text{if} \quad W_1 \in \mathcal{F}_{\mathcal{U}}\,. \end{cases}$$

The following lemma is a generalization of Lemma 9.21.

Lemma 9.35 *Let $\mathcal{F}$ be a z-ultrafilter on X containing $\mathcal{F}_0$, and suppose that h is a section of $H_{\mathcal{F}}$. Then there exists $f \in \{0,1\}^{\mathbb{N}}_{(\mathcal{F})}$ with $f \mid \mathbb{N}_{(\mathcal{U})} = h$.*

Proof Fix $g \in I^{\mathbb{N}}$ such that $h(a) = H_{\mathcal{F}}(a, [g]_{\mathcal{U}})\quad (a \in \mathbb{N}_{\mathcal{U}})$, and define

$$F : (k,z) \mapsto \sum_{j=1}^{\infty} \frac{z(j, g(k))}{2^j}, \quad X = \mathbb{N} \times Y \to \mathbb{I}\,,$$

so that, as in Lemma 9.21, $F \in C^b(X)$.

There are again two cases to consider.

Case 1 We suppose that $\pi_{\mathcal{F}}(F) \in \mathbb{D}_{(\mathcal{F})}$.

In this case, there is a $Q_{\mathcal{F}}$-partition $\{Z_j : j \in \mathbb{N}\}$ and a sequence (r_j) in $\mathbb{D}$ such that $F(Z_j) = \{r_j\} \quad (j \in \mathbb{N})$. By Lemma 9.30, we may suppose that $\{Z_j : j \in \mathbb{N}\}$ is well-placed.

Choose a sequence (m_j) in $\mathbb{N}$ such that $r_j \in \mathbb{D}_{m_j} \quad (j \in \mathbb{N})$. For each $(k, z) \in Z_j$, we have

$$\sum_{n=1}^{\infty} \frac{z(n, g(k))}{2^n} = r_j \in \mathbb{D}_{m_j} ,$$

and so the function $n \mapsto z(n, g(k))$ is uniquely determined by its value at the number $n = m_j + 1$ (noting that, as we explained in Chapter 8, an element of $\mathbb{D}$ has exactly two representations of the form $\sum_{n=1}^{\infty} \alpha_n / 2^n$, where (α_n) is a binary sequence). So, by again refining the partition, we may suppose that

$$z(n, g(k)) = w(n, g(k)) \quad (n \in \mathbb{N})$$

whenever (k, z) and (k, w) belong to the same set of the partition. Thus, for each $j \in \mathbb{N}$, we may define $F_j : \mathbb{N} \to \{0, 1\}$ by choosing $(k, z) \in Z_j$ (with k uniquely defined) and setting

$$F_j(n) = z(n, g(k)) \quad (n \in \mathbb{N}) .$$

For $j \in \mathbb{N}$, take $k_j \in \mathbb{N}$ such that $Z_j \subset Y_{k_j}$.

Now take $c \in \mathbb{N}_{(\mathcal{U})}$, say $c = [f_c]_{\mathcal{U}}$, where $f_c \in \mathbb{N}^{\mathbb{N}}$. By the definition of $H_{\mathcal{F}}$, for $\alpha \in \{0, 1\}$ we have $h(c) = \alpha$ if and only if

$$\{Z_j : F_j(f_c(k_j)) = \alpha\}$$

is a $Q_{\mathcal{F}}$-partition. But the function F_j is constant on $\{n \in \mathbb{N} : n \geq m_j + 1\}$, and so h is constant on $\{b \in \mathbb{N}_{(\mathcal{U})} : b \geq a + 1\}$, where $a \in \mathbb{N}_{(\mathcal{U})}$ is defined by the function $\Phi : Z_{\infty} \to \mathbb{N}$ such that $\Phi(Z_j) = \{m_j\} \quad (j \in \mathbb{N})$. Thus it suffices to find $f \in \{0, 1\}^{\mathbb{N}}_{(\mathcal{F})}$ such that

$$f \mid [1, a] \cap \mathbb{N}_{(\mathcal{U})} = h \mid [1, a] \cap \mathbb{N}_{(\mathcal{U})} .$$

For $j \in \mathbb{N}$, choose $x_j \in (\mathbb{Q} \cap \mathbb{I}) \setminus \mathbb{D}$ with $\varepsilon_{x_j} \restriction m_j = F_j \restriction m_j$, and choose $G \in C(\beta X)$ such that $G \mid Z_j = x_j$. Then $\pi_{\mathcal{F}}(G) \in (\mathbb{Q}_{(\mathcal{F})} \cap \mathbb{I}_{(\mathcal{F})}) \setminus \mathbb{D}_{(\mathcal{F})}$ and $\varepsilon_r^{(\mathcal{F})}(d) = h(d) \quad (d \in [1, a] \cap \mathbb{N}_{(\mathcal{U})})$, where $r = \pi_{\mathcal{F}}(G)$, and so $f = \varepsilon_r^{(\mathcal{F})}$ is the required function.

Case 2 We suppose that $\pi_{\mathcal{F}}(F) \notin \mathbb{D}_{(\mathcal{F})}$.

Set $r = \pi_{\mathcal{F}}(F)$. We shall prove that $\varepsilon_r^{(\mathcal{F})} \mid \mathbb{N}_{(\mathcal{U})} = h$, which suffices for the result. Fix $a \in \mathbb{N}_{(\mathcal{U})}$, say $a = [f]_{\mathcal{U}}$, where $f \in \mathbb{N}^{\mathbb{N}}$.

Define $F_1, F_2, G_1, G_2 \in C^b(X)$ as follows:

$$F_1 : (k, z) \mapsto \sum_{j=1}^{f(k)} z(j, g(k))/2^j ;$$

$$F_2 : (k, z) \mapsto \sum_{j=1}^{f(k)+1} z(j, g(k))/2^j ;$$

$$G_1 : (k, z) \mapsto F_1(k, z) + 2^{-f(k)} ;$$
$$G_2 : (k, z) \mapsto F_2(k, z) + 2^{-(f(k)+1)} .$$

Then $F_1 \leq F_2 \leq F \leq G_2 \leq G_1$ in $C^b(X)$, and, further, for $k \in \mathbb{N}$ and $x \in Y_k$:

$$G_1(x), F_1(x) \in \mathbb{D}_{f(k)} ; \qquad G_1(x) - F_1(x) = 2^{-f(k)} ;$$
$$G_2(x), F_2(x) \in \mathbb{D}_{f(k)+1} ; \quad G_2(x) - F_2(x) = 2^{-(f(k)+1)} .$$

Set $s_1 = \pi_{\mathcal{F}}(F_1)$, $t_1 = \pi_{\mathcal{F}}(G_1)$, $s_2 = \pi_{\mathcal{F}}(F_2)$, and $t_2 = \pi_{\mathcal{F}}(G_2)$, so that

$$s_1 \leq s_2 \leq r \leq t_2 \leq t_1 .$$

Consider the function F_1. The collection of sets

$$\{F_1^{-1}(\{t\}) \cap Y_k : t \in \mathbb{D}_{f(k)}, \ \ k \in \mathbb{N}\}$$

is a $Q_{\mathcal{F}}$-partition, and so $s_1 \in \mathbb{D}_a^{(\mathcal{F})}$. Similarly, $t_1 \in \mathbb{D}_a^{(\mathcal{F})}$, $s_2, t_2 \in \mathbb{D}_{a+1}^{(\mathcal{F})}$. However $t_1 - s_1 = 2^{-a}$ and $t_2 - s_2 = 2^{-(a+1)}$, and so, by Lemma 8.16, $\varepsilon_r^{(\mathcal{F})}(a+1) = 0$ if and only if $s_1 = s_2$ and $\varepsilon_r^{(\mathcal{F})}(a+1) = 1$ if and only if $t_1 = t_2$.

Suppose that $\varepsilon_r^{(\mathcal{F})}(a+1) = 0$. Then there exists a set $Z \in \mathcal{F}$ such that $F_1 \mid Z = F_2 \mid Z$. We have $z(f(k)+1, g(k)) = 0$ for $(k, z) \in Z$, and so it follows that $H_{\mathcal{F}}(a+1, [g]_{\mathcal{U}}) = 0$. Similarly, if $\varepsilon_r^{(\mathcal{F})}(a+1) = 1$, then it follows that $H_{\mathcal{F}}(a+1, [g]_{\mathcal{U}}) = 1$. Thus

$$h(a+1) = H_{\mathcal{F}}(a+1, [g]_{\mathcal{U}}) = \varepsilon_r^{(\mathcal{F})}(a+1) ,$$

and so $\varepsilon_r^{(\mathcal{F})} \mid \mathbb{N}_{(\mathcal{U})} = h$, as required. □

The following theorem summarizes the situation we have now reached.

Theorem 9.36 *Let $\mathcal{F}$ be a z-ultrafilter on X containing $\mathcal{F}_0$, and let J be an additively closed, proper initial segment of $\mathbb{N}_{(\mathcal{U})}$. Suppose that $\mathcal{F}_{\mathcal{U}}$ is J-complete. Then the unique prime ideal P in $C(\beta X)$ above $Q_{\mathcal{F}}$ such that $\mathbb{N}_{(P)} = J$ has the properties that:*

(i) *$h \mid J \in \{0,1\}^{\mathbb{N}}_{(P)}$ for each section h of $H_{\mathcal{F}}$;*

(ii) (CH) *$|\mathbb{N}_{(P)}| = |\Gamma_P| = \aleph_1$;*

(iii) *P is a cut prime and a valuation prime.*

Proof (i) The ideal $Q_{\mathcal{F}}$ is a valuation prime, and so, by Lemma 9.7(ii),

$$\begin{aligned} \{0,1\}^{\mathbb{N}}_{(P)} &= \{f \mid \mathbb{N}_{(P)} : f \in \{0,1\}^{<\mathbb{N}}_{(\mathcal{F})}\} \\ &= \{f \mid \mathbb{N}_{(P)} : f \in \{0,1\}^{\mathbb{N}}_{(\mathcal{F})}\} . \end{aligned}$$

Thus the result follows from Lemma 9.35.

(ii) By Theorem 8.34(ii), $|\mathbb{N}_{(P)}| = |\Gamma_P|$. We have

$$|\mathbb{N}_{(P)}| = |J| \leq |\mathbb{N}_{(\mathcal{U})}| = \aleph_1$$

with CH. By Theorem 4.28(iii), $|\Gamma_P| \geq \aleph_1$, and so $|\Gamma_P| = \aleph_1$.

(iii) This is immediate. □

We easily obtain from Theorem 9.36 a combinatorial condition which, if it be satisfied, yields a prime ideal P in $C(\beta X)$ such that $w(K_P) = \aleph_2$ and $|\Gamma_P| = \aleph_1$.

Theorem 9.37 (CH) *Let J be an additively closed, proper initial segment of $\mathbb{N}_{(\mathcal{U})}$. Assume that there exist $a_0 \in J$ and a z-ultrafilter $\mathcal{F}$ on X containing $\mathcal{F}_0$ such that $\mathcal{F}_{\mathcal{U}}$ is J-complete and*

$$|\{h \restriction a_0 : h \text{ is a section of } H_{\mathcal{F}}\}| = \aleph_2 . \tag{9.17}$$

Then the unique prime ideal P in $C(\beta X)$ such that $Q_{\mathcal{F}} \subsetneq P$ and $\mathbb{N}_{(P)} = J$ has the property that

$$|\mathbb{N}_{(P)}| = |\Gamma_P| = \aleph_1 \quad \text{and} \quad |\mathbb{M}_{(P)}| = w(K_P) \geq \aleph_2 . \tag{9.18}$$

Proof The first two equalities of (9.18) are given by Theorem 9.36(ii). By Theorem 8.34(i), $|\mathbb{M}_{(P)}| = w(K_P)$, and certainly $|\mathbb{M}_{(P)}| = |\mathbb{Q}_{(P)}|$. By Definition 8.38(iv),

$$\left|\{0,1\}^{a_0}_{(P)}\right| \leq |\mathbb{Q}_{(P)}| .$$

By Theorem 9.36(i), $h \restriction a_0 \in \{0,1\}^{a_0}_{(P)}$ for each section h of $H_{\mathcal{F}}$, and so, by hypothesis (9.17),

$$\aleph_2 \leq \left|\{0,1\}^{a_0}_{(P)}\right| .$$

Hence $\aleph_2 \leq |\mathbb{M}_{(P)}|$, and we have established (9.18). □

The final topic of this chapter is an attempt to construct a z-filter $\mathcal{F}$ to satisfy the conditions specified in the above theorem. As we explained, it may be that it is a theorem of the theory ZFC + CH that such a z-filter $\mathcal{F}$ exists, but we are only able to prove that it is relatively consistent with ZFC + CH that $\mathcal{F}$ exists.

The first technicality is to define a 'measure' for sets in $\mathbf{Z}(Y)_{(\mathcal{U})}$. For this we start by defining a measure for the sets in $\mathbf{Z}(Y)$: this is simply the product measure on $Y = \{0,1\}^{\mathbb{N}\times I}$ restricted to $\mathbf{Z}(Y)$. Indeed, let μ_0 be the measure on $\{0,1\}$ with

$$\mu_0(\{0\}) = \mu_0(\{1\}) = 1/2 ,$$

and let μ be the product measure from μ_0 defined on the Borel subsets of Y; in fact, we regard μ as being defined on $\mathbf{Z}(Y)$. Thus, for $Z \in \mathbf{Z}(Y)$, we have

$$\mu(Z) = \inf\left\{|\mathcal{R}_A[Z]|/2^{|A|} : A \in [\mathbb{N}\times I]^{<\omega}\right\} .$$

For each $k \in \mathbb{N}$, the measure induced by μ on $\mathbf{Z}(Y_k)$ is μ_k; specifically,

$$\mu_k(\{k\}\times Z) = \mu(Z) \quad (Z \in \mathbf{Z}(Y)) .$$

The 'measure' $\mu_{(\mathcal{U})} : \mathbf{Z}(Y)_{(\mathcal{U})} \to \mathbb{I}_{(\mathcal{U})}$ is defined from μ in the obvious fashion. Indeed, take $Z \in \mathbf{Z}(X)$, define

$$f : k \mapsto \mu_k(Z \cap Y_k), \quad \mathbb{N} \to \mathbb{I}, \quad \text{and} \quad \mu_{(\mathcal{U})}(Z_{\mathcal{U}}) = [f]_{\mathcal{U}} .$$

The function $\mu_{(\mathcal{U})}$ is well defined. The next few lemmas show that $\mu_{(\mathcal{U})}$ has the properties one would expect; the first is an immediate verification.

Lemma 9.38 *Let $Z, W \in \mathbf{Z}(X)$.*

(i) *Suppose that $Z \subset W$. Then $\mu_{(\mathcal{U})}(Z_{\mathcal{U}}) \leq \mu_{(\mathcal{U})}(W_{\mathcal{U}})$.*

(ii) $\mu_{(\mathcal{U})}(Z_{\mathcal{U}} \cup W_{\mathcal{U}}) \leq \mu_{(\mathcal{U})}(Z_{\mathcal{U}}) + \mu_{(\mathcal{U})}(W_{\mathcal{U}})$.

(iii) *Suppose that $Z \cap W = \emptyset$. Then*

$$\mu_{(\mathcal{U})}(Z_{\mathcal{U}} \cup W_{\mathcal{U}}) = \mu_{(\mathcal{U})}(Z_{\mathcal{U}}) + \mu_{(\mathcal{U})}(W_{\mathcal{U}}). \qquad \square$$

Lemma 9.39 *Let $Z, W \in \mathbf{Z}(X)$, and suppose that $Z \subset W$. For each $a \in \mathbb{N}_{(\mathcal{U})}$, there exists $V \in \mathbf{Z}(X)$ with $V \subset W \setminus Z$ and*

$$\mu_{(\mathcal{U})}(W_{\mathcal{U}}) \leq \mu_{(\mathcal{U})}(Z_{\mathcal{U}}) + \mu_{(\mathcal{U})}(V_{\mathcal{U}}) + a^{-1}.$$

Proof Fix $h \in \mathbb{N}^{\mathbb{N}}$ with $a = [h]_{\mathcal{U}}$, and, for each $k \in \mathbb{N}$, choose $V_k \in \mathbf{Z}(Y_k)$ with $V_k \subset W \setminus Z$ and

$$\mu_k(W \cap Y_k) \leq \mu_k(Z \cap Y_k) + \mu_k(V_k) + h(k)^{-1}.$$

Set $V = \bigcup V_k$. Then V has the required properties. $\square$

Lemma 9.40 *Suppose that $a \in \mathbb{N}_{(\mathcal{U})}$ and that $(W_b : b \leq a) \in \mathbf{Z}(Y)^a_{(\mathcal{U})}$. Then there exists $c \leq a$ such that*

$$\mu_{(\mathcal{U})}(W_c) \geq a^{-1}\mu_{(\mathcal{U})}\left(\bigcup\{W_b : b \leq a\}\right).$$

Proof First note that, by Lemma 9.31, there exists a set $Z \in \mathbf{Z}(X)$ with $Z_{\mathcal{U}} = \bigcup\{W_b : b \leq a\}$. Fix $h \in \mathbb{N}^{\mathbb{N}}$ with $a = [h]_{\mathcal{U}}$, and fix $g : \mathbb{N} \to \mathbf{Z}(X)$ with

$$g_{\mathcal{U}} \restriction a = (W_b : b \leq a),$$

and then define

$$\sigma = \left\{k \in \mathbb{N} : Z \cap Y_k = \bigcup\{g(j) \cap Y_k : j \leq h(k)\}\right\},$$

so that $\sigma \in \mathcal{U}$.

Choose $f \in \mathbb{N}^{\mathbb{N}}$ such that, for each $k \in \sigma$, $f(k) \leq h(k)$ and

$$\mu_k(g(f(k)) \cap Y_k) \geq \mu_k(Z_k)/h(k),$$

and set $c = [f]_{\mathcal{U}}$. Then c has the required properties. $\square$

Lemma 9.41 *Let $W \in \mathbf{Z}(Y)_{(\mathcal{U})}$ be such that $\mu_{(\mathcal{U})}(W) \geq a^{-1}$ for an element $a \in \mathbb{N}_{(\mathcal{U})}$. Suppose that $g_1, g_2 \in I^{\mathbb{N}}$ with $[g_1]_{\mathcal{U}} \neq [g_2]_{\mathcal{U}}$ and that $c \in \mathbb{N}_{(\mathcal{U})}$. Then there exists $d_0 \leq c$ such that*

$$\mu_{(\mathcal{U})}(\{F \in W : F(d_0, [g_1]_{\mathcal{U}}) \neq F(d_0, [g_2]_{\mathcal{U}}\}) \geq (a^{-1} - 2^{-c})/c.$$

Proof Fix $h \in \mathbb{N}^{\mathbb{N}}$ with $[h]_{\mathcal{U}} = c$. For $j, k \in \mathbb{N}$, set

$$Z_{j,k} = \{(k, f) \in \mathbb{N} \times Y = X : f(j, g_1(k)) \neq f(j, g_2(k))\},$$

so that $Z_{j,k} \subset Y_k$ and $Z_{j,k} \in \mathbf{Z}(X)$. Next, set

$$Z = \bigcup\{Z_{j,k} : j = 1, \ldots, h(k),\ k \in \mathbb{N}\} \quad \text{and} \quad V = X \setminus Z.$$

We have $Z, V \in \mathbf{Z}(X)$ and $\mu_k(V \cap Y_k) = 2^{-h(k)} \quad (k \in \mathbb{N})$, and this implies that $\mu_{(\mathcal{U})}(V_{\mathcal{U}}) = 2^{-c}$.

For $d \in \mathbb{N}_{(\mathcal{U})}$, define

$$W_d = \{F \in Y_{(\mathcal{U})} : F(d, [g_1]_{\mathcal{U}}) \neq F(d, [g_2]_{\mathcal{U}})\},$$

and set

$$W' = \bigcup \{W_d \cap W : d \leq c\},$$

so that $W' = W \cap Z_{\mathcal{U}} \in \mathbf{Z}(Y)_{(\mathcal{U})}$ and $W \subset W' \cup V_{\mathcal{U}}$. By Lemma 9.38(ii), $\mu_{(\mathcal{U})}(W) \leq \mu_{(\mathcal{U})}(W') + \mu_{(\mathcal{U})}(V_{\mathcal{U}})$, and so

$$\mu_{(\mathcal{U})}(W') \geq a^{-1} - 2^{-c}.$$

Define $g : j \mapsto \bigcup\{Z_{j,k} : k \in \mathbb{N}\}$, $\mathbb{N} \to \mathbf{Z}(X)$, and let

$$g_{\mathcal{U}} : \mathbb{N}_{(\mathcal{U})} \to \mathbf{Z}(Y)_{(\mathcal{U})}$$

be the induced function. Then $g_{\mathcal{U}}$ defines the sequence $(W_d : d \in \mathbb{N}_{(\mathcal{U})})$ in $\mathbf{Z}(Y)^{\mathbb{N}}_{(\mathcal{U})}$. By Lemma 9.40, there exists $d_0 \leq c$ such that

$$\mu_{(\mathcal{U})}(W_{d_0} \cap W) \geq c^{-1}\mu_{(\mathcal{U})}(W'),$$

and so the result follows. □

Lemma 9.42 *Let $(W_d : d \in \mathbb{N}_{(\mathcal{U})}) \in \mathbf{Z}(Y)^{\mathbb{N}}_{(\mathcal{U})}$.*

(i) *Suppose that $b \in \mathbb{N}_{(\mathcal{U})}$ and that $\mu_{(\mathcal{U})}(W_j) \geq b^{-1} \quad (j \in \mathbb{N})$. Then there exists $a \in \mathbb{N}_{(\mathcal{U})} \setminus \mathbb{N}$ such that $\mu_{(\mathcal{U})}(W_d) \geq b^{-1} \quad (d \leq a)$.*

(ii) *Suppose that $W_k \subset W_j$ for $j \leq k$ in $\mathbb{N}$. Then there exists $a \in \mathbb{N}_{(\mathcal{U})} \setminus \mathbb{N}$ such that $W_c \subset W_b$ whenever $b \leq c \leq a$ in $\mathbb{N}_{(\mathcal{U})}$.*

Proof (i) Fix $f \in \mathbb{N}^{\mathbb{N}}$ with $b = [f]_{\mathcal{U}}$, and fix a function $g : \mathbb{N} \to \mathbf{Z}(X)$ with $g_{\mathcal{U}} = (W_d : d \in \mathbb{N}_{(\mathcal{U})})$.

For each $k \in \mathbb{N}$, choose j_k to be the minimum $j \in \mathbb{N}$ with

$$\mu_k(g(j) \cap Y_k) < f(k)^{-1},$$

taking $j_k = k$ if there is no such j. Let $j \in \mathbb{N}$. Since $\mu_{(\mathcal{U})}(W_j) \geq b^{-1}$, it follows that

$$\{k \in \mathbb{N} : j_k > j\} \in \mathcal{U}.$$

Define $h : k \mapsto j_k$, $\mathbb{N} \to \mathbb{N}$, and set $a = [h]_{\mathcal{U}}$, so that $a \in \mathbb{N}_{(\mathcal{U})} \setminus \mathbb{N}$. For $d \in [1, a] \cap \mathbb{N}_{(\mathcal{U})}$, take $h_d \in \mathbb{N}^{\mathbb{N}}$ with $d = [h_d]_{\mathcal{U}}$. Then

$$\{k \in \mathbb{N} : h_d(k) < h(k)\} \in \mathcal{U},$$

and so

$$\{k \in \mathbb{N} : \mu_k(g(h_b(k)) \cap Y_k) \geq f(k)^{-1}\} \in \mathcal{U},$$

whence $\mu_{(\mathcal{U})}(W_d) \geq a^{-1}$.

(ii) This is a similar argument. □

Definition 9.43 *Let J be a multiplicatively closed initial segment of $\mathbb{N}_{(\mathcal{U})}$. Then*

$$\mathbb{P}_J = \{W \in \mathbf{Z}(Y)_{(\mathcal{U})} : \mu_{(\mathcal{U})}(W) \geq b^{-1} \ \text{ for some } \ b \in J\}.$$

The set $\mathbb{P}_J$ is a partially ordered set with respect to ordering by inclusion.

In the following lemmas, J is a fixed multiplicatively closed initial segment of $\mathbb{N}_{(\mathcal{U})}$.

Lemma 9.44 *Suppose that $W_1 \in \mathbf{Z}(Y)_{(\mathcal{U})}$, $W_2 \in \mathbb{P}_J$, and $W_1 \cap W_2 \notin \mathbb{P}_J$. Then there exists $W_3 \in \mathbb{P}_J$ with $W_3 \subset W_2 \setminus W_1$.*

Proof Fix $a \in J$ with $\mu_{(\mathcal{U})}(W_2) \geq a^{-1}$. By Lemma 9.39, there exists $V \in \mathbf{Z}(X)$ with $V_{\mathcal{U}} \subset W_2 \setminus W_1$ and

$$\mu_{(\mathcal{U})}(W_2) \leq \mu_{(\mathcal{U})}(W_1 \cap W_2) + \mu_{(\mathcal{U})}(V_{\mathcal{U}}) + (2a)^{-1}.$$

Since $W_1 \cap W_2 \notin \mathbb{P}_J$, we have $\mu_{(\mathcal{U})}(W_1 \cap W_2) \leq (4a)^{-1}$, and so

$$\mu_{(\mathcal{U})}(V_{\mathcal{U}}) \geq a^{-1} - (2a)^{-1} - (4a)^{-1} = (4a)^{-1},$$

whence $V_{\mathcal{U}} \in \mathbb{P}_J$. Set $W_3 = V_{\mathcal{U}}$ to obtain the required set. □

Lemma 9.45 *Suppose that $\operatorname{cof} J > \aleph_0$, and let (W_j) be a sequence in $\mathbb{P}_J$ with $W_k \subset W_j$ for $j \leq k$ in $\mathbb{N}$. Then there exists $W \in \mathbb{P}_J$ with $W \subset W_j \quad (j \in \mathbb{N})$.*

Proof There exists a sequence $(W_a : a \in \mathbb{N}_{(\mathcal{U})}) \in \mathbf{Z}(Y)_{(\mathcal{U})}$ taking the prescribed values W_j for $j \in \mathbb{N}$: take $G : \mathbb{N} \to \mathbf{Z}(X)$ with

$$G_{\mathcal{U}}(k) = W_k \quad (k \in \mathbb{N}),$$

and set $W_a = G_{\mathcal{U}}(a) \quad (a \in \mathbb{N}_{(\mathcal{U})})$.

For each $j \in \mathbb{N}$, there exists $b_j \in J$ with $\mu_{(\mathcal{U})}(W_j) \geq b_j^{-1}$. Since $\operatorname{cof} J > \aleph_0$, there exists $b \in J$ with $\mu_{(\mathcal{U})}(W_j) \geq b^{-1} \quad (j \in \mathbb{N})$. By Lemma 9.42(i), there exists $a_1 \in \mathbb{N}_{(\mathcal{U})} \setminus \mathbb{N}$ such that $\mu_{(\mathcal{U})}(W_d) \geq b^{-1} \quad (d \leq a_1)$. By Lemma 9.42(ii), there exists $a_2 \in \mathbb{N}_{(\mathcal{U})} \setminus \mathbb{N}$ with $W_d \subset W_c$ whenever $c \leq d \leq a_2$. Define $a = \min\{a_1, a_2\}$, and set $W = W_a$. Since $\mu_{(\mathcal{U})}(W) \geq b^{-1}$, we have $W \in \mathbb{P}_J$. Since $a >> \mathbb{N}$ in $\mathbb{N}_{(\mathcal{U})}$, we have $W \subset W_j \quad (j \in \mathbb{N})$. □

Lemma 9.46 *Let $a \in J$, and suppose that $(W_b : b \leq a) \in \mathbf{Z}(Y)^a_{(\mathcal{U})}$ and that $W \in \mathbb{P}_J$. Then either there exists $W' \in \mathbb{P}_J$ with $W' \subset W$ such that*

$$W' \cap W_b = \emptyset \quad (b \leq a)$$

or there exists $W' \in \mathbb{P}_J$ with $W' \subset W$ such that $W' \subset W_b$ for some $b \leq a$.

Proof Set $U = \bigcup\{W_b : b \leq a\}$. By Lemma 9.31, $U \in \mathbf{Z}(Y)_{(\mathcal{U})}$.

First, suppose that $U \cap W \notin \mathbb{P}_J$. By Lemma 9.44, there exists $W' \in \mathbb{P}_J$ with $W' \subset W \setminus U$, and hence such that $W' \cap W_b = \emptyset \quad (b \leq a)$.

Second, suppose that $U \cap W \in \mathbb{P}_J$, and fix $c \in J$ with $\mu_{(\mathcal{U})}(U \cap W) \geq c^{-1}$. By Lemma 9.40, there exists $b \leq a$ with $\mu_{(\mathcal{U})}(U \cap W \cap W_b) \geq (ca)^{-1}$. Set $W' = U \cap W \cap W_b \subset W_b$. Since J is multiplicatively closed, we have $ca \in J$, and so $W' \in \mathbb{P}_J$. □

We shall discuss the quite standard notions of dense subsets and filters on a partially ordered set. These concepts lie at the base of the theory of forcing. Indeed, let $(P, \leq)$ be a partially ordered set. A subset S of P is:

(i) *dense* in $(P, \leq)$ if, for each $a \in P$, there exists $b \in S$ with $b \leq a$;

(ii) a *prefilter* if, for each $a, b \in S$, there exists $c \in S$ with $c \leq a$ and $c \leq b$;

(iii) a *filter* if it is a prefilter and if, for each $a \in S$ and $b \in P$ with $a \leq b$, it follows that $b \in S$.

(iv) a *maximal* filter if S is maximal in the family of filters on $(P, \leq)$ with respect to inclusion.

We now continue with J a multiplicatively closed initial segment of $\mathbb{N}_{(\mathcal{U})}$, as in Definition 9.43, and consider the partially ordered set $(\mathbb{P}_J, \subset)$. We shall begin by defining some dense subsets of $\mathbb{P}_J$.

For $x, y \in I_{(\mathcal{U})}$ with $x \neq y$ and for $c \in J$, define

$$D^c_{x,y} = \{W \in \mathbb{P}_J : \text{there exists } \ d \leq c \ \text{ such that } \ f(d,x) \neq f(d,y) \quad (f \in W)\}.$$

Recall that, for a z-ultrafilter $\mathcal{F}$ on X containing $\mathcal{F}_0$,

$$\mathcal{F}_\mathcal{U} = \{Z_\mathcal{U} : Z \in \mathcal{F}\}$$

is a maximal $\mathbf{Z}(Y)_{(\mathcal{U})}$-filter on $Y_{(\mathcal{U})}$. A function $H_\mathcal{F} : \mathbb{N}_{(\mathcal{U})} \times I_{(\mathcal{U})} \to \{0,1\}$ has been defined. In the present situation, we see that

$$H_\mathcal{F}(d,x) \neq H_\mathcal{F}(d,y)$$

for some $d \leq c$ whenever there exists $W \in \mathcal{F}_{(\mathcal{U})}$ with $W \in D^c_{x,y}$.

Lemma 9.47 *For each $x, y \in I_{(\mathcal{U})}$ with $x \neq y$ and for each $c \in J$ with $2^c \notin J$, the set $D^c_{x,y}$ is dense in $\mathbb{P}_J$.*

Proof Take $W_0 \in \mathbb{P}_J$, and choose $a \in J$ with $\mu_{(\mathcal{U})}(W_0) \geq a^{-1}$. By Lemma 9.41, there exists $d_0 \leq c$ such that

$$\mu_{(\mathcal{U})}(W) \geq (a^{-1} - 2^{-c})/c,$$

where

$$W = \{f \in W_0 : f(d_0,x) \neq f(d_0,y)\}.$$

Certainly $W \subset W_0$. Since $2a \in J$ and $2^c \notin J$, we have $2a < 2^c$, and so $(a^{-1} - 2^{-c})/c \geq (2ac)^{-1}$. Since J is multiplicatively closed, $2ac \in J$, and so $W \in \mathbb{P}_J$, whence $W \in D^c_{x,y}$. This gives the result. □

For $a \in J$ and $s = (W_b : b \in \mathbb{N}_{(\mathcal{U})}) \in \mathbf{Z}(Y)^{\mathbb{N}}_{(\mathcal{U})}$, define

$$D_{a,s} = \{W \in \mathbb{P}_J : \ \text{ either } \ W \cap W_b = \emptyset \quad (b \leq a) \ \text{ or } \ W \subset W_b \ \text{ for some } \ b \leq a\}.$$

Lemma 9.48 *For each $s \in \mathbf{Z}(Y)^{\mathbb{N}}_{(\mathcal{U})}$ and each $a \in J$, $D_{a,s}$ is dense in $\mathbb{P}_J$.*

Proof This is immediate from Lemma 9.46. □

Definition 9.49 *The family* $\mathbb{D}_J$ *of subsets of* $\mathbb{P}_J$ *consists of the sets* $D^{\mathrm{c}}_{x,y}$ *for* $x, y \in I_{(\mathcal{U})}$ *with* $x \neq y$ *and* $c \in J$ *with* $2^{\mathrm{c}} \notin J$, *together with the sets* $D_{a,s}$ *for* $a \in J$ *and* $s \in \mathbf{Z}(Y)^{\mathbb{N}}_{(\mathcal{U})}$.

Theorem 9.50 *Let* J *be a multiplicatively closed initial segment of* $\mathbb{N}_{(\mathcal{U})}$ *such that* $2^{\mathrm{c}} \notin J$ *for some* $c \in J$. *Assume that* $\mathcal{G}$ *is a filter in* $\mathbb{P}_J$ *such that* $\mathcal{G} \cap D \neq \emptyset$ *for each* $D \in \mathbb{D}_J$. *Then there is a z-ultrafilter* $\mathcal{F}$ *on* X *containing* $\mathcal{F}_0$ *such that:*

(i) $\mathcal{G} = \mathcal{F}_{\mathcal{U}}$;

(ii) $\mathcal{F}_{\mathcal{U}}$ *is* J*-complete;*

(iii) *there is a unique prime ideal* P *in* $C(\beta X)$ *with* $Q_{\mathcal{F}} \subsetneq P$ *and* $\mathbb{N}_{(P)} = J$, *and* P *is both a cut prime and a valuation prime;*

(iv) (CH) $|\Gamma_P| = \left|\mathbb{N}_{(P)}\right| = \aleph_1$;

(v) $w(K_P) \geq |I|$.

Proof Take $W \in \mathbf{Z}(Y)_{(\mathcal{U})}$ with $W \notin \mathcal{G}$, and take

$$s = (W_b : b \in \mathbb{N}_{(\mathcal{U})}) \in \mathbf{Z}(Y)^{\mathbb{N}}_{(\mathcal{U})}$$

to be such that $W_1 = W$. Since $\mathcal{G} \cap D_{1,s} \neq \emptyset$, there exists $U \in \mathcal{G}$ such that either $U \cap W = \emptyset$ or $U \subset W$. Since $W \notin \mathcal{G}$, the alternative $U \cap W = \emptyset$ holds. It follows that $\mathcal{G}$ is a maximal $\mathbf{Z}(Y)_{(\mathcal{U})}$-filter.

There is a (unique) z-ultrafilter $\mathcal{F}$ on X containing $\mathcal{F}_0$ such that $\mathcal{G} = \mathcal{F}_{\mathcal{U}}$. Since $\mathcal{F}_{\mathcal{U}} \cap D_{a,s} \neq \emptyset$ for each $a \in J$ and $s \in \mathbf{Z}(Y)^{\mathbb{N}}_{(\mathcal{U})}$, it follows that $\mathcal{F}_{\mathcal{U}}$ is J-complete. By Lemma 9.34, the prime ideal P specified in (iii) exists; $Q_{\mathcal{F}} \neq P$ because $2^{\mathrm{c}} \notin J$. Clause (iv) follows as in Theorem 9.36(ii).

Since $Q_{\mathcal{F}}$ is a valuation prime,

$$\{0,1\}^{\mathbb{N}}_{(P)} = \{f \mid J : f \in \{0,1\}^{\mathbb{N}}_{(\mathcal{F})}\}$$

by Lemma 9.7. It follows from Lemma 9.35 that, for each section h of $H_{\mathcal{F}}$, there exists $f \in \{0,1\}^{\mathbb{N}}_{(\mathcal{F})}$ with $f \mid \mathbb{N}_{(\mathcal{U})} = h$, and so $h \restriction c \in \{0,1\}^{c}_{(P)}$.

For each $x \in I_{(\mathcal{U})}$, let

$$h_x(a) = H_{\mathcal{F}}(a, x) \quad (a \in \mathbb{N}_{(\mathcal{U})}),$$

so that h_x is a section of $H_{\mathcal{F}}$. If $x, y \in I_{(\mathcal{U})}$ with $x \neq y$, then $\mathcal{F}_{\mathcal{U}} \cap D^{\mathrm{c}}_{x,y} \neq \emptyset$, and so $h_x \restriction c \neq h_y \restriction c$. Hence

$$|I| \leq \left|\{0,1\}^{c}_{(P)}\right| \leq w(K_P),$$

giving clause (v). □

Let J be as in the above theorem, and assume that CH holds. To find an example of a super-real field K_P with $|\Gamma_P| = \aleph_1$ and $w(K_P) \geq \aleph_2$, it suffices to find a filter $\mathcal{G}$ in $\mathbb{P}_J$ with $\mathcal{G} \cap D \neq \emptyset$ for each $D \in \mathbb{D}_J$ in the case where $|I| = \aleph_2$. (In the case where $|I| = \aleph_1$, the filter $\mathcal{G}$ is easily constructed.) It follows from Theorem 4.28(ii) and Lemma 8.33(v) that the corresponding super-real field K_P is an η_1-field provided that $\operatorname{cof} J \geq \aleph_1$, and we shall restrict ourselves to this case. Even with this restriction, the construction of such a filter $\mathcal{G}$ is far from straightforward. We are in fact forced to use a forcing argument, and so we can only obtain the result that it is relatively consistent with the theory ZFC + CH that such a filter exists. The success of the forcing argument depends on the fact that the partially ordered set $\mathbb{P}_J$ satisfies the '$\aleph_2$-chain condition', as we shall now establish.

To prove this, we shall vary the index set I that arose in our previous constructions, and so we must modify our notation to indicate the dependence of the various structures that we have built on the set I; we write $X(I), Y(I)$, and $\mu_I^{(\mathcal{U})}$, for example.

Suppose, in fact, that σ and τ are index sets with $\sigma \subset \tau$. Then

$$\rho_{\tau,\sigma} : Y(\tau) \to Y(\sigma)$$

is the continuous map defined by

$$\rho_{\tau,\sigma}(f) = f \mid (\mathbb{N} \times \sigma) \quad (f \in Y(\tau)),$$

and

$$\pi_{\tau,\sigma} : X(\tau) \to X(\sigma)$$

is the corresponding continuous map. The product measure on $Y(\sigma)$ is μ_σ.

We start with some rather elementary observations. The first two are trivial verifications.

Lemma 9.51 *Suppose that σ is a subset of τ. Then:*

(i) *$\rho_{\tau,\sigma}(Z) \in \mathbf{Z}(Y(\sigma))$ for $Z \in \mathbf{Z}(Y(\tau))$ and also $\pi_{\tau,\sigma}(Z) \in \mathbf{Z}(X(\sigma))$ for $Z \in \mathbf{Z}(X(\tau))$;*

(ii) *$\mu_\tau(\rho_{\tau,\sigma}^{-1}(Z)) = \mu_\sigma(Z)$ for $Z \in \mathbf{Z}(Y(\sigma))$ and also $\mu_\sigma(\rho_{\tau,\sigma}(Z)) \geq \mu_\tau(Z)$ for $Z \in \mathbf{Z}(Y(\tau))$.* □

Let $\rho_{\tau,\sigma}^{(\mathcal{U})} : Y(\tau)_{(\mathcal{U})} \to Y(\sigma)_{(\mathcal{U})}$ be the function corresponding to $\rho_{\tau,\sigma}$; in fact,

$$\rho_{\tau,\sigma}^{(\mathcal{U})}(F) = F \mid \left(\mathbb{N}_{(\mathcal{U})} \times \sigma_{(\mathcal{U})}\right) \quad (F \in Y(\tau)_{(\mathcal{U})}).$$

Lemma 9.52 *Suppose that σ is a subset of τ. Then:*

(i) *$\rho_{\tau,\sigma}^{(\mathcal{U})}(Z_\mathcal{U}) = (\pi_{\tau,\sigma}(Z))_\mathcal{U}$ for $Z \in \mathbf{Z}(X(\tau))$;*

(ii) *$(\rho_{\tau,\sigma}^{(\mathcal{U})})^{-1}(Z_\mathcal{U}) = \left(\pi_{\tau,\sigma}^{-1}(Z)\right)_\mathcal{U}$ for $Z \in \mathbf{Z}(X(\sigma))$.* □

Lemma 9.53 *Suppose that σ is a subset of τ. Then:*

(i) *$\mu_\sigma^{(\mathcal{U})}\left(\rho_{\tau,\sigma}^{(\mathcal{U})}(W)\right) \geq \mu_\tau^{(\mathcal{U})}(W)$ for $W \in \mathbf{Z}(Y(\tau))_{(\mathcal{U})}$;*

(ii) *$\mu_\tau^{(\mathcal{U})}\left((\rho_{\tau,\sigma}^{(\mathcal{U})})^{-1}(W)\right) = \mu_\sigma^{(\mathcal{U})}(W)$ for $W \in \mathbf{Z}(Y(\sigma))_{(\mathcal{U})}$.*

Proof (i) Take $W \in \mathbf{Z}(Y(\tau))_{(\mathcal{U})}$, and choose $Z \in \mathbf{Z}(Y(\tau))$ with $Z_\mathcal{U} = W$. By Lemma 9.52(i), $\rho_{\tau,\sigma}^{(\mathcal{U})}(W) = (\pi_{\tau,\sigma}(Z))_\mathcal{U}$. Now $\mu_\tau^{(\mathcal{U})}(W) = [f]_\mathcal{U}$, where

$$f(k) = \mu_{\tau,k}(Z \cap Y_k(\tau)) \quad (k \in \mathbb{N}).$$

Set

$$h(k) = \mu_{\sigma,k}(\pi_{\tau,\sigma}(Z \cap Y_k(\tau))) \quad (k \in \mathbb{N}).$$

We have $f(k) \le h(k) \quad (k \in \mathbb{N})$, and so $[f]_{\mathcal{U}} \le [h]_{\mathcal{U}}$. Also

$$\mu_\sigma^{(\mathcal{U})}(\rho_{\tau,\sigma}^{(\mathcal{U})}(W)) = \mu_\sigma^{(\mathcal{U})}((\pi_{\tau,\sigma}(Z))_{\mathcal{U}}) = [h]_{\mathcal{U}},$$

and so $\mu_\tau^{(\mathcal{U})}(W) \le \mu_\sigma^{(\mathcal{U})}(\rho_{\tau,\sigma}^{(\mathcal{U})}(W))$.

(ii) This is similar. □

The next lemma is in essence simply an assertion of the standard fact that zero sets in $Y(\tau)$ or $X(\tau)$ have countable support.

Lemma 9.54 *Let τ be an index set.*

(i) *Suppose that $Z \in \mathbf{Z}(Y(\tau))$. Then there is a countable subset σ of τ such that $Z = \rho_{\tau,\sigma}^{-1}(\rho_{\tau,\sigma}(Z))$.*

(ii) *Suppose that $Z \in \mathbf{Z}(X(\tau))$. Then there is a countable subset σ of τ such that $Z = \pi_{\tau,\sigma}^{-1}(\pi_{\tau,\sigma}(Z))$.*

Proof (i) Every zero set in $Y(\tau)$ is a countable intersection of finite unions of sets of the form $\mathcal{N}_G$, where $G \in \mathrm{Fin}(\mathbb{N} \times \tau, \{0,1\})$. Thus

$$Z = \bigcap \left\{ \bigcup \{\mathcal{N}_{G_{i,j}} : i = 1, \dots, j\} : j \in \mathbb{N} \right\},$$

say, where each $G_{i,j} \in \mathrm{Fin}(\mathbb{N} \times \tau, \{0,1\})$. Set

$$\sigma = \bigcup \{\operatorname{dom} G_{i,j} : i = 1, \dots, j,\ j \in \mathbb{N}\}.$$

Then σ is countable, and $Z = \rho_{\tau,\sigma}^{-1}(\rho_{\tau,\sigma}(Z))$.

(ii) This follows from (i) because $X(\tau) = \mathbb{N} \times Y(\tau)$. □

The version of the above lemma relative to the ultrafilter $\mathcal{U}$ follows easily.

Lemma 9.55 *Let τ be an index set, and suppose that $W \in \mathbf{Z}(Y(\tau))_{(\mathcal{U})}$. Then there is a countable subset σ of τ such that*

$$W = (\rho_{\tau,\sigma}^{(\mathcal{U})})^{-1}(\rho_{\tau,\sigma}^{(\mathcal{U})}(W)).$$

Proof Choose $Z \in \mathbf{Z}(X(\tau))$ with $W = Z_{\mathcal{U}}$. By Lemma 9.54(ii), there is a countable set $\sigma \subset \tau$ with $Z = \pi_{\tau,\sigma}^{-1}(\pi_{\tau,\sigma}(Z))$. The result follows from Lemma 9.52. □

Lemma 9.56 *Let τ be an index set, and take subsets σ_1 and σ_2 of τ. Suppose that $Z_1, Z_2 \in \mathbf{Z}(Y(\tau))$ are such that*

$$Z_j = \rho_{\tau,\sigma_j}^{-1}(\rho_{\tau,\sigma_j}(Z_j)) \quad (j = 1, 2) \text{ and } \rho_{\tau,\sigma_1 \cap \sigma_2}(Z_1) = \rho_{\tau,\sigma_1 \cap \sigma_2}(Z_2).$$

Then $\mu_\tau(Z_1 \cap Z_2) \ge \mu_\tau(Z_1)\mu_\tau(Z_2)$.

Proof This is immediate because μ_τ is a product measure on $Y(\tau)$. □

Again, we require a version of the above lemma relative to the ultrafilter $\mathcal{U}$.

Lemma 9.57 *Let τ be an index set, and take subsets σ_1 and σ_2 of τ. Suppose that $W_1, W_2 \in \mathbf{Z}(Y(\tau))_{(\mathcal{U})}$ are such that*

$$W_j = (\rho^{(\mathcal{U})}_{\tau,\sigma_j})^{-1}(\rho^{(\mathcal{U})}_{\tau,\sigma_j}(W_j)) \quad (j = 1,2) \text{ and } \rho^{(\mathcal{U})}_{\tau,\sigma_1\cap\sigma_2}(W_1) = \rho^{(\mathcal{U})}_{\tau,\sigma_1\cap\sigma_2}(W_2)\,.$$

Then $\mu^{(\mathcal{U})}_\tau(W_1 \cap W_2) \geq \mu^{(\mathcal{U})}_\tau(W_1)\mu^{(\mathcal{U})}_\tau(W_2)$.

Proof For $j = 1, 2$, choose $Z_j \in \mathbf{Z}(X(\tau))$ such that $W_j = (Z_j)_{\mathcal{U}}$, and define $Z_j^k = Z_j \cap Y_k(\tau) \quad (k \in \mathbb{N})$. By Lemma 9.52,

$$(Z_j)_{\mathcal{U}} = \left(\pi^{-1}_{\tau,\sigma_j}\left(\pi_{\tau,\sigma_j}(Z_j)\right)\right)_{\mathcal{U}} \quad (j = 1,2)\,,$$

and so the set of $k \in \mathbb{N}$ such that

$$\pi^{-1}_{\tau,\sigma_j}\left(\pi_{\tau,\sigma_j}(Z_j^k)\right) = Z_j^k \quad (j = 1,2)$$

belongs to $\mathcal{U}$; we may suppose that this equality holds for all $k \in \mathbb{N}$. Set $\sigma = \sigma_1 \cap \sigma_2$. We may also suppose that $\pi_{\tau,\sigma}(Z_1^k) = \pi_{\tau,\sigma}(Z_2^k) \quad (k \in \mathbb{N})$.

By Lemma 9.56,

$$\mu_{\tau,k}(Z_1^k \cap Z_2^k) \geq \mu_{\tau,k}(Z_1^k)\mu_{\tau,k}(Z_2^k) \quad (k \in \mathbb{N})\,.$$

We recall that, for each $Z \in \mathbf{Z}(X(\tau))$, we have $\mu^{(\mathcal{U})}_\tau(Z_{\mathcal{U}}) = [f]_{\mathcal{U}}$, where

$$f(k) = \mu_{\tau,k}(Z \cap Y_k(\tau)) \quad (k \in \mathbb{N})\,.$$

Thus we have established the result. □

We shall require the following standard combinatorial lemma, called the Δ-*system lemma.*

Lemma 9.58 (CH) *Let I be an index set, and let $\mathcal{A}$ be a family of countable subsets of I with $|\mathcal{A}| = \aleph_2$. Then there is a countable subset σ of I and a subfamily $\mathcal{B}$ of $\mathcal{A}$, with $|\mathcal{B}| = \aleph_2$ such that, for each $\sigma_1, \sigma_2 \in \mathcal{B}$ with $\sigma_1 \neq \sigma_2$, necessarily $\sigma_1 \cap \sigma_2 \subset \sigma$.* □

We finally return to the partially ordered set $\mathbb{P}_J$; this set is now denoted by $\mathbb{P}_J^I$ when the index set which is implicit in its definition is I. We continue with a free ultrafilter $\mathcal{U}$ on $\mathbb{N}$.

Lemma 9.59 (CH) *Let J be a multiplicatively closed initial segment of $\mathbb{N}_{(\mathcal{U})}$, and let I be an index set. Suppose that S is a subset of $\mathbb{P}_J^I$ of cardinality $\aleph_2$. Then there exist $W_1, W_2 \in S$ such that $W_1 \cap W_2 \in \mathbb{P}_J^I$.*

Proof By Lemma 9.55, for each $W \in S$, there is a countable subset σ_W of I such that

$$W = \left(\rho^{(\mathcal{U})}_{I,\sigma_W}\right)^{-1}\left(\rho^{(\mathcal{U})}_{I,\sigma_W}(W)\right)\,.$$

Take $\mathcal{A} = \{\sigma_W : W \in S\}$, so that $|\mathcal{A}| \leq \aleph_2$. It follows from Lemma 9.58 in the case where $|\mathcal{A}| = \aleph_2$, and is immediate in the case where $|\mathcal{A}| < \aleph_2$, that there is a countable subset σ of I and a subset S' of S with $|S'| = \aleph_2$ such that $\sigma_{W_1} \cap \sigma_{W_2} \subset \sigma$ whenever $W_1, W_2 \in S'$ with $W_1 \neq W_2$.

By CH, $|\mathbf{Z}(Y(\sigma))_{(\mathcal{U})}| = \aleph_1$, and so

$$\left|\{\rho^{(\mathcal{U})}_{I,\sigma}(W) : W \in S'\}\right| \leq \aleph_1 .$$

Thus there exists $S'' \subset S'$ such that $|S''| = \aleph_2$ and

$$\rho^{(\mathcal{U})}_{I,\sigma}(W_1) = \rho^{(\mathcal{U})}_{I,\sigma}(W_2)$$

whenever $W_1, W_2 \in S''$ with $W_1 \neq W_2$.

By Lemma 9.57, we have

$$\mu^{(\mathcal{U})}_I(W_1 \cap W_2) \geq \mu^{(\mathcal{U})}_I(W_1)\mu^{(\mathcal{U})}_I(W_2) \quad (W_1, W_2 \in S'') .$$

Since J is multiplicatively closed, it follows that $W_1 \cap W_2 \in \mathbb{P}^I_J$ whenever $W_1, W_2 \in S''$, and this is a stronger result than the one we are proving. □

Let J be a multiplicatively closed initial segment of $\mathbb{N}_{(\mathcal{U})}$ such that $\operatorname{cof} J \geq \aleph_1$, and let I be an index set. We have established two key properties (in the theory ZFC + CH) of the partially ordered set $\mathbb{P}^I_J$.

• Suppose that $(W_j : j \in \mathbb{N})$ is a decreasing sequence in $\mathbb{P}^I_J$. Then there exists $W \in \mathbb{P}^I_J$ with $W \subset W_j \quad (j \in \mathbb{N})$.

• Suppose that S is a subset of $\mathbb{P}^I_J$ of cardinality $\aleph_2$. Then there exist $W_1, W_2 \in S$ such that $W_1 \cap W_2 \in \mathbb{P}^I_J$.

The first property says that the partially ordered set $\mathbb{P}^I_J$ is *ω-closed*; it was established in Lemma 9.45. The second property says that $\mathbb{P}^I_J$ satisfies the *$\aleph_2$-chain condition*; it was established in Lemma 9.59.

It is now an easy argument using the method of forcing to show from these two properties that it is relatively consistent with the theory ZFC + GCH that there is a filter $\mathcal{G} \subset \mathbb{P}^I_J$ such that $\mathcal{G} \cap D \neq \emptyset$ for each $D \in \mathbb{D}^I_J$. In the terminology of forcing, one simply forces with the partial order $\mathbb{P}^I_J$ itself. As we explained in the introduction, we do not give the details of forcing arguments. The generic filter so created is the desired filter $\mathcal{G}$. It follows from the above two properties that the index set I still has cardinality $\aleph_2$ in this generic extension (by GCH). The key point is that, since $\mathbb{P}^I_J$ is ω-closed, the ultrafilter $\mathcal{U}$ is still an ultrafilter in the generic extension, and the structures $\mathbb{P}^I_J$ and $\mathbb{D}^I_J$ defined in the generic extension are naturally isomorphic to the corresponding structures which were defined in the ground model. The spaces X and Y are different when calculated in this generic extension, but the corresponding spaces $C^b(X)$ and $C^b(Y)$ of bounded, continuous functions on X and Y are isomorphic to their respective original versions.

Thus, we have at last established the following result, provided that the above forcing argument be accepted. Let $\mathcal{U}$ be any free ultrafilter on $\mathbb{N}$. By Proposition 8.37(ii), there is a multiplicatively closed initial segment of $\mathbb{N}_{(\mathcal{U})}$ such that $\operatorname{cof} J = \aleph_1$ and such that there exists $c \in J$ with $2^c \notin J$. Let I be an index set of cardinality $\aleph_2$. The forcing argument provides a filter $\mathcal{G}$ in $\mathbb{P}^I_J$ such that $\mathcal{G} \cap D \neq \emptyset$ for each $D \in \mathbb{D}^I_J$. Thus, by Theorem 9.50, there is a prime ideal P in $C(\beta X)$ with the properties given in the final theorems of this chapter.

Theorem 9.60 *It is relatively consistent with the theory* ZFC + GCH *that there is a prime ideal P in a space of the form $C(\beta X)$ for a completely regular space X such that P is both a cut prime and a valuation prime and such that the super-real field K_P is an η_1-field with*

$$w(K_P) = \aleph_2 \quad \text{and} \quad |\Gamma_P| = \aleph_1 . \qquad \square$$

Theorem 9.61 *It is relatively consistent with the theory* ZFC + GCH *that there is a prime ideal P that is both a cut prime and a valuation prime such that K_P is a super-real η_1-field, but which is such that K_P is not isomorphic as a totally ordered group to any super-real field of the form K_Q, where Q is an exponential prime.*

Proof Let P be the prime ideal specified in Theorem 9.60, so that P is a cut prime and a valuation prime, K_P is an η_1-field, and $w(K_P) > |\Gamma_P|$. Assume that

$$(K_P, +) \simeq (K_Q, +),$$

where Q is an exponential prime. Then $w(K_Q) = w(K_P)$ and $|\Gamma_Q| = |\Gamma_P|$. But, by Theorem 6.29, $w(K_Q) = |\Gamma_Q|$, and so we have reached a contradiction. $\square$

Thus, in the language of Chapter 4, it cannot be proved in ZFC + GCH that the class **cut prime** $\boldsymbol{\eta_1}$ is contained in **exponential** $\boldsymbol{\eta_1}$, and, in particular, it cannot be proved in ZFC + GCH that the classes **exponential** $\boldsymbol{\eta_1}$ and **super-real** $\boldsymbol{\eta_1}$ are equal. *A fortiori*, it cannot be proved in ZFC + GCH that every super-real η_1-field is isomorphic to some hyper-real field.

We summarize some of the information that we have acquired at this stage in Fig. 9.2, on the following page.

Notes 9.62 The technique of forcing is extensively discussed in various texts. See, for example, Dales and Woodin (1987, Chapter 7), Jech (1978, Chapter 3), and Kunen (1980, Chapter VII).

The Δ-system lemma given in Lemma 9.58 is Kunen (1980, II, 1.6), for example.

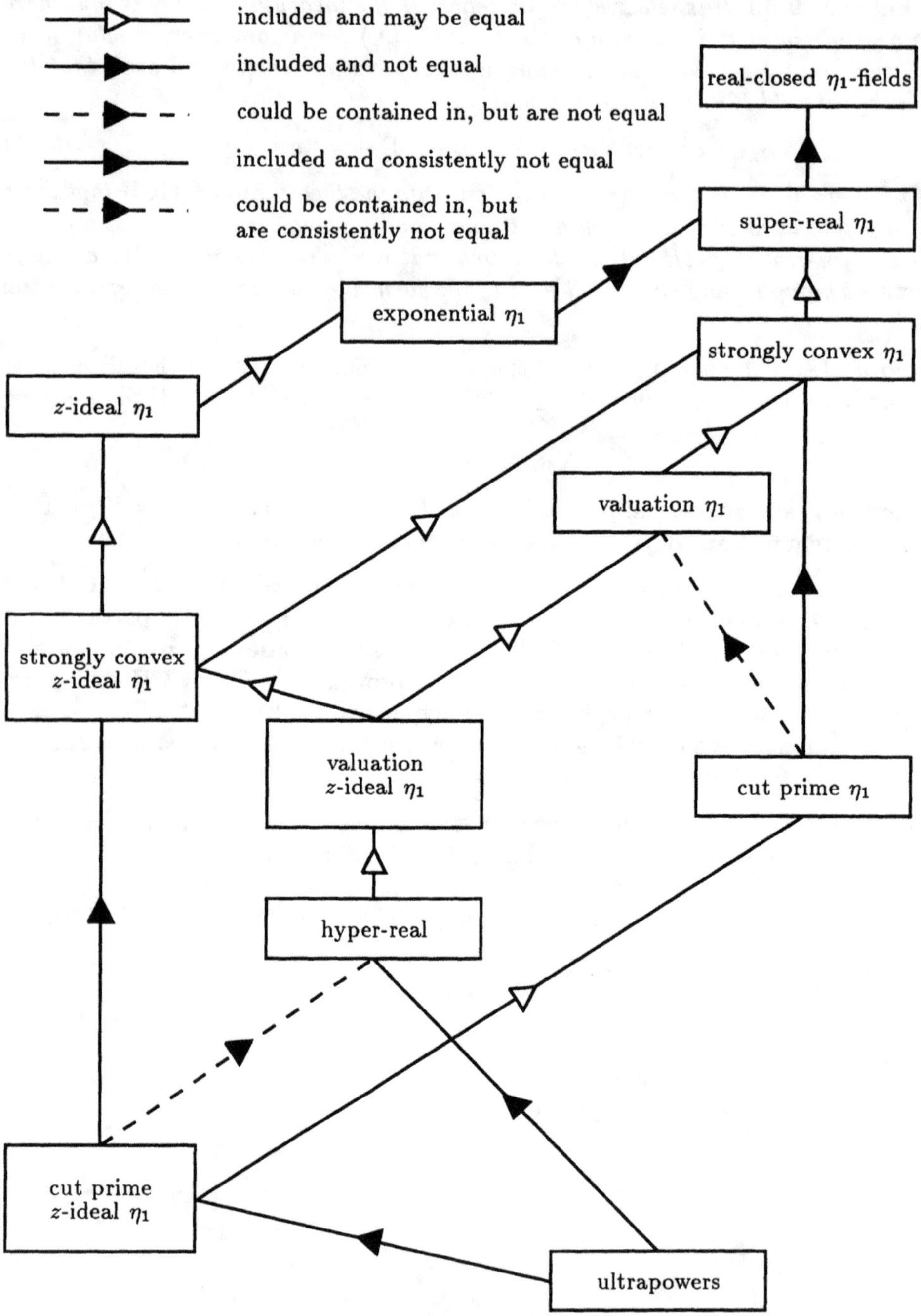

Fig. 9.2. The relationships between classes

10

Models and weak Cauchy completeness

Let P be a non-maximal, prime ideal in an algebra $C(\Omega)$, where Ω is a compact space, and let K_P be the corresponding super-real field. In the previous two chapters, we have associated various additional structures with the field K_P; for example, we introduced the structures $\mathbb{M}_{(P)}$, $\mathbb{N}_{(P)}$, $\mathbb{N}^{<\mathbb{N}}_{(P)}$, and $\{0,1\}^{<\mathbb{N}}_{(P)}$ in Chapter 8, and we studied the 'infinite' structure $\{0,1\}^{\mathbb{N}}_{(P)}$ in Chapter 9. Further, we have obtained a number of results by using these structures; in Chapter 8 we obtained the gap theorem for super-real semi-η_1-fields, Theorem 8.61, and in Chapter 9 we characterized Cauchy complete fields in Theorem 9.3 in terms of $\{0,1\}^{\mathbb{N}}_{(P)}$. It is a natural question to ask whether these additional structures are really necessary for an understanding of the super-real fields K_P. We shall show in Chapter 11 that these structures are indeed unique in some cases, and so a classification of super-real fields must presuppose a prior classification of these additional structures.

Our first step towards an understanding of the additional structures will be taken in this chapter, where we investigate $\{0,1\}^{\mathbb{N}}_{(P)}$. As we remarked, it is not always the case that $f_1 \vee f_2$ and $f_1 \wedge f_2$ belong to $\{0,1\}^{\mathbb{N}}_{(P)}$ whenever f_1 and f_2 belong to $\{0,1\}^{\mathbb{N}}_{(P)}$. This is a serious failure. To obviate this difficulty, we shall discuss a closely related structure ${}^*\{0,1\}^{\mathbb{N}}_{(P)}$. We shall prove in Theorem 10.23 that ${}^*\{0,1\}^{\mathbb{N}}_{(P)} \subset \{0,1\}^{\mathbb{N}}_{(P)}$, and in Theorem 10.24 that ${}^*\{0,1\}^{\mathbb{N}}_{(P)}$ is closed under the operations $\vee$ and $\wedge$. The possibility that ${}^*\{0,1\}^{\mathbb{N}}_{(P)} \neq \{0,1\}^{\mathbb{N}}_{(P)}$ is a manifestation of the difficulties that can arise because of the shortage of continuous functions on certain compact spaces Ω. It is natural to distinguish the prime ideals P such that we do have the equality

$$ {}^*\{0,1\}^{\mathbb{N}}_{(P)} = \{0,1\}^{\mathbb{N}}_{(P)} \,. \tag{10.1}$$

However ${}^*\{0,1\}^{\mathbb{N}}_{(P)}$ is not defined solely in terms of the field K_P, but rather in terms of a 'structure' $\mathfrak{N}_{(P)}$ associated with P. The structure $\mathfrak{N}_{(P)}$ is a model of a fragment of analysis, and so, to define $\mathfrak{N}_{(P)}$, we must first briefly discuss the notion of a 'model'.

We shall then say that the structure $\mathfrak{N}_{(P)}$ is weakly Cauchy complete if (10.1) holds; we shall show in Proposition 10.14 that, for a cut prime P, the structure $\mathfrak{N}_{(P)}$ is weakly Cauchy complete, but there are even hyper-real fields whose associated structure is not weakly Cauchy complete, as we shall show in Theorem 10.40. The examples of the final chapter, which will distinguish between some more of our classes, will be super-real fields derived from prime ideals P for which $\mathfrak{N}_{(P)}$ is weakly Cauchy complete.

To give a background in model theory, we shall discuss briefly two particular models, those of number theory (or arithmetic) and of analysis (or second-order number theory).

A model of number theory is, first, a structure

$$\mathfrak{N} = \langle \mathbb{N}_{(\mathfrak{N})}, +, \cdot \rangle ;$$

here, $\mathbb{N}_{(\mathfrak{N})} = \mathbb{Z}^+_{(\mathfrak{N})} \setminus \{0\}$, where

$$\langle \mathbb{Z}_{(\mathfrak{N})}, +, \cdot \rangle$$

is a totally ordered ring with identity 1 such that 1 is the minimum element of $\mathbb{N}_{(\mathfrak{N})}$. The model $\mathfrak{N}$ has associated with it a collection of sets, called the *definable sets* of $\mathfrak{N}$. A definable set is a subset of $(\mathbb{N}_{(\mathfrak{N})})^k$ for some $k \in \mathbb{N}$; these sets are defined either in the language of logic or as the minimal family satisfying various algebraic conditions. *Peano's axiom* (or the *axiom of induction*) for $\mathfrak{N}$ is the statement:

each definable subset of $\mathbb{N}_{(\mathfrak{N})}$ *has a minimum element.*

Thus, a *model of number theory* is defined to be a structure $\mathfrak{N} = \langle \mathbb{N}_{(\mathfrak{N})}, +, \cdot \rangle$, as above, together with its collection $\mathfrak{D}_{(\mathfrak{N})}$ of definable sets, such that Peano's axiom holds.

Two models $\mathfrak{N}$ and $\mathfrak{M}$ are defined to be *isomorphic*, written $\mathfrak{N} \cong \mathfrak{M}$, if there is a bijection $\psi : \mathbb{N}_{(\mathfrak{N})} \to \mathbb{N}_{(\mathfrak{M})}$ such that $\psi(a+b) = \psi(a) + \psi(b)$ and $\psi(a)\psi(b) = \psi(ab)$ for $a, b \in \mathbb{N}_{(\mathfrak{N})}$. Note that ψ induces a map of the definable sets $\mathfrak{D}_{(\mathfrak{N})}$ onto $\mathfrak{D}_{(\mathfrak{M})}$.

The model $\mathfrak{N}$ is a *standard model of number theory* if it is isomorphic to $\langle \mathbb{N}, +, \cdot \rangle$. Since $\mathbb{N}$ is countable, there are only countably many definable sets in this model. The definable sets of the standard model are the sets that can be defined by using first-order logic. A model $\mathfrak{N}$ is *non-standard* if it is not a standard model. For example, let $\mathcal{U}$ be a free ultrafilter on $\mathbb{N}$. Then

$$\langle \mathbb{N}^{\mathbb{N}}/\mathcal{U}, +, \cdot \rangle$$

is a non-standard model of number theory; this is, of course, the classical example of such a model.

Let P be a non-maximal, prime ideal in an algebra $C(\Omega)$. Then the structure

$$\langle \mathbb{N}_{(P)}, +, \cdot \rangle$$

may or may not be a model of number theory; a necessary condition for this is that $\mathbb{N}_{(P)}$ be multiplicatively closed, and then one must ascertain whether or not Peano's axiom holds for this structure. This structure is a model of number theory in the case where P is a z-ideal, which further reinforces the analogy between the classes **z-ideal η_1** and **ultrapowers**.

A model of analysis is, first, a structure

$$\mathfrak{N} = \langle \mathbb{N}_{(\mathfrak{N})}, +, \cdot, \mathcal{P}(\mathbb{N})_{(\mathfrak{N})} \rangle$$

such that $\langle \mathbb{N}_{(\mathfrak{N})}, +, \cdot \rangle$ is a model of number theory and $\mathcal{P}(\mathbb{N})_{(\mathfrak{N})}$ is a certain family of subsets of $\mathbb{N}_{(\mathfrak{N})}$, so that $\mathcal{P}(\mathbb{N})_{(\mathfrak{N})} \subset \mathcal{P}(\mathbb{N}_{(\mathfrak{N})})$. This structure $\mathfrak{N}$ also has associated with it a collection of sets, called the *definable sets* of $\mathfrak{N}$. In this case, a definable set is a subset of $(\mathcal{P}(\mathbb{N})_{(\mathfrak{N})}) \cup \mathbb{N}^k_{(\mathfrak{N})}$ for some $k \in \mathbb{N}$; again these sets are defined either in the language of logic or as the minimal family satisfying various algebraic conditions. (Our convention is that the relation

$$\{(a, A) \in \mathbb{N}_{(\mathfrak{N})} \times \mathcal{P}(\mathbb{N})_{(\mathfrak{N})} : a \in A\}$$

is always part of the structure.) The *axiom of comprehension* for $\mathfrak{N}$ is the statement:

$$A \in \mathcal{P}(\mathbb{N})_{(\mathfrak{N})} \textit{ for each definable subset } A \textit{ of } \mathbb{N}_{(\mathfrak{N})}\,.$$

The *axiom of induction* for $\mathfrak{N}$ is the statement:

$$\textit{each } A \in \mathcal{P}(\mathbb{N})_{(\mathfrak{N})} \textit{ has a minimum element.}$$

Thus, a *model of analysis* is defined to be a structure

$$\mathfrak{N} = \langle \mathbb{N}_{(\mathfrak{N})}, +, \cdot, \mathcal{P}(\mathbb{N})_{(\mathfrak{N})} \rangle$$

as above, together with its collection $\mathcal{C}_{(\mathfrak{N})}$ of definable sets, such that the axioms of comprehension and induction hold.

Two models of analysis $\mathfrak{N}$ and $\mathfrak{M}$ are *isomorphic*, written $\mathfrak{N} \cong \mathfrak{M}$, if there is an isomorphism of the models $\langle \mathbb{N}_{(\mathfrak{N})}, +, \cdot \rangle$ and $\langle \mathbb{N}_{(\mathfrak{M})}, +, \cdot \rangle$ which induces a bijection from $\mathcal{P}(\mathbb{N})_{(\mathfrak{N})}$ onto $\mathcal{P}(\mathbb{N})_{(\mathfrak{M})}$.

The model $\mathfrak{N}$ is a *standard model of analysis* if it is isomorphic to the model $\langle \mathbb{N}, +, \cdot, \mathcal{P}(\mathbb{N}) \rangle$. The definable sets of the standard model are the sets that can be defined by using second-order logic.

In the standard model of analysis, one can define many of the objects that are relevant to analysis on the real line; these objects include the algebra of all continuous functions on $\mathbb{R}$, the hierarchy of Borel sets, and Lebesgue measure on the Borel sets. However, whether or not one can define a well-ordering of the reals in the standard model of analysis is independent of ZFC; for example, in Gödel's constructible universe, a well-ordering of $\mathbb{R}$ is definable, but, if certain large cardinals exist in the universe of sets, then there can be no such well-ordering.

The model $\mathfrak{N}$ is *non-standard* if it is not a standard model. An example of a non-standard model of analysis is implicit in our earlier work. Indeed, let $\mathcal{U}$ be a non-principal ultrafilter on $\mathbb{N}$, and write $\mathbb{N}_{(\mathcal{U})}$ and $\{0,1\}^{\mathbb{N}}_{(\mathcal{U})}$ for $\mathbb{N}_{(Q)}$ and $\{0,1\}^{\mathbb{N}}_{(Q)}$, respectively, where Q is the prime ideal in $C(\beta\mathbb{N})$ corresponding to $\mathcal{U}$, so that $\mathbb{N}_{(\mathcal{U})} = \mathbb{N}^{\mathbb{N}}/\mathcal{U}$, as in Chapter 9. For an element $f \in \{0,1\}^{\mathbb{N}}_{(\mathcal{U})}$, set

$$A_f = \{a \in \mathbb{N}_{(\mathcal{U})} : f(a) = 1\} = f^{-1}(1)\,.$$

Then $\mathcal{P}(\mathbb{N})_{(\mathcal{U})}$ consists of the family

$$\left\{A_f : f \in \{0,1\}^{\mathbb{N}}_{(\mathcal{U})}\right\}$$

of subsets of $\mathbb{N}_{(\mathcal{U})}$, and it is easily checked by pointwise arguments that

$$\mathfrak{N}_{(\mathcal{U})} = \langle\, \mathbb{N}_{(\mathcal{U})}, +, \cdot, \mathcal{P}(\mathbb{N})_{(\mathcal{U})}\, \rangle$$

is a model of analysis; clearly $\mathfrak{N}_{(\mathcal{U})}$ is a non-standard model.

We shall generalize the above example by associating to an arbitrary non-maximal, prime ideal P in an algebra $C(\Omega)$ a structure

$$\mathfrak{N}_{(P)} = \langle\, \mathbb{N}_{(P)}, +, \cdot, \mathcal{P}(\mathbb{N})_{(P)}\, \rangle\,.$$

Here, $\mathcal{P}(\mathbb{N})_{(P)}$ is the collection of subsets of $\mathbb{N}_{(P)}$ whose characteristic functions are in $\{0,1\}^{\mathbb{N}}_{(P)}$. We must immediately caution that in general $\mathfrak{N}_{(P)}$ is not a model of analysis, just as, in general, $\mathbb{N}_{(P)}$ is not multiplicatively closed, and hence $\langle \mathbb{N}_{(P)}, +, \cdot\rangle$ is not a model of number theory; $\mathfrak{N}_{(P)}$ is a model of a weak fragment of analysis that we are not explicitly axiomatizing. It can be shown that, if $\mathfrak{N}_{(P)}$ is a model of analysis, then necessarily P is an exponential prime. The collection of primes P for which $\mathfrak{N}_{(P)}$ is a model of analysis is a very special class; it contains the class of primes which specify ultrapowers, but, for example, the proof in Theorem 9.9 essentially shows that the hyper-real field $\widetilde{\mathbf{R}}$ is not isomorphic to any super-real field K_P such that $\mathfrak{N}_{(P)}$ is a model of analysis. We showed in Theorem 9.26 that, with GCH, there is a prime z-ideal P in some $C(\Omega)$ such that $K_P \cong \widetilde{\mathbf{R}}$. Since P is a z-ideal, $\langle \mathbb{N}_{(P)}, +, \cdot\rangle$ is a model of number theory, but, in this case, $\langle \mathbb{N}_{(P)}, +, \cdot, \mathcal{P}(\mathbb{N})_{(P)}\rangle$ is not a model of analysis. The examples K_P, to be described in Chapter 11, are all such that $\mathfrak{N}_{(P)}$ is a model of analysis.

We begin with a formal definition of our model. In fact, we shall define two variants of our model corresponding to the difference between $\mathbb{M}_{(P)}$ and $\mathbb{N}_{(P)}$ that we discussed earlier.

Definition 10.1 *Let P be a non-maximal, prime ideal in an algebra $C(\Omega)$.*

(i) *The model $\mathfrak{N}_{(P)}$ is the structure*

$$\mathfrak{N}_{(P)} = \langle\, \mathbb{N}_{(P)}, +, \cdot, \mathcal{P}(\mathbb{N})_{(P)}\, \rangle\,,$$

where

$$\mathcal{P}(\mathbb{N})_{(P)} = \{\{a \in \mathbb{N}_{(P)} : f(a) = 1\} : f \in \{0,1\}^{\mathbb{N}}_{(P)}\}\,.$$

(ii) *The model $\mathfrak{M}_{(P)}$ is the structure*

$$\mathfrak{M}_{(P)} = \left\langle \mathbb{M}_{(P)}, \mathbb{N}_{(P)}, +, \cdot, \exp_{(P)}, \mathcal{P}(\mathbb{N})_{(P)}\right\rangle\,,$$

where

$$\exp_{(P)} : a \mapsto 2^a, \quad \mathbb{N}_{(P)} \to \mathbb{M}_{(P)}\,,$$

and $\mathcal{P}(\mathbb{N})_{(P)}$ is as in (i).

As before, in the structure $\mathfrak{N}_{(P)}$, the map $+$ is defined on all of $\mathbb{N}_{(P)} \times \mathbb{N}_{(P)}$, but the map $\cdot$ is a partial function which may be defined only on a proper subset of $\mathbb{N}_{(P)} \times \mathbb{N}_{(P)}$.

Definition 10.2 *Let P_1 and P_2 be non-maximal, prime ideals in algebras $C(\Omega_1)$ and $C(\Omega_2)$, respectively.*

(i) *The models $\mathfrak{N}_{(P_1)}$ and $\mathfrak{N}_{(P_2)}$ are* isomorphic, *written $\mathfrak{N}_{(P_1)} \cong \mathfrak{N}_{(P_2)}$, if there is a bijection $\psi : \mathbb{N}_{(P_1)} \to \mathbb{N}_{(P_2)}$ such that:*

$$\begin{aligned}
&\psi(a+b) = \psi(a) + \psi(b) \quad (a, b \in \mathbb{N}_{(P_1)})\,;\\
&a = bc \text{ if and only if } \psi(a) = \psi(b)\psi(c) \quad (a, b, c \in \mathbb{N}_{(P_1)})\,;\\
&\psi(A) \in \mathcal{P}(\mathbb{N})_{(P_2)} \quad (A \in \mathcal{P}(\mathbb{N})_{(P_1)})\,;\\
&\psi^{-1}(B) \in \mathcal{P}(\mathbb{N})_{(P_1)} \quad (B \in \mathcal{P}(\mathbb{N})_{(P_2)})\,.
\end{aligned}$$

(ii) *The models $\mathfrak{M}_{(P_1)}$ and $\mathfrak{M}_{(P_2)}$ are* isomorphic, *written $\mathfrak{M}_{(P_1)} \cong \mathfrak{M}_{(P_2)}$, if there is a bijection $\psi : \mathbb{M}_{(P_1)} \to \mathbb{M}_{(P_2)}$ such that:*

$$\begin{aligned}
&\psi(a+b) = \psi(a) + \psi(b) \quad (a, b \in \mathbb{M}_{(P_1)})\,;\\
&\psi(ab) = \psi(a)\psi(b) \quad (a, b \in \mathbb{M}_{(P_1)})\,;\\
&\psi(2^a) = 2^{\psi(a)} \quad (a \in \mathbb{N}_{(P_1)})\,;\\
&\psi(A) \in \mathcal{P}(\mathbb{N})_{(P_2)} \quad (A \in \mathcal{P}(\mathbb{N})_{(P_1)})\,;\\
&\psi^{-1}(B) \in \mathcal{P}(\mathbb{N})_{(P_1)} \quad (B \in \mathcal{P}(\mathbb{N})_{(P_2)})\,.
\end{aligned}$$

Let P be a non-maximal, prime ideal in an algebra $C(\Omega)$. Then we define

$$\mathcal{P}_{\text{fin}}(\mathbb{N})_{(P)} = \{A \cap [1, a]_{(P)} : a \in \mathbb{N}_{(P)},\ \ A \in \mathcal{P}(\mathbb{N})_{(P)}\}\,.$$

It follows from remarks in Chapter 8 that, if A and B belong to $\mathcal{P}_{\text{fin}}(\mathbb{N})_{(P)}$, then $A \cup B$, $A \cap B$, and $A \setminus B$ also belong to $\mathcal{P}_{\text{fin}}(\mathbb{N})_{(P)}$. However, the analogous questions for $\mathcal{P}(\mathbb{N})_{(P)}$ are more subtle. Let $A, B \in \mathcal{P}(\mathbb{N})_{(P)}$. Then clearly $\mathbb{N}_{(P)} \setminus A \in \mathcal{P}(\mathbb{N})_{(P)}$, but in general it is not true that $A \cup B \in \mathcal{P}(\mathbb{N})_{(P)}$ or that $A \cap B \in \mathcal{P}(\mathbb{N})_{(P)}$.

The next theorem shows that we have included enough information in the definition of $\mathfrak{M}_{(P)}$ to capture all the structures associated to the ideal P that we are interested in. However, this is really only a matter of cosmetics for us, for, if this were not the case, we would just add the relevant structure to the definition of the model.

Theorem 10.3 *Let $\psi : \mathbb{M}_{(P_1)} \to \mathbb{M}_{(P_2)}$ define an isomorphism of $\mathfrak{M}_{(P_1)}$ and $\mathfrak{M}_{(P_2)}$. Then there is a unique extension of ψ to a ring isomorphism*

$$\psi : K_{P_1} \to K_{P_2}$$

such that $\psi(\mathbb{D}_{(P_1)}) = \mathbb{D}_{(P_2)}$ and

$$\varepsilon_r^{(P_1)}(a) = \varepsilon_{\psi(r)}^{(P_2)}(\psi(a)) \quad (a \in \mathbb{N}_{(P_1)}, r \in \mathbb{I}_{(P_1)} \setminus \mathbb{D}_{(P_1)})\,.$$

Proof Clearly ψ extends to a unique isomorphism of rational fields

$$\psi_0 : \mathbb{Q}_{(P_1)} \to \mathbb{Q}_{(P_2)}\,.$$

It follows from the definitions of $\mathbb{D}_{(P_1)}$ and $\mathbb{D}_{(P_2)}$ that $\psi_0(\mathbb{D}_a^{(P_1)}) = \mathbb{D}_{\psi(a)}^{(P_2)}$ for each $a \in \mathbb{N}_{(P_1)}$, and hence that $\psi_0(\mathbb{D}_{(P_1)}) = \mathbb{D}_{(P_2)}$.

It is convenient to work with $\{0,1\}^{\mathbb{N}}_{(P_1)}$ and $\{0,1\}^{\mathbb{N}}_{(P_2)}$, rather than $\mathcal{P}(\mathbb{N})_{(P_1)}$ and $\mathcal{P}(\mathbb{N})_{(P_2)}$; we are free to do this because, for each $g : \mathbb{N}_{(P_2)} \to \{0,1\}$, $g \circ \psi \in \{0,1\}^{\mathbb{N}}_{(P_1)}$ if and only if $g \in \{0,1\}^{\mathbb{N}}_{(P_2)}$.

Take $r \in (\mathbb{I}_{(P_1)} \cap \mathbb{Q}_{(P_1)}) \setminus \mathbb{D}_{(P_1)}$, and define

$$g : b \mapsto \varepsilon_r^{(P_1)}(\psi^{-1}(b)), \quad \mathbb{N}_{(P_2)} \to \{0,1\}. \tag{10.2}$$

Then $g \in \{0,1\}^{\mathbb{N}}_{(P_2)}$. Since $\varepsilon_r^{(P_1)}$ is not eventually constant, g is not eventually constant, and so there exists $s \in \mathbb{I}_{(P_2)} \setminus \mathbb{D}_{(P_2)}$ with $\varepsilon_s^{(P_2)} = g$. Set

$$t = \psi_0(r) \in (\mathbb{I}_{(P_2)} \cap \mathbb{Q}_{(P_2)}) \setminus \mathbb{D}_{(P_2)}.$$

By a calculation involving Lemma 8.47 and using the fact that ψ is isotonic, it follows that $\varepsilon_s^{(P_2)} = \varepsilon_t^{(P_2)}$. By Lemma 8.46(ii), we have $s = t$. Thus we have shown that

$$\varepsilon_r^{(P_1)}(a) = \varepsilon_{\psi_0(r)}^{(P_2)}(\psi_0(a)) \quad (a \in \mathbb{N}_{(P_1)}, \quad r \in (\mathbb{I}_{(P_1)} \cap \mathbb{Q}_{(P_1)}) \setminus \mathbb{D}_{(P_1)}).$$

We now define $\psi_1 : \mathbb{I}_{(P_1)} \to \mathbb{I}_{(P_2)}$ to extend $\psi_0 \mid (\mathbb{I}_{(P_1)} \cap \mathbb{Q}_{(P_1)})$. Indeed, take $r \in \mathbb{I}_{(P_1)} \setminus \mathbb{D}_{(P_1)}$, and define g as in (10.2). Again there exists $s \in \mathbb{I}_{(P_2)} \setminus \mathbb{D}_{(P_2)}$ with $\varepsilon_s^{(P_2)} = g$. Set $\psi_1(r) = s$. Also define $\psi_1(r) = \psi_0(r) \quad (r \in \mathbb{D}_{(P_1)})$. We have defined $\psi_1 : \mathbb{I}_{(P_1)} \to \mathbb{I}_{(P_2)}$.

The range of ψ_1 certainly contains $\mathbb{Q}_{(P_2)}$. Fix $s \in \mathbb{I}_{(P_2)} \setminus \mathbb{Q}_{(P_2)}$, and define

$$f : a \mapsto \varepsilon_s^{(P_2)}(\psi_0(a)), \quad \mathbb{N}_{(P_1)} \to \{0,1\}.$$

Then $f \in \{0,1\}^{\mathbb{N}}_{(P_1)}$, and so there exists $r \in \mathbb{I}_{(P_1)} \setminus \mathbb{D}_{(P_1)}$ with $\varepsilon_r^{(P_1)} = f$. Clearly $\psi_1(r) = s$, and so ψ_1 is a surjection.

Fix $r_1, r_2 \in \mathbb{I}_{(P_1)} \setminus \mathbb{D}_{(P_1)}$, and set $s_1 = \psi_1(r_1)$ and $s_2 = \psi_1(r_2)$, so that $s_1, s_2 \in \mathbb{I}_{(P_2)} \setminus \mathbb{D}_{(P_2)}$. We have $r_1 < r_2$ if and only if $\varepsilon_{r_1}^{(P_1)} \prec \varepsilon_{r_2}^{(P_1)}$ and $s_1 < s_2$ if and only if $\varepsilon_{s_1}^{(P_2)} \prec \varepsilon_{s_2}^{(P_2)}$. But, by the definition of ψ_1, $\varepsilon_{r_1}^{(P_1)} \prec \varepsilon_{r_2}^{(P_2)}$ if and only if $\varepsilon_{s_1}^{(P_2)} \prec \varepsilon_{s_2}^{(P_2)}$. Thus $s_1 < s_2$ if and only if $r_1 < r_2$, and so ψ_1 is isotonic.

We now *claim* that

$$\psi_1(zr_1) = \psi_0(z)\psi_1(r) \quad (z \in \mathbb{I}_{(P_1)} \cap \mathbb{Q}_{(P_1)}, \quad r \in \mathbb{I}_{(P_1)}). \tag{10.3}$$

Assume towards a contradiction that there exist z and r such that (10.3) fails. Choose t_1, t_2 with $0 \le t_1 < rz < t_2 \le 1$ and such that

$$\psi_0(z)\psi_1(t) \notin [\psi_1(t_1), \psi_1(t_2)]_{(P_2)},$$

and choose $r_1, r_2 \in \mathbb{I}_{(P_1)} \cap \mathbb{Q}_{(P_1)}$ such that

$$r_1 \le r \le r_2 \quad \text{and} \quad t_1 < zr_1 \le zr_2 < t_2.$$

Then

$$\psi_0(z)\psi_1(r) \notin [\psi_1(zr_1), \psi_1(zr_2)]_{(P_2)}.$$

However $\psi_1(zr_1) = \psi_0(zr_1) = \psi_0(z)\psi_0(r_1)$, and so

$$\psi_1(zr_1) = \psi_0(z)\psi_1(r_1) \le \psi_0(z)\psi_1(r) \le \psi_0(z)\psi_1(r_2) = \psi_1(zr_2),$$

a contradiction. Thus (10.3) holds.

We now define $\psi : K_{P_1} \to K_{P_2}$. Take $r \in K_{P_1}$ and choose an element $x \in \mathbb{Q}_{(P_1)} \setminus \{0\}$ with $xr \in \mathbb{I}_{(P_1)}$. Define

$$\psi(r) = \psi_1(xr)/\psi_0(x).$$

To show that ψ is well defined, also choose $y \in \mathbb{Q}_{(P_1)} \setminus \{0\}$ with $yr \in \mathbb{I}_{(P_1)}$; we may suppose that $r > 0$ and that $xr \le yr$. Set $z = x/y \in \mathbb{I}_{(P_1)} \cap \mathbb{Q}_{(P_1)}$. Then

$$\psi_1(xr) = \psi_1(zyr) = \psi_0(z)\psi_1(yr)$$

by (10.3), and so $\psi_1(xr)/\psi_0(x) = \psi_0(z)\psi_1(yr)/\psi_0(x) = \psi_1(yr)/\psi_0(y)$. This shows that $\psi(r)$ is well defined.

Clearly $\psi \mid \mathbb{Q}_{(P_1)} = \psi_0$, $\psi \mid \mathbb{I}_{(P_1)} = \psi_1$, and ψ is isotonic. Now take $s \in K_{P_2}$, choose $y \in \mathbb{Q}_{(P_2)} \setminus \{0\}$ with $ys \in \mathbb{I}_{(P_2)}$, and set $x = \psi_0^{-1}(y)$ and $r = \psi_1^{-1}(ys)/x$, so that $x \in \mathbb{Q}_{(P_1)} \setminus \{0\}$ and $\psi(r) = \psi_1(xr)/\psi_0(x) = ys/y = s$. This shows that ψ is a surjection.

Since $\psi : K_{P_1} \to K_{P_2}$ is an isotonic bijection such that

$$\psi \mid \mathbb{Q}_{(P_1)} : \mathbb{Q}_{(P_1)} \to \mathbb{Q}_{(P_2)}$$

is a homomorphism and since $\mathbb{Q}_{(P_1)}$ is order-dense in K_{P_1}, ψ is a ring isomorphism uniquely specified by ψ_0, and hence by the original map ψ. □

We note that it is not necessarily the case that the map $\psi : K_{P_1} \to K_{P_2}$ defined in the above theorem is an isomorphism of the (real) *fields* K_{P_1} and K_{P_2}.

It is actually the case that, if $\mathfrak{N}_{(P_1)}$ and $\mathfrak{N}_{(P_2)}$ are isomorphic, then $\mathfrak{M}_{(P_1)}$ and $\mathfrak{M}_{(P_2)}$ are isomorphic, and so, by the theorem, K_{P_1} and K_{P_2} are isomorphic as rings. To see this, one can show that a model $\mathfrak{M}_{(P)}$ is naturally definable in the structure $\langle \mathbb{N}_{(P)}, +, \cdot, \mathcal{P}_{\text{fin}}(\mathbb{N})_{(P)} \rangle$, and then that the map $\exp_{(2)} : \mathbb{N}_{(P)} \to \mathbb{M}_{(P)}$ can be defined. We are omitting the details of this remark.

Before we begin the main work of this chapter, we wish to obtain some preliminary results related to the operational calculus which was introduced in Chapter 6. Once again, P is a non-maximal, prime ideal in an algebra $C(\Omega)$, and Q is the maximum prime z-ideal below P. We are concerned only with the case where $Q \ne P$. We must explore the relationship between $\mathbb{I}_{(P)}$ and $\mathbb{I}_{(Q)}$ more carefully; to do this, we construct a form of inverse to the canonical map

$$\pi_{PQ} : f + Q \mapsto f + P, \quad A_Q \to A_P.$$

Let $y \in \mathbb{I}_{(P)} \setminus \{0,1\}$. Then there exist $f, g \in C(\Omega)\backslash P$ with $0 < f < g < 1$ such that

$$y = \pi_P(f)/\pi_P(g). \tag{10.4}$$

Set

$$x = \pi_Q(f)/\pi_Q(g) \in \mathbb{I}_{(Q)} \subset K_Q, \tag{10.5}$$

where we note that x depends on the choice of f and g, and not just on y. Now set

$$J_y = \bigcap \left\{ \left(x - \frac{1}{a},\, x + \frac{1}{a} \right)_{(Q)} : a \in \mathbb{M}_{(P)} \right\} . \tag{10.6}$$

Then J_y is an interval in $\mathbb{I}_{(Q)}$. Suppose that also $y = \pi_P(f')/\pi_P(g')$, where $f', g' \in C(\Omega)\backslash P$ with $0 < f' < g' < 1$, and set $x' = \pi_Q(f')/\pi_Q(g')$. Then

$$x - x' = \pi_Q(fg' - f'g)/\pi_Q(gg'),$$

and so $|x - x'| < 1/a \quad (a \in \mathbb{M}_{(P)})$ because $f'g - fg' \in P$ and $gg' \notin P$. Thus $x' \in J_y$, and the interval defined by replacing x by x' in (10.6) is just J_y. Thus J_y depends only on y, and not on the choice of f and g.

Clearly, from the definition of $\mathbb{N}_{(P)}$ (Definition 8.32) and Lemma 8.33(iv), we have

$$J_y = \bigcap \{ (x - 2^{-a},\, x + 2^{-a})_{(Q)} : a \in \mathbb{N}_{(P)} \} . \tag{10.7}$$

We shall prove a technical lemma about x, y, and J_y, where x, y, and J_y are as in (10.5), (10.4), and (10.6), respectively. In particular, we have temporarily fixed f and g. We recall the relationships summarized in Figure 8.1. on page 183.

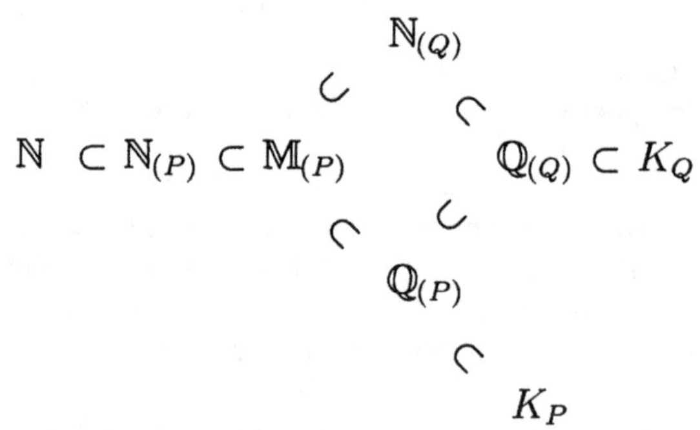

Fig. 10.1. The relationships between some sets

In particular, in the next result we are regarding $\mathbb{Q}_{(P)}$ as a subset of K_Q; recall from Chapter 8 that, if $a \in \mathbb{Q}_{(P)}$, then there exist $f \in C(\Omega)$ and $g \in C(\Omega)\backslash P$ such that $a = \pi_P(f)/\pi_P(g)$ in $\mathbb{Q}_{(P)}$ and $a = \pi_Q(f)/\pi_Q(g)$ in $\mathbb{Q}_{(Q)}$.

Lemma 10.4 (i) *Let $r \in \mathbb{Q}_{(P)}$. If $r < y$, then $r < x$; if $r > y$, then $r > x$.*

(ii) *The set $J_y \cap \mathbb{Q}_{(P)}$ contains at most one point.*

(iii) *Let $b \in \mathbb{Z}^+_{(P)}$. Then the following are equivalent:*

$$\text{(a) } J_y \cap \mathbb{D}_b^{(Q)} \neq \emptyset; \qquad \text{(b) } J_y \cap \mathbb{D}_b^{(P)} \neq \emptyset; \qquad \text{(c) } y \in J_y \cap \mathbb{D}_b^{(P)} .$$

(iv) *Let $S = \{b \in \mathbb{Z}^+_{(Q)} : J_y \cap \mathbb{D}_b^{(P)} \neq \emptyset\}$, and suppose that $S \cap \mathbb{Z}^+_{(P)} \neq \emptyset$. Then* $\min S$ *exists and belongs to $\mathbb{Z}^+_{(P)}$.*

(v) *Suppose that* $r, s \in J_y \setminus \mathbb{D}_{(Q)}$ *and that* $y \notin \mathbb{D}_{(P)}$. *Then*

$$\varepsilon_r^{(Q)} \restriction a = \varepsilon_s^{(Q)} \restriction a \quad (a \in \mathbb{N}_{(P)}).$$

(vi) *Suppose that* $y \notin \mathbb{D}_{(P)}$. *Then*

$$\varepsilon_y^{(P)} = \varepsilon_z^{(Q)} \mid \mathbb{N}_{(P)} \quad (z \in J_y \setminus \mathbb{D}_{(Q)}).$$

Proof (i) Take $f' \in C(\Omega)$ and $g' \in C(\Omega)\setminus P$ such that $r = \pi_Q(f')/\pi_Q(g')$. Suppose that $r \leq x$ in K_Q. Then there exists $Z \in \mathbf{Z}[Q]$ with

$$f'(x)g(x) \leq f(x)g'(x) \quad (x \in Z),$$

and so $\pi_P(f')\pi_P(g) \leq \pi_P(f)\pi_P(g')$ because π_{PQ} is an order-preserving epimorphism, i.e. $r \leq y$. Similarly, if $r \geq x$, then $r \geq y$.

(ii) This follows from the fact that $|r - s| < \delta$ for each $\delta \in \mathbb{Q}^+_{(P)}\setminus\{0\}$ whenever $r, s \in J_y$.

(iii) Since $\mathbb{D}_b^{(P)} \subset \mathbb{D}_b^{(Q)}$, it is sufficient to show that (a) implies (c).

To obtain a contradiction, first assume that $y \notin \mathbb{D}_b^{(P)}$. Then there exist $s, t \in \mathbb{D}_b^{(P)}$ with $s < y < t$ and $t - s = 2^{-b}$. Choose $c \in \mathbb{N}_{(P)}$ such that $2^{-c} < \min\{y - s, t - y\}$. By (i), we have

$$s < x - 2^{-c} < x < x + 2^{-c} < t,$$

and so $J_y \subset (s, t)_{(Q)}$. But $\mathbb{D}_b^{(Q)} \cap (s, t)_{(Q)} = \emptyset$, a contradiction of (a). Thus $y \in \mathbb{D}_b^{(P)}$.

Now assume that $y \notin J_y$. Then there exists $a \in \mathbb{N}_{(P)}$ with $|y - x| > 1/a$. If $y < x$, then $y < y + 1/a < x$ and $y + 1/a \in \mathbb{Q}_{(P)}$, a contradiction of (i). A similar contradiction arises in the case where $y > x$. Thus $y \in J_y$.

(iv) Take $b \in \mathbb{N}_{(P)}$ with $J_y \cap \mathbb{D}_b^{(Q)} \neq \emptyset$. By (iii), $y \in J_y \cap \mathbb{D}_b^{(P)}$. By Lemma 8.48, there is a minimum $a \in \mathbb{Z}^+_{(P)}$ with $y \in \mathbb{D}_a^{(P)}$. If $a = 0$, then certainly $a = \min S$. If $a \geq 1$, then $y \notin \mathbb{D}_{a-1}^{(P)}$, and so, by (iii), $J_y \cap \mathbb{D}_{a-1}^{(P)} = \emptyset$, showing that $a = \min S$.

(v) Let $a \in \mathbb{N}_{(P)}$. By (iii), $J_y \cap \mathbb{D}_a^{(Q)} = \emptyset$, and so $\varepsilon_r^{(Q)} \restriction a = \varepsilon_s^{(Q)} \restriction a$ by Lemma 8.15(i).

(vi) Take $b \in \mathbb{N}_{(P)}$ and $z \in J_y\setminus\mathbb{D}_{(Q)}$. Set

$$s_1 = \max(\mathbb{D}_b^{(P)} \cap [0, y]_{(P)}), \quad t_1 = \min(\mathbb{D}_b^{(P)} \cap [y, 1]_{(P)}),$$
$$s_2 = \max(\mathbb{D}_{b+1}^{(P)} \cap [0, y]_{(P)}), \quad t_2 = \min(\mathbb{D}_{b+1}^{(P)} \cap [y, 1]_{(P)}),$$

and choose $s, t \in \mathbb{Q}_{(P)}$ with

$$s_1 \leq s_2 < s < y < t < t_2 \leq t_1.$$

By (i), $s < x < t$, and so

$$s_1 \leq s_2 < z < t_2 \leq t_1.$$

But $t_1 - s_1 = 2^{-b}$ and $t_2 - s_2 = 2^{-(b+1)}$, and so, by Lemma 8.12(vi), we have $\mathbb{D}_b^{(Q)} \cap (s_1, t_1) = \emptyset$ and $\mathbb{D}_{b+1}^{(Q)} \cap (s_2, t_2) = \emptyset$. Hence

$$s_1 = \max(\mathbb{D}_b^{(Q)} \cap [0, z]_{(Q)}), \quad t_1 = \min(\mathbb{D}_b^{(Q)} \cap [z, 1]_{(Q)}),$$
$$s_2 = \max(\mathbb{D}_{b+1}^{(Q)} \cap [0, z]_{(Q)}), \quad t_2 = \min(\mathbb{D}_{b+1}^{(Q)} \cap [z, 1]_{(Q)}).$$

By Lemma 8.45, $\varepsilon_y^{(P)}(b+1) = \varepsilon_z^{(Q)}(b+1)$, and this gives the result. □

Recall that we defined an operational calculus in Chapter 6.

Lemma 10.5 *Suppose that $F \in \mathfrak{L}(\mathbb{I})$. Then there exist functions $f \in C(\Omega)$ and $g \in C(\Omega)\backslash P$ with*

$$F_{(P)}(y) = \pi_P(f)/\pi_P(g) \quad \text{and} \quad F_{(Q)}(x) = \pi_Q(f)/\pi_Q(g).$$

Proof This is immediate from the definition of the operational calculus. □

In Chapter 8, we defined $\varepsilon_y^{(P)}$ for $r \in \mathbb{I}_{(P)}\backslash\mathbb{D}_{(P)}$, but not for $y \in \mathbb{D}_{(P)}$. In fact, we expect from the analogous situation for $r \in \mathbb{D}$ that there will be two 'binary expansions' $\varepsilon_y^{+(P)}$ and $\varepsilon_y^{-(P)}$ for $y \in \mathbb{D}_{(P)}$, and this is indeed the case.

Thus, let $y \in \mathbb{D}_{(P)} \setminus \{0, 1\}$. Set $b = \min\{a \in \mathbb{N}_{(P)} : y \in \mathbb{D}_a^{(P)}\}$ and choose $z_1, z_2 \in (\mathbb{Q}_{(P)} \cap \mathbb{I}_{(P)})\backslash\mathbb{D}_{(P)}$ with

$$y - 2^{-(b+1)} < z_1 < y < z_2 < y + 2^{-(b+1)}.$$

Definition 10.6 *The functions $\varepsilon_y^{+(P)}$ and $\varepsilon_y^{-(P)}$ are defined on $\mathbb{N}_{(P)}$ by:*

$$\varepsilon_y^{+(P)}(a) = \begin{cases} \varepsilon_{z_2}^{(P)}(a) & (a \le b), \\ 0 & (a > b); \end{cases} \qquad \varepsilon_y^{-(P)}(a) = \begin{cases} \varepsilon_{z_1}^{(P)}(a) & (a \le b), \\ 1 & (a > b). \end{cases}$$

If $y = 0$, then $\varepsilon_y^{+(P)} = 0$; if $y = 1$, then $\varepsilon_y^{-(P)} = 1$.

It follows from Lemma 8.45 that $\varepsilon_y^{+(P)}$ and $\varepsilon_y^{-(P)}$ are well defined. By Definition 8.38(iii), we then have $\varepsilon_y^{+(Q)} \mid \mathbb{N}_{(P)} = \varepsilon_y^{+(P)}$ and $\varepsilon_y^{-(Q)} \mid \mathbb{N}_{(P)} = \varepsilon_y^{-(P)}$.

Lemma 10.7 *Let $y \in \mathbb{D}_{(P)}$, and take $z \in J_y\backslash\mathbb{D}_{(Q)}$.*

(i) *If $z > y$, then $\varepsilon_z^{(Q)} \mid \mathbb{N}_{(P)} = \varepsilon_y^{+(P)}$.*

(ii) *If $z < y$, then $\varepsilon_z^{(Q)} \mid \mathbb{N}_{(P)} = \varepsilon_y^{-(P)}$.*

Proof We prove (ii). Set $b = \min\{a \in \mathbb{N}_{(P)} : y \in \mathbb{D}_a^{(P)}\}$, and fix an element $z_1 \in (\mathbb{Q}_{(P)} \cap \mathbb{I}_{(P)}) \setminus \mathbb{D}_{(P)}$ with $y - 2^{-(b+1)} < z_1 < y$. Since $z < y$, we have $y - z < 2^{-a}$ $(a \in \mathbb{N}_{(P)})$, and so $z, z_1 \in (y - 2^{(b+1)}, y)$. Thus $(z_1, z) \cap \mathbb{D}_b^{(Q)} = \emptyset$. It follows from Lemma 8.14 that $\varepsilon_z^{(Q)} \restriction b = \varepsilon_{z_1}^{(Q)} \restriction b$. Take $a \in \mathbb{N}_{(P)}$ with $a \ge b$. Then

$$y = \min(\mathbb{D}_a^{(Q)} \cap (z, 1)_{(Q)}) = \min(\mathbb{D}_{a+1}^{(Q)} \cap (z, 1)_{(Q)}),$$

and so $\varepsilon_z^{(Q)}(a+1) = 1$ by Lemma 8.16. Thus $\varepsilon_z^{(Q)} \mid \mathbb{N}_{(P)} = \varepsilon_y^{-(P)}$. □

It follows from Lemma 10.4(vi) that $\varepsilon_y^{+(P)}$ and $\varepsilon_y^{-(P)}$ are indeed the natural definitions of the 'binary expansions' of the element y in the case where $y \in \mathbb{D}_{(P)}$.

Suppose that S is a subset of $\mathbb{Q} \times \mathbb{Q}$. Then for a prime z-ideal Q, $S_{(Q)}$ is the subset of $\mathbb{Q}_{(Q)} \times \mathbb{Q}_{(Q)}$ defined in the obvious pointwise manner using Q-partitions.

Lemma 10.8 *Suppose that $F \in \mathfrak{L}(\mathbb{I})$ with $F(\mathbb{D}_n) \subset \mathbb{D}_n \quad (n \in \mathbb{N})$. Then:*

(i) *for each $a \in \mathbb{N}_{(Q)}$, $F_{(Q)}(\mathbb{D}_a^{(Q)}) \subset \mathbb{D}_a^{(Q)}$;*

(ii) $F_{(Q)} \mid \mathbb{D}_{(P)} = F_{(P)} \mid \mathbb{D}_{(P)}$;

(iii) $(\operatorname{Gr} F \cap (\mathbb{D} \times \mathbb{D}))_{(Q)} = \operatorname{Gr} F_{(Q)} \cap (\mathbb{D}_{(Q)} \times \mathbb{D}_{(Q)})$.

Proof (i) Take $a \in \mathbb{N}_{(Q)}$ and $r \in \mathbb{D}_a^{(Q)}$. Then, as before, there are a Q-partition $\{Z_j : j \in \mathbb{N}\}$ and sequences (n_j) in $\mathbb{N}$ and (r_j) in $\mathbb{D}$ with $r_j \in \mathbb{D}_{n_j} \quad (j \in \mathbb{N})$ representing a and r, respectively.

Choose $g \in C(\Omega) \setminus Q$ such that

$$\lim_{j \to \infty} F(r_j)\, |g|_{Z_j} = \lim_{j \to \infty} r_j\, |g|_{Z_j} = 0\,;$$

this choice is possible because Q is a z-ideal. Next, choose $f \in C(\Omega)$ with

$$f(x) = F(r_j)g(x) \quad (x \in Z_j,\ j \in \mathbb{N})\,,$$

and define $s = \pi_Q(f)/\pi_Q(g)$. Since there exists $h \in C(\Omega)$ with

$$h(x) = r_j g(x) \quad (x \in Z_j,\ j \in \mathbb{N})\,,$$

it follows from the definition of $F_{(Q)}(r)$ in (6.1) that $s = F_{(Q)}(r)$. For each $j \in \mathbb{N}$, we have $F(r_j) \in F(\mathbb{D}_{n_j}) \subset \mathbb{D}_{n_j}$, and so, since Q is a z-ideal, it follows that $F_{(Q)}(r) \in \mathbb{D}_a^{(Q)}$.

(ii) Take $a \in \mathbb{N}_{(P)}$ and $r \in \mathbb{D}_a^{(P)} \subset \mathbb{D}_a^{(Q)}$, and set

$$x = F_{(Q)}(r), \quad y = F_{(P)}(r)\,.$$

By Lemma 10.5, there exist $f \in C(\Omega)$ and $g \in C(\Omega) \setminus P$ with

$$x = \pi_Q(f)/\pi_Q(g) \quad \text{and} \quad y = \pi_P(f)/\pi_P(g)\,.$$

By (i), $x \in \mathbb{D}_a^{(Q)}$. By the definition of J_y, we have $x \in J_y$, and so certainly $J_y \cap \mathbb{D}_a^{(Q)} \neq \emptyset$. By Lemma 10.4(iii), $y \in J_y \cap \mathbb{D}_a^{(P)}$. It follows that

$$|x - y| < 2^{-d} \quad (d \in \mathbb{N}_{(P)})\,,$$

and, in particular, $|x - y| < 2^{-a}$. But $x \in \mathbb{D}_a^{(Q)}$ and $y \in \mathbb{D}_a^{(P)} \subset \mathbb{D}_a^{(Q)}$, and so $x = y$. We have shown that $F_{(Q)} \mid \mathbb{D}_{(P)} = F_{(P)} \mid \mathbb{D}_{(P)}$.

(iii) This follows in a routine way from the various definitions. □

Suppose that $F \in C(\mathbb{I})$ and that $a \in \mathbb{I}_{(P)}$. Then we have defined $F_{(P)}(a)$ in the two cases: (i) $F \in \mathfrak{L}(\mathbb{I})$; (ii) P is a z-ideal. But there is no sensible definition of $F_{(P)}(a)$ in general; an example showed that such a definition is more than we can hope for. However there is a natural definition of $F_{(P)}(a)$ in the special case where $a \in \mathbb{I}_{(P)} \cap \mathbb{Q}_{(P)}$. Indeed, by an argument involving Q-partitions, there exist $f \in C(\Omega)$ and $g \in C(\Omega) \setminus P$ with $F_{(Q)}(a) = \pi_Q(f)/\pi_Q(g)$, and then we set

$$F_{(P)}(a) = \pi_P(f)/\pi_P(g)\,, \tag{10.8}$$

generalizing the known result in the case where $F \in \mathfrak{L}(\mathbb{I})$. It is easy to check that $F_{(P)}(a)$ is well defined and that it is independent of the choice of Q. In general, $F_{(P)} \mid \mathbb{I}_{(P)} \cap \mathbb{Q}_{(P)}$ is not continuous with respect to the order topology on $\mathbb{I}_{(P)}$.

We shall again require a coding device. Consider the map

$$\Psi : \mathbb{N} \times \{0,1\} \to \mathbb{N}$$

given by

$$\Psi(n,0) = 2n, \quad \Psi(n,1) = 2n-1 \qquad (n \in \mathbb{N})\,.$$

Then Ψ is a bijection, with

$$\Psi^{-1}(m) = \begin{cases} ((m+1)/2, 1) & (m \text{ odd})\,, \\ (m/2, 0) & (m \text{ even})\,. \end{cases}$$

Let $f : \mathbb{N} \to \{0,1\}$ be a function. Then $\operatorname{Gr} f$ is a subset of $\mathbb{N} \times \{0,1\}$, and so $\Psi(\operatorname{Gr} f)$ is a subset of $\mathbb{N}$. We define f^* to be the characteristic function of the set $\Psi(\operatorname{Gr} f)$, so that $f^* : \mathbb{N} \to \{0,1\}$ is a function. For example, if

$$f = (0,1,0,1,0,1,\ldots)\,, \quad \text{then} \quad f^* = (0,1,1,0,0,1,1,0,\ldots)\,.$$

In the general case, the function f is recoverable from the function f^* by an easily discerned formula. Indeed, the values of f on $\{1,\ldots,n\}$ determine f^* on $\{1,\ldots,2n\}$, and vice versa. A sequence (f_k) on $\mathbb{N}$ converges pointwise to f if and only if (f_k^*) converges pointwise to f^* on $\mathbb{N}$. Also, $f \prec g$ in the lexicographic order on $\mathbb{N}$ if and only if $f^* \prec g^*$.

We note a key triviality about f^* that follows from inspection of the coding function Ψ: for each $n \in \mathbb{N}$,

$$\{f^*(2n-1),\ f^*(2n)\} = \{0,1\}\,, \tag{10.9}$$

and this property characterizes the functions f^*. In particular, f^* is not eventually constant.

Similarly, for $f \in \{0,1\}^k$, we define $f^* \in \{0,1\}^{2k}$; for $f \in \{0,1\}^{\mathbb{N}}$, clearly $(f \restriction k)^* = f^* \restriction 2k$.

We now develop an analogue of the above coding device for a non-standard version $\mathbb{N}_{(P)}$ of $\mathbb{N}$. Thus take P to be a non-maximal, prime ideal in an algebra $C(\Omega)$, and consider the map $\Psi_{(P)} : \mathbb{N}_{(P)} \times \{0,1\} \to \mathbb{N}_{(P)}$ given by

$$\Psi_{(P)}(a,0) = 2a, \quad \Psi_{(P)}(a,1) = 2a-1 \quad (a \in \mathbb{N}_{(P)})\,.$$

Again, the map $\Psi_{(P)}$ is a bijection. In the case where P_1 and P_2 are non-maximal, prime ideals in $C(\Omega)$ with P_1 below P_2, we regard $\mathbb{N}_{(P_2)}$ as a subset of $\mathbb{N}_{(P_1)}$, as in Corollary 8.67, and we note that

$$\Psi_{(P_1)} \mid (\mathbb{N}_{(P_2)} \times \{0,1\}) = \Psi_{(P_2)}\,.$$

Let $f : \mathbb{N}_{(P)} \to \{0,1\}$ be a function. Then $\mathrm{Gr}\, f$ is a subset of $\mathbb{N}_{(P)} \times \{0,1\}$, and so $\Psi_{(P)}(\mathrm{Gr}\, f)$ is a subset of $\mathbb{N}_{(P)}$.

Definition 10.9 *Let P be a non-maximal, prime ideal in $C(\Omega)$.*

(i) *Let $f : \mathbb{N}_{(P)} \to \{0,1\}$ be a function. Then*

$$f^* : \mathbb{N}_{(P)} \to \{0,1\}$$

is the characteristic function of $\Psi_{(P)}(\mathrm{Gr}\, f)$.

(ii) *Let $a \in \mathbb{N}_{(P)}$, and let $f : [1,a] \cap \mathbb{N}_{(P)} \to \{0,1\}$ be a function. Then*

$$f^* : [1,2a] \cap \mathbb{N}_{(P)} \to \{0,1\}$$

is the characteristic function of $\Psi_{(P)}(\mathrm{Gr}\, f) \cap [1,2a]$.

Of course, given $f^* : \mathbb{N}_{(P)} \to \{0,1\}$, we can recover $f : \mathbb{N}_{(P)} \to \{0,1\}$. The triviality corresponding to (10.8) is that, for each $a \in \mathbb{N}_{(P)}$,

$$\{f^*(2a-1), f^*(2a)\} = \{0,1\}, \tag{10.10}$$

and this property characterizes the functions f^*. Also, it is clear that, for each $f : \mathbb{N}_{(P)} \to \{0,1\}$, we have $(f \upharpoonright a)^* = f^* \upharpoonright 2a$.

Lemma 10.10 *Let P be a non-maximal, prime ideal in an algebra $C(\Omega)$, and take $a \in \mathbb{N}_{(P)}$ and $f : [1,a] \cap \mathbb{N}_{(P)} \to \{0,1\}$. Then $f \in \{0,1\}^a_{(P)}$ if and only if $f^* \in \{0,1\}^{2a}_{(P)}$.*

Proof In the case where P is a z-ideal, this follows from the above remarks by standard pointwise arguments. The general case follows by the definition of $\varepsilon_r^{(P)}$ in Definition 8.38. □

Now let $f \in \{0,1\}^{\mathbb{N}}_{(P)}$. Then, for each $a \in \mathbb{N}_{(P)}$, we have $f \upharpoonright a \in \{0,1\}^a_{(P)}$, and so $f^* \upharpoonright 2a \in \{0,1\}^{2a}_{(P)}$. But it is not obvious that $f^* \in \{0,1\}^{\mathbb{N}}_{(P)}$; this is a real 'point at issue'. For this reason, we distinguish a subset of $\{0,1\}^{\mathbb{N}}_{(P)}$ in the following definition.

Definition 10.11 *Let P be a non-maximal, prime ideal in $C(\Omega)$. Then*

$${}^*\{0,1\}^{\mathbb{N}}_{(P)} = \{f : \mathbb{N}_{(P)} \to \{0,1\} : f^* \in \{0,1\}^{\mathbb{N}}_{(P)}\},$$

and

$${}^*\mathcal{P}(\mathbb{N})_{(P)} = \{\{a \in \mathbb{N}_{(P)} : f(a) = 1\} : f^* \in \{0,1\}^{\mathbb{N}}_{(P)}\}.$$

It follows that

$$\mathcal{P}_{\mathrm{fin}}(\mathbb{N})_{(P)} = \{A \cap [1,a] : a \in \mathbb{N}_{(P)},\ A \in {}^*\mathcal{P}(\mathbb{N})_{(P)}\}. \tag{10.11}$$

Definition 10.12 *Let P be a non-maximal, prime ideal in an algebra $C(\Omega)$. The structure $\mathfrak{N}_{(P)}$ is* weakly Cauchy complete *if*

$${}^*\{0,1\}^{\mathbb{N}}_{(P)} = \{0,1\}^{\mathbb{N}}_{(P)}.$$

We first note that weak Cauchy completeness is a property of the model $\mathfrak{N}_{(P)}$, and not necessarily of the field K_P: it is *a priori* possible that P_1 and P_2

are prime ideals in certain algebras with $K_{P_1} \cong K_{P_2}$, with $\mathfrak{N}_{(P_1)}$ weakly Cauchy complete, but with $\mathfrak{N}_{(P_2)}$ not weakly Cauchy complete, and this possibility does indeed arise.

Second, we remark that we shall eventually prove (in Theorem 10.23) that necessarily

$$^*\{0,1\}^{\mathbb{N}}_{(P)} \subset \{0,1\}^{\mathbb{N}}_{(P)},$$

and so the above 'point at issue' is the only point to be verified to show that $\mathfrak{N}_{(P)}$ is weakly Cauchy complete. Of course, we cannot yet assume this in our development of the theory.

Third, suppose that $\mathfrak{N}_{(P)}$ is a model of analysis. Then $\mathfrak{N}_{(P)}$ is weakly Cauchy complete; this is a routine consequence of the axiom of comprehension. In particular, structures associated to ultrapowers are weakly Cauchy complete.

The terminology of Definition 10.12 is justified by the following easy theorem, which is an immediate consequence of Theorem 9.3.

Theorem 10.13 *Let P be a non-maximal, prime ideal in $C(\Omega)$ for a compact space Ω, and suppose that K_P is Cauchy complete. Then $\mathfrak{N}_{(P)}$ is weakly Cauchy complete.* □

We now show that for a familiar class of prime ideals P, the structure $\mathfrak{N}_{(P)}$ is weakly Cauchy complete.

Proposition 10.14 *Let P be a cut prime in $C(\Omega)$. Then $\mathfrak{N}_{(P)}$ is weakly Cauchy complete.*

Proof Take P_1 and P_2 to be prime ideals in $C(\Omega)$ with $P_1 \subsetneq P_2$ and P_1 strongly convex, and fix $c \in \mathbb{N}_{(P_1)} \setminus \mathbb{N}_{(P_2)}$.

Let $f \in \{0,1\}^{\mathbb{N}}_{(P_2)}$. By Lemma 9.7(ii), there exists $g \in \{0,1\}^{c}_{(P_1)}$ such that $g \mid \mathbb{N}_{(P_2)} = f$. By Lemma 10.10, $g^* \in \{0,1\}^{2c}_{(P_1)}$, and so

$$f^* = g^* \mid \mathbb{N}_{(P_2)} \in \{0,1\}^{\mathbb{N}}_{(P_2)}.$$

Thus $f \in {}^*\{0,1\}^{\mathbb{N}}_{(P_2)}$, and so $\{0,1\}^{\mathbb{N}}_{(P_2)} \subset {}^*\{0,1\}^{\mathbb{N}}_{(P_2)}$.

The proof of the reverse inclusion is essentially the same. □

Thus there are many examples for which $\mathfrak{N}_{(P)}$ is weakly Cauchy complete, but K_P is not complete. However we shall in due course provide an example which shows that even a hyper-real field does not necessarily give a weakly Cauchy complete structure.

Our further analysis of the sets $\{0,1\}^{\mathbb{N}}_{(P)}$ and $^*\{0,1\}^{\mathbb{N}}_{(P)}$ first requires some classical real analysis. Recall that each sequence $f^* : \mathbb{N} \to \{0,1\}$ is not eventually constant, and so the real number

$$r = \sum_{n=1}^{\infty} \frac{f^*(n)}{2^n}$$

belongs to $\mathbb{I} \setminus \mathbb{D}$, and $\varepsilon_r = f^*$. However we do not obtain all elements of $\mathbb{I} \setminus \mathbb{D}$ in this way.

Definition 10.15 *The set U consists of those elements $r \in \mathbb{I} \setminus \mathbb{D}$ such that $\varepsilon_r = f^*$ for some $f : \mathbb{N} \to \{0,1\}$.*

We shall also require a specific subset V of U. Let $n \in \mathbb{N}$. Then V_n consists of those elements $r \in \mathbb{I} \setminus \mathbb{D}$ such that $\varepsilon_r = f^*$ for some $f : \mathbb{N} \to \{0,1\}$ with $f(m) = 1 \quad (m > n)$, and

$$V = \bigcup\{V_n : n \in \mathbb{N}\}.$$

If $r \in V_n$, then $\varepsilon_r(2m) = 0$ and $\varepsilon_r(2m-1) = 1$ whenever $m > n$. It follows that $V \subset U \cap \mathbb{Q}$; indeed, it is easy to see that each $r \in V_n$ has the form $2/3 \pm s$ for some $s \in \mathbb{D}_{2n}$.

Proposition 10.16 (i) *Let (r_k) be a sequence in $\mathbb{I} \setminus \mathbb{D}$, and let $r \in \mathbb{I} \setminus \mathbb{D}$ be such that $r_k \to r$. Then $\varepsilon_{r_k} \to \varepsilon_r$ pointwise on $\mathbb{N}$.*

(ii) *Let (r_k) be an increasing (respectively, decreasing) sequence in $\mathbb{I} \setminus \mathbb{D}$, and let $r \in \mathbb{D}$ be such that $r_k \to r$. Then $\varepsilon_{r_k} \to \varepsilon_r^-$ (respectively, $\varepsilon_{r_k} \to \varepsilon_r^+$) pointwise on $\mathbb{N}$.*

(iii) *The set U is a closed subset of $\mathbb{I}$ with $U \cap \mathbb{D} = \emptyset$.*

(iv) *For $r, s \in U$ with $|r - s| < 2^{-(n+2)}$, $\varepsilon_r \restriction n = \varepsilon_s \restriction n$.*

(v) *The subset V is dense in U.*

Proof (i) and (ii) These are clear.

(iii) Take (r_k) in U with $r_k \to r \in \mathbb{I}$. We may suppose, by passing to a subsequence, that either (r_k) is increasing or (r_k) is decreasing. By (i) or (ii), (ε_{r_k}) is pointwise convergent. Take (f_k) in $\{0,1\}^{\mathbb{N}}$ with $f_k^* = \varepsilon_{r_k} \quad (k \in \mathbb{N})$. Then (f_k) is pointwise convergent on $\mathbb{N}$, say $f_k \to f$. Take $s \in U$ with $\varepsilon_s = f^*$. Then $\varepsilon_{r_k} \to \varepsilon_s$ pointwise on $\mathbb{N}$, and so, for each $n \in \mathbb{N}$, $|r_k - s| < 2^{-n}$ eventually. This shows that $r_k \to s$ in $\mathbb{I}$, and so $r = s \in U$. Thus U is closed.

(iv) Take $r, s \in U$ with $r \neq s$, and suppose that $\varepsilon_r(n) = 1$ and $\varepsilon_s(n) = 0$, where $n = \delta(\varepsilon_r, \varepsilon_s)$.

By (10.9), ε_r and ε_s must each take the values 0 and 1 at one of the three points $n+1$, $n+2$, and $n+3$, and so

$$r - s \geq 2^{-n} + 2^{-(n+3)} - 2^{-(n+1)} - 2^{-(n+2)} = 3 \cdot 2^{-(n+3)} \geq 2^{-(n+2)}.$$

The result follows.

(v) Take $r \in U$, say $\varepsilon_r = f^*$, and take $m \in \mathbb{N}$. Define

$$g(n) = f(n) \quad (n = 1, \ldots, m), \quad g(n) = 1 \quad (n \geq m+1),$$

and take $s \in \mathbb{I}$ with $\varepsilon_s = g^*$, so that $s \in V$. We have $\varepsilon_r \restriction 2m = \varepsilon_s \restriction 2m$, and so $|s - r| < 2^{-2m}$. Thus V is dense in U. □

By a trivial variation of (v), above, we see that U has no isolated points, and so U is a compact, perfect, totally disconnected set which is homeomorphic to the Cantor set.

We now associate a canonical function Φ to the set U.

Definition 10.17 *For $r \in U$, take $f_r \in \{0,1\}^{\mathbb{N}}$ with $f_r^* = \varepsilon_r$, and define*

$$\Phi : r \mapsto \sum_{n=1}^{\infty} \frac{f_r(n)}{2^n}, \quad U \to \mathbb{I}. \tag{10.12}$$

Let $s \in \mathbb{I} \setminus \mathbb{D}$. Then there is a unique element $G(s) \in U$ with

$$\varepsilon_{G(s)} = \varepsilon_s^* .$$

Indeed we have

$$G(s) = \sum_{n=1}^{\infty} \frac{\varepsilon_s^*(n)}{2^n} \quad \text{and} \quad \Phi(G(s)) = \sum_{n=1}^{\infty} \frac{\varepsilon_s(n)}{2^n} = s . \tag{10.13}$$

This shows that $\Phi(U) \supset \mathbb{I} \setminus \mathbb{D}$ and that $G : \mathbb{I} \setminus \mathbb{D} \to U$ is a form of inverse to $\Phi : U \to \mathbb{I}$.

Let $r \in V_n$. Then

$$\Phi(r) = \sum_{k=1}^{n} \frac{f_r(k)}{2^k} + \sum_{k=n+1}^{\infty} \frac{1}{2^k} = \sum_{k=1}^{n} \frac{f_r(k)}{2^k} + \frac{1}{2^n},$$

and so $\Phi(r) \in \mathbb{D}_n$ and $\varepsilon^-_{\Phi(r)} = f_r$.

Take $r, s \in U$ with $f_r \prec f_s$ in the lexicographic order on $\mathbb{N}$. Then $f_r^* \prec f_s^*$, and so $r < s$. This shows that, if $r < s$ in U, then $f_r \prec f_s$, and hence that $\Phi(r) \leq \Phi(s)$. Thus Φ is increasing on U.

Next take $r, s \in U$ with $r \neq s$, and set

$$A = \{n \in \mathbb{N} : |r - s| \leq 2^{-(2n+2)}\},$$

a finite subset of $\mathbb{N}$. If $A \neq \emptyset$, set $n_0 = \max A$, so that $2^{-(2n_0+4)} < |r - s|$ and $\varepsilon_r \restriction 2n_0 = \varepsilon_s \restriction 2n_0$ by Proposition 10.16(iv), and hence

$$|\Phi(r) - \Phi(s)| \leq 2^{-n_0} \leq 4|r - s|^{1/2} .$$

If $A = \emptyset$, then $2^{-4} < |r - s|$ and $|\Phi(r) - \Phi(s)| \leq 1 \leq 4|r - s|^{1/2}$. It follows that

$$|\Phi(r) - \Phi(s)| \leq 4|r - s|^{1/2} \qquad (r, s \in U) .$$

In particular, $\Phi : U \to \mathbb{I}$ is continuous, and hence $\Phi(U) = \mathbb{I}$. Define $\Phi(0) = 0$ and $\Phi(1) = 1$. Let (r, s) be a complementary interval to the closed set U in $\mathbb{I}$. Then necessarily $\Phi(r) = \Phi(s)$; we extend Φ to have domain $\mathbb{I}$ by setting $\Phi(t) = \Phi(r)$ for $t \in (r, s)$ in this case. Clearly we now have

$$|\Phi(r) - \Phi(s)| \leq 4\,|r - s|^{1/2} \quad (r, s \in \mathbb{I}) .$$

We have proved the following result.

Proposition 10.18 *There is a function $\Phi \in \mathfrak{L}(\mathbb{I})$ such that:*

(i) *Φ is increasing, and $\Phi(0) = 0$ and $\Phi(1) = 1$;*

(ii) *$\Phi(U) = \Phi(\mathbb{I}) = \mathbb{I}$;*

(iii) *if $r \in U$ and $\Phi(r) \in \mathbb{I} \setminus \mathbb{D}$, then $\varepsilon^*_{\Phi(r)} = \varepsilon_r$;*

(iv) *if $r \in V_n$, then $\Phi(r) \in \mathbb{D}_n$ and $\varepsilon^{-*}_{\Phi(r)} = \varepsilon_r$.* □

Earlier, we defined the function $G : \mathbb{I} \setminus \mathbb{D} \to U$ so that $\varepsilon_{G(s)} = \varepsilon_s^*$ for each $s \in \mathbb{I} \setminus \mathbb{D}$ (see (10.13)). The function G is continuous, but it does not have a continuous extension to a map $G : \mathbb{I} \to \mathbb{I}$ because U is a closed and disconnected subset of $\mathbb{I}$. This 'failure' will have significance when we discuss weakly Cauchy complete structures a little later.

We now seek analogues related to a prime ideal of the set U of Definition 10.15, and of the associated function Φ, defined in Definition 10.17.

Definition 10.19 *The set $U_{(P)}$ consists of those elements $r \in \mathbb{I}_{(P)} \setminus \mathbb{D}_{(P)}$ such that $\varepsilon_r^{(P)} = f^*$ for some $f : \mathbb{N}_{(P)} \to \{0, 1\}$.*

The function f with $f^* = \varepsilon_r^{(P)}$ is uniquely specified, and certainly $f \restriction a \in \{0,1\}^a_{(P)}$ for each $a \in \mathbb{N}_{(P)}$.

The function Φ belongs to $\mathfrak{L}(\mathbb{I})$, and so Φ induces a function $\Phi_{(P)} : \mathbb{I}_{(P)} \to \mathbb{I}_{(P)}$ in the usual way (see Chapter 6). By Theorem 6.5(iii), $\Phi_{(P)}$ is increasing on $\mathbb{I}_{(P)}$. By Proposition 10.18(ii), $\varepsilon^*_{\Phi(r)} = \varepsilon_r$ whenever $r \in U$ and $\Phi(r) \in \mathbb{I} \setminus \mathbb{D}$, and so we would expect that

$$\varepsilon^{(P)*}_{\Phi_{(P)}(r)} = \varepsilon_r^{(P)} \tag{10.14}$$

whenever $r \in U_{(P)}$ and $\Phi_{(P)}(r) \in \mathbb{I}_{(P)} \setminus \mathbb{D}_{(P)}$. We shall prove that this is indeed the case. This is a typical example of the type of question that we must now be concerned with: we are given a function $F \in \mathfrak{L}(\mathbb{I}^m)$ with $F(\mathbb{I}^m) \subset \mathbb{I}$ for some $m \in \mathbb{N}$ such that F satisfies certain conditions, and we shall need to know that the induced function $F_{(P)}$ satisfies the analogous conditions.

We define a set $V_{(P)}$ to be the analogue of V. Let $a \in \mathbb{N}_{(P)}$. Then $V_a^{(P)}$ consists of those elements $r \in \mathbb{I}_{(P)} \setminus \mathbb{D}_{(P)}$ such that $\varepsilon_r^{(P)} = f^*$ for some function $f : \mathbb{N}_{(P)} \to \{0, 1\}$ with $f(b) = 1 \;\; (b > a)$, and

$$V_{(P)} = \bigcup \{V_a^{(P)} : a \in \mathbb{N}_{(P)}\}\,.$$

Take $r \in V_a^{(P)}$, and set $s = (2/3) \cdot 2^{-2a}$. Then $s \in \mathbb{Q}_{(P)} \setminus \mathbb{D}_{(P)}$ and

$$\varepsilon_s^{(P)}(b) = \varepsilon_r^{(P)}(b) \quad (b \geq 2a + 1)\,. \tag{10.15}$$

By Lemma 8.46(iii), $r - s \in \mathbb{D}_{2a}^{(P)}$, and so $V_{(P)} \subset U_{(P)} \cap \mathbb{Q}_{(P)}$.

Lemma 10.20 *Suppose that $r \in U_{(P)}$ and $s \in \mathbb{I}_{(P)}$ with $r < s$. Then there exists $t \in V_{(P)}$ such that $r \leq t < s$.*

Proof We may suppose that $s \in \mathbb{I}_{(P)} \setminus \mathbb{D}_{(P)}$ and that $r \notin V_{(P)}$.

Choose a function $f_r : \mathbb{N}_{(P)} \to \{0, 1\}$ with $\varepsilon_r^{(P)} = f_r^*$, so that f_r is not eventually equal to 1. There exists $b \in \mathbb{N}_{(P)}$ such that $\varepsilon_r^{(P)} \restriction 2b \prec \varepsilon_s^{(P)} \restriction 2b$, $f_r(b) = 0$, and $r + 2^{-b} < s$. Define $f : \mathbb{N}_{(P)} \to \{0, 1\}$ by

$$f(a) = \begin{cases} f_r(a) & (a \leq b)\,, \\ 1 & (a > b)\,. \end{cases}$$

Then $f, f^* \in \{0,1\}^{\mathbb{N}}_{(P)}$ by Lemma 9.2, where we note that $f_r \restriction b \in \{0,1\}^b_{(P)}$, and so there exists $t \in V_b^{(P)}$ with $\varepsilon_t^{(P)} = f^*$.

We have $\varepsilon_r^{(P)} \restriction 2b = \varepsilon_t^{(P)} \restriction 2b$, $\varepsilon_r^{(P)}(2b+1) = 0$, and $\varepsilon_t^{(P)}(2b+1) = 1$, and so $r < t$ and $t - r < 2^{-2b}$ by Lemma 8.46(ii). Thus $r \leq t < s$. □

It follows from Lemma 10.20 that $V_{(P)}$ is order-dense in $U_{(P)}$ and that $U_{(P)}$ is contained in the closure of $V_{(P)}$ in the order topology on $\mathbb{I}_{(P)}$.

Lemma 10.21 *Let Q be a non-maximal, prime z-ideal in $C(\Omega)$. Let $a \in \mathbb{N}_{(Q)}$, take $r \in V_a^{(Q)}$, and define $s = \Phi_{(Q)}(r)$. Then $s \in \mathbb{D}_a^{(Q)}$ and*

$$\varepsilon_s^{-(Q)*} = \varepsilon_r^{(Q)} .$$

Proof This follows from Proposition 10.18(iv) by another standard pointwise argument. □

Again P is a non-maximal, prime ideal in $C(\Omega)$, and Q is the maximum prime z-ideal below P.

Lemma 10.22 *Let $a \in \mathbb{N}_{(P)}$.*

(i) *Suppose that $r, s \in (\mathbb{Q}_{(P)} \cap \mathbb{I}_{(P)}) \setminus \mathbb{D}_{(P)}$ are such that $\varepsilon_r^{(P)} \restriction a = \varepsilon_s^{(P)*} \restriction a$. Then $\varepsilon_r^{(Q)} \restriction a = \varepsilon_s^{(Q)*} \restriction a$.*

(ii) $V_a^{(P)} \subset V_a^{(Q)}$.

(iii) $\Phi_{(P)} \mid V_{(P)} = \Phi_{(Q)} \mid V_{(P)}$.

Proof (i) Define $h : \mathbb{N}^5 \to \mathbb{N}$ as follows. Suppose that

$$n = (n_1, n_2, n_3, n_4, n_5) \in \mathbb{N}^5 ,$$

and set $r_1 = n_1/n_2$ and $r_2 = n_3/n_4$. If $r_1 \notin \mathbb{I} \setminus \mathbb{D}$ or if $r_2 \notin \mathbb{I} \setminus \mathbb{D}$, then $h(n) = n_5 + 1$. If both $r_1 \in \mathbb{I} \setminus \mathbb{D}$ and $r_2 \in \mathbb{I} \setminus \mathbb{D}$, define $h(n) = n_5 + 2$ in the case where $\varepsilon_{r_1} \restriction n_5 = \varepsilon^*_{r_2} \restriction n_5$ and otherwise define $h(n) = j$, where j is the minimum element of $\mathbb{N}$ with $\varepsilon_{r_1}(j) \neq \varepsilon^*_{r_2}(j)$.

Take $a_1, a_2, a_3, a_4 \in \mathbb{M}_{(P)}$ with $r = a_1/a_2$ and $s = a_3/a_4$.

Assume towards a contradiction that $\varepsilon_r^{(Q)} \restriction a \neq \varepsilon_s^{(Q)*} \restriction a$, and set

$$c = h_{(Q)}(a_1, a_2, a_3, a_4, a) .$$

By a standard pointwise argument, $c \leq a$ and $\varepsilon_r^{(Q)}(c) \neq \varepsilon_s^{(Q)*}(c)$. By the transfer principle, Corollary 8.30, $c \in \mathbb{M}_{(P)}$. Also $c \leq a \in \mathbb{N}_{(P)}$, and so, by Lemma 8.33(i), $c \in \mathbb{N}_{(P)}$. By Definition 8.38(iii), $\varepsilon_r^{(P)}(c) \neq \varepsilon_s^{(P)*}(c)$. But this contradicts the fact that $\varepsilon_r^{(P)} \restriction a = \varepsilon_s^{(P)*} \restriction a$. Hence $\varepsilon_r^{(Q)} \restriction a = \varepsilon_s^{(Q)*} \restriction a$.

(ii) Take $r \in V_a^{(P)}$, and set $s = (2/3) \cdot 2^{-2a}$. As above, $|r - s| \in \mathbb{D}_{2a}^{(P)}$, and so $|r - s| \in \mathbb{D}_{2a}^{(Q)}$. By Lemma 8.46(iii),

$$\varepsilon_s^{(Q)}(b) = \varepsilon_r^{(Q)}(b) \quad (b \geq 2a+1) ,$$

and so $r \in V_a^{(Q)}$.

(iii) Take $r \in V_a^{(P)}$, and set $x = \Phi_{(Q)}(r)$ and $y = \Phi_{(P)}(r)$. By Lemma 10.5, there exist $f \in C(\Omega)$ and $g \in C(\Omega) \setminus P$ with

$$x = \pi_Q(f)/\pi_Q(g) \quad \text{and} \quad y = \pi_P(f)/\pi_P(g),$$

as in (10.5) and (10.4), respectively.

By (ii), $r \in V_a^{(Q)}$, and so, by Lemma 10.21, $x \in \mathbb{D}_a^{(Q)}$. In the notation of Lemma 10.4, we have $x \in J_y \cap \mathbb{D}_a^{(Q)}$. It follows from Lemma 10.4(iii) that $y \in J_y \cap \mathbb{D}^{(P)}$, and so $x = y$ because $J_y \cap \mathbb{Q}_{(P)}$ contains at most one point by Lemma 10.4(ii). □

Theorem 10.23 *Let P be a non-maximal, prime ideal in an algebra $C(\Omega)$. Then*

$$^*\{0,1\}^{\mathbb{N}}_{(P)} \subset \{0,1\}^{\mathbb{N}}_{(P)}.$$

Proof As usual, Q is the maximum prime z-ideal below P.

Let $f \in {}^*\{0,1\}^{\mathbb{N}}_{(P)}$, and take $r \in \mathbb{I}_{(P)} \setminus \mathbb{D}_{(P)}$ with $f^* = \varepsilon_r^{(P)}$, so that $r \in U_{(P)}$. By Lemma 10.20, there is a decreasing sequence $(r_a : a \in \mathbb{N}_{(P)})$ in $V_{(P)}$ with $r_a \to r$ in the order-topology on $\mathbb{I}_{(P)}$. By Lemma 10.22(ii), $(r_a) \subset V_{(Q)}$, and so, for each $a \in \mathbb{N}_{(P)}$, there exist $f_a : \mathbb{N}_{(P)} \to \{0,1\}$ and $g_a : \mathbb{N}_{(Q)} \to \{0,1\}$ with $f_a^* = \varepsilon_{r_a}^{(P)}$ and $g_a^* = \varepsilon_{r_a}^{(Q)}$; we have $f_a = g_a \mid \mathbb{N}_{(P)}$.

Let $a \in \mathbb{N}_{(P)}$, and set $s_a = \Phi_{(P)}(r_a)$; by Lemma 10.22(iii), $s_a = \Phi_{(Q)}(r_a)$. The sequence (s_a) is decreasing, and $s_a \to s$, where $s = \Phi_{(P)}(r)$, because $\Phi_{(P)}$ is continuous with respect to the order-topology on $\mathbb{I}_{(P)}$. By Lemma 10.21, $s_a \in \mathbb{D}_{(Q)}$ and $g_a = \varepsilon_{s_a}^{-(Q)}$, and so $f_a = \varepsilon_{s_a}^{-(P)}$.

By Lemma 8.46(ii), $\varepsilon_{r_a}^{(P)} \to \varepsilon_r^{(P)}$ pointwise on $\mathbb{N}_{(P)}$, and so $f_a \to f$ pointwise on $\mathbb{N}_{(P)}$. Thus $\varepsilon_{s_a}^{-(P)} \to f$ pointwise on $\mathbb{N}_{(P)}$.

Set $s = \Phi_{(P)}(r)$. There are three possibilities:

(i) $s \in \mathbb{D}_{(P)}$ and $\varepsilon_{s_a}^{-(P)} \to \varepsilon_s^{-(P)}$ pointwise;

(ii) $s \in \mathbb{D}_{(P)}$ and $\varepsilon_{s_a}^{-(P)} \to \varepsilon_s^{+(P)}$ pointwise;

(iii) $s \notin \mathbb{D}_{(P)}$ and $\varepsilon_{s_a}^{-(P)} \to \varepsilon_s^{(P)}$ pointwise.

Thus either $s \in \mathbb{D}_{(P)}$ and $f = \varepsilon_s^{-(P)}$ or $s \in \mathbb{D}_{(P)}$ and $f = \varepsilon_s^{+(P)}$ or $s \notin \mathbb{D}_{(P)}$ and $f = \varepsilon_s^{(P)}$. In each case, $f \in \{0,1\}^{\mathbb{N}}_{(P)}$. □

As promised, we have proved that the model $\mathfrak{N}_{(P)}$ is weakly Cauchy complete if and only if

$$\{0,1\}^{\mathbb{N}}_{(P)} \subset {}^*\{0,1\}^{\mathbb{N}}_{(P)}.$$

In the course of the proof of Theorem 10.23, we proved that (10.14) does hold whenever $r \in U_{(P)}$ and $\Phi_{(P)}(r) \in \mathbb{I}_{(P)} \setminus \mathbb{D}_{(P)}$.

We now prove that $^*\{0,1\}^{\mathbb{N}}_{(P)}$ and $^*\mathcal{P}(N)_{(P)}$ have the required closure properties.

Theorem 10.24 *Let P be a non-maximal, prime ideal in an algebra $C(\Omega)$, and suppose that f_1 and f_2 belong to ${}^*\{0,1\}^{\mathbb{N}}_{(P)}$. Then $f_1 \vee f_2$, $f_1 \wedge f_2$, and $1 - f_1$ belong to ${}^*\{0,1\}^{\mathbb{N}}_{(P)}$.*

Proof Let U be as in Definition 10.15. Then U^2 is a closed subset of $\mathbb{I}^2$, and $(U \cap \mathbb{Q})^2$ is dense in U^2. For $r_1, r_2 \in U$, there exist $f_1, f_2 : \mathbb{N} \to \{0,1\}$ such that $\varepsilon_{r_1} = f_1^*$ and $\varepsilon_{r_2} = f_2^*$; we define $F(r_1, r_2) = s$, where $\varepsilon_s = (f_1 \vee f_2)^*$. Clearly

$$|F(r_1,r_2) - F(s_1,s_2)| \leq |r_1 - s_1| + |r_2 - s_2| \quad (r_1, r_2, s_1, s_2 \in U),$$

and so F extends to a function $F \in \mathfrak{L}(\mathbb{I}^2)$.

The theorem is proved by showing that $F_{(P)}$ acts on $U^2_{(P)}$ in the same way as F acts on U^2. First we work with $F_{(Q)}$ on $U^2_{(Q)}$, where Q is the maximum prime z-ideal below P.

The set $U_{(Q)} \cap \mathbb{Q}_{(Q)}$ is order-dense in $U_{(Q)}$. For $(x_1, x_2) \in (U_{(Q)} \cap \mathbb{Q}_{(Q)})^2$, choose $g_1, g_2 : \mathbb{N}_{(Q)} \to \{0,1\}$ such that $\varepsilon^{(Q)}_{x_1} = g_1^*$ and $\varepsilon^{(Q)}_{x_2} = g_2^*$. Then, by a pointwise argument, $F_{(Q)}(x_1, x_2) = z$, where $z \in \mathbb{I}_{(Q)}$ is taken to be such that $\varepsilon^{(Q)}_z = (g_1 \vee g_2)^*$. By a standard continuity argument, the same result holds whenever $(x_1, x_2) \in U^2_{(Q)}$.

We shall require the following remark. Let x and y be related as in (10.4) and (10.5), and suppose that $y \in U_{(P)}$. Then $U_{(Q)} \cap J_y \neq \emptyset$.

Now choose functions $f_1, f_2 \in {}^*\{0,1\}^{\mathbb{N}}_{(P)}$, say $f_j^* = \varepsilon^{(P)}_{y_j}$, where $y_j \in U_{(P)}$. For each $j = 1, 2$, take $\phi_j \in C(\Omega)^+$ and $\psi_j \in C(\Omega)^+ \setminus P$ with $y_j = \pi_P(\phi_j)/\pi_P(\psi_j)$, and set $x_j = \pi_Q(\phi_j)/\pi_Q(\psi_j)$, so that $x_j \in J_{y_j}$. Then there exist $\phi \in C(\Omega)^+$ and $\psi \in C(\Omega)^+ \setminus P$ with

$$\pi_P(\phi)/\pi_P(\psi) = F_{(P)}(y_1, y_2), \quad \pi_Q(\phi)/\pi_Q(\psi) = F_{(Q)}(x_1, x_2),$$

say $z = F_{(P)}(y_1, y_2)$ and $t = F_{(Q)}(x_1, x_2)$. Clearly $t \in J_z$. By the above remark, there exist $x_1' \in J_{y_1} \cap U_{(Q)}$ and $x_2' \in J_{y_2} \cap U_{(Q)}$. Set $t' = F_{(Q)}(x_1', x_2')$. Since $F \in \mathfrak{L}(\mathbb{I}^2)$, it follows that $|t - t'| < 2^{-a}$ $(a \in \mathbb{N}_{(P)})$, and so $t' \in J_z$.

For $j = 1, 2$, take $g_j : \mathbb{N}_{(Q)} \to \{0,1\}$ with $g_j^* = \varepsilon^{(Q)}_{x_j'}$, so that $f_j^* = g_j^* \mid \mathbb{N}_{(P)}$. Then $f_j = g_j \mid \mathbb{N}_{(P)}$ and

$$(f_1 \vee f_2)^* = (g_1 \vee g_2)^* \mid \mathbb{N}_{(P)}.$$

Since $x_j' \in U_{(Q)}$, we have $\varepsilon^{(Q)}_{t'} = (g_1 \vee g_2)^*$, and so $\varepsilon^{(Q)}_{t'} \mid \mathbb{N}_{(P)} = (f_1 \vee f_2)^*$. Assume that $z \in \mathbb{D}_{(P)}$. Then, by Lemma 10.7, either $\varepsilon^{(Q)}_{t'} \mid \mathbb{N}_{(P)} = \varepsilon_z^{+(P)}$ or $\varepsilon^{(Q)}_{t'} \mid \mathbb{N}_{(P)} = \varepsilon_z^{-(P)}$, a contradiction. Thus $z \notin \mathbb{D}_{(P)}$, and so $\varepsilon^{(P)}_z = \varepsilon^{(Q)}_{t'} \mid \mathbb{N}_{(P)}$ because $t' \in J_z$. It follows that $(f_1 \vee f_2)^* = \varepsilon^{(P)}_z$, and hence $(f_1 \vee f_2)^* \in \{0,1\}^{\mathbb{N}}_{(P)}$.

The other results are similar. □

Corollary 10.25 *Let P be a non-maximal, prime ideal in an algebra $C(\Omega)$, and let A and B belong to ${}^*\mathcal{P}(\mathbb{N})_{(P)}$. Then $A \cup B$, $A \cap B$, and $\mathbb{N}_{(P)} \setminus A$ belong to ${}^*\mathcal{P}(\mathbb{N})_{(P)}$.* □

The proof of Theorem 10.24 generalizes easily to show that ${}^*\{0,1\}^{\mathbb{N}}_{(P)}$ is a rich collection of functions closed under many operations; its closure properties are analogous to those of $\{0,1\}^{<\mathbb{N}}_{(P)}$. In fact, if $g_1, \dots, g_n$ belong to ${}^*\{0,1\}^{\mathbb{N}}_{(P)}$ and $h : \mathbb{N}_{(P)} \to \{0,1\}$ is any function which can be defined in $\langle \mathbb{N}_{(P)}, +, \cdot \rangle$ from $g_1, \dots, g_n$ by a formula involving only bounded quantifiers, then $h \in {}^*\{0,1\}^{\mathbb{N}}_{(P)}$.

We shall show below that there are prime ideals P such that $\mathfrak{N}_{(P)}$ is not weakly Cauchy complete. However, the following theorem, which will not be proved, shows that the set ${}^*\{0,1\}^{\mathbb{N}}_{(P)}$ is always quite substantial. For a subset S of $\mathbb{N}$, the set $S_{(Q)}$ was defined in Definition 8.10, and we set $S_{(P)} = S_{(Q)} \cap \mathbb{N}_{(P)}$.

Theorem 10.26 *Let P be a non-maximal, prime ideal in an algebra $C(\Omega)$, and let S be a subset of $\mathbb{N}$ such that $S_{(P)}$ is cofinal in $\mathbb{N}_{(P)}$. Let $f \in \{0,1\}^{\mathbb{N}}_{(P)}$ be such that, for each $a, b \in S_{(P)}$ with $a < b$, there exists $c \in [a,b] \cap \mathbb{N}_{(P)}$ with $f(c) = 0$. Then $f^* \in \{0,1\}^{\mathbb{N}}_{(P)}$.* □

In the case where $S = 2\mathbb{N}$, $S_{(P)}$ is indeed cofinal in $\mathbb{N}_{(P)}$ for each non-maximal, prime ideal P, and so the theorem implies that, for each $f \in \{0,1\}^{\mathbb{N}}_{(P)}$ such that $f^* \in \{0,1\}^{\mathbb{N}}_{(P)}$, necessarily $f^{**} \in \{0,1\}^{\mathbb{N}}_{(P)}$.

It would seem natural to those accustomed to working with ultrapowers to guess that at least every hyper-real field is weakly Cauchy complete. However this is not the case: we shall now present an example of a maximal ideal M in $C(\mathbb{N} \times \mathbb{I})$ such that the model $\mathfrak{N}_{(M)}$ is not weakly Cauchy complete (with CH). The construction is a variation of that given in Chapter 9. The result will not be required later.

The key to the production of the model $\mathfrak{N}_{(M)}$ is the exploitation of the failure referred to earlier: the function G such that $\varepsilon_{G(s)} = \varepsilon_s^*$ for each $s \in \mathbb{I} \setminus \mathbb{D}$ is continuous on $\mathbb{I} \setminus \mathbb{D}$, but does not have a continuous extension to $\mathbb{I}$. Indeed, suppose that $F \in C(\mathbb{I}, \mathbb{I})$ and that W is a non-empty, open subset of $\mathbb{I}$. Then there exist a non-empty, open subset V of W and $\varepsilon > 0$ with

$$|F(t) - G(t)| > \varepsilon \quad (t \in V \setminus \mathbb{D}). \tag{10.16}$$

To see this, choose $s \in W \setminus \mathbb{D}$ such that $F(s) \neq G(s)$, set $\varepsilon = |F(s) - G(s)|/3$, and choose $V \subset W$ with $s \in V$ such that

$$\max\{|F(t) - F(s)|, \ |G(t) - G(s)|\} < \varepsilon \quad (t \in V \setminus \mathbb{D}).$$

Let X be the completely regular topological space

$$X = \mathbb{N} \times \mathbb{I};$$

X is related to, but different from, the space which was called X in Chapter 9. As in Chapter 9, we fix a free ultrafilter $\mathcal{U}$ on $\mathbb{N}$; we recall the convention that we denote by $\mathbb{N}_{(\mathcal{U})}$, $\mathbb{I}_{(\mathcal{U})}$, etc., the objects $\mathbb{N}_{(Q)}$, $\mathbb{I}_{(Q)}$, etc., which are associated to the prime ideal $Q = l^\infty \cap \mathbf{Z}^{-1}[\mathcal{U}]$ in $C(\beta\mathbb{N})$; for example, for $h : \mathbb{N} \to \mathbb{I}$, the corresponding element $h_{\mathcal{U}} \in \mathbb{I}_{(\mathcal{U})} = \mathbb{I}^{\mathbb{N}}/\mathcal{U}$ is $[h]_{\mathcal{U}} = [(h(k))]_{\mathcal{U}}$. For $k \in \mathbb{N}$, the subset $\{k\} \times \mathbb{I}$ of X is denoted by $\mathbb{I}_k$, so that each $\mathbb{I}_k$ is homeomorphic to $\mathbb{I}$ and $X = \bigcup\{\mathbb{I}_k : k \in \mathbb{N}\}$. For each $\sigma \subset \mathbb{N}$,

$$\mathbb{I}_{[\sigma]} = \bigcup\{\mathbb{I}_k : k \in \sigma\},$$

so that $\mathbb{I}_{[\sigma]} \in \mathbf{Z}(X)$, and $\mathcal{F}_0$, defined to be

$$\mathcal{F}_0 = \{\mathbb{I}_{[\sigma]} : \sigma \in \mathcal{U}\},$$

is a z-filter on X. Essentially as in Chapter 9, we associate to each set $S \subset X$ a corresponding set $S_{\mathcal{U}}$ in $\mathbb{I}_{(\mathcal{U})}$. Indeed, there exists a sequence (S_k) such that $S_k \subset \mathbb{I}$ and $S \cap \mathbb{I}_k = \{k\} \times S_k$ for each $k \in \mathbb{N}$, and then

$$S_{\mathcal{U}} = \{h_{\mathcal{U}} = [h]_{\mathcal{U}} \ : \ h : \mathbb{N} \to \mathbb{I}, \ \ \{k \in \mathbb{N} : h(k) \in S_k\} \in \mathcal{U}\}.$$

Again it is clear that $\emptyset_{\mathcal{U}} = \emptyset$ and $Z_{\mathcal{U}} = \mathbb{I}_{(\mathcal{U})}$ for $Z \in \mathcal{F}_0$. Also $S_{\mathcal{U}} \cap T_{\mathcal{U}} = (S \cap T)_{\mathcal{U}}$, $S_{\mathcal{U}} \cup T_{\mathcal{U}} = (S \cup T)_{\mathcal{U}}$, and $(X \setminus S)_{\mathcal{U}} = \mathbb{I}_{(\mathcal{U})} \setminus S_{\mathcal{U}}$ for $S, T \subset X$. If S is open (respectively, closed, dense) in X, then $S_{\mathcal{U}}$ is open (respectively, closed, dense) in $\mathbb{I}_{(\mathcal{U})}$. Finally, we define

$$\mathbf{Z}(\mathbb{I})_{(\mathcal{U})} = \{Z_{\mathcal{U}} : Z \in \mathbf{Z}(X)\}.$$

We note that $\mathbf{Z}(\mathbb{I})_{(\mathcal{U})}$ contains each closed interval $[r, s]_{(\mathcal{U})}$ of $\mathbb{I}_{(\mathcal{U})}$ with $r \leq s$. For suppose that $r = [(r_k)]_{\mathcal{U}}$ and $s = [(s_k)]_{\mathcal{U}}$, and set $\sigma = \{k \in \mathbb{N} : r_k \leq s_k\}$. Then $\sigma \in \mathcal{U}$, and $[r, s]_{(\mathcal{U})} = Z_{\mathcal{U}}$, where

$$Z = \bigcup\{\{k\} \times [r_k, s_k] : k \in \sigma\}.$$

The *length* of an interval I in $\mathbb{I}$ or $\mathbb{I}_{(\mathcal{U})}$ is denoted by $l(I)$.

We fix some additional notation. Let $n \in \mathbb{N}$, and take $f \in \{0,1\}^n$. Then $r = \sum_{j=1}^{n} f(j)/2^j$ is the unique element in $\mathbb{D}_n$ with $\varepsilon_r^+ \restriction n = f$. We set

$$K_f = [r, r + 2^{-n}],$$

a closed subinterval of $\mathbb{I}$ with $l(K_f) = 2^{-n}$. For each $t \in \mathbb{I} \setminus \mathbb{D}$, we have $\varepsilon_t \restriction n = f$ if and only if $t \in K_f \setminus \mathbb{D}$.

Lemma 10.27 *Let $F \in C(\mathbb{I}, \mathbb{I})$, and let U be a non-empty, open subset of $\mathbb{I}$. Then there exists $f \in \{0,1\}^{<\mathbb{N}}$ such that $K_f \subset U$ and $F(K_f) \cap K_{f^*} = \emptyset$.*

Proof Let V be a non-empty, open subset of U and $\varepsilon > 0$ be such that (10.16) holds. Choose $k \in \mathbb{N}$ and $f \in \{0,1\}^k$ with $2^{-k} < \varepsilon$ and with $K_f \subset V$.

Assume towards a contradiction that there exists $t \in F(K_f) \cap K_{f^*}$. Then there exists $s \in K_f \setminus \mathbb{D}$ with $|F(s) - t| < \varepsilon - 2^{-k}$. We have $G(s) \in K_{f^*}$. But $l(K_{f^*}) = 2^{-2k}$, and so $|G(s) - t| \leq 2^{-2k}$, whence

$$|F(s) - G(s)| < \varepsilon - 2^{-k} + 2^{-2k} < \varepsilon,$$

a contradiction. Thus $F(K_f) \cap K_{f^*} = \emptyset$. □

Now let $a \in \mathbb{N}_{(\mathcal{U})}$, and take $f \in \{0,1\}^a_{(\mathcal{U})}$. There is a unique $r \in \mathbb{D}_a^{(\mathcal{U})}$ with $\varepsilon_r^{+(\mathcal{U})} \restriction a = f$; set

$$K_f^{(\mathcal{U})} = [r, r + 2^{-a}],$$

a closed subinterval of $\mathbb{I}_{(\mathcal{U})}$ with $l(K_f^{(\mathcal{U})}) = 2^{-a}$. Thus, for each $t \in \mathbb{I}_{(\mathcal{U})} \setminus \mathbb{D}_{(\mathcal{U})}$, we have $\varepsilon_t^{(\mathcal{U})} \upharpoonright a = f$ if and only if $t \in K_f^{(\mathcal{U})} \setminus \mathbb{D}_{(\mathcal{U})}$. The following lemma about the intervals $K_f^{(\mathcal{U})}$ is immediate.

Lemma 10.28 *Let $a, b \in \mathbb{N}_{(\mathcal{U})}$, let $f \in \{0,1\}^a_{(\mathcal{U})}$, and let $g \in \{0,1\}^b_{(\mathcal{U})}$. Then:*

(i) $K_f^{(\mathcal{U})} \subset K_g^{(\mathcal{U})}$ *if and only if $b \leq a$ and $g = f \upharpoonright b$;*

(ii) *either $K_f^{(\mathcal{U})} \cap K_g^{(\mathcal{U})} = \emptyset$ or $K_f^{(\mathcal{U})} \subset K_g^{(\mathcal{U})}$ or $K_g^{(\mathcal{U})} \subset K_f^{(\mathcal{U})}$ or there exists $r \in \mathbb{D}_{(\mathcal{U})}$ with $K_f^{(\mathcal{U})} \cap K_g^{(\mathcal{U})} = \{r\}$.* □

We associate to each $F \in C(X, \mathbb{I})$ a function $F_{\mathcal{U}} : \mathbb{I}_{(\mathcal{U})} \to \mathbb{I}_{(\mathcal{U})}$ and an open, dense subset $V_{\mathcal{U}}^F$ as follows.

For $k \in \mathbb{N}$, define $F_k : t \mapsto F(k,t)$, $\mathbb{I} \to \mathbb{I}$, so that (F_k) is a sequence in $C(\mathbb{I}, \mathbb{I})$. Then

$$F_{\mathcal{U}} : [(t_k)]_{\mathcal{U}} \mapsto [F_k(t_k)]_{\mathcal{U}}, \quad \mathbb{I}^{\mathbb{N}}/\mathcal{U} \to \mathbb{I}^{\mathbb{N}}/\mathcal{U},$$

defines the function $F_{\mathcal{U}}$ via our canonical identification of $\mathbb{I}^{\mathbb{N}}/\mathcal{U}$ with $\mathbb{I}_{(\mathcal{U})}$. It is easy to see that $F_{\mathcal{U}}$ is continuous relative to the order topology on $\mathbb{I}_{(\mathcal{U})}$, and that $\mathbf{Z}(\mathbb{I})_{(\mathcal{U})}$ is a family of zero sets, in the sense that

$$\mathbf{Z}(\mathbb{I})_{(\mathcal{U})} = \{F_{\mathcal{U}}^{-1}(0) : F \in C(X, \mathbb{I})\} = \{F_{\mathcal{U}}^{-1}(Z_{\mathcal{U}}) : Z \in \mathbf{Z}(X),\ F \in C(X, \mathbb{I})\}.$$

Again, a filter on $\mathbb{I}_{(\mathcal{U})}$ contained in $\mathbf{Z}(\mathbb{I})_{(\mathcal{U})}$ is a $\mathbf{Z}(\mathbb{I})_{(\mathcal{U})}$*-filter* .

Next, for each $k \in \mathbb{N}$, take a maximal family $\mathcal{C}_k$ of pairwise disjoint intervals of the form K_f for which $F_k(K_f) \cap K_{f^*} = \emptyset$, and set

$$V_k = \bigcup \{\operatorname{int} K_f : K_f \in \mathcal{C}_k\};$$

by Lemma 10.27, V_k is dense in $\mathbb{I}$. We define

$$V^F = \bigcup \{\{k\} \times V_k : k \in \mathbb{N}\}.$$

Then V^F is indeed an open, dense subset of X. The associated subset $V_{\mathcal{U}}^F$ of $\mathbb{I}_{(\mathcal{U})}$ is open and dense in $\mathbb{I}_{(\mathcal{U})}$.

Lemma 10.29 *Let $F \in C(X, \mathbb{I})$, and suppose that K is a non-empty, closed interval of $\mathbb{I}_{(\mathcal{U})}$ such that $K \subset V_{\mathcal{U}}^F$. Then there exists $f \in \{0,1\}^{<\mathbb{N}}_{(\mathcal{U})}$ such that $K \subset K_f^{(\mathcal{U})}$ and*

$$F_{\mathcal{U}}(K_f^{(\mathcal{U})}) \cap K_{f^*}^{(\mathcal{U})} = \emptyset.$$

Proof Associate (F_k) and (V_k) to F as above.

Take I to be an open interval in $\mathbb{I}_{(\mathcal{U})}$ which is a maximal subinterval of $V_{\mathcal{U}}^F$ with $I \supset K$, and choose $h : \mathbb{N} \to \mathbb{Q} \cap \mathbb{I}$ with $[h]_{\mathcal{U}} \in K$. Set

$$\sigma = \{k \in \mathbb{N} : h(k) \in V_k\};$$

since $[h]_{\mathcal{U}} \in V_{\mathcal{U}}^F$, necessarily $\sigma \in \mathcal{U}$. For each $k \in \sigma$, take I_k to be the component of V_k containing $h(k)$, so that $I_k = \operatorname{int} K_{f_k}$ for some $f_k \in \{0,1\}^{<\mathbb{N}}$ such that

$$F(K_{f_k}) \cap K_{f_k^*} = \emptyset.$$

Set $V = \bigcup\{\{k\} \times I_k : k \in \sigma\}$; $V_{\mathcal{U}}$ is then a non-empty, open interval in $\mathbb{I}_{(\mathcal{U})}$ such that $V_{\mathcal{U}} \subset V_{\mathcal{U}}^F$ and $[h]_{\mathcal{U}} \in V_{\mathcal{U}}$, and so $V_{\mathcal{U}} \subset I$.

The sequence $(f_k : k \in \sigma)$ defines an element, say f, in $\{0,1\}^{<\mathbb{N}}_{(\mathcal{U})}$. We have $\overline{V}_{\mathcal{U}} = K_f^{(\mathcal{U})}$ and $F_{\mathcal{U}}(K_f^{(\mathcal{U})}) \cap K_{f^*}^{(\mathcal{U})} = \emptyset$. Neither of the two end-points of the closed interval $K_f^{(\mathcal{U})}$ belongs to $V_{\mathcal{U}}^F$, and so we must have $I = V_{\mathcal{U}}$, and then $K \subset I \subset \overline{V}_{\mathcal{U}} = K_f^{(\mathcal{U})}$. □

We now return to a construction analogous to that in Chapter 9. The version of Definition 9.14 applicable to our present situation is the following.

Definition 10.30 *Let $H : \mathbb{N}_{(\mathcal{U})} \to \{0,1\}$ be such that $H \restriction a \in \{0,1\}^{(a)}_{(\mathcal{U})}$ for each $a \in \mathbb{N}_{(\mathcal{U})}$. Then*

$$\mathcal{F}_H = \{Z \in \mathbf{Z}(X) : Z_{\mathcal{U}} \cap K^{(\mathcal{U})}_{H\restriction a} \neq \emptyset \quad (a \in \mathbb{N}_{(\mathcal{U})})\}\,.$$

Let $a \in \mathbb{N}_{(\mathcal{U})}$, and take $Z \in \mathbf{Z}(X)$ with $Z_{\mathcal{U}} = K^{(\mathcal{U})}_{H\restriction a}$. Then it follows from Lemma 10.28(i) that $Z \in \mathcal{F}_H$.

The following result is a copy of Lemma 9.15, applicable in our new interpretation of the symbols; the proof is similar to the previous one.

Lemma 10.31 *Let $\mathcal{F}_H$ be as above.*

(i) *Suppose that $Z \in \mathcal{F}_H$ and that $W \in \mathbf{Z}(X)$ with $W \supset Z$. Then $W \in \mathcal{F}_H$.*

(ii) *Suppose that $Z_1, \ldots, Z_m \in \mathcal{F}_H$. Then $\bigcap_{j=1}^m Z_j \neq \emptyset$.*

(iii) *For $Z \in \mathbf{Z}(X)$ either $Z \in \mathcal{F}_H$ or there exists $W \in \mathcal{F}_H$ with $Z \cap W = \emptyset$.*

Proof (i) and (iii) are immediate.

(ii) Assume towards a contradiction that $\bigcap_{j=1}^m Z_j = \emptyset$. For each $k \in \mathbb{N}$, there exists $n_k \in \mathbb{N}$ such that, for each closed subinterval J of $\mathbb{I}$ with $l(J) \leq 2^{-n_k}$, we have $(\{k\} \times J) \cap Z_j = \emptyset$ for some $j \in \{1, \ldots, m\}$. The sequence (n_k) defines an element $a \in \mathbb{N}_{(\mathcal{U})}$, and $K^{(\mathcal{U})}_{H\restriction a}$ is a closed subinterval of $\mathbb{I}_{(\mathcal{U})}$ with $l(K^{(\mathcal{U})}_{H\restriction a}) = 2^{-a}$. Fix $W \in \mathbf{Z}(X)$ with $K^{(\mathcal{U})}_{H\restriction a} = W_{\mathcal{U}}$, and take $(W_k) \subset \mathbf{Z}(\mathbb{I})$ with

$$W \cap \mathbb{I}_k = \{k\} \times W_k \quad (k \in \mathbb{N})\,.$$

Then $\sigma \in \mathcal{U}$, where σ is the set of $k \in \mathbb{N}$ such that W_k is a closed interval of $\mathbb{I}$ with $l(W_k) \leq 2^{-n_k}$. For each $k \in \sigma$, choose $j_k \in \{1, \ldots, m\}$ such that $(\{k\} \times W_k) \cap Z_{j_k} = \emptyset$. Since $\mathcal{U}$ is an ultrafilter, there exist $j \in \{1, \ldots, m\}$ such that $\{k \in \mathbb{N} : j_k = j\} \in \mathcal{U}$, and then $(Z_j)_{\mathcal{U}} \cap K^{(\mathcal{U})}_{H\restriction a} = \emptyset$, a contradiction of the fact that $Z_j \in \mathcal{F}_H$. □

The following result follows from Lemma 10.31 exactly as Theorem 9.16 followed from Lemma 9.15.

Theorem 10.32 *Let $\mathcal{F}_H$ be as above. Then $\mathcal{F}_H$ is a z-ultrafilter on X containing $\mathcal{F}_0$.* □

Let $H : \mathbb{N}_{(\mathcal{U})} \to \{0,1\}$ be such that $H \restriction a \in \{0,1\}^a_{(\mathcal{U})}$ for each $a \in \mathbb{N}_{(\mathcal{U})}$, and let Q_H be the valuation prime z-ideal in $C(\beta X)$ specified by the z-ultrafilter

$\mathcal{F}_H$, so that K_{Q_H} is a hyper-real field; by Theorem 4.29, K_{Q_H} is a real-closed η_1-field. As in Chapter 9, there is a natural embedding of $\mathbb{N}_{(\mathcal{U})}$ into $\mathbb{N}_{Q_H}$, and we identify $\mathbb{N}_{(\mathcal{U})}$ with its image under this embedding. The following result is the analogue of Theorem 9.19, and its proof is similar.

Theorem 10.33 *With the above notation,* $\mathbb{N}_{(\mathcal{U})} = \mathbb{N}_{(Q_H)}$.

Proof As before, it is sufficient to prove the following: for each Q_H-partition $\{S_j : j \in \mathbb{N}\}$ of βX with $S_j \subset X \quad (j \in \mathbb{N})$, there exists $S \in \mathbf{Z}[Q_H]$ such that, for each $k \in \mathbb{N}$,

$$S \cap S_j \cap Y_k \neq \emptyset$$

for at most one $j \in \mathbb{N}$.

Let $\{S_j : j \in \mathbb{N}\}$ be such a partition, and take $S_0 \in \mathbf{Z}(\beta X)$ such that the partition has limit set contained in S_0. Since $S_0 \notin \mathbf{Z}[Q_H]$, we have $S_0 \cap X \notin \mathcal{F}_H$, and so, by Lemma 10.31(iii), there exists $Z \in \mathcal{F}_H$ with $S_0 \cap X \cap Z = \emptyset$.

Let $k \in \mathbb{N}$. Then $\{S_j \cap Z \cap \mathbb{I}_k : j \in \mathbb{N}\}$ is a pairwise disjoint subfamily of $\mathbf{Z}(\mathbb{I}_k)$, and only finitely many of these sets are non-empty because $S_0 \cap Z \cap \mathbb{I}_k = \emptyset$. Thus there exists $n_k \in \mathbb{N}$ such that

$$\{j \in \mathbb{N} : S_j \cap Z \cap (\{k\} \times K) \neq \emptyset\}$$

contains at most one element for each closed subinterval K of $\mathbb{I}$ with $l(K) \leq 2^{-n_k}$.

Set $b = [(n_k)]_{\mathcal{U}}$, and take $W \in \mathcal{F}_H$ with $W_{\mathcal{U}} = K^{(\mathcal{U})}_{H \restriction b}$, say

$$W \cap \mathbb{I}_k = \{k\} \times W_k \quad (k \in \mathbb{N}).$$

We may suppose that, for each $k \in \mathbb{N}$, W_k is a closed interval in $\mathbb{I}$ such that $l(W_k) = 2^{-n_k}$. For each $k \in \mathbb{N}$,

$$\{j \in \mathbb{N} : S_j \cap Z \cap W \cap \mathbb{I}_k \neq \emptyset\}$$

contains at most one element. But $Z \cap W \in \mathcal{F}_H$, and so

$$\{S_j \cap Z \cap W \cap \mathbb{I}_k : j \in \mathbb{N},\, k \in \mathbb{N}\}$$

is a Q_H-partition. Let S be the closure in βX of the union of these sets, so that $S \in \mathcal{F}_H$. Then S has the required properties. □

It follows easily from Theorem 10.33 that $\{0,1\}^{<\mathbb{N}}_{(\mathcal{U})} = \{0,1\}^{<\mathbb{N}}_{(Q_H)}$ and that $\{0,1\}^{\mathbb{N}}_{(\mathcal{U})} \subset \{0,1\}^{\mathbb{N}}_{(Q_H)}$.

We continue with a function $H : \mathbb{N}_{(\mathcal{U})} \to \{0,1\}$ such that

$$H \restriction a \in \{0,1\}^{a}_{(\mathcal{U})} \quad (a \in \mathbb{N}_{(\mathcal{U})}),$$

and we define

$$\mathcal{G}_H = \{Z_{\mathcal{U}} : Z \in \mathcal{F}_H\},$$

so that

$$\mathcal{G}_H = \{W \in \mathbf{Z}(\mathbb{I})_{(\mathcal{U})} : W \cap K^{(\mathcal{U})}_{H \restriction a} \neq \emptyset \quad (a \in \mathbb{N}_{(\mathcal{U})})\};$$

since $\mathcal{F}_H$ is a z-ultrafilter on X, $\mathcal{G}_H$ is a maximal $\mathbf{Z}(\mathbb{I})_{(\mathcal{U})}$-filter on $\mathbb{I}_{(\mathcal{U})}$. (The family $\mathcal{G}_H$ would be denoted by $(\mathcal{F}_H)_{\mathcal{U}}$ in notation analogous to that of Definition 9.29.) We have

$$K^{(\mathcal{U})}_{H\restriction a} \in \mathcal{G}_H \quad (a \in \mathbb{N}_{(\mathcal{U})})\,.$$

Recall that $\mathbf{Z}(\mathbb{I})_{(\mathcal{U})}$ contains all the closed intervals of $\mathbb{I}_{(\mathcal{U})}$ and that, by Theorem 10.33, $\mathbb{Q}_{(\mathcal{U})} = \mathbb{Q}_{(Q_H)}$.

Lemma 10.34 *Let $r \in \mathbb{Q}_{(\mathcal{U})} \cap \mathbb{I}_{(\mathcal{U})}$ and $F \in C(X, \mathbb{I})$. Then $\pi_{Q_H}(F) \leq r$ if and only if $F^{-1}_{\mathcal{U}}([0, r]_{(\mathcal{U})}) \in \mathcal{G}_H$.*

Proof Take $h : \mathbb{N} \to \mathbb{Q} \cap \mathbb{I}$ with $r = [h]_{\mathcal{U}}$, and define $\widetilde{h} : (k, t) \mapsto h(t)$, so that $\widetilde{h} \in C(X, \mathbb{I})$ and $\pi_{Q_H}(\widetilde{h}) = r$ by the definition of $\mathcal{F}_H$. Clearly $\pi_{Q_H}(F) \leq \pi_{Q_H}(\widetilde{h})$ if and only if $Z \in \mathcal{F}_H$, where $Z = \{x \in X : F(x) \leq \widetilde{h}(x)\}$. The result follows because $Z_{\mathcal{U}} = F^{-1}_{\mathcal{U}}([0, r]_{(\mathcal{U})})$. □

Let $a \in \mathbb{N}_{(\mathcal{U})}$, and take $f \in \{0, 1\}^a_{(\mathcal{U})}$. Then we have defined the interval $K^{(\mathcal{U})}_f = [r, s]_{(\mathcal{U})}$ in $\mathbb{I}_{(\mathcal{U})}$ for certain $r, s \in \mathbb{D}^{(\mathcal{U})}_a$. We have $r, s \in \mathbb{D}^{(Q_H)}_a$; the corresponding interval in $\mathbb{I}_{(Q_H)}$ is

$$K^{(Q_H)}_f = [r, s]_{(Q_H)}\,.$$

It is immediate from Lemma 10.34 that, for $F \in C(X, \mathbb{I})$ and $f \in \{0, 1\}^a_{(\mathcal{U})}$, we have $\pi_{Q_H}(F) \in K^{(Q_H)}_f$ if and only if $F^{-1}_{\mathcal{U}}(K^{(\mathcal{U})}_f) \in \mathcal{G}_H$. Further, for each $z \in \mathbb{I}_{(Q_H)} \setminus \mathbb{D}_{(Q_H)}$, we have $\varepsilon^{(Q_H)}_z \restriction a = f$ if and only if $z \in K^{(Q_H)}_f \setminus \mathbb{D}_{(Q_H)}$.

Lemma 10.35 *Let H be as above. Then $H \in \{0, 1\}^{\mathbb{N}}_{(Q_H)}$.*

Proof Define $F : (k, t) \mapsto t$, $X \to \mathbb{I}$, so that $F_{\mathcal{U}}$ is the function

$$F_{\mathcal{U}} : z \mapsto z\,, \quad \mathbb{I}_{(\mathcal{U})} \to \mathbb{I}_{(\mathcal{U})}\,.$$

If $H \in \{0, 1\}^{\mathbb{N}}_{(\mathcal{U})}$, then certainly $H \in \{0, 1\}^{\mathbb{N}}_{(Q_H)}$. So we may suppose that $H \notin \{0, 1\}^{\mathbb{N}}_{(\mathcal{U})}$; set $r = \pi_{Q_H}(F)$. Assume that $r \in \mathbb{D}_{(\mathcal{U})}$. Then, from the definition of $\mathcal{G}_H$, $\{r\} = F^{-1}_{\mathcal{U}}(r) \in \mathcal{G}_H$, and so either $H = \varepsilon^{-(\mathcal{U})}_r$ or $H = \varepsilon^{+(\mathcal{U})}_r$, a contradiction of the fact that $H \notin \{0, 1\}^{\mathbb{N}}_{(\mathcal{U})}$. Thus $r \in \mathbb{I}_{(\mathcal{U})} \setminus \mathbb{D}_{(\mathcal{U})}$.

Now set $H' = \varepsilon^{(Q_H)}_r \in \{0, 1\}^{\mathbb{N}}_{(Q_H)}$, and assume that $H' \neq H$. For each $a \in \mathbb{N}_{(\mathcal{U})}$, we have $r \in K^{(Q_H)}_{H'\restriction a} \setminus \mathbb{D}_{(\mathcal{U})}$, and so $F^{-1}_{\mathcal{U}}(K^{(\mathcal{U})}_{H'\restriction a}) \in \mathcal{G}_H$, i.e. $K^{(\mathcal{U})}_{H'\restriction a} \in \mathcal{G}_H$. But $K^{(\mathcal{U})}_{H\restriction a} \in \mathcal{G}_H$, and so

$$K^{(\mathcal{U})}_{H'\restriction a} \cap K^{(\mathcal{U})}_{H\restriction a} \in \mathcal{G}_H \quad (a \in \mathbb{N}_{(\mathcal{U})})\,.$$

Take $a \in \mathbb{N}_{(\mathcal{U})}$ such that $H' \restriction a \neq H \restriction a$. Since H is not eventually constant, it follows from Lemma 10.28 that $K^{(\mathcal{U})}_{H'\restriction a} \cap K^{(\mathcal{U})}_{H\restriction a} = \emptyset$, a contradiction. Thus $H' = H$, and so $H \in \{0, 1\}^{\mathbb{N}}_{(Q_H)}$. □

We now come to the key lemma in the proof of the existence of a structure $\mathfrak{N}_{(M)}$ which is not weakly Cauchy complete.

Lemma 10.36 *Let H be as above, and suppose that H is not eventually constant. Take $F \in C(X, \mathbb{I})$ and set $z = \pi_{Q_H}(F)$. Suppose that $z \notin \mathbb{D}_{(\mathcal{U})}$ and that $K_{H\restriction a}^{(\mathcal{U})} \subset V_{\mathcal{U}}^F$ for some $a \in \mathbb{N}_{(\mathcal{U})}$. Then $\varepsilon_z^{(Q_H)} \neq H^*$.*

Proof By Lemma 10.29, there exists $b \in \mathbb{N}_{(\mathcal{U})}$ and $f \in \{0,1\}_{(\mathcal{U})}^b$ such that

$$K_{H\restriction a}^{(\mathcal{U})} \subset K_f^{(\mathcal{U})} \quad \text{and} \quad F_{\mathcal{U}}(K_f^{(\mathcal{U})}) \cap K_{f^*}^{(\mathcal{U})} = \emptyset .$$

By Lemma 10.28(i), we have $b \leq a$ and $f = H \restriction b$, and so

$$K_{H\restriction b}^{(\mathcal{U})} \cap F_{\mathcal{U}}^{-1}(K_{f^*}^{(\mathcal{U})}) = \emptyset .$$

Thus $F_{\mathcal{U}}^{-1}(K_{f^*}^{(\mathcal{U})}) \notin \mathcal{G}_H$, and hence $\pi_{Q_H}(F) \notin K_{f^*}^{(Q_H)}$. Now $f^* = H^* \restriction 2b$, and so we have shown that

$$\varepsilon_z^{(Q_H)} \restriction 2b \neq H^* \restriction 2b ;$$

this implies the result. □

Definition 10.37 *Let H be as above. Then H is* generic *if, for each dense, open subset U of X, there exists $a \in \mathbb{N}_{(\mathcal{U})}$ such that $K_{H\restriction a}^{(\mathcal{U})} \subset U_{\mathcal{U}}$.*

Lemma 10.38 *Let H be as above, and assume that H is generic. Then:*

(i) *H is not eventually constant;*

(ii) *$H^* \notin \{0,1\}_{(Q_H)}^{\mathbb{N}}$;*

(iii) *$\mathfrak{N}_{(Q_H)}$ is not weakly Cauchy complete.*

Proof (i) Let $a \in \mathbb{N}_{(\mathcal{U})}$; we shall prove that there exists $b \in \mathbb{N}_{(\mathcal{U})}$ with $b \geq a$ such that $H(b) \neq H(b+1)$. Take $h : \mathbb{N} \to \mathbb{N}$ with $a = [h]_{\mathcal{U}}$.

For each $m \in \mathbb{N}$, there is a dense, open subset U_m of $\mathbb{I}$ such that, for each component I of U_m, necessarily $\overline{I} = K_f$, where $f \in \{0,1\}^n$ for some $n > m$ and $f(k) \neq f(k+1)$ for some $k \in \{m, \cdots, n-1\}$. Set

$$U = \bigcup \{\{k\} \times U_{h(k)} : k \in \mathbb{N}\} ,$$

so that U is a dense, open subset of X.

Since H is generic, there exists $d \in \mathbb{N}_{(\mathcal{U})}$ with $K_{H\restriction d}^{(\mathcal{U})} \subset U_{\mathcal{U}}$. Take I to be the maximal open subinterval of $\mathbb{I}_{(\mathcal{U})}$ with $K_{H\restriction d}^{(\mathcal{U})} \subset I \subset U_{\mathcal{U}}$. Then there exist $c > a$ and $f \in \{0,1\}_{(\mathcal{U})}^c$ with $\overline{I} = K_f^{(\mathcal{U})}$ and such that $f(b) \neq f(b+1)$ for some $b \in [a, c-1] \cap \mathbb{N}_{(\mathcal{U})}$. We have $K_{H\restriction d}^{(\mathcal{U})} \subset K_f^{(\mathcal{U})}$, and so, by Lemma 10.28(i), $c \leq d$ and $f = H \restriction c$. But now $H(b) \neq H(b+1)$, as required.

(ii) Assume towards a contradiction that $H^* \in \{0,1\}_{(Q_H)}^{\mathbb{N}}$. By (i), there exists $z \in \mathbb{I}_{(Q_H)} \setminus \mathbb{D}_{(Q_H)}$ with $H^* = \varepsilon_z^{(Q_H)}$. Since Q_H is a valuation prime in $C(\beta X)$, there exists $F \in C(X, \mathbb{I})$ with $\pi_{Q_H}(F) = z$. The set V^F associated with F is dense and open in X, and so there exists $a \in \mathbb{N}_{(\mathcal{U})}$ with $K_{H\restriction a}^{(\mathcal{U})} \subset V_{\mathcal{U}}^F$. By Lemma 10.36, $\varepsilon_z^{(Q_H)} \neq H^*$, a contradiction. Thus $H^* \notin \{0,1\}_{(Q_H)}^{\mathbb{N}}$.

(iii) By Lemma 10.35, $H \in \{0,1\}^{\mathbb{N}}_{(Q_H)}$. By (ii), $H^* \notin \{0,1\}^{\mathbb{N}}_{(Q_H)}$, and so $H \notin {}^*\{0,1\}^{\mathbb{N}}_{(Q_H)}$. □

To produce our example, it remains to find a function H which is generic. At this point, we require CH.

Lemma 10.39 (CH) *There is a generic function $H : \mathbb{N}_{(\mathcal{U})} \to \{0,1\}$ such that $H \restriction a \in \{0,1\}^a_{(\mathcal{U})}$ for each $a \in \mathbb{N}_{(\mathcal{U})}$.*

Proof The topological space X is separable, and so there are only $\aleph_1$ open sets in X (with CH); let $(U_\alpha : \alpha < \omega_1)$ be an enumeration of the dense, open subsets of X, and let $(b_\alpha : \alpha < \omega_1)$ be an increasing, cofinal sequence in $\mathbb{N}_{(\mathcal{U})}$.

We shall construct by induction sequences $(a_\alpha : \alpha < \omega_1)$ and $(h_\alpha : \alpha < \omega_1)$, where $a_\alpha \in \mathbb{N}_{(\mathcal{U})}$ and $h_\alpha \in \{0,1\}^{a_\alpha}_{(\mathcal{U})}$, such that:

(i) $a_\alpha < a_\beta$ and $h_\beta \restriction \alpha = h_\alpha$ whenever $\alpha < \beta < \omega_1$;

(ii) $a_\alpha \geq b_\alpha \quad (\alpha < \omega_1)$;

(iii) $K^{(\mathcal{U})}_{h_\alpha} \subset (U_\alpha)_\mathcal{U} \quad (\alpha < \omega_1)$.

Indeed, take $\beta < \omega_1$ and assume that a_α and h_α have been defined for each $\alpha < \beta$. By Lemma 8.19 (which applies because K_{Q_H} is an η_1-field), there exists $a \in \mathbb{N}_{(\mathcal{U})}$ with $a > b_\beta$ and $g \in \{0,1\}^a_{(\mathcal{U})}$ with $a_\alpha < a$ and $g \restriction a_\alpha = h_\alpha$ whenever $\alpha < \beta$. Fix $c \in \mathbb{N}_{(\mathcal{U})}$ with $c > b_\beta$ such that there exist $r, s \in \mathbb{D}^{(\mathcal{U})}_c$ with $s = r + 2^{-c}$ and

$$[r,s]_{(\mathcal{U})} \subset (U_\beta)_\mathcal{U} \cap K^{(\mathcal{U})}_g ;$$

such an element c exists because $(U_\beta)_\mathcal{U}$ is a dense, open subset of $\mathbb{I}_{(\mathcal{U})}$. Next, choose an element $z \in [r,s]_{(\mathcal{U})} \setminus \mathbb{D}_{(\mathcal{U})}$, and set $h = \varepsilon^{(\mathcal{U})}_z \restriction c$, so that $z \in K^{(\mathcal{U})}_h$ and $g = h \restriction a$. We have $K^{(\mathcal{U})}_h \subset [r,s]_{(\mathcal{U})} \subset (U_\beta)_\mathcal{U}$; set $a_\beta = c$ and $h_\beta = h$ to continue the inductive construction.

Define $H : \mathbb{N}_{(\mathcal{U})} \to \{0,1\}$ by the formula

$$H \restriction a_\alpha = h_\alpha \quad (\alpha < \omega_1).$$

Then H is well defined and $H \restriction a \in \{0,1\}^a_{(\mathcal{U})}$ for each $a \in \mathbb{N}_{(\mathcal{U})}$. By (iii), the function H is generic. □

We see that, by combining Lemmas 10.38 and 10.39, we have obtained the following result.

Theorem 10.40 (CH) *There is a maximal ideal M in the algebra $C(\mathbb{N} \times \mathbb{I})$ such that the structure $\mathfrak{N}_{(M)}$ associated to the hyper-real field $C(\mathbb{N} \times \mathbb{I})/M$ is not weakly Cauchy complete.* □

A similar method to the above can be used to produce a family of related examples. For example, we have a proof of the following theorem.

Theorem 10.41 (CH) *There is a maximal ideal M in the algebra $C(\mathbb{N} \times \mathbb{I} \times \mathbb{I})$ such that the family $\mathcal{P}(\mathbb{N})_{(M)}$ is not closed under finite unions.* □

Notes 10.42 For a discussion of model theory, see Chang and Keisler (1990).

Let P be a non-maximal, prime ideal in an algebra $C(\Omega)$. The structure $\mathfrak{N}_{(P)}$ can in fact be recovered from the more elementary structure

$$\langle \mathbb{N}_{(P)}, +, \mathcal{P}(\mathbb{N})_{(P)} \rangle .$$

This result follows from standard results of computational complexity concerning the definition of multiplication of binary strings, transcribed to our setting. Indeed most of the structures that we have associated with the prime ideal P are recoverable from this simpler structure; these structures include the set $\mathbb{D}_{(P)}$, the maps $\varepsilon_r^{(P)}$ (for $r \in \mathbb{I}_{(P)} \setminus \mathbb{D}_{(P)}$), and the ring K_P.

Experts in forcing will notice that, in the production of a structure $\mathfrak{N}_{(M)}$ which is not weakly Cauchy complete, the generic function H constructed in Lemma 10.39 is really a 'generic Cohen real over $\mathbb{N}_{(\mathcal{U})}$', in the language of forcing.

The hyper-real field $C(\mathbb{N} \times \mathbb{I})/M$ of Theorem 10.40 has cardinality $\aleph_1$, and so, by the isomorphism theorem, Theorem 2.34, it is isomorphic to an ultrapower, whose associated structure is weakly Cauchy complete. The techniques of Chapter 11, below, can probably be adapted to produce a hyper-real field K with the property that $\mathfrak{N}_{(P)}$ is not weakly Cauchy complete for any prime ideal P such that $K \simeq K_P$.

11

Rigid fields and solid structures

The main work of this final chapter will be to show that there are super-real fields K_P of weight $\aleph_1$ for which the associated structures $\mathbb{N}_{(P)}$ and $\{0,1\}^{\mathbb{N}}_{(P)}$ are *invariants* of the field K_P. Thus any classification of such super-real fields must involve a prior classification of these structures. The basic method will also be used to show, in Theorem 11.49, that it is relatively consistent with ZFC + GCH that there is a prime ideal P with corresponding super-real K_P an η_1-field of weight $\aleph_1$ such that P is both an exponential prime and a cut prime, but such that K_P is not even *order-isomorphic* to any super-real field of the form K_Q, where Q is a prime z-ideal in some $C(\Omega)$. Thus we shall separate the classes **exponential η_1** and **z-ideal** of Chapter 4, even when we restrict ourselves to order-dense subfields of $\widetilde{\mathbf{R}}$. In fact we shall prove that it is relatively consistent with ZFC + GCH that the pairs of classes **strongly convex η_1** and **strongly convex z-ideal η_1**, **valuation** and **valuation z-ideal**, and **cut prime η_1** and **cut prime z-ideal η_1** are distinct in the category of totally ordered sets.

This will conclude all that we can say about the separation of classes in the fundamental diagram. The situation, as far as we can determine it, will be summarized at the end of the chapter; a list of the remaining open questions will be given in Chapter 12.

The notions which form the backbone of the theory of this chapter are those of rigid fields and solid structures. Essentially a rigid super-real field K_P is one in which the associated structure $\mathfrak{N}_{(P)}$ is uniquely determined by the order structure of K_P, and a solid structure $\mathfrak{N}_{(P)}$ is one in which each continuous function on an open set agrees with a continuous internal function on some non-empty open subset. The proof of our main theorem, Theorem 11.6, that a solid model $\mathfrak{N}_{(P)}$ gives a rigid field K_P, requires us to determine addition and multiplication in some ordered fields merely from knowledge of the order structure in the field; the proof involves the construction of some specific functions in $\mathfrak{L}(\mathbb{I})$ and $\mathfrak{L}(\mathbb{I}^2)$. Because we have to deal with a wide class of prime ideals, including primes which are not exponential, the construction is complex (but elementary); it is given in perhaps unusual detail. In the latter part of the chapter, both for the proof that, with GCH, there are solid models $\mathfrak{N}_{(P)}$ and in the proof of the relative consistency of the separation of classes, we shall require greater sophistication and background in logic than we have assumed heretofore, and we confess that our final 'proofs' will only be briefly sketched, and will require knowledge of forcing.

We start with the definition of a rigid super-real field. The definition of when two models $\mathfrak{N}_{(P_1)}$ and $\mathfrak{N}_{(P_2)}$ are isomorphic, written $\mathfrak{N}_{(P_1)} \cong \mathfrak{N}_{(P_2)}$, was given in Definition 10.2(i).

Definition 11.1 *Let Ω_0 be a compact space, and let P_0 be a non-maximal, prime ideal in $C(\Omega_0)$. The super-real field K_{P_0} is* rigid *if $\mathfrak{N}_{(P)} \cong \mathfrak{N}_{(P_0)}$ whenever P is a non-maximal, prime ideal in $C(\Omega)$ for some compact space Ω and the ordered sets $(K_P, \le)$ and $(K_{P_0}, \le)$ are order-isomorphic.*

We must stress the key fact that, in the above definition, the fields K_P and K_{P_0} are only required to be isomorphic as ordered sets; they are not required *a priori* to be isomorphic as fields. Of course, by Theorem 10.3 and the following remark, K_P and K_{P_0} are isomorphic as rings, but not necessarily as fields, whenever K_{P_0} is rigid. Rigidity is a property of the ordered set $(K_{P_0}, \le)$.

In this context, we first make a trivial remark about order-isomorphisms of fields. Let $(K, \le)$ be a real-closed field. Then the map

$$x \mapsto \frac{2x-1}{x(1-x)}, \quad (0,1) \to K,$$

is an order-isomorphism. Thus, $(K_P, \le)$ and $(K_{P_0}, \le)$ are order-isomorphic if and only if $(\mathbb{I}_{(P)}, \le)$ and $(\mathbb{I}_{(P_0)}, \le)$ are order-isomorphic.

At first sight, it seems unlikely that rigid super-real fields exist. Indeed, the first proposition shows that the unique real-closed β_1-η_1-field $\mathbf{R}$ is not rigid.

Proposition 11.2 (CH) *The super-real field $\mathbf{R}$ is not rigid.*

Proof The field $\mathbf{R}$ has the form $C(\beta\mathbb{N})/Q$, where Q is a valuation prime z-ideal in $C(\beta\mathbb{N})$. Let I be an additively closed initial segment of $\mathbb{N}_{(Q)}$ such that I has cofinality $\aleph_1$, but such that I is not multiplicatively closed. By Proposition 8.36(ii), there is a prime ideal P in $C(\beta\mathbb{N})$ with P above Q such that $\mathbb{N}_{(P)} = I$. By Lemma 8.33(v), $\delta(K_P) = \operatorname{cof} \mathbb{N}_{(P)} = \aleph_1$, and so, by Theorem 4.28(ii), K_P is an η_1-field. Since $|K_P| = |\mathbf{R}| = \aleph_1$, it follows from the fundamental isomorphism theorem, Theorem 1.15, that the ordered sets $(K_P, \le)$ and $(\mathbf{R}, \le)$ are order-isomorphic. The structures $\mathfrak{N}_{(P)}$ and $\mathfrak{N}_{(Q)}$ are not isomorphic because $\mathbb{N}_{(Q)}$ is multiplicatively closed, but $\mathbb{N}_{(P)}$ is not. Thus $\mathbf{R}$ is not rigid. □

We note that, with CH, each real-closed η_1-field which is order-isomorphic to $\mathbf{R}$ is isomorphic to $\mathbf{R}$ (as a field); this is immediate from Corollary 2.34(ii). (We do not know if this result requires CH.) Thus rigidity is a stronger property than the property that isomorphism follows from order-isomorphism.

The next collection of η_1-fields in which one would search for rigid super-real fields is the class of super-real subfields K of the special field $\widetilde{\mathbf{R}}$ such that $\mathbf{R} \subset K \subset \widetilde{\mathbf{R}}$ and $\mathbf{R} \ncong K$. Under the assumption of CH, these are precisely the super-real fields of weight $\aleph_1$ and cardinality at least $\aleph_2$. We shall prove that, in the theory ZFC + GCH, *there are* rigid super-real fields in this class. The only method we know for establishing that a super-real field K_{P_0} is rigid is to work through a related concept, called solidity, and so we must now introduce this notion.

Suppose that P is a non-maximal, prime ideal in $C(\Omega)$, and that P is an exponential prime in the sense of Definition 4.45. We shall define ${}^*C(\mathbb{I}^n, \mathbb{I})_{(P)}$ for each $n \in \mathbb{N}$ as a collection of functions from $\mathbb{I}^n_{(P)}$ into $\mathbb{I}_{(P)}$ which are continuous with respect to the product topology on $\mathbb{I}^n_{(P)}$ obtained from the order topology on $\mathbb{I}_{(P)}$. The definition of ${}^*C(\mathbb{I}^n, \mathbb{I})_{(P)}$ requires some preliminary remarks, which we make for an arbitrary prime ideal.

As on page 20, we denote by $[S]^\omega$ the family of countable subsets a set of S. Of course, in the case where S is countable, $[S]^\omega = \mathcal{P}(S)$.

Now suppose again that P is a non-maximal, prime ideal in an algebra $C(\Omega)$, and take $n \in \mathbb{N}$. We form the set $\mathbb{D}^n_{(P)}$, as usual; we shall define ${}^*[\mathbb{D}^n]^\omega_{(P)}$ as a collection of subsets of $\mathbb{D}^n_{(P)}$. Each set in ${}^*[\mathbb{D}^n]^\omega_{(P)}$ will be the range of a function from $\mathbb{N}_{(P)}$ into $\mathbb{D}^n_{(P)}$, and so it will not necessarily be the case that $\mathbb{D}^n_{(P)}$ itself belongs to ${}^*[\mathbb{D}^n]^\omega_{(P)}$; for example, in the case where $|\Gamma_P| < w(K_P)$, we have $|\mathbb{N}_{(P)}| < \left|\mathbb{D}^n_{(P)}\right|$ by Theorem 8.34, and so $\mathbb{D}^n_{(P)} \notin {}^*[\mathbb{D}^n]^\omega_{(P)}$. In the case where P is an exponential prime, it can be shown by using Theorem 10.26 with S taken to be $2\mathbb{N}$ that $\mathbb{D}^n_{(P)} \in {}^*[\mathbb{D}^n]^\omega_{(P)}$.

We associate to each $f \in \{0,1\}^{\mathbb{N}}_{(P)}$ and each $n \in \mathbb{N}$ a subset $A^{(P)}_{f,n}$ of $\mathbb{D}^n_{(P)}$ as follows. Take $(r_1, \dots, r_n) \in \mathbb{D}^n_{(P)}$. Then

$$(r_1, \dots, r_n) \in A^{(P)}_{f,n} \tag{11.1}$$

if there exist $b_1, \dots, b_{n+2}$ in $\mathbb{N}_{(P)}$ with $b_1 < \cdots < b_{n+2}$ such that:

(i) $f(4b_1) = f(4b_{n+2}) = 1$;

(ii) $f(4b_2) = f(4b_3) = \cdots = f(4b_n) = 0$;

(iii) for $b \in [b_1, b_{n+2}] \cap \mathbb{N}_{(P)}$, $b \in \{b_1, \dots, b_{n+2}\}$ if and only if $f(4b+1) = 0$;

(iv) for $i = 1, \dots, n$, we have $f(4b_i + 2) = 1$ if and only if $r_i = 1$;

(v) for each $i \in \{1, \dots, n\}$ such that $r_i \neq 1$, necessarily $r_i \in \mathbb{D}^{(P)}_{b_{i+1}-b_i}$ and

$$\varepsilon^{+(P)}_{r_i}(c) = f(4b_i + 3 + 4c) \quad (c \in \{1, \dots, b_{i+1} - b_i\}).$$

Thus (11.1) holds if the n-tuple $(r_1, \dots, r_n)$ is determined by the restriction

$$f \mid ([4b_1, 4b_{n+2}] \cap \mathbb{N}_{(P)})$$

in a certain fashion.

Suppose that $b_1, \dots, b_{n+2} \in \mathbb{N}_{(P)}$ with $b_1 < \cdots < b_{n+2}$ are such that clauses (i)–(iii), above, are satisfied. Then $f \mid \left([4b_1, 4b_{n+2}] \cap \mathbb{N}_{(P)}\right)$ uniquely determines *some* element $(r_1, \dots, r_n) \in \mathbb{D}^n_{(P)}$. For take $i \in \{1, \dots, n\}$, and set $d = b_{i+1} - b_i$. If $f(4b_i + 2) = 1$, set $r_i = 1$. Otherwise, define

$$g : c \mapsto f(4b_i + 3 + 4c), \quad [1, d] \cap \mathbb{N}_{(P)} \to \{0, 1\};$$

it follows by standard arguments that $g \in \{0,1\}^d_{(P)}$, and so $g = \varepsilon_s^{+(P)} \upharpoonright d$ for some unique $s \in \mathbb{D}_d^{(P)}$, and in this case we set $r_i = s$.

Definition 11.3 *Let P be a non-maximal, prime ideal in $C(\Omega)$, and let $k \in \mathbb{N}$. Then*

$$^*[\mathbb{D}^n]^\omega_{(P)} = \left\{A^{(P)}_{f,n} : f \in {}^*\{0,1\}^{\mathbb{N}}_{(P)}\right\}.$$

We caution that the sets in $^*[\mathbb{D}^n]^\omega_{(P)}$ may be rather strange. For example, suppose that $K_P \cong \widetilde{\mathbf{R}}$ and that CH holds. Then, for each $a \in \mathbb{N}_{(P)} \setminus \mathbb{N}$, $^*[\mathbb{D}]^\omega_{(P)}$ contains every initial segment of uncountable cofinality of $(\mathbb{D}_a^{(P)}, \leq)$, and, in particular, $^*[\mathbb{D}]^\omega_{(P)}$ contains subsets of $(\mathbb{D}_a^{(P)}, \leq)$ with no maximum element.

We continue towards the definition of $^*C(\mathbb{I}^n, \mathbb{I})_{(P)}$ in the case where P is an exponential prime.

Definition 11.4 *Let P be a non-maximal, exponential prime ideal in $C(\Omega)$, and let $n \in \mathbb{N}$. Then $^*C(\mathbb{I}^n, \mathbb{I})_{(P)}$ is the set of all continuous functions $G : \mathbb{I}^n_{(P)} \to \mathbb{I}_{(P)}$ such that $A_G \in {}^*[\mathbb{D}^{2n+2}]^\omega_{(P)}$, where*

$$A_G = \{(a_1, b_1, \ldots, a_n, b_n, a, b) \in \mathbb{D}^{2n+2}_{(P)} : a_1 < b_1, \ \ldots, \ a_n < b_n, \ a < b, \\ G([a_1, b_1] \times \cdots \times [a_n, b_n]) \subset (a, b)\}.$$

The functions G of $^*C(\mathbb{I}^n, \mathbb{I})_{(P)}$ are the continuous internal *functions.*

In the case where $\mathfrak{N}_{(P)}$ is a model of analysis, it can be shown that the space $^*C(\mathbb{I}, \mathbb{I})_{(P)}$ is precisely the collection of all continuous functions from $\mathbb{I}_{(P)}$ into $\mathbb{I}_{(P)}$ which are definable in the model.

We now come to the key definition of solidity; it shows that, in the case where a model $\mathfrak{N}_{(P)}$ is solid, an arbitrary continuous function agrees with a continuous, internal function on a non-empty subinterval.

Definition 11.5 *Let P be a non-maximal, exponential prime ideal in $C(\Omega)$ for which the model $\mathfrak{N}_{(P)}$ is weakly Cauchy complete. Then the model $\mathfrak{N}_{(P)}$ is* solid *if, for every $n \in \mathbb{N}$, for every non-empty open subset U of $\mathbb{I}^n_{(P)}$, and for every continuous function $F \in C(U, \mathbb{I}_{(P)})$, there exist a non-empty, open subset V of U and a function $G \in {}^*C(\mathbb{I}^n, \mathbb{I})_{(P)}$ such that $G \mid V = F \mid V$.*

It follows easily from this definition that we can 'match finitely many functions' in the following sense. Take $n \in \mathbb{N}$. Suppose that, for $i = 1, \ldots, k$, we have $n_i \in \{1, \ldots, n\}$ and $F_i \in C(\mathbb{I}^{n_i}_{(P)}, \mathbb{I}_{(P)})$, and suppose that $I_1, \ldots, I_n$ are non-empty, open subintervals of $\mathbb{I}_{(P)}$. Then there exist non-empty, open subintervals $J_1, \ldots, J_n$ such that $J_m \subset I_m$ $(m = 1, \ldots, n)$ and $G_i \in {}^*C(\mathbb{I}^{n_i}, \mathbb{I})_{(P)}$ for $i = 1, \ldots, k$ such that

$$G_i \mid J_1 \times \cdots \times J_{n_i} = F_i \mid J_1 \times \cdots \times J_{n_i} \quad (i = 1, \ldots, k).$$

We now state our main theorem relating solidity and rigidity.

Theorem 11.6 *Let P_0 be a non-maximal, prime ideal in $C(\Omega_0)$ such that $\mathfrak{N}_{(P_0)}$ is a model of analysis. Suppose that $\mathfrak{N}_{(P_0)}$ is solid. Then K_{P_0} is rigid.*

Before beginning the proof of this theorem, we make two remarks.

First, recall that, in the case where $\mathfrak{N}_{(P_0)}$ is a model of analysis, P_0 is necessarily an exponential prime and $\mathfrak{N}_{(P_0)}$ is weakly Cauchy complete.

Second, we state without proof a theorem that illustrates a consequence of the hypothesis that $\mathfrak{N}_{(P_0)}$ be a model of analysis. For an exponential prime P, set

$$^*\mathbf{Z}(\mathbb{I})_{(P)} = \{F^{-1}(0) : F \in {}^*C(\mathbb{I}, \mathbb{I})_{(P)}\} ;$$

each $Z \in {}^*\mathbf{Z}(\mathbb{I})_{(P)}$ is a closed subset of $\mathbb{I}_{(P)}$.

Theorem 11.7 *Let P_0 be a non-maximal, prime ideal in $C(\Omega_0)$ such that $\mathfrak{N}_{(P_0)}$ is a model of analysis. Suppose that $Z \in {}^*\mathbf{Z}(\mathbb{I})_{(P_0)}$ and that $Z \cap (0,1]_{(P_0)}$ is a discrete subset with no minimum element. Then $Z \cap (0,1]_{(P_0)}$ is coinitial in $(0,1]_{(P_0)}$, and there exists $H \in {}^*C(\mathbb{I}, \mathbb{I})_{(P_0)}$ such that*

$$Z = \{H(1/a) : a \in \mathbb{N}_{(P_0)}\} \cup \{0\} .$$

Further, for each such H and each $W \in {}^*\mathbf{Z}(\mathbb{I})_{(P_0)}$,

$$\{a \in \mathbb{N}_{(P_0)} : H(1/a) \in W\} \in {}^*\mathcal{P}(\mathbb{N})_{(P_0)} . \qquad \square$$

In fact the requirement in Theorem 11.6 that $\mathfrak{N}_{(P_0)}$ be a model of analysis is unnecessarily strong; it could probably be replaced by the requirement that all the conclusions of Theorem 11.7 hold.

The proof of Theorem 11.6 involves the detailed construction of a triple (Λ, Ξ, Π) of functions in $\mathfrak{L}(\mathbb{I})$ with a quite remarkable property, which we now explain. We write Υ for the specific function

$$\Upsilon : (s,t) \mapsto (s-t)^2, \quad \mathbb{I}^2 \to \mathbb{I} . \tag{11.2}$$

Let P be a non-maximal, prime ideal in $C(\Omega)$, and suppose that V_1 and V_2 are non-empty, open subintervals of $\mathbb{I}_{(P)}$. First, in the case where P is not an exponential prime, we shall compute from the *ordered set* $(\mathbb{I}_{(P)}, \leq)$, together with the information

$$\Pi_{(P)} \mid V_1 \quad \text{and} \quad \Upsilon_{(P)} \mid V_1 \times V_2 ,$$

the ordered set $(\mathbb{N}_{(P)}, \leq)$ (up to order-isomorphism) together with the subset $\{a \in \mathbb{N}_{(P)} : 2^a \in \mathbb{N}_{(P)}\}$. Second, in the case where P is an exponential prime, we shall compute from the ordered set $(\mathbb{I}_{(P)}, \leq)$, together with the information

$$\Lambda_{(P)} \mid V_1 \quad \text{and} \quad \Upsilon_{(P)} \mid V_1 \times V_2 ,$$

the structure $\langle \mathbb{N}_{(P)}, +, \cdot \rangle$ up to isomorphism. Finally, in the case where P is an exponential prime and $\langle \mathbb{N}_{(P)}, +, \cdot \rangle$ is a model of number theory, we shall compute the full structure $\mathfrak{N}_{(P)} = \langle \mathbb{N}_{(P)}, +, \cdot, \mathcal{P}(\mathbb{N})_{(P)} \rangle$ from the information

$$\Lambda_{(P)} \mid V_1, \quad \Xi_{(P)} \mid V_1, \quad \text{and} \quad \Upsilon_{(P)} \mid V_1 \times V_2 .$$

Thus our specific functions Λ, Ξ, and Π are used to 'code' information that takes us from the order-structure $(\mathbb{I}_{(P)}, \leq)$ to the full structure $\mathfrak{N}_{(P)}$.

To construct our specific functions Λ, Ξ, and Π, we first describe some subsets of $\mathbb{D}$ and some functions in $C(\mathbb{I}, \mathbb{I})$.

First, for each $s \in \mathbb{D}$, denote by $n(s)$ the minimum element $n \in \mathbb{Z}^+$ such that $s \in \mathbb{D}_n$. Let $s, t \in \mathbb{D} \setminus \{0, 1\}$. Then clearly $st \in \mathbb{D} \setminus \{0, 1\}$ and

$$n(st) = n(s) + n(t)\,.$$

If, further, $s + t < 1$, then

$$n(s+t) \leq \max\{n(s), n(t)\}$$

and

$$n(s+t) = \max\{n(s), n(t)\}$$

in the case where $n(s) \neq n(t)$.

We define a total order $\preccurlyeq$ on $\mathbb{D}$ as follows: for $r, s \in \mathbb{D}$, set $r \preccurlyeq s$ if $n(r) < n(s)$ or if $n(r) = n(s)$ and $r \leq s$. Then there is an isotonic bijection

$$\Gamma : (\mathbb{N}, \leq) \to (\mathbb{D}, \preccurlyeq)\,. \tag{11.3}$$

All our functions will be built on certain subsets of $\mathbb{D}$. A closed subset L of $\mathbb{I}$ is a *dyadic set* if $\{0, 1\} \subset L \subset \mathbb{D}$ and if, for each complementary interval (s, t) of L (i.e. for each component interval (s, t) of $\mathbb{I} \setminus L$), we have

$$t - s = 2^{-n}, \quad \text{where} \quad n = \max\{n(s), n(t)\}\,. \tag{11.4}$$

For example, the subset $\{0, 1/2, 7/8, 1\}$ is not a dyadic set, but $\{0, 1/2, 3/4, 7/8, 1\}$ is a dyadic set. A dyadic set may contain infinitely many points: for example, $\{0\} \cup \{2^{-n} : n \in \mathbb{Z}^+\}$ is such a set. Suppose that L is a dyadic set with complementary interval (s, t). Then $n(s) \neq n(t)$; say $n = \max\{n(s), n(t)\}$. Set $r = (s+t)/2$, the mid-point of (s, t). Then $n(r) = n + 1$, and the set $L \cup \{r\}$ is a dyadic set. In fact the dyadic sets are precisely the closed sets $L \subset \mathbb{D}$ which are generated from $\{0, 1\}$ by successively adding mid-points to some complementary intervals of the existing set.

The *mesh* of a dyadic set L is the maximum of the lengths of the complementary intervals of L in $\mathbb{I}$.

Let L_0 and L_1 be dyadic sets. We *add copies of* L_1 *to* L_0 by taking some or all of the complementary intervals (a, b) of L_0, and adding $\gamma_{a,b}(L_1)$ to L_0, where

$$\gamma_{a,b} : t \mapsto a + t(b - a)\,, \quad \mathbb{I} \to [a, b]\,. \tag{11.5}$$

Lemma 11.8 *Let L_0 and L_1 be dyadic sets, and form L by adding copies of L_1 to L_0. Then L is a dyadic set.*

Proof Take (s, t) to be a complementary interval of L_1, say $t - s = 2^{-m}$; in fact, suppose that $m = n(s)$. Let (a, b) be a complementary interval of L_0 to which L_1 is added, and set $\gamma = \gamma_{a,b}$ and $n = \max\{n(a), n(b)\}$. Then $n(a) \leq n$ and $n(2^{-n}s) = m + n$, and so

$$n(\gamma(s)) = n(a + 2^{-n}s) = m + n\,.$$

Also $n(\gamma(t)) \leq m+n$ and $\gamma(t) - \gamma(s) = 2^{-(m+n)}$, and so the interval $(\gamma(s), \gamma(t))$ satisfies the analogue of (11.4). This is sufficient for the result. □

The functions that we shall define will be built by strategically placing triangles on certain dyadic sets. The details of the positions and sizes of these triangles will encode information that will be extracted later.

The *slope* of a function $F \in \mathrm{Lip}_1\mathbb{I}$ is

$$s(F) = \sup\left\{\frac{|F(s) - F(t)|}{|s-t|} : s, t \in \mathbb{I}, \quad s \neq t\right\}.$$

Take $\delta \geq 0$. A *triangle of height* δ on a subinterval $[s,t]$ of $\mathbb{I}$ is a function F on $[s,t]$ such that $F(s) = F(t) = 0$, $F((s+t)/2) = \delta$, and F is linear on $[s,(s+t)/2]$ and on $[(s+t)/2, t]$; the triangle is *special* if $\delta = (t-s)/2$, so that the slopes of the sides of a special triangle are ± 1.

Let L be a dyadic set. A *dyadic triangle function on* L is a function $F \in \mathrm{Lip}_1\mathbb{I}$ with

$$F(\mathbb{D}_n) \subset \mathbb{D}_n \quad (n \in \mathbb{N}) \tag{11.6}$$

and such that, for each complementary interval I of L, $F \mid \overline{I}$ is a triangle of height δ_I; for such a function F,

$$s(F) = 2\sup\{\delta_I / l(I)\}.$$

A *special dyadic triangle function* for L is a function $F \in \mathrm{Lip}_1\mathbb{I}$ such that, for each complementary interval I of L, $F \mid \overline{I}$ is a triangle function of height $\ell(I)/2$ or 0 (so that $s(F) = 1$ if $F \neq 0$); the special dyadic triangle function is *full for* L if, in each case, $\delta_I = l(I)/2$. There is a unique such function F which is full for L. Each special dyadic triangle function F satisfies condition (11.6), and so is a dyadic triangle function.

A *dyadic line function* for L is a function $f \in \mathrm{Lip}_1\mathbb{I}$ satisfying (11.6) such that, for each complementary interval I of L, the restriction $F \mid \overline{I}$ is linear.

We make a trivial remark. Let L be a dyadic set of mesh ε, and let L_1 be the subset of $\mathbb{I}$ formed by adding to L the mid-point of every complementary interval of L. Then L_1 is a dyadic set of mesh $\varepsilon/2$. Suppose, further, that F is a dyadic triangle function of slope m for L. Then F is a dyadic line function of slope m for L_1.

Let L_0, L_1, and L be as in Lemma 11.8, and suppose that F is a dyadic triangle function of slope m for L_1. For a complementary interval (a,b) of L_0, let $F_{a,b}$ be the scaled copy of F on $[a,b]$, so that

$$F_{a,b}(t) = (b-a)F(\gamma_{a,b}^{-1}(t)) \quad (t \in [a,b]).$$

Form a new function G on $\mathbb{I}$ by requiring that

$$G \mid [a,b] = F_{a,b} \mid [a,b]$$

whenever $\gamma_{a,b}(L_1)$ was added to L_0, and setting $G = 0$ otherwise. Then G is a dyadic triangle function of slope m for L.

Recall that, for each $\alpha \in (0,1]$, $(\mathrm{Lip}_\alpha\mathbb{I}, \|\cdot\|_\alpha)$ is a Banach space.

Lemma 11.9 *Let F be a dyadic triangle function of slope m on a dyadic set of mesh ε, and take $\alpha \in (0,1]$. Then $F \in \mathrm{Lip}_\alpha \mathbb{I}$ and*

$$\|F\|_\alpha \leq 2m\varepsilon^{1-\alpha}.$$

Proof We have $|F|_{\mathbb{I}} \leq m\varepsilon/2 \leq m\varepsilon^{1-\alpha}$. Take $s, t \in \mathbb{I}$ with $s \neq t$. If $|s-t| \leq \varepsilon$, then

$$\frac{|F(s)-F(t)|}{|s-t|^\alpha} \leq \frac{m\,|s-t|}{|s-t|^\alpha} \leq m\varepsilon^{1-\alpha}.$$

If $|s-t| \geq \varepsilon$, then

$$\frac{|F(s)-F(t)|}{|s-t|^\alpha} \leq \frac{1}{2}\frac{m\varepsilon}{|s-t|^\alpha} \leq m\varepsilon^{1-\alpha}.$$

The result follows. □

We shall consider pairs (F, L), where L is a dyadic set of mesh ε and F is a dyadic line function of slope m for L. A *refinement* of (F, L) is a pair $(\widetilde{F}, \widetilde{L})$ satisfying the following conditions. First, L' is a dyadic set containing L; the mesh of L' is at most ε, and F is also a dyadic line function of slope m for L'. Now let (s,t) be a complementary interval of L' such that $F(s) \neq F(t)$, and set

$$s_1 = \frac{1}{4}(3s+t), \quad s_2 = \frac{1}{2}(s+t), \quad s_3 = \frac{3}{4}(s+3t),$$

so that we have divided $[s,t]$ into four equal subintervals. Note that

$$n(s_1) = n(s_3) = n(s_2) + 1 = \max\{n(s), n(t)\} + 2. \tag{11.7}$$

The set $\widetilde{L}$ consists of L', together with these three intermediate points for every complementary subinterval (s,t), so that $\widetilde{L}$ is a dyadic set of mesh at most $\varepsilon/4$. The function $\widetilde{F}$ is formed by building platforms into the graph of F. Indeed, let (s,t) be some complementary interval of L', take s_1, s_2, s_3 as above, and set $x = (F(t)-F(s))/(t-s)$. Suppose that $x > 0$. Then:

$$\widetilde{F}(r) = \begin{cases} F(r) & (s \leq r \leq s_1), \\ F(s_1) + 2(r-s_1)x & (s_1 \leq r \leq s_2), \\ F(s_3) & (s_2 \leq r \leq s_3), \\ F(r) & (s_3 \leq r \leq t). \end{cases}$$

Suppose that $x < 0$. Then:

$$\widetilde{F}(r) = \begin{cases} F(r) & (s \leq r \leq s_2), \\ F(s_2) & (s_2 \leq r \leq s_3), \\ F(s_1) + 2(r-s_3)x & (s_3 \leq r \leq t). \end{cases}$$

In both cases, the function $\widetilde{F}$ is constant on the *third* subinterval $[s_2, s_3]$ of the interval $[s,t]$, and so an affine transformation from one such complementary subinterval of L' to another will preserve the intervals on which platforms are built. The line of height $\widetilde{F}(s_2)$ on this third subinterval is a *platform*. (It is not required that a platform be built into the graph on every complementary subinterval of L'.)

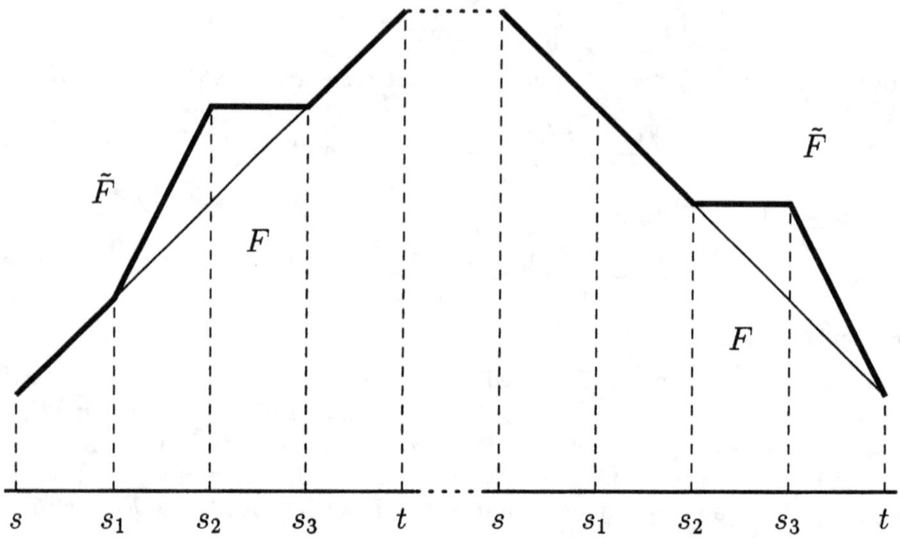

Fig. 11.1. Refinement of a pair (F, L)

We see that $\widetilde{F}$ satisfies (11.6), using the fact that L is a dyadic set. Thus $\widetilde{F}$ is a dyadic line function of slope at most $2m$ for $\widetilde{L}$. Further, the function $\widetilde{F} - F$ is a dyadic triangle function of slope at most m for a dyadic set. Note that $\widetilde{F} \mid L = F \mid L$. A refinement $(\widetilde{F}, \widetilde{L})$ of (F, L) is *full* if platforms are inserted in the graph of F for every complementary subinterval (s, t) of L, so that (F, L) has a unique full refinement with $L' = L$.

There are three further notions, to be used later in our construction of the code, that we require before we begin our description of the algorithm. Take $a < b$, and let $G : [a, b] \to \mathbb{R}$ be a continuous function.

First, a subinterval $[c, d]$ with $c < d$ of $[a, b]$ is a *peak interval* for G if

$$G(a) = G(c) = G(d) = G(b),$$
$$G(t) \geq G(a) \quad (t \in [a, b]), \quad \text{and} \quad G(t) > G(a) \quad (t \in (c, d)).$$

The family of peak intervals for G is denoted by

$$\mathfrak{P}\mathfrak{I}(G).$$

For example, let G be a dyadic triangle function for a dyadic set L. Then $\mathfrak{P}\mathfrak{I}(G)$ is the family of all closed subintervals $\overline{I}$ of $\mathbb{I}$ for complementary intervals I to L for which $\delta_I > 0$.

Second, a subinterval $[c, d]$ of $[a, b]$ is a *twin-peak interval* for G if both

$$[c, (c + d)/2] \quad \text{and} \quad [(c + d)/2, d]$$

are peak intervals for $G \mid [c,d]$. Thus $[c,d]$ is a twin-peak interval if and only if

$$G(c) = G((c+d)/2) = G(d) \quad \text{and} \quad G(t) > G(c) \quad (t \in (c,d) \setminus \{(c+d)/2\}).$$

The family of twin-peak intervals for G is denoted by

$$\mathfrak{TP}(G).$$

Finally, a *flat component* for G is a maximal subinterval $[s,t]$ of $[a,b]$ such that $s < t$ and G is constant on $[s,t]$; the *flat part* of G, denoted by

$$\mathfrak{FP}(G),$$

is the union of the flat components for G.

Lemma 11.10 *Suppose that L_0 is a dyadic set and F_0 is a special dyadic triangle function for L_0. Suppose also that (L_k) and (F_k) are sequences such that $F_1 = F_0$ and, for each $k \in \mathbb{N}$:*

(i) *L_k is a dyadic set;*

(ii) *F_k is a dyadic line function for L_k;*

(iii) *$L_{k-1} \subset L_k$;*

(iv) *(F_{2k}, L_{2k}) is a refinement of (F_{2k-1}, L_{2k-1});*

(v) *$F_{2k+1} - F_{2k}$ is a special dyadic triangle function for a dyadic set containing L_{2k} with* $\operatorname{supp}(F_{2k+1} - F_{2k}) \subset \mathfrak{FP}(F_{2k})$.

Define $\widehat{F} : \mathbb{I} \to \mathbb{I}$ by

$$\widehat{F}(r) = \lim_{k\to\infty} F_k(r) = \sup\{F_k(r) : k \in \mathbb{N}\} \quad (r \in \mathbb{I}).$$

Then $\widehat{F} \in \mathfrak{L}(\mathbb{I})$ and $\widehat{F}(\mathbb{D}_n) \subset \mathbb{D}_n \quad (n \in \mathbb{N})$. Further, for each $k \in \mathbb{N}$ and each flat component $[a,b]$ for F_k,

$$\mathfrak{PI}(\widehat{F} \mid [a,b]) = \bigcup\{\mathfrak{PI}(F_j \mid [a,b]) : j \geq k\} \tag{11.8}$$

and

$$\mathfrak{TP}(\widehat{F} \mid [a,b]) = \bigcup\{\mathfrak{TP}(F_j \mid [a,b]) : j \geq k\}. \tag{11.9}$$

Proof For each $k \in \mathbb{N}$, define $\varepsilon_k > 0$ to be the mesh of L_k and set $m_k = s(F_k)$ and $H_k = F_{k+1} - F_k$. Since H_{2k} is a special dyadic triangle function, we have $m_{2k+1} \leq m_{2k} + 1$. Also $m_2 \leq 2m_1$, $\varepsilon_2 \leq 1/4$, and $m_{2k+2} \leq 2m_{2k+1}$ and $\varepsilon_{2k+2} \leq \varepsilon_{2k+1}/4$. Thus

$$m_{2k} \leq 3^k \overline{m}_1, \quad m_{2k+1} \leq 3^k \overline{m}_1 + 1, \quad \text{and} \quad \varepsilon_{2k+1} \leq \varepsilon_{2k} \leq 4^{-k},$$

where $\overline{m}_1 = \max\{m_1, 2\}$. Let $\alpha \in (0,1]$. By (v) and Lemma 11.9,

$$\|H_{2k}\|_\alpha \leq 2\varepsilon_{2k}^{1-\alpha} \leq 2 \cdot 4^{-k(1-\alpha)}.$$

Since H_{2k+1} is a dyadic triangle function of slope at most m_{2k+1} on a dyadic set of mesh at most $\varepsilon_{2k+1}/2$, we have, again by Lemma 11.9, that

$$\|H_{2k+1}\|_\alpha \leq 2m_{2k+1}(\varepsilon_{2k+1}/2)^{1-\alpha} = O\left(3^k \cdot 4^{-k(1-\alpha)}\right) \quad \text{as } k \to \infty.$$

Choose $\alpha \in (0,1]$ with $3 < 4^{1-\alpha}$, and set $\eta = 3/4^{1-\alpha} < 1$. Then

$$\|H_k\|_\alpha = O\left(\eta^{k/2}\right) \quad \text{as } k \to \infty,$$

and so (F_k) converges in $(\mathrm{Lip}_\alpha \mathbb{I}, \|\cdot\|_\alpha)$. But $F_k \to \widehat{F}$ pointwise on $\mathbb{I}$, and so

$$\widehat{F} \in \mathrm{Lip}_\alpha \mathbb{I} \subset \mathfrak{L}(\mathbb{I}).$$

It is immediate from (iii) and (iv) that $\bigcup\{L_k : k \in \mathbb{N}\} = \mathbb{D}$. For each $k \in \mathbb{N}$, we have $F_{k+1} \mid L_k = F_k \mid L_k$, and so $\widehat{F} \mid L_k = F_k \mid L_k$. It follows that $\widehat{F}(\mathbb{D}_n) \subset \mathbb{D}_n \quad (n \in \mathbb{N})$. A moment's consideration shows that conditions (11.8) and (11.9) are satisfied. □

Let (L_k) and (F_k) be sequences satisfying the conditions of Lemma 11.10, and again take $\widehat{F} = \lim F_k$, so that $\widehat{F} \in \mathfrak{L}(\mathbb{I})$. We shall require some further properties of $\widehat{F}$.

Let $[a, b]$ correspond to a platform of a function F_{2k} (each such interval is a flat component for F_{2k}, but there may be other flat components). Then

$$\widehat{F}(t) \geq \widehat{F}(a) = \widehat{F}(b) \quad (t \in [a, b]).$$

Also, for each $\varepsilon > 0$, there exists

$$t \in (a - \varepsilon, a) \cup (b, b + \varepsilon) \quad \text{with} \quad \widehat{F}(t) < \widehat{F}(a).$$

The peak intervals of $\widehat{F} \mid [a, b]$ are of the form $[c, d]$, where $[c, d] \subset [a, b]$, $c, d \in \mathbb{D}$, and $d = c + 2^{-k}$, where $k = \max\{n(c), n(d)\}$.

Now suppose that $[a_1, b_1]$ and $[a_2, b_2]$ are two distinct intervals specifying platforms arising in the construction. Then either these intervals are disjoint or one is contained in the other; if $[a_2, b_2] \subset [a_1, b_1]$, then $\widehat{F}(a_1) < \widehat{F}(a_2)$.

Lemma 11.11 *Suppose that $n \in \mathbb{Z}^+$ and that $a, b \in \mathbb{D}_n$ with $a < b$.*

(i) *Set $u = \min\{\widehat{F}(t) : t \in [a, b]\}$, and let a_0 and b_0 be the minimum and maximum, respectively, of $\{t \in [a, b] : \widehat{F}(t) = u\}$. Then $a_0, b_0 \in \mathbb{D}_{n+1}$.*

(ii) *Suppose that there exists $c_0 \in (a, b)$ with*

$$\widehat{F}(c_0) < \widehat{F}(t) \quad (t \in [a, b] \setminus \{c_0\}).$$

Set

$$r = \sup\{t \in [0, c_0) : \widehat{F}(t) \leq \widehat{F}(c_0)\}, \quad s = \inf\{t \in (c_0, 1] : \widehat{F}(t) \leq \widehat{F}(c_0)\}.$$

Then $c_0 \in \mathbb{D}_n$ and either $r \in \mathbb{D}_n$ or $s \in \mathbb{D}_n$.

Proof (i) We shall prove that $a_0 \in \mathbb{D}_{n+1}$; the argument that $b_0 \in \mathbb{D}_{n+1}$ is similar. We may suppose that $a_0 \in (a, b)$, for otherwise the result is trivial.

By inspection of the construction of $\widehat{F}$, we see that there exists $[c, d]$ corresponding to a platform with $a_0 \in [c, d]$ and $\widehat{F}(c) = \widehat{F}(d) = u$. Set

$$m = \max\{n(c), n(d)\},$$

so that $d = c + 2^{-m}$ and $|n(c) - n(d)| = 1$ by (11.7). It is clear that either $a \in [c, d]$ or $b \in [c, d]$.

If $a \in [c, d]$, then there exists a peak interval $[c', a_0]$ of $\widehat{F} \mid [c, d]$ such that $a \in (c', a_0)$. We have $c', a_0 \in \mathbb{D}_k$ and $a_0 = c' + 2^{-k}$, where $k = \max\{n(c'), n(a_0)\}$. It cannot be that $a \in \mathbb{D}_k$, and so $k < n(a) \leq n$, whence $a_0 \in \mathbb{D}_n$.

Now suppose that $a \notin [c, d]$ and $b \in [c, d]$. Then $a_0 = c$ and $b \in (c, d]$. If $b \in (c, d)$, then $m < n$ and $a_0 \in \mathbb{D}_n$. If $b = d$, then $n(d) \leq n$, and so $n(c) \leq n + 1$ because $|n(c) - n(d)| = 1$. Thus $a_0 \in \mathbb{D}_{n+1}$.

(ii) By inspection of the construction, we see that either $[r, c_0]$ or $[c_0, s]$ (or both) is the base of an added triangle, say this is true for $[r, c_0]$. Then r and c_0 are adjacent points of a dyadic set, and so $c_0 = r + 2^{-m}$, say. But $a \in (r, c_0)$ and $a \in \mathbb{D}_n$, and so $m < n$. Thus $r, c_0 \in \mathbb{D}_n$. □

There is a particular way in which the sequences (L_k) and (F_k) arise; this method will be used in the construction of the functions Λ and Π. We start with a dyadic set L and the special dyadic triangle function F which is full for L. The pair (L, F) is the *seed* of the construction. We construct sequences (L_k) of dyadic sets and (F_k) of dyadic line functions for L_k as follows. First, set $L_0 = L$ and $F_0 = F$. Then L_1 is formed from L_0 by adding the mid-point of every complementary interval of L_0 to L_0, and $F_1 = F_0$, regarded as a dyadic line function of slope 1 on L_1. Now suppose that the odd-numbered terms L_{2k-1} and F_{2k-1} have been constructed. Then (F_{2k}, L_{2k}) is the full refinement $(\widetilde{F}_{2k-1}, \widetilde{L}_{2k-1})$ of (F_{2k-1}, L_{2k-1}), so that the graph of F_{2k} includes platforms; the intervals corresponding to these platforms are exactly the flat components for F_{2k}. The pair (F_{2k+1}, L_{2k+1}) is formed by building a scaled version of the pair (F_1, L_1) on each platform: for each flat component $[a, b]$ for F_{2k}, add a copy of L_1 to L_{2k} on $[a, b]$, so forming the dyadic set L_{2k+1}, and add $F_{a,b}$, the scaled version of the original function F, to F_{2k} on each such interval $[a, b]$ to form the new dyadic line function F_{2k+1} for L_{2k+1}.

The sequences (L_k) and (F_k) so constructed satisfy the conditions of Lemma 11.10. Let $\widehat{F} = \lim F_k$, so that, by Lemma 11.10, $\widehat{F} \in \mathfrak{L}(\mathbb{I})$. The function $\widehat{F}$ is the function *germinated* by the seed (L, F). In this case, we have

$$\mathfrak{P}\mathfrak{I}(\widehat{F}) = \mathfrak{P}\mathfrak{I}(F), \tag{11.10}$$

and so the initial seed (L, F) is uniquely determined by $\widehat{F}$.

Let $\mathcal{I}(\widehat{F})$ be the union of the families of flat components for the various functions F_{2k} occurring in the above construction, together with the interval $\mathbb{I}$ itself. On each interval $[a, b]$ in $\mathcal{I}(\widehat{F})$, the function $\widehat{F}$ replicates itself, in the sense that

$$\widehat{F} \mid [a, b] = \widehat{F}(a) + \widehat{F}_{a,b} \mid [a, b]. \tag{11.11}$$

It follows that $\widehat{F}$ has the following fractal-like property: for each non-empty, open subset U of $\mathbb{I}$, there exists an interval $[a, b] \subset U$ such that $\widehat{F}$ repeats itself on $[a, b]$, in the sense that (11.11) holds.

Lemma 11.12 *Suppose that $\varphi : \mathbb{N} \to \mathbb{N}$ is a strictly increasing function such that, for each $n \in \mathbb{N}$ and $a, b \in \mathbb{D}_n \cap L_1$ with $a < b$, there exist $a', b' \in \mathbb{D}_{\varphi(n)} \cap L_1$ with $a \le a' < b' \le b$ and with a', b' adjacent points in L_1. Then, for each $n \in \mathbb{N}$ and $a, b \in \mathbb{D}_n$ with $a < b$, there exists $[c, d] \in \mathcal{I}(\widehat{F})$ with $a \le c < d \le b$ and $c, d \in \mathbb{D}_{\varphi(n)+2}$.*

Proof We first prove the special case of the result in which $a, b \in L_1$. By hypothesis, there exist $a', b' \in \mathbb{D}_{\varphi(n)} \cap L_1$ with $a \le a' < b' \le b$ and a', b' adjacent points in L_1, and then, because of the way in which platforms were inserted, there exists $[c, d] \in \mathcal{I}(\widehat{F})$ with $a' < c < d < b'$ and $c, d \in \mathbb{D}_{\varphi(n)+2}$. This special case proves the lemma for $n = 1$.

We shall now prove this lemma by induction on $n \in \mathbb{N}$ simultaneously for the specified function φ and seed (L, F) and for the function $\varphi' : n \mapsto n$ and the seed (L', F'), where $L' = \{0, 1\}$. We have the result in the case where $n = 1$.

We now assume inductively that the claimed result holds for $n \in \mathbb{N}$ both for the given φ, L and for φ', L', and shall prove the result for $n + 1$ and φ, L; an easier version of the argument handles the case for $n + 1$ and φ', L'. Take $a, b \in \mathbb{D}_{n+1}$ with $a < b$; we may suppose that $a \notin L_1$ or $b \notin L_1$.

Suppose first that there exists $[r, s] \in \mathcal{I}(\widehat{F}) \setminus \{\mathbb{I}\}$ with $a \in (r, s)$, and set

$$m = \max\{n(r), n(s)\},$$

so that $m < n(a) \le n + 1$. By reducing b to s, if necessary, we may suppose that $b \le s$. Take $a', b' \in \mathbb{I}$ with $\gamma_{r,s}(a') = a$ and $\gamma_{r,s}(b') = b$, so that $a', b' \in \mathbb{D}_{n+1-m}$. By the inductive hypothesis, there exists $[c', d'] \in \mathcal{I}(\widehat{F})$ with $a' \le c' < d' \le b'$ and $c', d' \in \mathbb{D}_{\varphi(n+1-m)+2}$. Set $c = \gamma_{r,s}(c')$ and $d = \gamma_{r,s}(d')$, so that $a \le c < d \le b$. Then $[c, d] \in \mathcal{I}(\widehat{F})$ and

$$\max\{n(c), n(d)\} \le \varphi(n + 1 - m) + 2 \le \varphi(n + 1) + 2,$$

as required. A similar argument covers the case where $b \in (r, s)$.

The remaining case is that where $a, b \notin (r, s)$ for each $[r, s] \in \mathcal{I}(\widehat{F}) \setminus \{\mathbb{I}\}$; we may suppose that $a \notin L_1$. Let (u, v) be the complementary interval of L_1 with $a \in (u, v)$; we may also suppose that $b \le v$. Set

$$m = \max\{n(u), n(v)\},$$

so that $m < n(a) \le n + 1$. Define $\gamma : [u, v] \to [0, 1/2]$ to be the scaling map. By the inductive hypothesis for the points $\gamma(a), \gamma(b)$ and the function φ', there exists $[c', d'] \in \mathcal{I}(\widehat{F}')$ with $\gamma(a) \le c' < d' \le \gamma(b)$ and $\max\{n(c'), n(d')\} = k$, where

$$k \le \varphi'(n + 1 - m) + 2 = n + 3 - m;$$

we choose $[c', d']$ to minimize k. Set $c = \gamma^{-1}(c')$ and $d = \gamma^{-1}(d')$, so that $a \le c < d \le b$. The key point is that $[c, d] \in \mathcal{I}(\widehat{F})$ because $a \notin (r, s)$, $b \notin (r, s)$, and k is minimal. Also

$$\max\{n(c), n(d)\} \le k + m \le n + 3 \le \varphi(n + 1) + 2.$$

The induction continues. □

Our next task is to determine appropriate dyadic sets L with the special dyadic triangle functions F which are full for L to act as a seed for the germination of $\widehat{F}$, and hence to build two of our required functions, Λ and Π. Throughout, we denote by $\mathcal{J}$ the family of closed subintervals $[a, b]$ of $\mathbb{I}$ with $a < b$.

An intermediate set T and function g We recall that we defined a bijection $\Phi : \mathbb{N} \times \mathbb{N} \to \mathbb{N}$ in Chapter 8, with the property, given in (8.8), that $m \vee n \leq \Phi(m, n) \leq (m \vee n)^2$ for $m, n \in \mathbb{N}$.

We define a set $\mathbf{T}$ by first including the points 2^{-n} for $n \in \mathbb{Z}^+$, and also the point 0. We define $\mathbf{g}$ on each interval $[2^{-n},\ 2^{-n+1}]$ of $\mathbb{I}$ (for $n \in \mathbb{N}$). On the 'second half' $[2^{-n} + 2^{-n-1},\ 2^{-n+1}]$ of this interval, $\mathbf{g}$ is the special triangle of height 2^{-n-2}, so that $[2^{-n} + 2^{-n-1},\ 2^{-n+1}] \in \mathfrak{P}\mathfrak{I}(\mathbf{g})$. On the 'second quarter'

$$I_n = [2^{-n} + 2^{-n-2},\ 2^{-n} + 2^{-n-1}]$$

of this interval, $\mathbf{g}$ is 0. We write

$$\mathcal{K} = \{I_n : n \in \mathbb{N}\}$$

for the family of these second quarters; the family $\mathcal{K}$ is discrete in $(0, 1]$ in the sense that its members are pairwise disjoint and its union is closed in $(0, 1]$. We write $\gamma(I_n)$ for the corresponding first quarter, so that

$$\gamma(I_n) = [2^{-n},\ 2^{-n} + 2^{-n-2}] \quad (n \in \mathbb{N}),$$

and we have a map $\gamma : \mathcal{K} \to \mathcal{J}$. For $I \in \mathcal{K}$, $\gamma(I)$ is characterized as that interval $J \in \mathcal{J}$ with $J \leq I$, $l(J) = l(I)$, and $|J \cap I| = 1$.

Divide the interval $\gamma(I_n)$ into subintervals

$$J_{n,k} = [2^{-n} + 2^{-n-2-k},\ 2^{-n} + 2^{-n-2-k+1}]$$

for $k \in \mathbb{N}$, and divide $J_{n,k}$ into 2^m equal subintervals, where

$$m = \Phi(n, k+3) - (n + k + 2) \in \mathbb{N};$$

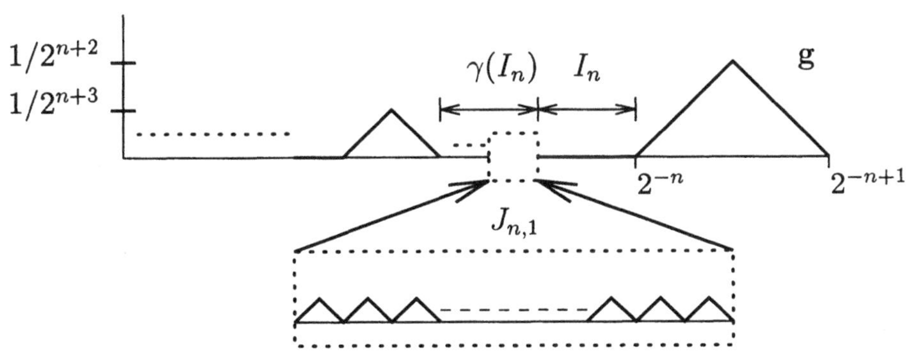

Fig. 11.2. The function **g**

each of these subintervals has length $2^{-\Phi(n,k+3)}$, and on this subinterval $\mathbf{g}$ is the special triangle of height $2^{-\Phi(n,k+3)-1}$. For example, $J_{1,1} = [9/16, 5/8]$, and $J_{1,1}$ is divided into eight subintervals, each of height $1/256$, because $\Phi(1,4) = 7$. We shall need to notice, first, that, for each $n \in \mathbb{N}$, there are such subintervals of $\gamma(I_n)$ with arbitrarily short length, and, second, that, if two such subintervals are distinct, they have different lengths.

Finally, $\mathbf{g}(2^{-n}) = \mathbf{g}(0) = 0$. It is clear that $\mathbf{g} \in \mathrm{Lip}_1\mathbb{I}$.

The set $\mathbf{T}$ consists of the end-points of the four quarter subintervals of $[2^{-n}, 2^{-n+1}]$, all the end-points of the intervals $J_{n,k}$, and all the end-points of the 2^m specified subintervals of $J_{n,k}$, for each $n \in \mathbb{N}$. Certainly $\mathbf{T}$ is a dyadic set, and $\mathbf{g}$ satisfies (11.6), but $\mathbf{g}$ is not full for $\mathbf{T}$ because $\mathbf{g}$ is equal to 0 on some complementary intervals of $\mathbf{T}$. In fact, $\mathbf{T}$ is the minimum dyadic set for which $\mathbf{g}$ is a special dyadic triangle function on $\mathbf{T}$.

The sets $\mathbf{K_1}$ and $\mathbf{K_2}$ These two sets are formed from $\mathbf{T}$ by inserting points in each of the intervals I_n, defined above.

Take $n \in \mathbb{N}$, and divide I_n into four equal subintervals, called successively (from right to left) $I_{n,1}, I_{n,2}, I_{n,3}$, and $I_{n,4}$. We shall define $m_1, m_2, m_3, m_4 \in \mathbb{N}$ (the numbers depending on n), and then divide $I_{n,j}$ into 2^{m_j} equal subintervals, each of length 2^{-n-4-m_j}, for $j = 1, 2, 3, 4$.

The dyadic set $\mathbf{K_1}$ is the union of $\mathbf{T}$ and the end-points of these subintervals for certain choices of m_1, m_2, m_3, and m_4. For $n \in \mathbb{N}$, there are uniquely defined numbers $i, j \in \mathbb{N}$ with $\Phi(i,j) = n$. We define:

$m_1 = 1$;

$m_2 = \Phi(i+j, k+3) - (n+4)$, where k is the minimum element of $\mathbb{N}$ with $\Phi(i+j, k+3) \geq n+5$;

$m_3 = \Phi(j, k+3) - (n+4)$, where k is the minimum element of $\mathbb{N}$ with $\Phi(j, k+3) \geq n+5$;

$m_4 = \Phi(i, k+3) - (n+4)$, where k is the minimum element of $\mathbb{N}$ with $\Phi(i, k+3) \geq n+5$.

The set $\mathbf{K_2}$ is defined in exactly the same way, save that the number m_2 is replaced by m_2', defined by the condition:

$m_2' = \Phi(ij, k+3) - (n+4)$, where k is the minimum element of $\mathbb{N}$ with $\Phi(ij, k+3) \geq n+5$.

The special dyadic triangle functions which are full for $\mathbf{K_1}$ and $\mathbf{K_2}$ are $\mathbf{f_1}$ and $\mathbf{f_2}$, respectively.

It will be guessed that these numbers m_j and m_j' are to be used in some future coding of addition and multiplication. To see how to do this, take $m, n, p, q \in \mathbb{N}$.

Then $m + n = p$ and $q = \Phi(m,n)$ if and only if there exist R_1, R_2, R_3 and S_1, S_2, S_3 in $\mathcal{J}$ such that $l(R_j) = l(S_j)$ $(j = 1,2,3)$ and, further:

$$\begin{aligned}&R_1 \text{ is a peak interval for } \mathbf{f_1} \mid \gamma(I_m)\,;\\&R_2 \text{ is a peak interval for } \mathbf{f_1} \mid \gamma(I_n)\,;\\&R_3 \text{ is a peak interval for } \mathbf{f_1} \mid \gamma(I_p)\,;\\&S_1, S_2, S_3 \text{ are peak intervals for } \mathbf{f_1} \mid I_q\,.\end{aligned}$$

(The number q in the above remarks acts as a 'marker', showing where to look for the number $p = m + n$.) Thus the function which represents addition is

$$\sigma : (I_m, I_n) \mapsto I_{m+n}, \quad \mathcal{K} \to \mathcal{K}\,. \tag{11.12}$$

A similar condition to the above determines when $mn = p$ and $q = \Phi(m,n)$ for $m, n, p, q \in \mathbb{N}$, and hence we obtain the product function

$$\pi : (I_m, I_n) \mapsto I_{mn}, \quad \mathcal{K} \to \mathcal{K}\,. \tag{11.13}$$

The set $\mathbf{K_3}$ We first associate a set $K(r)$ with each $r \in \mathbb{D} \setminus \{0,1\}$. Indeed, set

$$K(r) = \mathbb{D}_{n(r)} \cup \{r + 2^{-n(r)-1}\}\,.$$

Thus we add to $\mathbb{D}_{n(r)}$ the mid-point of the complementary subinterval of $\mathbb{D}_{n(r)}$ which is immediately to the right of r. The special dyadic triangle function which is full for $K(r)$ consists of $2^{n(r)} - 1$ triangles of height $2^{-n(r)-1}$ and two triangles of height $2^{-n(r)-2}$.

The set $\mathbf{K_3}$ is also obtained from the intermediate set $\mathbf{T}$ by inserting points in each of the intervals I_n. Thus take $n \in \mathbb{N}$, and set $r = \Gamma(n+2)$, where $\Gamma : \mathbb{N} \to \mathbb{D}$ is the bijection specified in (11.3), and add a scaled copy of $K(r)$ to I_n. We note that, trivially, $n(\Gamma(n+2)) \le n$ and so $\mathbf{K_3} \cap I_n \subset \mathbb{D}_{(n+2)^2}$. For example, with $n = 1$, we have $I_1 = [5/8, 3/4]$ and $r = 1/2$; we add one triangle of height $1/32$ to the interval $[5/8, 11/16]$ and two adjacent triangles, each of

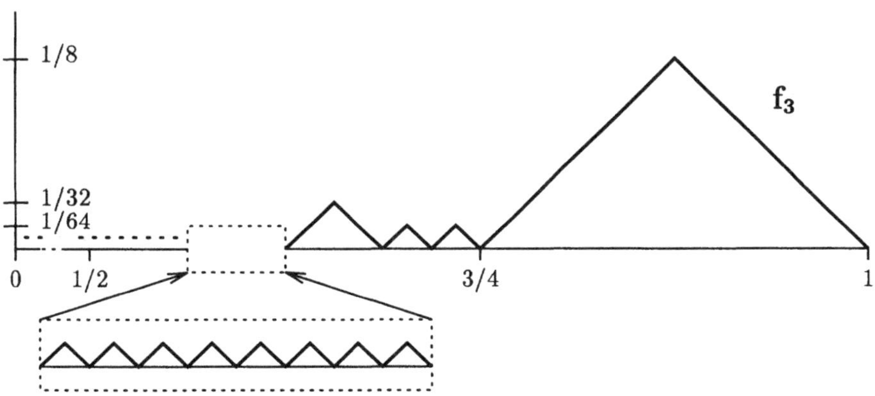

Fig. 11.3. The function $\mathbf{f_3}$

height $1/64$, to the interval $[11/16, 3/4]$. (We recall that there are eight triangles on the interval $[9/16, 5/8]$.) Certainly the set $\mathbf{K_3}$ so formed is a dyadic set.

The special dyadic triangle function which is full for $\mathbf{K_3}$ is $\mathbf{f_3}$.

We can now define the first two of our special functions.

The function Λ The dyadic set $\mathbf{L}$ consists of a scaled version of $\mathbf{K_1}$ on $[0, 1/4]$, of $\mathbf{K_2}$ on $[1/4, 1/2]$, and of $\mathbf{K_3}$ on $[1/2, 1]$.

The special dyadic triangle function which is full for $\mathbf{L}$ is $\mathbf{F}$, and the function Λ is the corresponding function $\widehat{\mathbf{F}}$ germinated from the seed $(\mathbf{L}, \mathbf{F})$ by the process described above; by Lemma 11.10, $\Lambda \in \mathfrak{L}(\mathbb{I})$.

The function Π This function is easily defined; we again build a dyadic set L. First let L include the points 2^{-n} for $n \in \mathbb{Z}^+$, and also the point 0. Each interval $[2^{-n}, 2^{-n+1}]$ of $\mathbb{I}$ is further divided as follows. In the case where $n \neq 2^m$ for any $m \in \mathbb{N}$, the interval is divided into two equal subintervals by the insertion of the mid-point $2^{-n} + 2^{-n-1}$. In the case where $n = 2^m$ for some $m \in \mathbb{N}$, the interval is divided into three subintervals by the insertion of the mid-point $2^{-n} + 2^{-n-1}$ and by the 'three-quarters' point $2^{-n} + 2^{-n-1} + 2^{-n-2}$.

The special dyadic function which is full for L is F, and the function Π is the function $\widehat{F}$ germinated from the seed (L, F) by the process described above; by Lemma 11.10, $\Pi \in \mathfrak{L}(\mathbb{I})$.

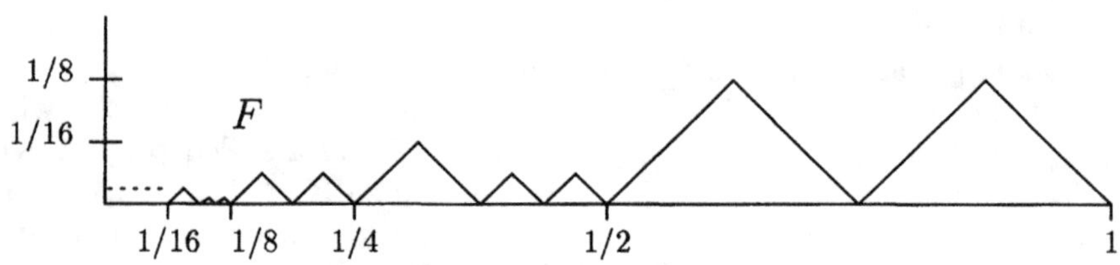

Fig. 11.4. The seed function F for Π

We apply Lemma 11.12 to our functions Λ and Π to obtain the following estimates.

Lemma 11.13 *Suppose that $n \in \mathbb{N}$ and $a, b \in \mathbb{D}_n$ with $a < b$.*

(i) *There exists $[c, d] \in \mathcal{I}(\Lambda)$ with $a \leq c < d \leq b$ and $c, d \in \mathbb{D}_{n^2+8}$.*

(ii) *There exists $[c, d] \in \mathcal{I}(\Pi)$ with $a \leq c < d \leq b$ and $c, d \in \mathbb{D}_{n+5}$.*

Proof (i) By inspection of the dyadic set L_1 arising from $\mathbf{L}$, we see that Lemma 11.12 applies with the choice $\varphi(n) = n^2 + 6$.

(ii) This is similar, with the choice $\varphi(n) = n + 3$. □

Before giving the construction of the final function Ξ, we introduce some further notation connected with peak intervals. Let G be a continuous function on an interval $[a, b]$. We have defined the family $\mathfrak{PI}(G)$ of peak intervals of G.

Now it may be that two or more distinct peak intervals have the same length; all these intervals are deleted from $\mathfrak{PI}(G)$ to form a new family

$$\mathfrak{PI}'(G).$$

The members of $\mathfrak{PI}'(G)$ are the *individual peak intervals.*

For example, let L be the set described in the above construction of Π, and let F be the special dyadic function which is full for L. Many of the peak intervals of F are of the same length as another peak interval, and so are deleted; the only peak intervals that survive and become individual peak intervals for F are those of the form $[2^{-n},\ 2^{-n}+2^{-n-1}]$ with $n=2^m$ for some $m \in \mathbb{N}$.

We now come to the construction of the remaining function Ξ. Our intent is to code the equivalence relation on $(0,1/2)$:

$$x \sim y \quad \text{if} \quad x = 2^k y \quad \text{for some} \quad k \in \mathbb{Z}.$$

The need for this will become clear when we prove Theorem 11.6. The function Ξ is the limit of a sequence of dyadic line functions satisfying the conditions set forth in Lemma 11.10. However, in distinction to the above constructions of Λ and Π, the sequence for Ξ is not generated by a single seed.

It will be convenient to recall the following notation: for a subset S of $\mathbb{R}$, for $\alpha \in \mathbb{R}$, and $a \in \mathbb{R}$,

$$\alpha S = \{\alpha s : s \in S\} \quad \text{and} \quad S \pm a = \{s \pm a : s \in S\}.$$

We shall construct by induction on k a strictly increasing sequence (m_k) and an increasing sequence (n_k) in $\mathbb{N}$, a dyadic line function G_k on $\mathbb{D}_{m_k}$ such that the sequence $((G_k, \mathbb{D}_{m_k}))$ satisfies the conditions imposed on $((F_k, L_k))$ in Lemma 11.10, and a sequence $(\mathcal{I}_k)$ such that, for each $k \in \mathbb{N}$, $\mathcal{I}_k$ is a finite subset of $\mathcal{I}(\Lambda)$, $\mathcal{I}_k \subset \mathcal{I}_{k+1}$, and the following conditions hold.

First, let $m_1 = 4$, and let G_1 be the dyadic line function on $\mathbb{D}_4$ obtained by placing special triangles on each of the seven subintervals of $\mathbb{I}$ of the form $[j/2^3, (j+1)/2^3]$ for $j = 0, 1, 2, 3, 5, 6$, and 7. (There is no triangle on $[1/2, 5/8]$.) Also set $\mathcal{I}_1 = \{\mathbb{I}\}$; the condition that n_1 must satisfy will arise in (3), below.

We shall use the following notation in the case where $k \in \mathbb{N}$ is odd. Let $[c,d]$ be an interval in $\mathcal{I}_k$. Then

$$d^{(k)} = (c+d)/2 + 2^{1-m_k},$$

so that the interval $[(c+d)/2, d^{(k)}]$ is a (small) interval with left-hand end-point the mid-point of $[c,d]$.

The following inductive conditions are imposed for each odd $k \in \mathbb{N}$. (There are no corresponding conditions in the case where k is even.)

(1) Suppose that $[c,d]$ belongs to $\mathcal{I}_k$. Then:

(1.1) $G_k(c) = G_k((c+d)/2) = G_k(d^{(k)}) = G_k(d)$ and

$$G_k(t) \geq G_k(c) \quad (t \in [c,d]);$$

(1.2) the flat components for G_k are exactly the intervals $[(c+d)/2, d^{(k)}]$;

(1.3) for each $I \in \mathfrak{PI}'(G_k \mid [(c+d)/2, d])$, there is exactly one twin-peak interval J of $G_k \mid [c, (c+d)/2]$ with $l(J) = l(I)$;

(1.4) for each pair $\{I_1, I_2\}$ of abutting intervals, each of which is an individual peak interval of $G_k \mid [(c+d)/2, d]$ (with $I_1 \leq I_2$), there exist twin-peak intervals J_1, J_2 of $G_k \mid [c, (c+d)/2]$ with $J_2 \subset [(3c+d)/4, (c+d)/2]$, with $l(J_1) = l(I_1)$ and $l(J_2) = l(I_2)$, and with $J_2 - c = 2^n(J_1 - c)$ for some $n \in \mathbb{N}$.

(2) Suppose that $[c_1, d_1]$ and $[c_2, d_2]$ are distinct elements of $\mathcal{I}_k$. Then:

$$[(c_1 + d_1)/2,\ d_1^{(k)}] \cap [(c_2 + d_2)/2,\ d_2^{(k)}] = \emptyset\,.$$

(3) Suppose that $p \in \mathbb{N}$ with $n_p \leq k$ and that $a, b \in \mathbb{D}_p$ with $a < b$. Then:

(3.1) there exists $I \in \mathcal{I}_k$ with $I \subset (a, b)$;

(3.2) whenever $[c, d]$ belongs to $\mathcal{I}_k$ with $[a, b] \cap (c, (3c+d)/4) \neq \emptyset$, there exists a pair $\{I_1, I_2\}$ of abutting intervals, each of which is an individual peak interval of $G_k \mid [(c+d)/2, d]$, and a twin-peak interval J_1 of $G_k \mid [c, (c+d)/2]$ such that $J_1 \subset (a, b)$ and $l(J_1) = l(I_1)$.

We must also force the new sequence $((G_k, \mathbb{D}_{m_k}))$ to satisfy the conditions (i)–(v) of Lemma 11.10; of these, (i)–(iii) are satisfied automatically, and so the following conditions must hold for each $k \in \mathbb{N}$:

(4) $(G_{2k}, \mathbb{D}_{m_{2k}})$ is a refinement of $(G_{2k-1}, \mathbb{D}_{m_{2k-1}})$ with $m_{2k} = m_{2k-1} + 2$;

(5) $G_{2k+1} - G_{2k}$ is a special dyadic triangle function for a dyadic set containing $\mathbb{D}_{m_{2k}}$ with $\operatorname{supp}(G_{2k+1} - G_{2k}) \subset \mathfrak{FP}(G_{2k})$.

Our task is to construct sequences satisfying the above inductive conditions. We have already fixed G_1, m_1, and $\mathcal{I}_1$. Now suppose that k is odd and assume that we have constructed $G_1, \ldots, G_k$, $m_1, \ldots, m_k$, and $\mathcal{I}_1, \ldots, \mathcal{I}_k$, and also that we have defined $n_1, \ldots, n_p$ for some $p \in \mathbb{N}$ with $n_p \leq k$. Our goal is to define n_{p+1}, and to then define $G_{k+1}, \ldots, G_l$, $m_{k+1}, \ldots, m_l$, and $\mathcal{I}_{k+1}, \ldots, \mathcal{I}_l$ for some odd $l \geq k$ with $l \geq n_{p+1}$.

The first remark is that, for each $n \in \mathbb{N}$, there is a so-called *simple extension*, defining G_{k+1}, G_{k+2}, m_{k+1}, m_{k+2}, and $\mathcal{I}_{k+1}$, $\mathcal{I}_{k+2}$ such that (1)–(5) still hold and that $m_{k+2} \geq n$. In fact, define $G_{k+1} = G_k$ and $m_{k+1} = m_k + 2$ (so that $(G_{k+1}, \mathbb{D}_{m_{k+1}})$ is a refinement of $(G_k, \mathbb{D}_{m_k})$ without adding any platforms), and define $\mathcal{I}_{k+1} = \mathcal{I}_{k+2} = \mathcal{I}_k$. Define G_{k+2} by adding triangles to the flat components of G_k as follows. By (1.2), the flat components for G_k are the intervals $[(c+d)/2, d^{(k)}]$ for $[c, d] \in \mathcal{I}_k$, and each of these intervals has length 2^{-m_k+1}. Divide each such interval into 2^{n+1} subintervals, each of length 2^{-m_k-n}, and place a special triangle on all of these subintervals save for the one with left-hand end-point $(c+d)/2$. Define $m_{k+2} = m_{k+1} + n - 1 = m_k + n + 1$ (so that

$$d^{(k+2)} = (c+d)/2 + 2^{1-m_{k+2}}$$

is the left-hand end-point of the block of added triangles). Note that, for each $[c, d] \in \mathcal{I}_k$,

$$\mathfrak{PI}'(G_{k+2} \mid [(c+d)/2, d]) = \mathfrak{PI}'(G_k \mid [(c+d)/2, d])\,,$$

and so (1.3) and (1.4) hold with k replaced by $k+2$. The other conditions are clearly satisfied.

The main effect of this simple extension is the following. The family $\mathcal{I}_k$ is finite, and so the collection

$$M_k = \{(c+d)/2 : [c,d] \in \mathcal{I}_k\}$$

of mid-points is a finite subset of $\mathbb{I}$. Suppose that S is a closed subset of $\mathbb{I}$ such that $S \cap M_k = \emptyset$. Then, by taking a simple extension with m_{k+2} sufficiently large, we may further suppose that

$$S \cap [(c+d)/2, d^{(k+2)}] = \emptyset$$

for each $[c,d] \in \mathcal{I}_k$. Thus, by (1.2), $S \cap \mathfrak{FP}(G_{k+2}) = \emptyset$, and so we can modify G_{k+2} on both S and on $\mathfrak{FP}(G_{k+2})$ without there being any 'interference' between these two separate modifications.

We now turn to the definition of n_{p+1}. In defining n_{p+1}, we may suppose that condition (3) fails for the definition $n_{p+1} = k$, for otherwise the next step of the construction would be obtained trivially.

We first arrange that (3.1) be satisfied. Take $a, b \in \mathbb{D}_{p+1}$ with $a < b$. There exist $x, y \in \mathbb{D}$ with $x < y$, $[x,y] \subset (a,b)$, and $[x,y] \cap M_k = \emptyset$. By taking a simple extension, we may suppose that $[x,y] \cap \mathfrak{FP}(G_{k+2}) = \emptyset$ and that $x, y \in \mathbb{D}_{m_{k+2}}$; also, by modifying x, we may suppose that $y = x + 2^{-m_{k+2}}$.

Write m for m_{k+2}, set $m_{k+3} = m+2$, and define G_{k+3} by adding a platform to the graph of G_{k+2} within the interval $[x,y]$; the closed interval corresponding to this platform is J, say, so that $l(J) = 2^{-m-2}$ and $\mathfrak{FP}(G_{k+3}) = \mathfrak{FP}(G_{k+2}) \cup \{J\}$. Set $\mathcal{I}_{k+3} = \mathcal{I}_k$.

To define G_{k+4} from G_{k+3}, choose $I \in \mathcal{I}$ with $I \subset J$, say $I = [r,s]$, and set $n = \max\{n(r), n(s)\}$, so that $r = s + 2^{-n}$. The end-points of J belong to $\mathbb{D}_n$. Divide J into 2^{n-m+1} subintervals, each of length 2^{-n-3}; necessarily I is exactly equal to the union of eight of these subintervals. Place a special triangle on all of these subintervals save for the one corresponding to the fifth subinterval of I from the left, so that we have inserted $2^{n-m+1} - 1$ triangles on J.

The family

$$\{[(c+d)/2, d^{(k+2)}] : [c,d] \in \mathcal{I}_k\}$$

is pairwise disjoint (by (2)), and each interval of the family is disjoint from J and has length 2^{-m+1}; we divide each interval into 2^{n+4-m} subintervals, each of length 2^{-n-1}, and place a special triangle on each of these subintervals save for the one with left-hand end-point $(c+d)/2$. Continue by defining $m_{k+4} = n+4$ and $\mathcal{I}_{k+4} = \mathcal{I}_{k+3} \cup \{I\}$.

We see that $G_1, \ldots, G_{k+4}$, $m_1, \ldots, m_{k+4}$, $\mathcal{I}_1, \ldots, \mathcal{I}_{k+4}$, and $n_1, \ldots, n_p$ satisfy (1)–(5), and, further, that $[a,b]$ does not provide a counter-example to assertion (3.1) when n_{p+1} is defined to be $k+4$. By repeating this process, we can eliminate all possible counter-examples to (3.1), noting that there are only finitely many potential counter-examples. Thus (with a new definition of k), we may suppose that G_k, m_k, and $\mathcal{I}_k$ have been constructed so that (3.1) holds in the case where $n_{p+1} = k$.

We now arrange that (3.2) be satisfied. Choose $a, b \in \mathbb{D}_{p+1}$ with $a < b$ and allow the possibility that there exists $[c_0, d_0] \in \mathcal{I}_k$ with

$$[a, b] \cap (c_0, (3c_0 + d_0)/4) \neq \emptyset,$$

witnessing the failure of (3.2) when $n_{p+1} = k$. Then there exist $x_1, y_1, x_2, y_2 \in \mathbb{D}$ such that:

(i) $x_2 < y_2$ and $[x_2, y_2] \subset ((3c_0 + d_0)/4, (c_0 + d_0)/2)$;

(ii) $x_1 < y_1$, $[x_1, y_1] \subset (a, b)$ and there exists $i \in 2\mathbb{N}$ with

$$x_1 = c_0 + 2^{-i}(x_2 - c_0), \quad y_1 = c_0 + 2^{-i}(y_2 - c_0);$$

(iii) $([x_1, y_1] \cup [x_2, y_2]) \cap M_k = \emptyset$.

By taking a simple extension, we obtain G_{k+2} and m_{k+2} such that $x_2, y_2 \in \mathbb{D}_{m_{k+2}}$ and:

(iii)$'$ $([x_1, y_1] \cup [x_2, y_2]) \cap \mathfrak{FP}(G_{k+2}) = \emptyset$.

Write m for m_{k+2}. By modifying x_1 and x_2, we may suppose both that $y_2 = x_2 + 2^{-m}$ and that (ii), above, still holds.

Set $m_{k+3} = m + i + 2$, and define G_{k+3} by adding a platform to the graph of G_{k+2} within each of the intervals $[x_1, y_1]$ and $[x_2, y_2]$. The closed intervals corresponding to these two platforms are J_1 and J_2, respectively, say, so that $l(J_1) = 2^{-i-m-2}$ and $l(J_2) = 2^{-m-2}$ and

$$\mathfrak{FP}(G_{k+3}) = \mathfrak{FP}(G_{k+2}) \cup \{J_1, J_2\}.$$

Set $\mathcal{I}_{k+3} = \mathcal{I}_k$.

The process of forming G_{k+4} from G_{k+3} is similar to the earlier one.

First, consider the intervals J_1 and J_2; divide each of these into two equal subintervals, of length 2^{-i-m-3} and 2^{-m-3} in the two cases. Place special triangles of base length 2^{-i-m-3} and 2^{-m-3} on each of the two subintervals of J_1 and J_2, respectively. Note that $J_1, J_2 \in \mathfrak{TP}(G_{k+4} \mid [c_0, (c_0 + d_0)/2])$ with

$$J_2 \subset [(3c_0 + d_0)/4, (c_0 + d_0)/2]$$

and $J_2 - c_0 = 2^i(J_1 - c_0)$, and also that J_1 and J_2 are the unique intervals in

$$\mathfrak{TP}(G_{k+4} \mid [c_0, (c_0 + d_0)/2])$$

of lengths 2^{-i-m-2} and 2^{-m-2}, respectively.

Second, consider the special subinterval $[(c_0 + d_0)/2, d_0^{(k+2)}]$. Place a special triangle of base length $l(J_2) = 2^{-m-2}$ on an interval, say I_2, with right-hand end-point $d_0^{(k+2)}$, and place a special triangle of base length $l(J_1) = 2^{-i-m-2}$ on the interval, say I_1, which abuts I_2 on the left. Divide the remaining part of $[(c_0 + d_0)/2, d_0^{(k+2)}]$ into subintervals of length 2^{-i-m-3} (which is less than $l(I_1)$), and place a special triangle on all of these subintervals save for the one with left-hand end-point $(c_0 + d_0)/2$. Note that in this last step we have added

$$2^{i+m+3}\left(2^{-m+1} - 2^{-m-2} - 2^{-i-m-2}\right) - 1 = 14 \cdot 2^i - 3 \geq 2$$

triangles. Also note that I_1 and I_2 are both individual peak intervals of the function $G_{k+4} \mid [(c_0 + d_0)/2, d_0]$, that I_1 and I_2 are abutting intervals, that $l(I_1) = l(J_1)$ and $l(I_2) = l(J_2)$, and that

$$\mathfrak{PI}'(G_{k+4} \mid [(c_0 + d_0)/2, d_0]) = \mathfrak{PI}'(G_{k+2} \mid [(c_0 + d_0)/2, d_0]) \cup \{I_1, I_2\}.$$

Finally, we consider the intervals $[(c + d)/2, d^{(k+2)}]$, where $[c, d] \in \mathcal{I}_k$ and $[c, d] \neq [c_0, d_0]$. Divide each such interval into 2^{i+4} subintervals, each of length 2^{-i-m-3}, and place a special triangle on all of these subintervals save for the one with left-hand end-point $(c + d)/2$. Define $m_{k+4} = m + i + 4$ and $\mathcal{I}_{k+4} = \mathcal{I}_{k+3}$. Note that

$$\mathfrak{PI}'(G_{k+4} \mid [(c + d)/2, d]) = \mathfrak{PI}'(G_{k+2} \mid [(c + d)/2, d]).$$

We have defined $G_1, \ldots, G_{k+4}$, $m_1, \ldots, m_{k+4}$, $\mathcal{I}_1, \ldots, \mathcal{I}_{k+4}$ and $n_1, \ldots, n_p$. The notes that we have inserted as the construction proceeded are sufficient to show that (1)–(5) are satisfied, and, further, that $[a, b]$ does not provide a counter-example to (3.2) when n_{p+1} is defined to be $k + 4$. By repeating this process, we can eliminate all possible counter-examples to (3.2). Thus (with a new definition of k), we may suppose that G_k, m_k, and $\mathcal{I}_k$ have been constructed so that both (3.1) and (3.2), and hence all of (3), hold in the case where $n_{p+1} = k$.

This concludes the description of the inductive construction of the sequences to satisfy (1)–(5) for all $n \in \mathbb{N}$.

The function Ξ Having established that the above inductive construction is possible, we define the sequences to satisfy (1)–(5) in a special way. First, choose $n_1 \in \mathbb{N}$ and $G_1, \ldots, G_{n_1}$, $m_1, \ldots, m_{n_1}$, and $\mathcal{I}_1, \ldots, \mathcal{I}_{n_1}$ to satisfy the conditions (1)–(5) for $k = 1, \ldots, n_1$ in such a way that m_{n_1} attains its minimum possible value. Second, choose $n_2 \in \mathbb{N}$ such that $n_2 > n_1$ and choose $G_{n_1+1}, \ldots, G_{n_2}$, $m_{n_1+1}, \ldots, m_{n_2}$, and $\mathcal{I}_{n_1+1}, \ldots, \mathcal{I}_{n_2}$ to satisfy the conditions (1)–(5) for $k = 1, \ldots, n_2$ in such a way that m_{n_2} attains its minimum possible value. Continue in this way. The point of this procedure is that the sequence (m_{n_k}) is recursive, in the sense that a bound on the size of m_{n_k} as a function of k can be given by an explicit formula, which we are not calculating.

The sequence $((G_k, \mathbb{D}_{m_k}))$ so defined satisfies the conditions imposed on the sequence $((L_k, F_k))$ in Lemma 11.10. The corresponding limit $\widehat{F}$ provided by this lemma is the required function Ξ. We have $\Xi \in \mathfrak{L}(\mathbb{I})$, and Ξ satisfies (11.6).

Note that Ξ is not self-replicating in the way that Λ and Π are, but the key features of Ξ recur within arbitrarily small open intervals of $\mathbb{I}$.

We now define a relation $\mathcal{R}$ on the set $\mathcal{J}$ (of all subintervals $[a, b]$ of $\mathbb{I}$ with $a < b$). Indeed, $\mathcal{R}$ is the subset of $\mathcal{J} \times \mathcal{J}$ consisting of the pairs (J_1, J_2) for which there is another pair (I_1, I_2) in $\mathcal{J} \times \mathcal{J}$ such that:

J_1 and J_2 are twin-peak intervals of $\Xi \mid [0, 1/2]$;
I_1 and I_2 are individual peak intervals of $\Xi \mid [1/2, 1]$;
$l(I_1) = l(J_1)$ and $l(I_2) = l(J_2)$;
$I_1 \leq I_2$ and $|I_1 \cap I_2| = 1$.

In the case where $(J_1, J_2) \in \mathcal{R}$, necessarily $J_1 \subset (0, 1/4)$, $J_2 \subset (1/4, 1/2)$, and $J_2 = 2^n J_1$ for some (uniquely specified) $n \in \mathbb{N}$. The point of our long construction above, is that we can find $(J_1, J_2) \in \mathcal{R}$ with J_1 contained in an arbitrary (small) open interval: if $r, s \in \mathbb{D}$ with $0 < r < s < 1/4$, then there exists $(J_1, J_2) \in \mathcal{R}$ with $J_1 \subset (r, s)$. This is verified by considering the construction.

The following *key claim* follows. Let $x, y \in (0, 1/4)$. Then $x = 2^k y$ for some $k \in \mathbb{Z}$ (i.e. $x \sim y$) if and only if, for each $\delta > 0$, there exist (J_1, J_2) and (J_3, J_4) in $\mathcal{R}$ such that

$$J_1 \subset (x - \delta,\, x + \delta),\quad J_3 \subset (y - \delta,\, y + \delta),\quad J_2 \subset J_4\,.$$

The claim is verified by considering the construction. Indeed, a local version of this claim is true and will be required. Set

$$S^{\Xi} = \{(c, d) \in \mathbb{D} \times \mathbb{D} : [c, d] \in \bigcup \mathcal{I}_k\}\,.$$

(Thus S^{Ξ} is a family of ordered pairs, not of open intervals.) For each $a, b \in \mathbb{I}$ with $a < b$, there exists $(c, d) \in S^{\Xi}$ with $a < c < d < b$. We can now scale the definition of the relation $\mathcal{R}$ to $[c, d]$ by using $S^{\Xi} \mid [c, d]$; we obtain the relation $\mathcal{R}^{[c,d]}$, say, on the scaled version of $\mathcal{J}$ on $[c, d]$. Then, for elements $x, y \in (c, (3c + d)/4)$, the relation

$$x - c = 2^k (y - c) \quad \text{for some} \quad k \in \mathbb{Z}$$

can be expressed in terms of pairs of elements of $\mathcal{R}^{[c,d]}$ in an analogous way to the above.

Having completed a season of standard real analysis, we return to super-real fields. Again, let Ω be a compact space, and let P be a non-maximal, prime ideal in $C(\Omega)$; the maximal prime z-ideal below P is Q. Let $F \in \mathfrak{L}(\mathbb{I})$, so that, by the operational calculus of Chapter 6, we have the associated functions $F_{(P)}$ and $F_{(Q)}$ on $\mathbb{I}_{(P)}$ and $\mathbb{I}_{(Q)}$, respectively. Many of the properties of $F_{(Q)}$ that we shall be interested in now are properties of

$$\operatorname{Gr} F_{(Q)} \cap (\mathbb{D}_{(Q)} \times \mathbb{D}_{(Q)})\,;$$

by Lemma 10.8(iii), this set is equal to $(\operatorname{Gr} F \cap (\mathbb{D} \times \mathbb{D}))_{(Q)}$, and so the properties can be easily established by standard pointwise arguments using Q-partitions. The somewhat tedious task is to transfer these properties to $F_{(P)}$.

As before, we take $\widehat{F}$ to be a function in $\mathfrak{L}(\mathbb{I})$ obtained through the limiting process of Lemma 11.10 from certain sequences (L_k) and (F_k) satisfying the hypotheses of that lemma. By Lemma 11.10, $\widehat{F}$ satisfies (11.6), and so it follows from Lemma 10.8(ii) that

$$\widehat{F}_{(Q)} \mid \mathbb{D}_{(P)} = \widehat{F}_{(P)} \mid \mathbb{D}_{(P)}\,.$$

We maintain these hypotheses during the next three lemmas; the purpose of the lemmas is to allow us to transfer known results about $\widehat{F}_{(Q)}$ to $\widehat{F}_{(P)}$.

Lemma 11.14 *Let $x, y \in \mathbb{D}_{(P)}$ with $x < y$. Then there exist $u \in \mathbb{D}_{(P)}$ and $r_0, s_0 \in \mathbb{D}_{(P)}$ such that*

$$u = \min\{\widehat{F}_{(Q)}(z) : z \in [x,y]_{(Q)}\}\,,$$
$$r_0 = \min\{z \in [x,y]_{(Q)} : \widehat{F}_{(Q)}(z) = u\}\,,$$
$$s_0 = \max\{z \in [x,y]_{(Q)} : \widehat{F}_{(Q)}(z) = u\}\,.$$

Proof For $r, s \in \mathbb{D}$ with $r < s$, define $u(r,s) = \min\{\widehat{F}(t) : t \in [r,s]\}$ and

$$H(r,s) = \min\{t \in [r,s] : \widehat{F}(t) = u(r,s)\}\,;$$

also, set $H(r,s) = 0$ if $r \geq s$. It follows from Lemma 11.11(i) that

$$H(\mathbb{D}_n \times \mathbb{D}_n) \subset \mathbb{D}_{n+1} \quad (n \in \mathbb{N})\,.$$

By standard arguments, u is well defined in K_Q and $H_{(Q)}(x,y) = r_0 \in \mathbb{D}_{a+1}^{(Q)}$, where $a \in \mathbb{N}_{(P)}$ is such that $x, y \in \mathbb{D}_a^{(Q)}$.

By a familiar argument from the transfer principle, we see that

$$H_{(Q)}(\mathbb{D}_{(P)} \times \mathbb{D}_{(P)}) \subset \mathbb{D}_{(P)}\,,$$

and so r_0 is well defined and $r_0 \in \mathbb{D}_{(P)}$. The same result applies to s_0. Finally $F_{(Q)}(r_0) = F_{(P)}(r_0)$ and $F_{(Q)}(r_0) = u$, and so $u \in \mathbb{D}_{(P)}$. □

We have defined peak intervals, twin-peak intervals, and flat components for a continuous function on a subinterval $[a,b]$ of $\mathbb{I}$. Formally the same definitions apply to continuous functions on subintervals $[x,y]_{(P)}$ of $\mathbb{I}_{(P)}$ and $[x,y]_{(Q)}$ of $\mathbb{I}_{(Q)}$.

Lemma 11.15 *Let $x, y \in \mathbb{D}_{(P)}$ with $x < y$.*

(i) *Suppose that $[r,s]_{(P)}$ is a peak interval for $\widehat{F}_{(P)} \mid [x,y]_{(P)}$. Then $r, s \in \mathbb{D}_{(P)}$ and $[r,s]_{(Q)}$ is a peak interval for $\widehat{F}_{(Q)} \mid [x,y]_{(Q)}$.*

(ii) *Suppose that $[r,s]_{(Q)}$ is a peak interval for $\widehat{F}_{(Q)} \mid [x,y]_{(Q)}$ and that $(r,s)_{(Q)} \cap \mathbb{D}_{(P)} \neq \emptyset$. Then $r, s \in \mathbb{D}_{(P)}$ and $[r,s]_{(P)}$ is a peak interval for $\widehat{F}_{(P)} \mid [x,y]_{(P)}$.*

(iii) *Suppose that there exists $z_0 \in (x,y)_{(Q)}$ with*

$$\widehat{F}_{(Q)}(z_0) < \widehat{F}_{(Q)}(t) \quad (t \in [x,y]_{(Q)} \setminus \{z_0\})\,.$$

Set

$$r = \sup\{t \in [0,z_0)_{(Q)} : \widehat{F}_{(Q)}(t) \leq \widehat{F}_{(Q)}(z_0)\}\,,$$
$$s = \inf\{t \in (z_0,1]_{(Q)} : \widehat{F}_{(Q)}(t) \leq \widehat{F}_{(Q)}(z_0)\}\,.$$

Then $z_0 \in \mathbb{D}_{(P)}$ and either $r \in \mathbb{D}_{(P)}$ or $s \in \mathbb{D}_{(P)}$.

Proof We have $\widehat{F}_{(P)}(x) = \widehat{F}_{(Q)}(x)$ and $\widehat{F}_{(P)}(y) = \widehat{F}_{(Q)}(y)$.

(i) By hypothesis, $\widehat{F}_{(P)}(x) = \widehat{F}_{(P)}(r) = \widehat{F}_{(P)}(s) = \widehat{F}_{(P)}(y)$, and also

$$\widehat{F}_{(P)}(z) \geq \widehat{F}_{(P)}(x) \quad (z \in [x,y]_{(P)}), \quad \widehat{F}_{(P)}(z) > \widehat{F}_{(P)}(x) \quad (z \in (r,s)_{(P)})\,.$$

Fix $t \in (r,s)_{(P)} \cap \mathbb{D}_{(P)}$, and set

$$\begin{aligned} u &= \min\{\widehat{F}_{(Q)}(z) : z \in [x,t]_{(Q)}\}, \\ v &= \min\{\widehat{F}_{(Q)}(z) : z \in [t,y]_{(Q)}\}, \\ r_0 &= \max\{z \in [x,t]_{(Q)} : \widehat{F}_{(Q)}(z) = u\}, \\ s_0 &= \min\{z \in [t,y]_{(Q)} : \widehat{F}_{(Q)}(z) = v\}. \end{aligned}$$

By Lemma 11.14, u, v, r_0, and s_0 are well defined, and $r_0, s_0 \in \mathbb{D}_{(P)}$. We have $\widehat{F}_{(P)}(r_0) = \widehat{F}_{(Q)}(r_0) = u$ and $\widehat{F}_{(P)}(x) = \widehat{F}_{(Q)}(x) \geq u$. But $\widehat{F}_{(P)}(r_0) \geq \widehat{F}_{(P)}(x)$, and so $\widehat{F}_{(P)}(x) = u$. Similarly $\widehat{F}_{(P)}(y) = v$, and so $u = v$. Further $r = r_0$ and $s = s_0$, and hence $r, s \in \mathbb{D}_{(P)}$. Thus

$$\widehat{F}_{(Q)}(z) \geq \widehat{F}_{(Q)}(x) = \widehat{F}_{(Q)}(y) \quad (z \in [x,y]_{(Q)}),$$

and, for $z \in (r,t]_{(Q)} \cup [t,s)_{(Q)} = (r,s)_{(Q)}$, we have $\widehat{F}_{(Q)}(z) > u$. Therefore $[r,s]_{(Q)}$ is a peak interval for $\widehat{F}_{(Q)} \mid [x,y]_{(Q)}$.

(ii) Fix $t \in (r,s)_{(Q)} \cap \mathbb{D}_{(P)}$, and let u, v, r_0, and s_0 be defined with respect to this t as in (i). Since $[r,s]_{(Q)}$ is a peak interval for $\widehat{F}_{(Q)} \mid [x,y]_{(Q)}$, we have $r = r_0$ and $s = s_0$, and so, by Lemma 11.14, $r, s \in \mathbb{D}_{(P)}$.

We *claim* that

$$\widehat{F}_{(P)}(z) \geq \widehat{F}_{(P)}(x) = \widehat{F}_{(P)}(y) \quad (z \in [x,y]_{(P)}).$$

The inequality holds for $z \in [x,y]_{(Q)}$, and hence for $z \in [x,y]_{(P)} \cap \mathbb{D}_{(P)}$. By Lemma 8.40, the latter set is order-dense in $[x,y]_{(P)}$, and, by Theorem 6.5(ii), $\widehat{F}_{(P)}$ is continuous with respect to the order-topology on $[x,y]_{(P)}$, and so the claim holds.

We also *claim* that

$$\widehat{F}_{(P)}(z) > \widehat{F}_{(P)}(x) \quad (z \in (r,s)_{(P)}).$$

Fix $t_1, t_2 \in (r,s)_{(P)} \cap \mathbb{D}_{(P)}$ with $t_1 < t_2$, and set

$$\begin{aligned} u_1 &= \min\{\widehat{F}_{(Q)}(z) : z \in [t_1,t_2]_{(Q)}\}, \\ r_1 &= \min\{z \in [t_1,t_2]_{(Q)} : \widehat{F}_{(Q)}(z) = u_1\}; \end{aligned}$$

again it follows from Lemma 11.14 that u_1 and r_1 are well defined and $r_1 \in \mathbb{D}_{(P)}$. Since $\widehat{F}_{(Q)}(z) > \widehat{F}_{(Q)}(x)$ for $z \in (r,s)_{(Q)}$, we have $u_1 > \widehat{F}_{(Q)}(x) = \widehat{F}_{(P)}(x)$. As in the proof of the first claim, it follows that

$$\widehat{F}_{(P)}(z) \geq u_1 > \widehat{F}_{(P)}(x) \quad (z \in [t_1,t_2]_{(P)}).$$

The second claim follows because $(r,s)_{(P)} \cap \mathbb{D}_{(P)}$ is order-dense in $(r,s)_{(P)}$.

We have shown that $[r,s]_{(P)}$ is a peak interval for $\widehat{F}_{(P)} \mid [x,y]_{(P)}$.

(iii) The proof of this result is similar to that of Lemma 11.14, using Lemma 11.11(ii). □

Lemma 11.16 (i) *Suppose that* $x, y \in \mathbb{I}_{(P)}$ *and* $[x, y]_{(P)}$ *is a twin-peak interval for* $\widehat{F}_{(P)}$. *Then* $x, y \in \mathbb{D}_{(P)}$ *and* $[x, y]_{(Q)}$ *is a twin-peak interval for* $\widehat{F}_{(Q)}$.

(ii) *Suppose that* $x, y \in \mathbb{I}_{(Q)}$ *and* $[x, y]_{(Q)}$ *is a twin-peak interval for* $\widehat{F}_{(Q)}$, *and suppose that*

$$(x, (x+y)/2)_{(Q)} \cap \mathbb{D}_{(P)} \neq \emptyset, \quad ((x+y)/2, y)_{(Q)} \cap \mathbb{D}_{(P)} \neq \emptyset.$$

Then $x, y \in \mathbb{D}_{(P)}$ *and* $[x, y]_{(P)}$ *is a twin-peak interval for* $\widehat{F}_{(P)}$.

Proof (i) Fix $r, s \in \mathbb{D}_{(P)}$ with $x < r < (x+y)/2 < s < y$, and set

$$\begin{aligned} u &= \min\{\widehat{F}_{(Q)}(t) : t \in [r, s]_{(Q)}\}, \\ r_0 &= \min\{t \in [r, s]_{(Q)} : \widehat{F}_{(Q)}(t) = u\}, \\ s_0 &= \max\{t \in [r, s]_{(Q)} : \widehat{F}_{(Q)}(t) = u\}. \end{aligned}$$

By Lemma 11.14, u, r_0, and s_0 are well defined and $r_0, s_0 \in \mathbb{D}_{(P)}$. But clearly $r_0 = s_0 = (x+y)/2$, and so

$$\widehat{F}_{(Q)}((x+y)/2) < \widehat{F}_{(Q)}(t) \quad (t \in [r, s]_{(Q)} \setminus \{(x+y)/2\}).$$

Set

$$\begin{aligned} r_1 &= \sup\{t \in [0, (x+y)/2)_{(Q)} : \widehat{F}_{(Q)}(t) \leq \widehat{F}_{(Q)}((x+y)/2)\}, \\ s_1 &= \inf\{t \in ((x+y)/2, 1]_{(Q)} : \widehat{F}_{(Q)}(t) \leq \widehat{F}_{(Q)}((x+y)/2)\}. \end{aligned}$$

By Lemma 11.15(iii), either $r_1 \in \mathbb{D}_{(P)}$ or $s_1 \in \mathbb{D}_{(P)}$. Suppose that $r_1 \in \mathbb{D}_{(P)}$. Since $\widehat{F}_{(P)} \mid \mathbb{D}_{(P)} = \widehat{F}_{(Q)} \mid \mathbb{D}_{(P)}$, we have

$$\widehat{F}_{(P)}(t) \geq \widehat{F}_{(P)}(r_1) \quad (t \in [r_1, x]_{(P)})$$

by the continuity of $\widehat{F}_{(P)}$ and the definition of r_1. Hence $\widehat{F}_{(P)}(r_1) = \widehat{F}_{(P)}(x)$, and so $[x, (x+y)/2]_{(P)}$ is a peak interval of $\widehat{F}_{(P)} \mid [r_1, (x+y)/2]_{(P)}$. By Lemma 11.15(i), $x \in \mathbb{D}_{(P)}$ (and so $r_1 = x$). Similarly, in the case where $s_1 \in \mathbb{D}_{(P)}$, it follows that $y \in \mathbb{D}_{(P)}$.

It now follows from the fact that $(x+y)/2 \in \mathbb{D}_{(P)}$ that both x and y belong to $\mathbb{D}_{(P)}$.

Next set

$$\begin{aligned} v &= \min\{\widehat{F}_{(Q)}(t) : t \in [x, r]_{(Q)}\}, \\ r_2 &= \max\{t \in [x, r]_{(Q)} : \widehat{F}_{(Q)}(t) = v\}. \end{aligned}$$

By Lemma 11.14 again, $r_2 \in \mathbb{D}_{(P)}$, and so $r_2 = x$. Thus

$$\widehat{F}_{(Q)}(t) > \widehat{F}_{(Q)}(x) \quad (t \in (x, r]_{(Q)}).$$

Similarly, this inequality holds for $t \in (s, y]_{(Q)}$, and so $[x, y]_{(Q)}$ is a twin-peak interval for $\widehat{F}_{(Q)}$.

(ii) Fix $r \in (x, (x+y)/2)_{(Q)} \cap \mathbb{D}_{(P)}$ and $s \in ((x+y)/2, y)_{(Q)} \cap \mathbb{D}_{(P)}$. By Lemma 11.15(iii), $(x+y)/2 \in \mathbb{D}_{(P)}$ and also either $r_1 \in \mathbb{D}_{(P)}$ or $s_1 \in \mathbb{D}_{(P)}$, where r_1 and s_1 are as defined above.

Clearly $r_1 = x$ and $s_1 = y$, and so $x \in \mathbb{D}_{(P)}$ or $y \in \mathbb{D}_{(P)}$. Again since $(x+y)/2 \in \mathbb{D}_{(P)}$, both x and y belong to $\mathbb{D}_{(P)}$.

By Lemma 11.15(ii), both $[x,(x+y)/2]_{(P)}$ and $[(x+y)/2,y]_{(P)}$ are peak-intervals of $\widehat{F}_{(P)} \mid [x,y]_{(P)}$, and so $[x,y]_{(P)}$ is a twin-peak interval for $\widehat{F}_{(P)}$, as required. □

Take $x, y \in \mathbb{D}_{(Q)}$ with $x < y$, and let $\gamma_{x,y}^{(Q)}$ be the Q-version of the scaling function $\gamma_{x,y}$ of (11.5); the scaled version of a function F on $[x,y]_{(Q)}$ is $F_{x,y}^{(Q)}$, so that

$$F_{x,y}^{(Q)}(z) = (y-x)\left(F \circ (\gamma_{x,y}^{(Q)})^{-1}\right)(z) \quad (z \in [x,y]_{(Q)})\,.$$

In this way, we define

$$\Lambda_{x,y}^{(Q)}, \quad \Pi_{x,y}^{(Q)}, \quad \text{and} \quad \Xi_{x,y}^{(Q)}\,.$$

Similarly, we have P-versions of all these functions.

Lemma 11.17 *Let $x, y \in \mathbb{D}_{(P)}$ with $x < y$.*

(i) *There exist $u, v \in \mathbb{D}_{(P)}$ with $x \le u < v \le y$ and*

$$\Pi_{(P)} \mid [u,v]_{(P)} = \Pi_{(P)}(u) + \Pi_{u,v}^{(P)}\,.$$

(ii) *Suppose that P is an exponential prime. Then there exist $u, v \in \mathbb{D}_{(P)}$ with $x \le u < v \le y$ and*

$$\Lambda_{(P)} \mid [u,v]_{(P)} = \Lambda_{(P)}(u) + \Lambda_{u,v}^{(P)}\,.$$

Proof (i) For $a, b \in \mathbb{D}$ with $a < b$, set $n = \max\{n(a), n(b)\}$. By Lemma 11.13(ii), there exist $c, d \in \mathbb{D}_{n+5}$ with $a \le c < d \le b$ and

$$\Pi \mid [c,d] = \Pi(c) + \Pi_{c,d}\,.$$

We set $H(a,b) = (c,d)$; in the case where $a \ge b$, set $H(a,b) = (0,0)$. We have thus defined a function $H : \mathbb{D} \times \mathbb{D} \to \mathbb{D} \times \mathbb{D}$. By an argument using the transfer principle, Corollary 8.30, we see that $H_{(Q)}(\mathbb{D}_{(P)} \times \mathbb{D}_{(P)}) \subset \mathbb{D}_{(P)} \times \mathbb{D}_{(P)}$. Now define

$$(u,v) = H_{(Q)}(x,y)\,,$$

so that $u, v \in \mathbb{D}_{(P)}$ and $x \le u < v \le y$.

By a pointwise argument,

$$\Pi_{(Q)} \mid [u,v]_{(Q)} = \Pi_{(Q)}(a) + \Pi_{u,v}^{(Q)}\,.$$

But $\Pi_{(P)} \mid \mathbb{D}_{(P)} = \Pi_{(Q)} \mid \mathbb{D}_{(P)}$, and so

$$\Pi_{(P)} \mid \left([u,v]_{(P)} \cap \mathbb{D}_{(P)}\right) = \Pi_{(P)}(u) + \Pi_{u,v}^{(P)} \mid \mathbb{D}_{(P)}\,.$$

Since $\Pi_{(P)}$ is continuous, the result follows.

(ii) This is similar to the proof of (i), but uses Lemma 11.13(i); the bound '$n+5$' of (i) is replaced by 'n^2+8', and so we must use the fact that P is an exponential prime to see that $\mathbb{N}_{(P)}$ is multiplicatively closed. □

The analogue of the sets $\mathcal{I}(\Lambda)$ and $\mathcal{I}(\Pi)$ for the function Ξ is the set S^{Ξ}. There is a function $\varphi : \mathbb{N} \to \mathbb{N}$ such that, for each $n \in \mathbb{N}$ and $a, b \in \mathbb{D}_n$ with $a < b$, there exists $(c, d) \in S^{\Xi}$ with $a \leq c < d \leq b$ and $c, d \in \mathbb{D}_{\varphi(n)}$. It is more than tedious to calculate a reasonable formula for φ. However, for our applications, it suffices that there is such a function φ with a recursive definition, and that it is provable in number theory that this recursive definition defines a function on all of $\mathbb{N}$.

Lemma 11.18 *Let $x, y \in \mathbb{D}_{(P)}$ with $x < y$. Suppose that the structure*

$$\langle \mathbb{N}_{(P)}, +, \cdot \rangle$$

is a model of number theory. Then there exist $u, v \in \mathbb{D}_{(P)}$ with $(u, v) \in S^{\Xi}_{(Q)}$ and $x \leq u < u \leq y$.

Proof Define $H : \mathbb{D} \times \mathbb{D} \to \mathbb{D} \times \mathbb{D}$ essentially as in Lemma 11.17, save that now $H(a, b) \in \mathbb{D}_{\varphi(n)} \times \mathbb{D}_{\varphi(n)}$ for $a, b \in \mathbb{D}_n$, where φ is the recursive function described above. The recursive definition which defines φ defines in $\langle \mathbb{N}_{(Q)}, +, \cdot \rangle$ the function $\varphi_{(Q)}$. Since $\langle \mathbb{N}_{(P)}, +, \cdot \rangle$ is also a model of number theory, the recursive definition of φ defines in $\langle \mathbb{N}_{(P)}, +, \cdot \rangle$ a function $\psi : \mathbb{N}_{(P)} \to \mathbb{N}_{(P)}$. By using the transfer principle, which shows a certain closure of $\mathbb{N}_{(P)}$ in $\mathbb{N}_{(Q)}$, we see that $\psi = \varphi_{(Q)} \mid \mathbb{N}_{(P)}$. In particular, $\varphi_{(Q)}(a) \in \mathbb{N}_{(P)}$ $(a \in \mathbb{N}_{(P)})$. From this, it follows that $H_{(Q)}(\mathbb{D}_{(P)} \times \mathbb{D}_{(P)}) \subset \mathbb{D}_{(P)} \times \mathbb{D}_{(P)}$, and then the result follows, as before. □

We can now, at last, turn to the proof of Theorem 11.6 itself.

Proof of Theorem 11.6 The hypothesis in this theorem is that P_0 is a non-maximal, prime ideal in $C(\Omega_0)$ such that $\mathfrak{N}_{(P_0)}$ is a model of analysis and $\mathfrak{N}_{(P_0)}$ is solid (in the sense of Definition 11.5). Let P be a non-maximal, prime ideal in $C(\Omega)$ for some compact space Ω such that $(K_P, \leq)$ and $(K_{P_0}, \leq)$ are order-isomorphic. We must show that the models $\mathfrak{N}_{(P_0)}$ and $\mathfrak{N}_{(P)}$ are isomorphic.

We begin with the definitions of various structures.

First $\mathcal{J}_{(P)}$ is the family of all closed intervals of $\mathbb{I}_{(P)}$. Next, for $a \in \mathbb{N}_{(P)}$, we set

$$I_a^{(P)} = [2^{-a} + 2^{-a-2},\, 2^{-a} + 2^{-a-1}]_{(P)} = \left[5l(I_a^{(P)}),\, 6l(I_a^{(P)})\right]_{(P)},$$

so that $I_a^{(P)}$ is the P-version of the set I_n, defined earlier. The family of these intervals is

$$\mathcal{K}_{(P)} = \{I_a^{(P)} : a \in \mathbb{N}_{(P)}\},$$

so that $\mathcal{K}_{(P)}$ is a discrete family in $(0, 1]_{(P)}$. We write $\gamma_{(P)}(I_a^{(P)})$ for the corresponding interval 'to the left of $I_a^{(P)}$':

$$\gamma_{(P)}(I_a^{(P)}) = [2^{-a},\, 2^{-a} + 2^{-a-2}]_{(P)} \quad (a \in \mathbb{N}_{(P)}),$$

defining the map $\gamma_{(P)} : \mathcal{K}_{(P)} \to \mathcal{J}_{(P)}$. There is an obvious notion of addition and multiplication of the intervals of $\mathcal{K}_{(P)}$: define

$$\left.\begin{array}{l} \sigma_{(P)} : (I_a^{(P)}, I_b^{(P)}) \mapsto I_{a+b}^{(P)} \\ \pi_{(P)} : (I_a^{(P)}, I_b^{(P)}) \mapsto I_a^{(P)} \end{array}\right\}, \quad \mathcal{K}_{(P)} \times \mathcal{K}_{(P)} \to \mathcal{K}_{(P)}\,.$$

Clearly we have an isomorphism of structures

$$\iota_{(P)} : \langle \mathbb{N}_{(P)}, +, \cdot \rangle \to \langle K_{(P)}, \sigma_{(P)}, \pi_{(P)} \rangle\,.$$

The order $\leq$ on $\mathcal{K}_{(P)}$ is defined by the map $\sigma_{(P)}$. Indeed, $I_a < I_b$ if there exists $I_c \in \mathcal{K}_{(P)}$ with $\sigma_{(P)}(I_a^{(P)}, I_c^{(P)}) = I_b^{(P)}$.

Next we define

$$\mathcal{P}_{(P)} = \{\iota_{(P)}(A) : A \in \mathcal{P}(\mathbb{N})_{(P)}\}\,,$$

so that we have an isomorphism of structures

$$\iota_{(P)} : \langle \mathbb{N}_{(P)}, +, \cdot, \mathcal{P}(\mathbb{N})_{(P)}) \rangle \to \langle \mathcal{K}_{(P)}, \sigma_{(P)}, \pi_{((P)}, \mathcal{P}_{(P)} \rangle\,.$$

The analogue of the function ε is called E : for $x \in \mathbb{I}_{(P)} \setminus \mathbb{D}_{(P)}$, we define

$$E_{(P)}(x) : I_a \mapsto \varepsilon_x^{(P)}(a), \quad \mathcal{K}_{(P)} \to \{0,1\}\,.$$

As a tool for understanding $E_{(P)}$, we shall require the notation for an equivalence relation: for $x, y \in \mathbb{I}_{(P)}$, set

$$x \sim_{(P)} y \quad \text{if } x = 2^a y \quad \text{for some} \quad a \in \mathbb{Z}_{(P)}\,.$$

We shall also require the notation for a 'local version' of all the above structures. Take $x, y \in \mathbb{I}_{(P)}$, and let $\gamma_{x,y}^{(P)}$ be the corresponding scaling function corresponding to γ specified in (11.5). For a function F on $\mathbb{I}_{(P)}$, the scaled copy of F on $[x,y]_{(P)}$ is $F_{x,y}$. We also transfer the above structures and maps to $[x,y]_{(P)}$. Thus, for example, we have

$$\mathcal{K}_{x,y}^{(P)} = \{\gamma_{x,y}^{(P)}(I) : I \in \mathcal{K}_{(P)}\}\,,$$

$$\sigma_{x,y}^{(P)} : (\gamma_{x,y}^{(P)}(I), \gamma_{x,y}^{(P)}(J)) \mapsto (\gamma_{x,y}^{(P)} \circ \sigma)(I,J), \quad \mathcal{K}_{x,y}^{(P)} \times \mathcal{K}_{x,y}^{(P)} \to \mathcal{K}_{x,y}^{(P)}\,.$$

The image of the equivalence relation $\sim_{(P)}$ under $\gamma_{x,y}^{(P)}$ is $\sim_{x,y}^{(P)}$, so that

$$r \sim_{x,y}^{(P)} s \quad \text{if} \quad r - x = 2^a(s - x) \quad \text{for some } a \in \mathbb{Z}_{(P)}\,,$$

where $r, s \in [x,y]_{(P)}$. We shall see that the functions $\Pi_{(P)}$ and $\Lambda_{(P)}$ are self-replicating on many intervals $[x,y]_{(P)} \in \mathcal{J}_{(P)}$ and that, whilst this is not true for $\Xi_{(P)}$, it is true that the qualitative features of $\Xi_{(P)}$ reproduce themselves on many such intervals. This will be the key to our proof. Our procedure to prove Theorem 11.6 will be to see how to determine all of

$$\left\langle \mathcal{K}_{x,y}^{(P)}, \sigma_{x,y}^{(P)}, \pi_{x,y}^{(P)}, \mathcal{P}_{x,y}^{(P)} \right\rangle \quad \text{and} \quad E_{x,y}^{(P)}$$

from the order-structure on $[x,y]_{(P)}$, the restrictions to $[x,y]_{(P)}$ of $\Pi_{(P)}, \Lambda_{(P)}$, and $\Xi_{(P)}$, and the restriction of $\Upsilon_{(P)}$ to a suitable small square in $\mathbb{I}^2_{(P)}$. We shall succeed for sufficiently many intervals $[x,y]_{(P)}$.

We begin with the special case in which P is a z-ideal, and we write Q for P in this case.

Recall that the function $\Upsilon : (s,t) \mapsto (s-t)^2$ was defined in (11.2). The Q-version of Υ is $\Upsilon_{(Q)} : \mathbb{I}^2_{(Q)} \to \mathbb{I}_{(Q)}$. The first remark is that addition can be defined as a partial function from $\mathbb{I}^2_{(Q)}$ to $\mathbb{I}_{(Q)}$ by using $\Upsilon_{(Q)}$ and the order on $\mathbb{I}_{(Q)}$. Indeed, let $x,y,z \in \mathbb{I}_{(Q)}$. Then:

$$x+y=z \quad \text{if and only if} \quad \Upsilon_{(Q)}(x,0)=\Upsilon_{(Q)}(z,y) \quad \text{and} \quad y \le z\,.$$

From addition, we can define translation on $\mathcal{J}_{(Q)}$, so constructing the equivalence relation

$$I \leftrightarrow J \quad (I,J \in \mathcal{J}_{(Q)})\,.$$

Indeed, suppose that $I=[x_1,y_1]_{(Q)}$ and $J=[x_2,y_2]_{(Q)}$. Then $I \leftrightarrow J$ in the case where $x_1 \le x_2$ if there exists $z \in \mathbb{I}_{(Q)}$ with $x_1+z=x_2$. Translation leads to the definition of many relations between I and J of $\mathcal{J}_{(Q)}$ from the given information. For example, we may define: '$l(I)=l(J)$'; 'I is the first half of J'; 'I is the fourth quarter of J'; etc.

We next *claim* that $\mathcal{K}_{(Q)}$ can be defined from $\leftrightarrow$ and the function $(\mathbf{f_1})_{(Q)}$. Indeed, $\mathcal{K}_{(Q)}$ is the family of intervals $I=[x,y]_{(Q)}$ in $\mathcal{J}_{(Q)}$ such that there is an interval $[y,z]_{(Q)}$ in $\mathcal{J}_{(Q)}$ with $z=y+2l(I)$, with $[y,z]_{(Q)}$ the fourth quarter of $[0,z]_{(Q)}$, and with $[y,z]_{(Q)} \in \mathfrak{PI}((\mathbf{f_1})_{(Q)})$. To see that this recipe does define $\mathcal{K}_{(Q)}$ uniquely, one must inspect carefully the construction of the dyadic set $\mathbf{K_1}$.

We see that $\gamma_{(Q)}(I)=J$ for $I \in \mathcal{K}_{(Q)}$ and $J \in \mathcal{J}_{(Q)}$ if and only if $J \le I$, $I \leftrightarrow J$, and $|I \cap J|=1$, so that $\gamma_{(Q)}$ is also definable from $\leftrightarrow$ and $(\mathbf{f_1})_{(Q)}$.

We now have enough preliminary information to see that we can define $\sigma_{(Q)}$ from $\leftrightarrow$ and $(\mathbf{f_1})_{(Q)}$. For take $a,b,c,d \in \mathbb{N}_{(Q)}$. Then $a+b=c$ and $d=\Phi_{(Q)}(c,d)$ if and only if there exist R_1,R_2,R_3 and S_1,S_2,S_3 in $\mathcal{J}_{(Q)}$ with $R_j \leftrightarrow S_j$ for $j=1,2,3$, and, further:

$$\begin{aligned}
&R_1 \text{ is a peak interval for } (\mathbf{f_1})_{(Q)} \mid \gamma_{(Q)}(I_a^{(Q)})\,;\\
&R_2 \text{ is a peak interval for } (\mathbf{f_1})_{(Q)} \mid \gamma_{(Q)}(I_b^{(Q)})\,;\\
&R_3 \text{ is a peak interval for } (\mathbf{f_1})_{(Q)} \mid \gamma_{(Q)}(I_c^{(Q)})\,;\\
&S_1,S_2,S_3 \text{ are peak intervals for } (\mathbf{f_1})_{(Q)} \mid I_d^{(Q)}\,.
\end{aligned}$$

This follows from an earlier remark by pointwise arguments. Thus we can define $\sigma_{(Q)} : \mathcal{K}_{(Q)} \times \mathcal{K}_{(Q)} \to \mathcal{K}_{(Q)}$.

In exactly the same way we can define $\pi_{(Q)} : \mathcal{K}_{(Q)} \times \mathcal{K}_{(Q)} \to \mathcal{K}_{(Q)}$ from the information given by $\leftrightarrow$ and $(\mathbf{f_2})_{(Q)}$.

We next bring into play the function Ξ. Our claim is that the equivalence relation $\sim_{(Q)}$ can be defined from $\leftrightarrow$ and $\Xi_{(Q)}$. The verification of this claim is similar to the above arguments. Let $I \in \mathcal{J}_{(Q)}$. First note that the families $\mathfrak{PI}'(\Xi_{(Q)} \mid I)$ of individual peak intervals and $\mathfrak{TP}(\Xi_{(Q)} \mid I)$ of twin-peak intervals are easily defined from $\mathfrak{PI}(\Xi_{(Q)} \mid I)$. For example, $\mathfrak{PI}'(\Xi_{(Q)} \mid I)$ is the family of intervals J in $\mathfrak{PI}(\Xi_{(Q)} \mid I)$ such that $J' = J$ whenever $J' \in \mathfrak{PI}(\Xi_{(Q)} \mid I)$ and $J' \leftrightarrow J$. Let $\mathcal{R}_{(Q)}$ be the subset of $\mathcal{J}_{(Q)} \times \mathcal{J}_{(Q)}$ consisting of the pairs (J_1, J_2) for which there is another pair (I_1, I_2) in $\mathcal{J}_{(Q)} \times \mathcal{J}_{(Q)}$ such that:

$$\begin{aligned}
&J_1, J_2 \text{ are twin-peak intervals of } \Xi_{(Q)} \mid [0,1/2]_{(Q)}\,;\\
&I_1, I_2 \text{ are individual peak intervals of } \Xi_{(Q)} \mid [1/2,1]_{(Q)}\,;\\
&I_1 \leftrightarrow J_1 \text{ and } I_2 \leftrightarrow J_2\,;\\
&I_1 \leq I_2 \text{ and } |I_1 \cap I_2| = 1\,.
\end{aligned}$$

We see by pointwise arguments from remarks about $\mathcal{R}$ that, in the case where $(J_1, J_2) \in \mathcal{R}_{(Q)}$, we have $J_1 \subset (0,1/4)_{(Q)}$, $J_2 \subset (1/4,1/2)_{(Q)}$, and $J_2 = 2^a J_1$ for a (uniquely specified) $a \in \mathbb{N}_{(Q)}$. Further, if $r, s \in \mathbb{D}_{(Q)}$ with $0 < r < s < 1/4$, then there exists $(J_1, J_2) \in \mathcal{R}_{(Q)}$ with $J_1 \subset (r,s)_{(Q)}$. Now take $x, y \in (0,1/4)_{(Q)}$. We see that $x \sim_{(Q)} y$ in the sense of the above definition if and only if, for each $\delta > 0$ in $\mathbb{I}_{(Q)}$, there exist (J_1, J_2) and (J_3, J_4) in $\mathcal{R}_{(Q)}$ such that

$$J_1 \subset (x-\delta, x+\delta)_{(Q)}, \quad J_3 \subset (y-\delta, y+\delta)_{(Q)}, \quad J_2 \subset J_4\,.$$

It is easy to define $\sim_{(Q)}$ on $(0,1/2)_{(Q)}$ from $\sim_{(Q)}$ on $(0,1/4)_{(Q)}$ by using translation. Thus the relation $\sim_{(Q)}$ is definable from $\leftrightarrow$ and $\Xi_{(Q)}$.

Finally, we consider $(\mathbf{f_3})_{(Q)}$; we shall utilize the maps $\Gamma_{(Q)} : \mathbb{N}_{(Q)} \to \mathbb{D}_{(Q)}$ and $n_{(Q)} : \mathbb{D}_{(Q)} \to \mathbb{Z}^+_{(Q)}$, which are the induced maps corresponding to the bijection Γ of (11.3) and the map $n : \mathbb{D} \to \mathbb{Z}^+$, respectively. Since Q is an exponential prime, $\Gamma_{(Q)}$ is also a bijection. For $r \in \mathbb{D}_{(Q)} \setminus \{0,1\}$, set

$$K_{(Q)}(r) = \mathbb{D}_d^{(Q)} \cup \{r + 2^{-d-1}\}\,,$$

where $d = n_{(Q)}(r)$. Let $c \in \mathbb{N}_{(Q)}$, set $r_c = \Gamma_{(Q)}(c+2)$ and $d_c = n_{(Q)}(\Gamma_{(Q)}(c+2))$; we note that $d_c \leq c+2$. Then we see that $\mathfrak{PI}\left((\mathbf{f_3})_{(Q)} \mid I_c^{(Q)}\right)$ is the set of closed intervals $[a,b]_{(Q)}$ such that a, b are adjacent points in a scaled copy of $K_{(Q)}(r_c)$ on $I_c^{(Q)}$; each of these intervals has length 2^{-d_c-c-2}, save for the two intervals which have $2^{-c} + 2^{-c-2} + r_c \cdot 2^{-c-2}$ as an end-point, these two intervals having length 2^{-d_c-c-3}.

From the information given by $\leftrightarrow$ and $(\mathbf{f_3})_{(Q)}$, we can define the function

$$I_c^{(Q)} \mapsto 2^{-c} + 2^{-c-2} + r_c \cdot 2^{-c-2} = l(I_c^{(Q)})(5 + r_c), \quad \mathcal{K}_{(Q)} \to \mathbb{I}_{(Q)}\,,$$

just because the information about peak intervals picks out the image of $I_c^{(Q)}$. But now we can define the function

$$U_{(Q)} : I_c^{(Q)} \mapsto r_c \cdot 2^{-c-2}, \quad \mathcal{K}_{(Q)} \to \mathbb{I}_{(Q)}\,.$$

For each $c \in \mathbb{N}_{(Q)}$, $r_c = 2(a - 1/2)$, where $a \in (1/2,1)_{(Q)}$ is the unique element with $a \sim_{(Q)} r_c \cdot 2^{-c-2}$. Thus, with the extra information given by $\Xi_{(Q)}$ (recall that we used $\Xi_{(Q)}$ to define $\sim_{(Q)}$), we can define

$$V_{(Q)} : I_c^{(Q)} \mapsto \Gamma_{(Q)}(c), \quad \mathcal{K}_{(Q)} \to \mathbb{I}_{(Q)}\,.$$

The set $\mathbb{D}_{(Q)}$ is exactly the range of $\Gamma_{(Q)}$, and hence it is the range of $V_{(Q)}$ on $\mathcal{K}_{(Q)}$. Thus we have now defined $\mathbb{D}_{(Q)}$ from our permitted information.

We next *claim* that the map $E_{(Q)}$ is indeed definable from the permitted information. For let S be the set of triples (m, n, p) in $\mathbb{N}^3$ with $\Gamma(m) < \Gamma(n)$ and

$$\varepsilon_x(p) = 1 \quad (x \in (\Gamma(m), \Gamma(n)))\,.$$

Then S is definable in $\langle \mathbb{N}, +, \cdot \rangle$, and so the corresponding Q-version $S_{(Q)}$ is definable in $\langle \mathbb{N}_{(Q)}, +, \cdot \rangle$. The set

$$\{(I_a^{(Q)}, I_b^{(Q)}, I_c^{(Q)}) \in \mathcal{K}_{(Q)}^3 : (a, b, c) \in S_{(Q)}\}$$

is definable in $\langle \mathcal{K}_{(Q)}, \sigma_{(Q)}, \pi_{(Q)} \rangle$, and the map $E_{(Q)}$ is easily definable from this set and $V_{(Q)}$: indeed, $E_{(Q)}(x)(I_c^{(Q)}) = 1$ if and only if there exists a triple $(I_a^{(Q)}, I_b^{(Q)}, I_c^{(Q)})$ in this set with $V_{(Q)}(I_a^{(Q)}) < x < V_{(Q)}(I_b^{(Q)})$.

The set $\mathcal{P}_{(Q)}$ can be defined from the function $E_{(Q)}$ in exactly the way that $\mathcal{P}(\mathbb{N})_{(Q)}$ was defined from the function $x \mapsto \varepsilon_x^{(Q)}$; first we defined $\{0,1\}_{(Q)}^{\mathbb{N}}$ in Definition 9.1 and then $\mathcal{P}(\mathbb{N})_{(Q)}$ in Definition 10.1. In defining $\mathcal{P}_{(Q)}$, we must take account of the 'eventually constant' functions, as we did before, but this is straightforward.

Let $(\mathbf{L}, \mathbf{F})$ be the seed from which the function Λ was germinated. Thus

$$\mathbf{F} \mid [0, 1/4] = (\mathbf{f_1})_{0,1/4}, \quad \mathbf{F} \mid [1/4, 1/2] = (\mathbf{f_2})_{1/4,1/2}, \quad \mathbf{F} \mid [1/2, 1] = (\mathbf{f_3})_{1/2,1}\,.$$

The definition of $\mathcal{K}_{(Q)}$, $\sigma_{(Q)}$, $\pi_{(Q)}$, and $V_{(Q)}$ required only the notion of $\leftrightarrow$ and the sets of intervals

$$\mathfrak{P}\mathfrak{I}((\mathbf{f_1})_{(Q)}), \quad \mathfrak{P}\mathfrak{I}((\mathbf{f_2})_{(Q)}), \quad \text{and} \quad \mathfrak{P}\mathfrak{I}((\mathbf{f_3})_{(Q)}),$$

and these three families are easily defined from $\mathfrak{P}\mathfrak{I}(\mathbf{F}_{(Q)})$. By (11.7), we have $\mathfrak{P}\mathfrak{I}(\mathbf{F}) = \mathfrak{P}\mathfrak{I}(\Lambda)$, and a pointwise argument shows that $\mathfrak{P}\mathfrak{I}(\mathbf{F}_{(Q)}) = \mathfrak{P}\mathfrak{I}(\Lambda_{(Q)})$.

In summary, starting from $(\mathbb{I}_{(Q)}, \leq)$ and $\Upsilon_{(Q)}$ and $\Lambda_{(Q)}$, we can define

$$\mathcal{K}_{(Q)}, \quad \sigma_{(Q)}, \quad \pi_{(Q)}, \quad \text{and} \quad V_{(Q)}\,,$$

and, with the further information from $\Xi_{(Q)}$, we can define all of

$$\langle \mathcal{K}_{(Q)}, \sigma_{(Q)}, \pi_{(Q)}, \mathcal{P}_{(Q)} \rangle \quad \text{and} \quad E_{(Q)}\,.$$

We now turn to the definition of the local versions of these structures from the permitted information. Since the function Λ has the self-replicating property given in (11.11), for many intervals $[x, y]_{(Q)}$ in $\mathcal{J}_{(Q)}$ we have

$$\Lambda_{(Q)} \mid [x, y]_{(Q)} = \Lambda_{(Q)}(x) + \Lambda_{x,y}^{(Q)}\,. \tag{11.14}$$

In this case, $\mathcal{K}_{x,y}^{(Q)}$, $\sigma_{x,y}^{(Q)}$, $\pi_{x,y}^{(Q)}$, and $V_{x,y}^{(Q)}$ can be defined from the ordered set $([x, y]_{(Q)}, \leq)$ with the help of information given by $\Lambda_{(Q)} \mid [x, y]_{(Q)}$ and $\leftrightarrow$.

We shall obtain the equivalence relation $\sim_{x,y}^{(Q)}$ from $\Xi_{(Q)} \mid [x, y]_{(Q)}$ for suitable $x, y \in \mathbb{I}_{(Q)}$. Let S^{Ξ} be as specified in the construction of Ξ, and let $S_{(Q)}^{\Xi}$ be the

Q-version of S^{Ξ}, defined pointwise. Then $\sim_{x,y}^{(Q)}$ is definable from $\Xi_{(Q)} \mid [x,y]_{(Q)}$ provided that $(x,y) \in S^{\Xi}_{(Q)}$, and in this case (11.14) also holds. We note that, for each $x, y \in \mathbb{I}_{(Q)}$ with $x < y$, there exists $(x', y') \in S^{\Xi}_{(Q)}$ with $x \leq x' < y' \leq y$.

Translation can also be defined locally. Indeed, suppose that $I_1 = [r_1, s_1]_{(Q)}$ and $I_2 = [r_2, s_2]_{(Q)}$ are closed subintervals of $[x,y]_{(Q)}$ and that $x', y' \in \mathbb{I}_{(Q)}$ with

$$y' - x' = y - x \quad \text{and} \quad [x,y]_{(Q)} \cap [x',y']_{(Q)} = \emptyset\,. \tag{11.15}$$

Then $I_1 \leftrightarrow I_2$ if and only if there exists $z \in [x', y']_{(Q)}$ with

$$\Upsilon_{(Q)}(r_1, z) = \Upsilon_{(Q)}(r_2, z) \quad \text{and} \quad \Upsilon_{(Q)}(s_1, z) = \Upsilon_{(Q)}(s_2, z)\,,$$

and so $E_{x,y}^{(Q)}$ can be defined from

$$\Upsilon_{(Q)} \mid [x,y]_{(Q)} \times [x',y']_{(Q)}\,.$$

Thus we can define the local structure $\mathcal{P}_{x,y}^{(Q)}$.

We are now at last in a position to invoke the hypothesis that $\mathfrak{N}_{(P_0)}$ be solid. There is an order-isomorphism

$$\beta : (\mathbb{I}_{(P_0)}, \leq) \to (\mathbb{I}_{(Q)}, \leq)\,;$$

let $(\Lambda_{(Q)})^{\beta}$, $(\Xi_{(Q)})^{\beta}$, and $(\Upsilon_{(Q)})^{\beta}$ be the transferred functions on $\mathbb{I}_{(P)}$, so that $(\Lambda_{(Q)})^{\beta} = \beta^{-1} \circ \Lambda_{(Q)} \circ \beta$, for example. Then $(\Lambda_{(Q)})^{\beta}$, $(\Upsilon_{(Q)})^{\beta} \in C(\mathbb{I}_{(P_0)}, \mathbb{I}_{(P_0)})$ and $(\Upsilon_{(Q)})^{\beta} \in C(\mathbb{I}^2_{(P_0)}, \mathbb{I}_{(P_0)})$. Since $\mathfrak{N}_{(P_0)}$ is solid, there exist non-empty, open intervals V_1 and V_2 in $\mathbb{I}_{(P_0)}$ and functions

$$G_1, G_2 \in {}^*C(\mathbb{I}, \mathbb{I})_{(P_0)} \quad \text{and} \quad G_3 \in {}^*C(\mathbb{I}^2, \mathbb{I})_{(P_0)}$$

such that

$$G_1 \mid V_1 = (\Lambda_{(Q)})^{\beta} \mid V_1, \quad G_2 \mid V_2 = (\Xi_{(Q)})^{\beta} \mid V_2\,,$$
$$\text{and} \quad G_3 \mid V_1 \times V_2 = (\Upsilon_Q)^{\beta} \mid V_1 \times V_2\,.$$

Let $x, y, x', y' \in \mathbb{I}_{(Q)}$ be such that $\beta(V_1) = (x,y)_{(Q)}$ and $\beta(V_2) = (x', y')_{(Q)}$; by reducing V_1 and V_2, we may suppose that $(x,y) \in S^{\Xi}_{(Q)}$ and that (11.15) is satisfied. Since $(x,y) \in S^{\Xi}_{(Q)}$, equation (11.14) is also satisfied, and thus

$$\left\langle \mathcal{K}_{x,y}^{(Q)},\, \sigma_{x,y}^{(Q)},\, \pi_{x,y}^{(Q)},\, \mathcal{P}_{x,y}^{(Q)} \right\rangle \quad \text{and} \quad E_{x,y}^{(Q)}$$

can be defined from

$$\Lambda_{(Q)} \mid [x,y]_{(Q)}, \quad \Xi_{(Q)} \mid [x,y]_{(Q)}, \quad \text{and} \quad \Upsilon_{(Q)} \mid [x,y]_{(Q)} \times [x',y']_{(Q)}\,.$$

Let $\langle \mathcal{K}^{\beta}, \sigma^{\beta}, \pi^{\beta}, \mathcal{P}^{\beta} \rangle$ be the naturally transferred version of the above structure, so that $\mathcal{K}^{\beta}$ is a family of closed subintervals of $[\beta^{-1}(x), \beta^{-1}(y)]_{(P_0)}$, for example. Let E^{β} be the corresponding image of $E_{x,y}^{(Q)}$. Thus

$$\langle \mathcal{K}^{\beta}, \sigma^{\beta}, \pi^{\beta}, \mathcal{P}^{\beta} \rangle \quad \text{and} \quad E^{\beta}$$

can be defined from

$$G_1 \mid V_1, \quad G_2 \mid V_2, \quad \text{and} \quad G_3 \mid V_1 \times V_2\,.$$

We now define an order $\prec$ on $\mathcal{K}^\beta$. Note first that the family $\mathcal{K}^\beta$ is discrete in the interval $(\beta^{-1}(x), \beta^{-1}(y)]_{(P_0)}$. For $I, J \in \mathcal{K}^\beta$, set $I \prec J$ if there exists $K \in \mathcal{K}^\beta$ with $\sigma^\beta(J, K) = I$, so that $\prec$ is a strict partial order on $\mathcal{K}^\beta$. Thus $I \prec J$ if and only if $J << I$ in $[\beta^{-1}(x), \beta^{-1}(y)]_{(P_0)}$.

Since $\mathfrak{N}_{(P_0)}$ is a model of analysis and the structure $\langle \mathcal{K}^\beta, \sigma^\beta, \pi^\beta \rangle$ is definable in $\mathfrak{N}_{(P_0)}$, the totally ordered set $(\mathcal{K}^\beta, \preceq)$ is definable in $\mathfrak{N}_{(P_0)}$, and so there is an order-isomorphism

$$\rho : (\mathbb{N}_{(P_0)}, \leq) \to (\mathcal{K}^\beta, \preceq).$$

From this it follows that ρ defines an isomorphism of the structures $\langle \mathbb{N}_{(P_0)}, +, \cdot \rangle$ and $\langle \mathcal{K}^\beta, \sigma^\beta, \pi^\beta \rangle$. A final application of the fact that $\mathfrak{N}_{(P_0)}$ is a model of analysis, now using the fact that E^β is definable in $\mathfrak{N}_{(P_0)}$, shows that ρ defines an isomorphism of $\langle \mathbb{N}_{(P_0)}, +, \cdot, \mathcal{P}(\mathbb{N})_{(P_0)} \rangle$ and $\langle \mathcal{K}^\beta, \sigma^\beta, \pi^\beta, \mathcal{P}^\beta \rangle$. It follows that $\mathfrak{N}_{(P_0)} \cong \mathfrak{N}_{(Q)}$, and so we have completed the proof of Theorem 11.6 in the special case where $P = Q$ is a z-ideal.

We now turn to the general case in the proof of Theorem 11.6: P is a non-maximal, prime ideal in $C(\Omega)$ for some compact space Ω and $(K_P, \leq)$ and $(K_{P_0}, \leq)$ are order-isomorphic. Let

$$\beta : (\mathbb{I}_{(P_0)}, \leq) \to (\mathbb{I}_{(P)}, \leq)$$

be an isotonic map. As above, let $(\Pi_{(P)})^\beta$, $(\Lambda_{(P)})^\beta$, $(\Xi_{(P)})^\beta$, and $(\Upsilon_{(P)})^\beta$ be the transferred functions on $\mathbb{I}_{(P)}$ corresponding to $\Pi_{(P)}$, $\Lambda_{(P)}$, $\Xi_{(P)}$, and $\Upsilon_{(P)}$, respectively. As before, it follows from the fact that $\mathfrak{N}_{(P_0)}$ is solid that there exist non-empty, open intervals V_1 and V_2 in $\mathbb{I}_{(P)}$ and functions

$$G_1, G_2, G_3 \in {}^*C(\mathbb{I}, \mathbb{I})_{(P_0)} \quad \text{and} \quad G_4 \in {}^*C(\mathbb{I}^2, \mathbb{I})_{(P_0)}$$

such that

$$G_1 \mid V_1 = (\Pi_{(P)})^\beta \mid V_1, \quad G_2 \mid V_1 = (\Lambda_{(P)})^\beta \mid V_1, \quad G_3 \mid V_1 = (\Xi_{(P)})^\beta \mid V_1,$$

$$\text{and} \quad G_4 \mid V_1 \times V_2 = (\Upsilon_{(P)})^\beta \mid V_1 \times V_2 .$$

We shall prove that $\mathfrak{N}_{(P_0)} \cong \mathfrak{N}_{(P)}$ in three steps. First, we shall use the functions G_1 and G_4 to prove that P is necessarily an exponential prime. Second, we shall use the functions G_2 and G_4 to prove that the structures $\langle \mathbb{N}_{(P_0)}, +, \cdot \rangle$ and $\langle \mathbb{N}_{(P)}, +, \cdot \rangle$ are isomorphic. Since $\mathfrak{N}_{(P_0)}$ is a model of analysis, this will prove that $\langle \mathbb{N}_{(P)}, +, \cdot \rangle$ is a model of number theory. Finally, by repeating the argument for the special case in which P was a z-ideal, we shall prove that $\mathfrak{N}_{(P_0)} \cong \mathfrak{N}_{(P)}$; the earlier argument given for a z-ideal will apply to the case where P is an exponential prime because it has been established that $\langle \mathbb{N}_{(P)}, +, \cdot \rangle$ is a model of number theory.

We begin by showing that P is necessarily an exponential prime. Take Q to be the maximum z-ideal below P. To obtain a contradiction, assume that P is not an exponential prime. We set

$$\mathcal{L}_{(P)} = \{ I_a^{(P)} : a = 2^b \text{ for some } b \in \mathbb{N}_{(P)} \},$$

so that $\mathcal{L}_{(P)}$ is a subset of $\mathcal{K}_{(P)}$ with no maximum element and which is not cofinal in $(\mathcal{K}_{(P)}, \le)$. Certainly the corresponding families $\mathcal{K}_{(Q)}$ and $\mathcal{L}_{(Q)}$ can be defined from $\leftrightarrow$ and $\mathfrak{PI}(\Pi_{(Q)})$; to see that $\mathcal{K}_{(P)}$ and $\mathcal{L}_{(P)}$ can be defined from $\mathfrak{PI}(\Pi_{(P)})$, we need to note that $\Pi_{(P)} \mid \mathbb{D}_{(P)} = \Pi_{(Q)} \mid \mathbb{D}_{(P)}$ and to apply (i) and (ii) of Lemma 11.15 to see that peak intervals are maintained. Let $x, y, x', y' \in \mathbb{I}_{(P)}$ be such that $\beta(V_1) = (x, y)_{(P)}$ and $\beta(V_2) = (x', y')_{(P)}$; by Lemma 11.17(i), we may suppose that $x, y \in \mathbb{D}_{(P)}$ and that

$$\Pi_{(P)} \mid [x, y]_{(P)} = \Pi_{(P)}(x) + \Pi^{(P)}_{x,y}\,.$$

We may also suppose that $y' - x' = y - x$ and that $[x, y]_{(P)} \cap [x', y']_{(P)} = \emptyset$. The scaled versions $\mathcal{K}^{(P)}_{x,y}$ and $\mathcal{L}^{(P)}_{x,y}$ of $\mathcal{K}_{(P)}$ and $\mathcal{L}_{(P)}$ can be defined from $\leftrightarrow$ and $\Pi_{(P)} \mid [x, y]_{(P)}$, and hence from

$$\Pi_{(P)} \mid [x, y]_{(P)} \quad \text{and} \quad \Upsilon_{(P)} \mid [x, y]_{(P)} \times [x', y']_{(P)}\,.$$

Again, the naturally transferred versions $\mathcal{K}^{\beta}$ and $\mathcal{L}^{\beta}$ are definable from

$$G_1 \mid V_1 \quad \text{and} \quad G_4 \mid V_1 \times V_2\,.$$

The family $\mathcal{K}^{\beta}$ is a discrete family of closed intervals in V_1 and $\mathcal{L}^{\beta}$ is a subfamily. Since $\mathcal{L}_{(P)}$ is not cofinal in $\mathcal{K}_{(P)}$, $\mathcal{L}^{\beta}$ is not cofinal in $\mathcal{K}^{\beta}$. But the families $\mathcal{K}^{\beta}$ and $\mathcal{L}^{\beta}$ of intervals are definable in $\mathfrak{N}_{(P)}$, which is a model of analysis. This is a contradiction, and so P is an exponential prime.

Now that we know that P is an exponential prime, we can apply Lemma 11.17(ii) to see that we may suppose that $\beta(V_1) = (x, y)_{(P)}$, where $x, y \in \mathbb{D}_{(P)}$ and

$$\Lambda_{(P)} \mid [x, y]_{(P)} = \Lambda_{(P)}(x) + \Lambda^{(P)}_{x,y}\,.$$

Again, we may suppose that $\beta(V_2) = (x', y')_{(P)}$, where x', y' satisfy the same conditions as before. Since $\Lambda_{(Q)} \mid \mathbb{D}_{(P)} = \Lambda_{(P)} \mid \mathbb{D}_{(P)}$, it follows from (i) and (ii) of Lemma 11.15 that $\mathcal{K}_{(P)}$, $\sigma_{(P)}$, and $\pi_{(P)}$ are definable from $\leftrightarrow$ and $\Lambda_{(P)}$. Hence $\mathcal{K}^{(P)}_{x,y}$, $\sigma^{(P)}_{x,y}$, and $\pi^{(P)}_{x,y}$ are definable from

$$\Lambda_{(P)} \mid [x, y]_{(P)} \quad \text{and} \quad \Upsilon_{(P)} \mid [x, y]_{(P)} \times [x', y']_{(P)}\,,$$

and so the transferred structure $\langle \mathcal{K}^{\beta}, \sigma^{\beta}, \pi^{\beta} \rangle$ is definable from

$$G_2 \mid V_1 \quad \text{and} \quad G_4 \mid V_1 \times V_2\,.$$

As in the case where P is a prime z-ideal, it follows that $\langle \mathbb{N}_{(P_0)}, +, \cdot \rangle$ and $\langle \mathcal{K}^{\beta}, \sigma^{\beta}, \pi^{\beta} \rangle$ are isomorphic, and so both are isomorphic to $\langle \mathbb{N}_{(P)}, +, \cdot \rangle$, which is thus a model of number theory.

It now follows from Lemma 11.18 that we may suppose that $\beta(V_1) = (x, y)_{(P)}$, where $x, y \in \mathbb{D}_{(P)}$ and $(x, y) \in S^{\Xi}_{(Q)}$; we also choose x', y' to satisfy the usual conditions. Thus

$$\left\langle \mathcal{K}^{(Q)}_{x,y}, \sigma^{(Q)}_{x,y}, \pi^{(Q)}_{x,y}, \mathcal{P}^{(Q)}_{x,y} \right\rangle \quad \text{and} \quad E^{(Q)}_{x,y}$$

can be defined from

$$\Lambda_{(Q)} \mid [x,y]_{(Q)}, \quad \Xi_{(Q)} \mid [x,y]_{(Q)}, \quad \text{and} \quad \Upsilon_{(Q)} \mid [x,y]_{(Q)} \times [x',y']_{(Q)}.$$

By (i) and (ii) of both Lemmas 11.15 and 11.16, it follows that

$$\left\langle \mathcal{K}^{(P)}_{x,y},\, \sigma^{(P)}_{x,y},\, \pi^{(P)}_{x,y},\, \mathcal{P}^{(P)}_{x,y} \right\rangle \quad \text{and} \quad E^{(P)}_{x,y}$$

can be defined from

$$\Lambda_{(P)} \mid [x,y]_{(P)}, \quad \Xi_{(P)} \mid [x,y]_{(P)}, \quad \text{and} \quad \Upsilon_{(P)} \mid [x,y]_{(P)} \times [x',y']_{(P)}.$$

Hence the transferred structures $\langle \mathcal{K}^\beta, \sigma^\beta, \pi^\beta \rangle$ and E^β are definable from

$$G_2 \mid V_1, \quad G_3 \mid V_1, \quad \text{and} \quad G_4 \mid V_1 \times V_2.$$

Our situation is now identical to that which arose in the case where P was a z-ideal, and we conclude as before that $\mathfrak{N}_{(P_0)} \cong \mathfrak{N}_{(P)}$.

This concludes our proof of Theorem 11.6 in the general case. □

To show that there is an example of a rigid super-real field K_{P_0}, and hence an example for which the associated structure $\mathfrak{N}_{(P)} = \langle \mathbb{N}_{(P)}, +, \cdot, \mathcal{P}(\mathbb{N})_{(P)} \rangle$ is an invariant of the field, we should now exhibit a prime ideal P_0 in $C(\Omega_0)$ such that $\mathfrak{N}_{(P_0)}$ is a model of analysis and is solid. Unfortunately this requires a more sophisticated background than we have cared to assume so far. Our aim in this book has been to prove all results in a manner accessible to non-logicians, and, in particular, to those unacquainted with the techniques of forcing. We have essentially achieved this so far, save that the proof of Theorem 9.60 required an elementary forcing argument, and we were not able to present the proof of Theorem 7.16. It is also true that we may have fallen a little short of our aim at some points in the present chapter. However, for the final few pages we are forced to abandon our previous policy on forcing, and we shall now use the notation and ideas of this subject without any pretence of giving an explanation. Indeed, even within this framework, we shall omit some details of the proofs.

Theorem 11.19 (GCH) *There is a compact space Ω_0 and a non-maximal, prime z-ideal P_0 in $C(\Omega_0)$ such that K_{P_0} is a hyper-real field, $w(K_{P_0}) = \aleph_1$, $|K_{P_0}| = \aleph_2$, and $\mathfrak{N}_{(P_0)}$ is a model of analysis and is solid.*

The proof of this theorem will emerge from some discussion and some lemmas.

Let ZC* be the collection of Zermelo Axioms for set theory together with the following formulation of the Axiom of Choice: for every set S, there is an ordinal α and a bijection $\pi : S \to \alpha$. The theory ZC* is in essence a 'reformulation' of ZC. We shall write $\mathbf{V} = \bigcup \mathbf{V}_\alpha$ for the universe of all sets. Suppose that α is an infinite ordinal such that $|\mathbf{V}_\alpha| = \alpha$ (so that α is a cardinal). The axioms of ZFC guarantee that such ordinals α exist; with GCH, these ordinals are exactly the ordinals α such that $\aleph_\alpha = \alpha$. Then $\mathbf{V}_\alpha \models$ ZC* .

We take $\mathcal{M} = (M, E)$ to be a model of ZC*. We extend our notational conventions, for example by setting

$$\mathfrak{N}_{(\mathcal{M})} = \langle \mathbb{N}_{(\mathcal{M})}, +, \cdot, \mathcal{P}(\mathbb{N})_{(\mathcal{M})} \rangle$$

for the structure $\langle \mathbb{N}, +, \cdot, \mathcal{P}(\mathbb{N})\rangle$ as computed in $\mathcal{M}$. Since $\mathcal{M}$ is a model of ZC*, $\mathfrak{N}_{(\mathcal{M})}$ is a model of analysis.

Now let $\mathfrak{M}$ be a model of analysis. Then $\mathbb{Q}_{(\mathfrak{M})}$ and $\mathbb{R}_{(\mathfrak{M})}$ are the rational fields defined by applying within $\mathfrak{M}$ the standard constructions of $\mathbb{Q}$ from $\mathbb{N}$ and $\mathbb{R}$ from $\mathcal{P}(\mathbb{N})$, respectively. We note that $\mathbb{R}_{(\mathfrak{M})}$ is not a field because no map from $\mathbb{R}$ into $\mathbb{R}_{(\mathfrak{M})}$ is specified (although there are non-canonical embeddings of $\mathbb{R}$ into $\mathbb{R}_{(\mathfrak{M})}$ with respect to which $\mathbb{R}_{(\mathfrak{M})}$ is a field in the case where $\mathbb{R}_{(\mathfrak{M})}$ is an η_1-set). Again, for each $n \in \mathbb{N}$, ${}^*C(\mathbb{I}^n, \mathbb{I})_{(\mathfrak{M})}$ denotes the family of continuous functions from $\mathbb{I}^n_{(\mathfrak{M})}$ to $\mathbb{I}_{(\mathfrak{M})}$ which are *internal* to the model; these functions can also be obtained by applying Definition 11.4 with $\mathfrak{N}_{(P)}$ replaced by $\mathfrak{M}$. In the case where $\mathcal{M}$ is a model of ZC* and $\mathfrak{M} = \mathfrak{N}_{(\mathcal{M})}$, we write $\mathbb{Q}_{(\mathcal{M})}$, $\mathbb{R}_{(\mathcal{M})}$, and ${}^*C(\mathbb{I}^n, \mathbb{I})_{(\mathcal{M})}$ for $\mathbb{Q}_{(\mathfrak{M})}$, $\mathbb{R}_{(\mathfrak{M})}$, and $C^*(\mathbb{I}^n, \mathbb{I})_{(\mathfrak{M})}$, respectively.

We begin with an application of the constructions given in Chapter 9. Suppose that $\mathcal{U}$ is a free ultrafilter on $\mathbb{N}$, and consider the model

$$\mathcal{M} = (M, E) = \mathbf{V}_\alpha^{\mathbb{N}}/\mathcal{U},$$

the ultrapower of $\mathbf{V}_\alpha$ by $\mathcal{U}$, where α is an infinite ordinal as described above. By Los's theorem, $\mathcal{M} \models$ ZC*. It is clear that the structure $\mathfrak{N}_{(\mathcal{M})}$ is precisely the structure $\mathfrak{N}_{(\mathcal{U})}$ which we previously associated to $\mathcal{U}$. We suppose that $\mathbb{P} \in M$ and that

$$\mathcal{M} \models \text{ `}\mathbb{P} \text{ is a partial order'}.$$

Further, suppose that $\mathcal{G}$ is a filter contained in $\mathbb{P}$ and that $\mathcal{G}$ is $\mathcal{M}$-generic. Then $\mathcal{G}$ can be used to define a generic extension of the model $\mathcal{M}$; this extension is denoted by

$$\mathcal{M}[\mathcal{G}] = (M[\mathcal{G}], E).$$

We return to the terminology

$$X = \mathbb{N} \times Y = \mathbb{N} \times \{0,1\}^{\mathbb{N}\times I}$$

for the topological space of Chapter 9.

Lemma 11.20 *Let $\mathcal{M}[\mathcal{G}]$ be as above. Then there is a prime z-ideal P_0 in $C(\beta X)$ such that K_{P_0} is a hyper-real field and*

$$\mathfrak{N}_{(P_0)} \cong \langle \mathbb{N}_{(\mathcal{M}[\mathcal{G}])}, +, \cdot, \mathcal{P}(\mathbb{N})_{(\mathcal{M}[\mathcal{G}])}\rangle .$$

Proof We use notation taken from Chapter 9. There is a limit ordinal β with $\beta < \alpha$ such that $\mathbb{P} \in \mathbf{V}_\beta^{\mathbb{N}}/\mathcal{U}$. Set $I = \mathbf{V}_\beta$, so that $I_{(\mathcal{U})} = \mathbf{V}_\beta^{\mathbb{N}}/\mathcal{U}$, and let J be the set of all $z \in M$ such that

$$\mathcal{M} \models \text{ `}z \subset \mathbb{P} \times \mathbb{N}\text{'}.$$

Let $F \in M$ be a function such that

$$\mathcal{M} \models \text{ `}J = \operatorname{ran} F\text{'}.$$

Then each element of J is naturally a term in $\mathcal{M}$ for a subset of $\mathbb{N}_{(\mathcal{M})} = \mathbb{N}_{(\mathcal{U})}$. Let $\tau_{\mathcal{G}} : J \to \mathcal{P}(\mathbb{N}_{(\mathcal{U})})$ be a map which makes this interpretation, so that $\tau_{\mathcal{G}}$

associates to each $z \in J$ the subset t of $\mathbb{N}_{(\mathcal{U})}$ defined as follows: for $a \in \mathbb{N}_{(\mathcal{U})}$, we have $a \in t$ if there exists $q \in \mathcal{G}$ such that

$$\mathcal{M} \models \text{`}(q, a) \in z\text{'}.$$

It is an elementary fact about forcing extensions that the range of the function $\tau_{\mathcal{G}}$ is precisely the set $\mathcal{P}(\mathbb{N})_{(\mathcal{M}[\mathcal{G}])}$.

We now aim to define a function $H : \mathbb{N}_{(\mathcal{U})} \times I_{(\mathcal{U})} \to \{0,1\}$ satisfying the condition of Definition 9.14. Indeed, in the case where $x \in J$, we define

$$H(a,x) = \begin{cases} 1 & \text{if} \quad a \in \tau_{\mathcal{G}}(x)\,, \\ 0 & \text{if} \quad a \notin \tau_{\mathcal{G}}(x)\,. \end{cases}$$

If $x \notin J$, we set $H(a,x) = 0 \quad (a \in \mathbb{N}_{(\mathcal{U})})$. Then, for each $A \in [\mathbb{N} \times \mathbb{I}]^{<\omega}_{(\mathcal{U})}$ (the latter set was defined in (9.11)), we have

$$H \mid A \in \mathrm{Fin}\,(\mathbb{N} \times I, \{0,1\})\,,$$

the required condition. Thus, as in Definition 9.14, $\mathcal{F}_H$ is defined to be a family of subsets of $\mathbf{Z}(X)$; in Theorem 9.16 it was shown that $\mathcal{F}_H$ is a z-ultrafilter on X. The associated ideal Q_H was defined in Definition 9.17; Q_H is a prime z-ideal in $C(\beta X)$, and Q_H is the required ideal P_0. Certainly

$$K_{P_0} \cong C(X)/\mathcal{F}_H$$

is a hyper-real field. It is easily checked that $\mathcal{P}(\mathbb{N})_{(P_0)} \subset \mathcal{M}[\mathcal{G}]$.

The functions $h : \mathbb{N}_{(\mathcal{U})} \to \{0,1\}$ which are the sections of H, as in Definition 9.20, are precisely the functions which belong to $\{0,1\}^{\mathbb{N}}_{(\mathcal{M}[\mathcal{G}])}$, and so we have $\mathfrak{N}_{(P_0)} \cong \mathfrak{N}_{(\mathcal{M}[\mathcal{G}])}$. This proves the lemma. □

We shall now prove three technical lemmas about Cohen extensions of non-standard models.

For each ordinal α, let $\mathbb{P}_\alpha$ be the partial order for adding α Cohen reals to $\mathbf{V}$. Thus $\mathbb{P}_\alpha$ is the partial order defined on the set of finite partial functions from $\mathbb{N} \times \alpha$ to $\{0,1\}$ by extension.

Let $(\omega_2^{<\omega}, \leq)$ be the partial order of finite sequences from ω_2 ordered by extension. In the case where $\mathcal{M} = (M, E)$ is a model of ZC*, we take $\omega_2^{(\mathcal{M})}$ to be the set of elements $a \in M$ such that

$$\mathcal{M} \models \text{`}a \quad \text{is an ordinal with} \quad a < \omega_2\text{'}.$$

Then $(\omega_2^{(\mathcal{M})}, \leq)$ is a totally ordered set, where $\leq$ is the order on $\omega_2^{(\mathcal{M})}$ given by E. The set $\omega_2^{<\omega}$ as computed in $\mathcal{M}$ is denoted by $(\omega_2^{<\omega})_{(\mathcal{M})}$, and

$$\left((\omega_2^{<\omega})_{(\mathcal{M})}, \leq\right)$$

is the corresponding partially ordered set. It is straightforward to show that $(\mathbb{Q}_{(\mathcal{M})}, \leq)$ is an η_1-set in the case where the above partially ordered set is ω-closed.

For each $a \in M$ such that

$$\mathcal{M} \models \text{ `}a \text{ is an ordinal'},$$

we let $\mathbb{P}_a^{(\mathcal{M})}$ denote the partial order $\mathbb{P}$ as computed in $\mathcal{M}$ relative to a; this is naturally a partial order in $\mathbf{V}$.

Lemma 11.21 (CH) *Let $\mathcal{M}$ be a model of* ZC* *such that the partially ordered set $((\omega_2^{<\omega})_{(\mathcal{M})}, \leq)$ is ω-closed and $|\mathcal{P}(\mathbb{N})_{(\mathcal{M})}| = 2^{\aleph_0}$. Suppose that (a_k) is an increasing sequence in $(\omega_2^{(\mathcal{M})}, \leq)$ and $(\mathcal{G}_k)$ is a sequence of $\mathcal{M}$-generic filters such that, for each $k \in \mathbb{N}$:*

(i) $\mathcal{G}_k \subset \mathbb{P}_{a_k}^{(\mathcal{M})}$;

(ii) $\mathcal{G}_{k+1} \mid \mathbb{P}_{a_k}^{(\mathcal{M})} = \mathcal{G}_k$.

Then there exist $a \in \omega_2^{(\mathcal{M})}$ and $\mathcal{G} \subset \mathbb{P}_a^{(\mathcal{M})}$ such that $\mathcal{G}$ is $\mathcal{M}$-generic and such that $\mathcal{G} \mid \mathbb{P}_{a_k}^{(\mathcal{M})} = \mathcal{G}_k \quad (k \in \mathbb{N})$.

Proof Fix $a \in \omega_2^{(\mathcal{M})}$ such that $a_k < a \quad (k \in \mathbb{N})$; the element a exists because $(\omega_2^{(\mathcal{M})}, \leq)$ has uncountable cofinality. The set

$$\{b \in \omega_2^{(\mathcal{M})} : a_k < b < a \quad (k \in \mathbb{N})\}$$

has coinitiality $\aleph_1$ because $|\mathcal{P}(\mathbb{N})_{(\mathcal{M})}| = \aleph_1$ (with CH), and the partial order $\mathbb{P}_a^{(\mathcal{M})}$ is ω-closed because this is true of $((\omega_2^{<\omega})_{(\mathcal{M})}, \leq)$.

It is now straightforward to construct by induction on $\alpha < \omega_1$ a decreasing sequence $\langle p_\alpha : \alpha < \omega_1 \rangle$ of conditions in $\mathbb{P}_a^{(\mathcal{M})}$ such that

$$p_\alpha \mid \mathbb{P}_{a_k}^{(\mathcal{M})} \in \mathcal{G}_k \quad (k \in \mathbb{N},\ \alpha < \omega_1)$$

and such that the filter in $\mathbb{P}_a^{(\mathcal{M})}$ generated by $\{p_\alpha : \alpha < \omega_1\}$ is an $\mathcal{M}$-generic filter; this is the required filter $\mathcal{G}$. □

Again, suppose that $\mathcal{M} = (M, E)$ is a model of ZC*, and let $\mathcal{M}[\mathcal{G}]$ be a generic extension of $\mathcal{M}$. The extension $\mathcal{M}[\mathcal{G}]$ is a *Cohen extension* if there exists $a \in M$ such that

$$\mathcal{M} \models \text{ `}a \text{ is an ordinal'}$$

and if $\mathcal{G}$ is an $\mathcal{M}$-generic filter for the partially ordered set $\mathbb{P}_a^{(\mathcal{M})}$. It is an elementary fact about Cohen extensions that, in the case where $\mathcal{M}[\mathcal{G}]$ is a Cohen extension of $\mathcal{M}$ and

$$(\mathcal{M}[\mathcal{G}])[\mathcal{H}] = \mathcal{M}[\mathcal{G}][\mathcal{H}]$$

is a Cohen extension of $\mathcal{M}[\mathcal{G}]$, also $\mathcal{M}[\mathcal{G}][\mathcal{H}]$ is itself a Cohen extension of the model $\mathcal{M}$.

Lemma 11.22 (CH) *Let $\mathcal{M}$ be a model of* ZC* *for which $(\mathbb{Q}_{(\mathcal{M})}, \leq)$ is an η_1-set and such that $|\mathcal{P}(\mathbb{N})_{(\mathcal{M})}| = 2^{\aleph_0}$.*

(i) *Suppose that* $f : \mathbb{N}_{(\mathcal{M})} \to \{0,1\}$ *is a function with*

$$f \restriction a \in \{0,1\}^a_{(\mathcal{M})} \quad (a \in \mathbb{N}_{(\mathcal{M})}) \quad \text{and} \quad f \notin \{0,1\}^{\mathbb{N}}_{(\mathcal{M})}\,.$$

Then there is an $\mathcal{M}$*-generic filter* $\mathcal{G}$ *in* $\mathbb{P}_1^{(\mathcal{M})}$ *such that, for each Cohen extension* $\mathcal{M}[\mathcal{G}][\mathcal{H}]$ *of* $\mathcal{M}[\mathcal{G}]$, *we have*

$$f \notin \{0,1\}^{\mathbb{N}}_{(\mathcal{M}[\mathcal{G}][\mathcal{H}])}\,.$$

(ii) *Let* $n \in \mathbb{N}$. *Suppose that* $F : \mathbb{D}^n_{(\mathcal{M})} \to \mathbb{I}_{(\mathcal{M})}$ *is a continuous function with respect to the order topology and that* V *is a non-empty, open subset of* $\mathbb{I}^n_{(\mathcal{M})}$ *such that* $\{z \in \mathbb{D}^n_{(\mathcal{M})} \cap V : G(z) \neq F(z)\}$ *is dense in* V *for each* $G \in {}^*C(\mathbb{I}^n, \mathbb{I})_{(\mathcal{M})}$. *Then there exists an* $\mathcal{M}$*-generic filter* $\mathcal{G}$ *in* $\mathbb{P}_n^{(\mathcal{M})}$ *such that, for each Cohen extension* $\mathcal{M}[\mathcal{G}][\mathcal{H}]$ *of* $\mathcal{M}[\mathcal{G}]$, F *cannot be extended to a continuous function*

$$F^* : \mathbb{I}^n_{(\mathcal{M}[\mathcal{G}][\mathcal{H}])} \to \mathbb{I}_{(\mathcal{M}[\mathcal{G}][\mathcal{H}])}\,.$$

Proof (i) We need only consider the case where there is a Cohen extension $\mathcal{M}[\mathcal{G}_0]$ of $\mathcal{M}$ for which $f \in \{0,1\}_{(\mathcal{M}[\mathcal{G}_0])}$, for otherwise $\mathcal{M}[\mathcal{G}]$ may be chosen to be any Cohen extension of $\mathcal{M}$.

By standard properties of Cohen extensions, we may suppose that $\mathcal{G}_0 \subset \mathbb{P}_1^{(\mathcal{M})}$ and that $\mathcal{M}[\mathcal{G}_0] = \mathcal{M}[f]$, where $\mathcal{M}[f]$ is the forcing extension of $\mathcal{M}$ given by f. Thus it suffices to find a Cohen extension $\mathcal{M}[\mathcal{G}]$ of $\mathcal{M}$ such that $\mathcal{G}_0 \notin \mathcal{M}[\mathcal{G}][\mathcal{H}]$ whenever $\mathcal{M}[\mathcal{G}][\mathcal{H}]$ is a Cohen extension of $\mathcal{M}[\mathcal{G}]$. In fact, this is easily done using the facts that $|\mathcal{P}(\mathbb{N})_{(\mathcal{M})}| = \aleph_1$ (with CH) and $(\mathbb{Q}_{(\mathcal{M})}, \leq)$ is an η_1-set.

(ii) We sketch the argument in the case where $n = 1$; the general case is similar. It is convenient to work with a variation of $\mathbb{P}_1$. Let $\mathbb{O}_1$ be the partially ordered family of non-empty, open intervals in $(0,1)$ with rational end-points, ordered by inclusion; it is easy to see that there is an isotonic map from $\mathbb{P}_1$ onto a dense subset of $\mathbb{O}_1$. The $\mathcal{M}$-version of $\mathbb{O}_1$ is denoted by $\mathbb{O}_1^{(\mathcal{M})}$. Since $(\mathbb{Q}_{(\mathcal{M})}, \leq)$ is an η_1-set, the partially ordered set $\mathbb{O}_1^{(\mathcal{M})}$ is ω-closed.

Fix $x, y \in \mathbb{D}_{(\mathcal{M})}$ such that $x < y$ and $(x,y)_{(\mathcal{M})} \subset V$. By a routine construction involving diagonalization against $\mathbb{R}_{(\mathcal{M})}$ (by CH, $|\mathbb{R}_{(\mathcal{M})}| = \aleph_1$), one can construct an $\mathcal{M}$-generic filter $\mathcal{G}_0 \subset \mathbb{O}_1^{(\mathcal{M})}$ with $(x,y)_{(\mathcal{M})} \in \mathcal{G}_0$ and such that $\langle A, B\rangle$ is a Cauchy gap in $(\mathbb{I}_{(\mathcal{M}[\mathcal{G}])}, \leq)$, where

$$\begin{aligned} A &= \{z \in (0,1)_{(\mathcal{M})} : F(I \cap \mathbb{D}_{(\mathcal{M})}) \subset (z,1)_{(\mathcal{M})} \quad \text{for some} \quad I \in \mathcal{G}_0\}\,, \\ B &= \{z \in (0,1)_{(\mathcal{M})} : F(I \cap \mathbb{D}_{(\mathcal{M})}) \subset (0,z)_{(\mathcal{M})} \quad \text{for some} \quad I \in \mathcal{G}_0\}\,. \end{aligned}$$

Since $\langle A, B\rangle$ is a Cauchy gap, it defines a function $f : \mathbb{N}_{(\mathcal{M})} \to \{0,1\}$ as follows. For each $a \in \mathbb{N}_{(\mathcal{M})}$, choose $r \in A \cap \mathbb{D}_a^{(\mathcal{M})}$ and $s \in B \cap \mathbb{D}_a^{(\mathcal{M})}$ with $s = r + 2^{-a}$, choose $t \in (r,s)_{(\mathcal{M})} \setminus \mathbb{D}_{(\mathcal{M})}$, and set $f \restriction a = \varepsilon_t^{(\mathcal{M})} \restriction a$. By Lemma 8.46(ii), f is well defined. We see that

$$f \restriction a \in \{0,1\}^a_{(\mathcal{M}[\mathcal{G}_0])} \quad (a \in \mathbb{N}_{(\mathcal{M})})\,,$$

but $f \notin \{0,1\}^{\mathbb{N}}_{(\mathcal{M}[\mathcal{G}_0])}$.

By standard arguments, $\mathbb{N}_{(\mathcal{M})} = \mathbb{N}_{(\mathcal{M}[\mathcal{G}_0])}$, and it follows that

$$|\mathcal{P}(\mathbb{N})_{(\mathcal{M}[\mathcal{G}_0])}| = \aleph_1$$

because $|\mathcal{P}(\mathbb{N})_{(\mathcal{M})}| = \aleph_1$ (with CH). By (i), there is an $\mathcal{M}[\mathcal{G}_0]$-generic filter $\mathcal{G}_1$ in $\mathbb{P}_1^{(\mathcal{M})}$ such that, for each Cohen extension $\mathcal{M}[\mathcal{G}_0][\mathcal{G}_1][\mathcal{H}]$ of $\mathcal{M}[\mathcal{G}_0][\mathcal{G}_1]$,

$$f \notin \{0,1\}^{\mathbb{N}}_{(\mathcal{M}[\mathcal{G}_0][\mathcal{G}_1][\mathcal{H}])}. \tag{11.16}$$

Let $\mathcal{G}$ be an $\mathcal{M}$-generic filter in $\mathbb{P}_1^{(\mathcal{M})}$ such that $\mathcal{M}[\mathcal{G}] = \mathcal{M}[\mathcal{G}_0][\mathcal{G}_1]$. We *claim* that $\mathcal{G}$ is the required filter. For take $\mathcal{M}[\mathcal{G}][\mathcal{H}]$ to be a Cohen extension of $\mathcal{M}[\mathcal{G}]$. Assume towards a contradiction that

$$F^* : \mathbb{I}_{(\mathcal{M}[\mathcal{G}][\mathcal{H}])} \to \mathbb{I}_{(\mathcal{M}[\mathcal{G}][\mathcal{H}])}$$

is a continuous function extending F, and take $x_0 \in \mathbb{I}_{(\mathcal{M}[\mathcal{G}_0])}$ to be the unique element such that $x_0 \in (r,s)_{(\mathcal{M}[\mathcal{G}_0])}$ for each $(r,s)_{(\mathcal{M})} \in \mathcal{G}_0$. Since $\langle A, B\rangle$ defines a Cauchy pregap in $(\mathbb{I}_{(\mathcal{M}[\mathcal{G}][\mathcal{H}])}, \leq)$ and F^* is continuous, it follows that

$$A << F^*(x_0) << B.$$

Set $y_0 = F^*(x_0)$. Then $f = \varepsilon_{y_0}^{(\mathcal{M}[\mathcal{G}][\mathcal{H}])}$, a contradiction of (11.16). Thus there is no such function F^*. □

Proposition 11.23 (GCH) *Let $\mathcal{M}$ be a model of* ZC* *such that the partially ordered set $((\omega_2^{<\omega})_{(\mathcal{M})}, \leq)$ is ω-closed and such that $|\mathcal{P}(\mathbb{N})_{(\mathcal{M})}| = 2^{\aleph_0}$. Suppose that $(a_\alpha : \alpha < \omega_2)$ is a strictly increasing sequence in $(\omega_2^{(\mathcal{M})}, \leq)$ such that, for each $\alpha < \omega_2$ with* $\operatorname{cof}\alpha = \aleph_1$, $a_\alpha = \sup\{a_\beta : \beta < \alpha\}$ *and*

$$\mathcal{M} \models \text{`}\operatorname{cof} a_\alpha = \aleph_1\text{'}.$$

Then there is an $\mathcal{M}$-generic filter $\mathcal{G} \subset \mathbb{P}_a^{(\mathcal{M})}$, where $\mathcal{M} \models$ '$a = \omega_2$', such that $\mathfrak{N}_{(\mathcal{M}[\mathcal{G}])}$ is solid.

Proof We *claim* that there is an inductive construction of an increasing sequence $(\mathcal{G}_\alpha : \alpha < \omega_2)$ of $\mathcal{M}$-generic filters such that the following conditions are satisfied for all $\alpha < \omega_2$:

(i) $\mathcal{G}_\alpha \subset \mathbb{P}_{a_\alpha}^{(\mathcal{M})}$;

(ii) $\mathcal{G}_\alpha \mid \mathbb{P}_{a_\beta}^{(\mathcal{M})} = \mathcal{G}_\beta \quad (\beta < \alpha)$;

(iii) for each $n \in \mathbb{N}$, each continuous function $F : \mathbb{D}^n_{(\mathcal{M}[\mathcal{G}_\alpha])} \to \mathbb{I}_{(\mathcal{M}[\mathcal{G}_\alpha])}$, and each non-empty, open subset V of $\mathbb{I}^n_{(\mathcal{M}[\mathcal{G}_\alpha])}$ either there exist $\beta < \omega_2$ and $G \in {}^*C(\mathbb{I}^n, \mathbb{I})_{(\mathcal{M}[\mathcal{G}_\beta])}$ such that

$$\{z \in \mathbb{D}^n_{(\mathcal{M}[\mathcal{G}_\beta])} \cap V : G(z) \neq F(z)\}$$

is not dense in V, or there exists $\beta < \omega_2$ such that, for each Cohen extension $\mathcal{M}[\mathcal{G}_\beta][\mathcal{H}]$ of $\mathcal{M}[\mathcal{G}_\beta]$, the function F cannot be extended to a continuous function

$$F^* : \mathbb{I}^n_{(\mathcal{M}[\mathcal{G}_\beta][\mathcal{H}])} \to \mathbb{I}_{(\mathcal{M}[\mathcal{G}_\beta][\mathcal{H}])}.$$

In the proof of this claim, Lemma 11.21 is used to construct $\mathcal{G}_\alpha$ in the case where $\operatorname{cof}\alpha = \aleph_0$ and Lemma 11.22(ii) is used at successor steps to achieve condition (iii). It remains to explain how the construction is carried out in the case where $\operatorname{cof}\alpha = \aleph_1$. Take α to be such an ordinal, assume that $(\mathcal{G}_\beta : \beta < \alpha)$ has been constructed, and set $\mathcal{G}_\alpha = \bigcup\{\mathcal{G}_\beta : \beta < \alpha\}$. Since $a_\alpha = \sup\{a_\beta : \beta < \alpha\}$ and since $\mathcal{M} \models$ '$\operatorname{cof} a_\alpha = \aleph_1$', it follows that $\mathcal{G}_\alpha$ is $\mathcal{M}$-generic, and this is sufficient for the inductive construction to continue.

Let

$$\mathcal{G} = \bigcup\{\mathcal{G}_\alpha : \alpha < \omega_2\}.$$

Then $\mathcal{G} \in \mathbb{P}_a^{(\mathcal{M})}$, where $\mathcal{M} \models$ '$a = \omega_2$', and $\mathcal{G}$ is an $\mathcal{M}$-generic filter. By standard properties of Cohen extensions, $\mathcal{M}[\mathcal{G}]$ is a Cohen extension of $\mathcal{M}[\mathcal{G}_\alpha]$ for each $\alpha < \omega_2$. It follows from (iii) that $\mathfrak{N}_{(\mathcal{M}[\mathcal{G}])}$ is solid. □

As an immediate corollary of Proposition 11.23, we obtain the following theorem.

Theorem 11.24 (GCH) *Let $\mathcal{U}$ be a free ultrafilter on $\mathbb{N}$, let α be an infinite ordinal such that $|\mathbf{V}_\alpha| = \alpha$, and set $\mathcal{M} = \mathbf{V}_\alpha^{\mathbb{N}}/\mathcal{U}$. Then there is an $\mathcal{M}$-generic filter $\mathcal{G} \subset \mathbb{P}_a^{(\mathcal{M})}$, where $\mathcal{M} \models$ '$a = \omega_2$', such that $\mathfrak{N}_{(\mathcal{M}[\mathcal{G}])}$ is solid.*

Proof For each $\alpha < \omega_2$, let $a_\alpha \in \mathcal{M}$ be the element defined by the function $f_\alpha : \mathbb{N} \to \omega_2$ which takes the constant value α. Then $(a_\alpha : \alpha < \omega_2)$ is a strictly increasing sequence in $(\omega_2^{(\mathcal{M})}, \leq)$ such that, for each $\alpha < \omega_2$ with $\operatorname{cof}\alpha = \aleph_1$, $a_\alpha = \sup\{a_\beta : \beta < \alpha\}$ and

$$\mathcal{M} \models \text{ `}\operatorname{cof} a_\alpha = \aleph_1\text{'}.$$

By a pointwise argument, we see that $((\omega_2^{<\omega})_{(\mathcal{M})}, \leq)$ is ω-closed. With CH, $|\mathcal{P}(\mathbb{N})_{(\mathcal{U})}| = \aleph_1$, and so $|\mathcal{P}(\mathbb{N})_{(\mathcal{M})}| = \aleph_1$. Thus Proposition 11.23 applies to give the result. □

Proof of Theorem 11.19 Let $\mathcal{U}$ be a free ultrafilter on $\mathbb{N}$, and take $\mathcal{M}$ to be $\mathbf{V}_\alpha^{\mathbb{N}}/\mathcal{U}$, where α is as above. By Theorem 11.24, there is an $\mathcal{M}$-generic filter $\mathcal{G}$ such that $\mathfrak{N}_{(\mathcal{M}[\mathcal{G}])}$ is solid. Since $\mathcal{M}$ is a model of ZC*, $\mathfrak{N}_{(\mathcal{M}[\mathcal{G}])}$ is a model of analysis.

By Lemma 11.20, there is a prime z-ideal P_0 in $C(\beta X)$ such that K_{P_0} is a hyper-real field and $\mathfrak{N}_{(P_0)} \cong \mathfrak{N}_{(\mathcal{M}[\mathcal{G}])}$.

We have $|\mathbb{Q}_{(\mathcal{M})}| = |\mathbb{Q}_{(\mathcal{U})}| = \aleph_1$ (with CH). Also $\mathbb{Q}_{(\mathcal{M})} = \mathbb{Q}_{(\mathcal{M}[\mathcal{G}])}$, which is order-dense in $\mathbb{R}_{(\mathcal{M}[\mathcal{G}])}$, and so $w(K_{P_0}) = \aleph_1$. Necessarily $|K_{P_0}| \leq 2^{\aleph_1} = \aleph_2$ with GCH.

This concludes the proof of Theorem 11.19. □

By combining Theorems 11.6 and 11.19, we succeed in finding a rigid super-real field.

Theorem 11.25 (GCH) *There is a rigid hyper-real field K such that $w(K) = \aleph_1$ and $|K| = \aleph_2$.* □

We pointed out in Chapter 4 that we began our study of super-real fields with the intention of characterizing these fields algebraically in the class of all totally ordered fields; at least we might be able to distinguish the subfields of $\widetilde{\mathbf{R}}$ that are super-real. Perhaps 'having an exponentiation on a convex subgroup' and a few similar properties might suffice? Theorem 11.28, below, shows that there is a subfield K of $\widetilde{\mathbf{R}}$ which is the rational field associated with a certain model

$$\mathfrak{N} = \langle \mathbb{N}_{(\mathfrak{N})}, +, \cdot, \mathcal{P}(\mathbb{N})_{(\mathfrak{N})} \rangle$$

of analysis, and hence such that K has all of the properties of a super-real field that one could imagine, but is still not itself a super-real field; our sought-for characterization is seen to be evanescent. Any characterization of super-real η_1-fields must involve a classification of the structures $\mathfrak{N}_{(P)}$ arising from non-maximal, prime ideals P. There is evidence, which we shall discuss in Chapter 12, that such a classification is possible, at least if we restrict ourselves to the structures $\mathfrak{N}_{(P)}$ where K_P is an η_1-field with $w(K_P) = \aleph_1$, i.e. where K_P is an η_1-subfield of $\widetilde{\mathbf{R}}$ containing $\mathbf{R}$.

We require as a preliminary a result which is a slight generalization of Theorem 11.6. Let $\mathfrak{M}$ be a model of analysis. Then there is a natural generalization of Definition 11.5 that defines *solidity* for $\mathfrak{M}$.

Theorem 11.26 *Let $\mathfrak{M}$ be a model of analysis which is solid, and suppose that P is a non-maximal, prime ideal in $C(\Omega)$ such that $(K_P, \leq)$ and $(\mathbb{R}_{(\mathfrak{M})}, \leq)$ are order-isomorphic. Then $\mathfrak{N}_{(P)} \cong \mathfrak{N}_{(\mathfrak{M})}$.* □

We shall produce in Theorem 11.28 an example of a model of analysis whose associated field $\mathbb{R}_{(\mathfrak{M})}$ is an η_1-rational field of weight $\aleph_1$ which is not a super-real field. The proof requires the following proposition. The notation

$$\mathrm{Th}\,\langle M, +, \cdot \rangle ,$$

the *theory* of $\langle M, +, \cdot \rangle$, denotes the collection of sentences of the appropriate language that are true in a structure $\langle M, +, \cdot \rangle$.

Proposition 11.27 *Let Q be a non-maximal, prime z-ideal in $C(\Omega)$. Then*

$$\mathrm{Th}\,\langle \mathbb{N}_{(Q)}, +, \cdot \rangle = \mathrm{Th}\,\langle \mathbb{N}, +, \cdot \rangle .$$

Proof This is easily proved by a pointwise argument using Q-partitions. □

We make the following technical observation. Let P be a non-maximal, prime ideal, and let Q be a prime z-ideal below P. The above proposition shows that $\mathrm{Th}\,\langle \mathbb{N}_{(Q)}, +, \cdot \rangle = \mathrm{Th}\,\langle \mathbb{N}, +, \cdot \rangle$. By the transfer principle, Corollary 8.30, the Σ_1^0 sentences which are true in $\langle \mathbb{N}_{(P)}, +, \cdot \rangle$ are exactly the Σ_1^0 sentences true in $\langle \mathbb{N}, +, \cdot \rangle$; this fact also follows from Matijacevič's theorem.

Theorem 11.28 (GCH) *There is a model $\mathfrak{M}$ of analysis such that $\mathbb{R}_{(\mathfrak{M})}$ is an η_1-rational field of weight $\aleph_1$ and such that $(\mathbb{R}_{(\mathfrak{M})}, \leq)$ is not order-isomorphic to $(K, \leq)$ for any super-real field K.*

Proof Take $\mathcal{M} = (M, E)$ to be a model such that

$$\mathcal{M} \models \text{ZC}^* + \varphi,$$

where φ is the Σ_1^0 sentence which expresses the statement that the theory ZC* is inconsistent. We suppose that $\mathcal{M}$ satisfies the conditions of Proposition 11.23. It is a standard construction to build such a model from a countable model, say $\mathcal{M}_0$, of ZC* + φ (such a model $\mathcal{M}_0$ exists by Gödel's incompleteness theorem).

By Proposition 11.23, there is a Cohen extension $\mathcal{M}[\mathcal{G}]$ of $\mathcal{M}$ such that $\mathfrak{N}_{(\mathcal{M}[\mathcal{G}])}$ is solid. Let $\mathfrak{M} = \mathfrak{N}_{(\mathcal{M}[\mathcal{G}])}$, so that $\mathfrak{M}$ is solid and

$$\langle \mathbb{N}_{(\mathfrak{M})}, +, \cdot \rangle = \langle \mathbb{N}_{(\mathcal{M})}, +, \cdot \rangle ;$$

it follows that $w(\mathbb{R}_{(\mathfrak{M})}) = \aleph_1$ because $|\mathcal{P}(\mathbb{N})_{(\mathcal{M})}| = \aleph_1$.

Assume towards a contradiction that $(\mathbb{R}_{(\mathfrak{M})}, \leq)$ is order-isomorphic to the ordered set $(K_P, \leq)$ for a non-maximal, prime ideal P in some $C(\Omega)$. Then it follows from Theorem 11.26 that $\mathfrak{N}_{(\mathfrak{M})} \cong \mathfrak{N}_{(P)}$, and so we have the isomorphism $\langle \mathbb{N}_{(\mathcal{M})}, +, \cdot \rangle \cong \langle \mathbb{N}_{(P)}, +, \cdot \rangle$. Therefore

$$\langle \mathbb{N}_{(P)}, +, \cdot \rangle \models \varphi,$$

a contradiction since $\langle \mathbb{N}, +, \cdot \rangle \models \neg\varphi$. □

The final labour in this book is to show that it is relatively consistent with ZFC + GCH that various pairs of classes of super-real fields are distinct in the category of totally ordered sets. Essentially, we are seeking a super-real η_1-field which does not belong to ***z*-ideal η_1**.

Suppose that $\mathfrak{N}$ and $\mathfrak{M}$ are models of analysis such that

$$\langle \mathbb{N}_{(\mathfrak{N})}, +, \cdot \rangle = \langle \mathbb{N}_{(\mathfrak{M})}, +, \cdot \rangle \quad \text{and} \quad \mathcal{P}(\mathbb{N})_{(\mathfrak{N})} \subset \mathcal{P}(\mathbb{N})_{(\mathfrak{M})} .$$

Then $\mathfrak{M}$ is a *Cohen extension* of $\mathfrak{N}$ if, for each $A \in \mathcal{P}(\mathbb{N})_{(\mathfrak{M})}$, there is an $\mathfrak{N}$-generic filter $\mathcal{G} \subset \mathbb{P}_1^{(\mathfrak{N})}$ with $A \in \mathcal{P}(\mathbb{N})_{(\mathfrak{N}[\mathcal{G}])}$. The connection with the earlier version of this notion is the following. Let $\mathcal{M}$ be a model of ZFC, and let $\mathcal{M}[\mathcal{G}]$ be a Cohen extension of $\mathcal{M}$. Then $\mathfrak{N}_{(\mathcal{M}[\mathcal{G}])}$ is a Cohen extension of $\mathfrak{N}_{(\mathcal{M})}$.

The following lemma is standard.

Lemma 11.29 *Let $\mathfrak{N}$ be a model of analysis, and let $\mathcal{S}$ be a collection of subsets of $\mathbb{N}_{(\mathfrak{N})}$ such that, for each $A, B \in \mathcal{S}$, there is an $\mathfrak{N}$-generic filter $\mathcal{G} \subset \mathbb{P}_1^{(\mathfrak{N})}$ with $A, B \in \mathcal{P}(\mathbb{N})_{(\mathfrak{N}[\mathcal{G}])}$ and $\mathcal{P}(\mathbb{N})_{(\mathfrak{N}[\mathcal{G}])} \subset \mathcal{S}$. Then $\langle \mathbb{N}_{(\mathfrak{N})}, +, \cdot, \mathcal{S} \rangle$ is a model of analysis which is a Cohen extension of $\mathfrak{N}$.* □

We now fix a free ultrafilter $\mathcal{U}$ on $\mathbb{N}$. Once again we also fix an index set I (which will eventually be a set of cardinality $\aleph_2$) and set $Y = \{0,1\}^{\mathbb{N}\times I}$ and $X = \mathbb{N} \times Y$. Recall from page 217 that $Y_{(\mathcal{U})}$ is a collection of functions from $\mathbb{N}_{(\mathcal{U})} \times I_{(\mathcal{U})}$ into $\{0,1\}$ determined by functions $h : \mathbb{N} \to \{0,1\}^{\mathbb{N}\times I}$, and from page 220 that $\mathbf{Z}(Y)_{(\mathcal{U})}$ is a family of subsets of $Y_{(\mathcal{U})}$.

On page 216, we defined a structure $S_{(\mathcal{U})}$ associated with a set S; in fact, there is a canonical identification of $S_{(\mathcal{U})}$ with $S^{\mathbb{N}}/\mathcal{U}$. We now apply this successively in the cases where $S = (\mathbb{N} \times I)^{\mathbb{N}}$, where $S = [\mathbb{N} \times I]^{\omega}$ (the family of countable subsets of $\mathbb{N} \times I$), where $S = \mathcal{P}(\mathbb{N})^{\mathbb{N}}$, and where $S = \bigcup\{\mathcal{P}(\{0,1\}^k) : k \in \mathbb{N}\}$ to obtain

$$(\mathbb{N} \times I)^{\mathbb{N}}_{(\mathcal{U})}, \quad [\mathbb{N} \times I]^{\omega}_{(\mathcal{U})}, \quad \mathcal{P}(\mathbb{N})^{\mathbb{N}}_{(\mathcal{U})}, \quad \left(\bigcup\{\mathcal{P}(\{0,1\}^k) : k \in \mathbb{N}\}\right)_{(\mathcal{U})},$$

respectively. In an analogous way to that in Chapter 9, we interpret these sets in the following way:

$(\mathbb{N} \times I)^{\mathbb{N}}_{(\mathcal{U})}$ is a collection of functions from $\mathbb{N}_{(\mathcal{U})}$ into $\mathbb{N}_{(\mathcal{U})} \times I_{(\mathcal{U})}$;

$[\mathbb{N} \times I]^{\omega}_{(\mathcal{U})}$ is a collection of subsets of $\mathbb{N}_{(\mathcal{U})} \times I_{(\mathcal{U})}$;

$\mathcal{P}(\mathbb{N})^{\mathbb{N}}_{(\mathcal{U})}$ is a collection of functions from $\mathbb{N}_{(\mathcal{U})}$ into $\mathcal{P}(\mathbb{N})_{(\mathcal{U})}$;

$(\bigcup\{\mathcal{P}(\{0,1\}^k) : k \in \mathbb{N}\})_{(\mathcal{U})}$ is a family of sets, each of which is a subset of $\{0,1\}^c_{(\mathcal{U})}$ for some $c \in \mathbb{N}_{(\mathcal{U})}$.

On page 230 we explained that we would also consider an initial segment J of the structure $\mathbb{N}_{(\mathcal{U})}$. We shall associate to such an initial segment J a certain partial order $\mathbb{O}_J$ with

$$\mathbb{O}_J \subset (\mathbb{N} \times I)^{\mathbb{N}}_{(\mathcal{U})} \times \mathbf{Z}(Y)_{(\mathcal{U})};$$

this partial order $\mathbb{O}_J$ is analogous to the partial order $\mathbb{P}_J$ which was defined in Definition 9.43. As a preliminary, we define two related partial orders, $\mathbb{O}^c_J$ and $\mathbb{O}^{\mathbb{N}}_J$.

Let $W \in (\bigcup\{\mathcal{P}(\{0,1\})^k : k \in \mathbb{N}\})_{(\mathcal{U})}$, and take $c \in \mathbb{N}_{(\mathcal{U})}$ with $W \subset \{0,1\}^c_{(\mathcal{U})}$. The *support* of W, written $\operatorname{supp} W$, is the set of elements $a \in [1,c]_{(\mathcal{U})} \cap \mathbb{N}_{(\mathcal{U})}$ such that there exist $f \in W$ and $g \in \{0,1\}^c_{(\mathcal{U})} \setminus W$ with $f(b) = g(b)$ for each $b \in \mathbb{N}_{(\mathcal{U})}$ with $b \leq c$ and $b \neq a$. (In the cases where $W = \emptyset$ or $W = \{0,1\}^c_{(\mathcal{U})}$, we take $\operatorname{supp} W$ to be $\emptyset$.) We have

$$\operatorname{supp} W \in [\mathbb{N}]^{<\omega}_{(\mathcal{U})}.$$

Now take $c \in \mathbb{N}_{(\mathcal{U})} \setminus J$. We define $\mathbb{O}^c_J$ to be the family of all non-empty subsets W of $\{0,1\}^c_{(\mathcal{U})}$ such that $W \in \left(\bigcup\{\mathcal{P}(\{0,1\}^k) : k \in \mathbb{N}\}\right)_{(\mathcal{U})}$ and such that there exists $b \in \mathbb{N}_{(\mathcal{U})} \setminus J$ with $b \leq c$ for which

$$\operatorname{supp}\{f \restriction b : f \in W\} \subset J.$$

The set $\mathbb{O}^c_J$ is clearly a partially ordered set with respect to inclusion.

We also define $\mathbb{O}^{\mathbb{N}}_J$, a closely related partial order. For $W \in \mathbf{Z}(\{0,1\}^{\mathbb{N}})_{(\mathcal{U})}$, the *support* of W, $\operatorname{supp} W$, is the set of elements $a \in \mathbb{N}_{(\mathcal{U})}$ such that there exists $f \in W$ and $g \in \{0,1\}^{\mathbb{N}})_{(\mathcal{U})} \setminus W$ with

$$\{a\} = \{b \in \mathbb{N}_{(\mathcal{U})} : f(b) \neq g(b)\}.$$

We have $\operatorname{supp} W \in [\mathbb{N}]^{\omega}_{(\mathcal{U})}$. Now $\mathbb{O}^{\mathbb{N}}_J$ is the family of all non-empty sets W in $\mathbf{Z}(\{0,1\}^{\mathbb{N}})_{(\mathcal{U})}$ such that there exists $b \in \mathbb{N}_{(\mathcal{U})} \setminus J$ with $\operatorname{supp}\{f \restriction b : f \in W\} \subset J$. Again, $\mathbb{O}^{\mathbb{N}}_J$ is a partially ordered set with respect to inclusion.

Finally, suppose that $W \in \mathbf{Z}(Y)_{(\mathcal{U})}$, so that $W \subset Y_{(\mathcal{U})}$. In this case the *support* of W, also written $\operatorname{supp} W$, is the set of elements $(c, x) \in \mathbb{N}_{(\mathcal{U})} \times I_{(\mathcal{U})}$ such that there exist $f \in W$ and $g \in Y_{(\mathcal{U})} \setminus W$ with

$$\{(c,x)\} = \{(d,y) \in \mathbb{N}_{(\mathcal{U})} \times I_{(\mathcal{U})} : f(d,y) \neq g(d,y)\}.$$

We have $\operatorname{supp} W \in [\mathbb{N} \times I]^{\omega}_{(\mathcal{U})}$. It is easily verified by pointwise arguments that, if $f \in W$ and $g \in Y_{(\mathcal{U})} \setminus W$, then there exists $(c, x) \in \operatorname{supp} W$ with $f(c, x) \neq g(c, x)$.

Let $W \in \mathbf{Z}(Y)_{(\mathcal{U})}$ and $f \in (\mathbb{N} \times I)^{\mathbb{N}}_{(\mathcal{U})}$. We define

$$W_f = \{h \circ f : h \in W\},$$

so that W_f is a subset of $\{0,1\}^{\mathbb{N}}_{(\mathcal{U})}$, and it is easily verified that $W_f \in \mathbf{Z}(\{0,1\}^{\mathbb{N}})_{(\mathcal{U})}$. For each $a \in \mathbb{N}_{(\mathcal{U})}$, we set

$$W_{(f,a)} = \{g \restriction a : g \in W_f\}.$$

The partial order $\mathbb{O}_J$ consists, as a set, of all pairs (f, W) satisfying the following conditions:

(i) $W \in \mathbf{Z}(Y)_{(\mathcal{U})}$;

(ii) $f : \mathbb{N}_{(\mathcal{U})} \to \mathbb{N}_{(\mathcal{U})} \times I_{(\mathcal{U})}$ is an injection with $f \in (\mathbb{N} \times I)^{\mathbb{N}}_{(\mathcal{U})}$;

(iii) $\operatorname{supp} W \subset f(\mathbb{N}_{(\mathcal{U})})$;

(iv) there exist $a \in \mathbb{N}_{(\mathcal{U})} \setminus J$ and $b \in J$ such that $\operatorname{supp} W_{(f,a)} \subset [1, b]_{(\mathcal{U})} \cap \mathbb{N}_{(\mathcal{U})}$.

Condition (iv) is equivalent to the assertion that $W_f \in \mathbb{O}^{\mathbb{N}}_J$.

The partial order on $\mathbb{O}_J$ is defined by saying that

$$(f_1, W_1) \leq (f_2, W_2)$$

if $W_1 \subset W_2$ and there exists $h \in \mathbb{N}^{\mathbb{N}}_{(\mathcal{U})}$ with $h(J) \subset J$ and $f_1 \circ h = f_2$. It is easy to check that $(\mathbb{O}_J, \leq)$ is indeed a partially ordered set.

Lemma 11.30 *Let J be an initial segment of $\mathbb{N}_{(\mathcal{U})}$ such that* $\operatorname{cof} J \geq \aleph_1$ *and* $\operatorname{coi}(\mathbb{N}_{(\mathcal{U})} \setminus J) \geq \aleph_1$. *Then:*

(i) $(\mathbb{O}_J, \leq)$ *is ω-closed;*

(ii) (CH) *for each subset $\mathcal{A}$ of $\mathbb{O}_J$ for which $|\mathcal{A}| = \aleph_2$, there exist $(f_1, W_1) \in \mathcal{A}$, $(f_2, W_2) \in \mathcal{A}$, and $(f, W) \in \mathbb{Q}_J$ such that $(f, W) < (f_j, W_j)$ $(j = 1, 2)$.*

Proof This proof is a routine adaptation of earlier arguments: the proofs for (i) and (ii) are essentially the same as the proofs of Lemma 9.45 and Lemma 9.59, respectively. □

We are seeking a prime ideal P such that K_P is an η_1-field of weight $\aleph_1$, but such that K_P is not order-isomorphic to K_Q for any prime z-ideal Q. This example will be obtained from a sufficiently generic filter $\mathcal{G}$ in $\mathbb{O}_J$, where J is chosen to have various special properties and the index set I that occurs in the definition of the topological space X is a set of cardinality $\aleph_2$. Our situation will be quite similar to that in Theorems 9.37 and 9.50.

As before, let J be an initial segment of $\mathbb{N}_{(\mathcal{U})}$, with the extra requirement that J be additively closed. Then we define

$$\mathfrak{N}_J^{(\mathcal{U})} = \langle \mathbb{N}_{(\mathcal{U})}, +, \cdot, J, \mathcal{P}(\mathbb{N})_{(\mathcal{U})} \rangle .$$

We set $\mathcal{P}(J)_{(\mathcal{U})} = \{A \cap J : A \in \mathcal{P}(\mathbb{N})_{(\mathcal{U})}\}$. We recall from Proposition 8.36(ii) that there is a unique prime ideal P in $C(\beta\mathbb{N})$ above the prime ideal given by $\mathcal{U}$ with $\mathbb{N}_{(P)} = J$. We now call this prime P_J, so that $\mathcal{P}(J)_{(\mathcal{U})} = \mathcal{P}(\mathbb{N})_{(P_J)}$ and

$$\mathfrak{N}_{(P_J)} = \langle J, +, \cdot, \mathcal{P}(J)_{(\mathcal{U})} \rangle .$$

The structure $\mathfrak{N}_J^{(\mathcal{U})}$ is complicated, but $\mathfrak{N}_{(P_J)}$ is easily defined within $\mathfrak{N}_J^{(\mathcal{U})}$.

Let us discuss a natural example of the above situation. We take J to be $\mathbb{N}$ itself. Then $\mathfrak{N}_{(P_J)}$ is simply the standard model of analysis $\langle \mathbb{N}, +, \cdot, \mathcal{P}(\mathbb{N}) \rangle$. In general, the structure $\mathfrak{N}_J^{(\mathcal{U})}$ is a much richer structure than this standard model, but it does satisfy the following condition. Suppose that a subset A of J is definable in the structure $\mathfrak{N}_J^{(\mathcal{U})}$ from various parameters. Then $A \in \mathcal{P}(J)_{(\mathcal{U})}$ because, in this specific case, $\mathcal{P}(J)_{(\mathcal{U})}$ contains all the subsets of J.

Define

$$\mathcal{P}(\mathbb{N})_{(\mathcal{U})}^{<J} = \bigcup \{\mathcal{P}(\mathbb{N})_{(\mathcal{U})}^{a} : a \in J\} .$$

A subset A of $\mathcal{P}(\mathbb{N})_{(\mathcal{U})}^{<J}$ is *closed under initial segments* if, for each $a \in J$ and each $f \in A$ with $f \in \mathcal{P}(\mathbb{N})_{(\mathcal{U})}^{a}$, $f \restriction b \in A$ for each $b \in [1, a]_{(\mathcal{U})} \cap \mathbb{N}_{(\mathcal{U})}$. An element $f \in A$ is *maximal* if, for each $g \in A$ with $f \neq g$ and each $a \in J$, we have $f \neq g \restriction a$.

Definition 11.31 *Let J be an additively closed initial segment of $\mathbb{N}_{(\mathcal{U})}$. Then J is a* saturated cut *in $\mathfrak{N}_{(\mathcal{U})}$ if, for each non-empty subset A of $\mathcal{P}(\mathbb{N})_{(\mathcal{U})}^{<J}$ such that A is closed under initial segments and has no maximal elements and such that A is definable from parameters in the structure $\mathfrak{N}_J^{(\mathcal{U})}$, there is a function $h : \mathbb{N}_{(\mathcal{U})} \to \mathcal{P}(\mathbb{N})_{(\mathcal{U})}$ such that $h \in \mathcal{P}(\mathbb{N})_{(\mathcal{U})}^{\mathbb{N}}$ and $h \restriction a \in A$ for each $a \in J$.*

For example, $\mathbb{N}$ is a saturated cut in $\mathbb{N}_{(\mathcal{U})}$; this is implied by the well-known assertion that $\mathfrak{N}_{(\mathcal{U})}$ is a countably saturated model in the usual sense of model theory.

The following lemma is an easy exercise; we note that, for each function $f : J \to \mathcal{P}(\mathbb{N})_{(\mathcal{U})}$ which is definable from parameters in $\mathcal{M}_J$, there is a function $h : \mathbb{N}_{(\mathcal{U})} \to \mathcal{P}(\mathbb{N})_{(\mathcal{U})}$ such that $h \in \mathcal{P}(\mathbb{N})_{(\mathcal{U})}^{\mathbb{N}}$ and $h \mid J = f$.

Lemma 11.32 *Let J be a saturated cut in $\mathbb{N}_{(\mathcal{U})}$. Then P_J is an exponential prime in $C(\beta\mathbb{N})$ and $\mathfrak{N}_{(P_J)}$ is a model of analysis.* □

Theorem 11.33 (CH) *Let $\mathcal{U}$ be a free ultrafilter on $\mathbb{N}$. Then there is a saturated cut J in $\mathbb{N}_{(\mathcal{U})}$ such that* $\operatorname{cof} J = \operatorname{coi}(\mathbb{N}_{(\mathcal{U})} \setminus J) = \aleph_1$, *and*

$$\operatorname{Th}\langle J, +, \cdot \rangle \neq \operatorname{Th}\langle \mathbb{N}, +, \cdot \rangle .$$

Proof Suppose that $\mathfrak{N} = \langle \mathbb{N}_{(\mathcal{U})}, +, \cdot, \mathcal{P}(\mathbb{N})_{(\mathcal{U})} \rangle$ is a model of analysis and that I is an additively closed initial segment of $\mathbb{N}_{(\mathcal{U})}$. Define

$$\mathfrak{N}_I = \left\langle \mathbb{N}_{(\mathcal{U})}, +, \cdot, I, \mathcal{P}(\mathbb{N})_{(\mathcal{U})} \right\rangle .$$

We generalize Definition 11.31 in the obvious fashion to define 'I is a saturated cut in $\mathfrak{N}$'. In the case where $\mathfrak{N}$ is countably saturated, $\mathbb{N}$ is a saturated cut in $\mathfrak{N}$.

We begin with the following preliminary remark. Let T_0 be the set of Π^0_1 sentences φ such that

$$\langle \mathbb{N}, +, \cdot \rangle \models \varphi .$$

Suppose that T is a theory in the language of analysis containing the axioms for analysis and such that $T \cup T_0$ is consistent, and suppose that $\mathcal{U}$ is a free ultrafilter on $\mathbb{N}$. Then it is easily proved using finite games and an overspill argument that, under the assumption of CH, there is an initial segment L of $\mathbb{N}_{(\mathcal{U})}$ with $\operatorname{cof} L = \aleph_1$ such that

$$\left\langle L, +, \cdot, \mathcal{P}(L)_{(\mathcal{U})} \right\rangle \models T ,$$

where

$$\mathcal{P}(L)_{(\mathcal{U})} = \{A \cap L : A \in \mathcal{P}(\mathbb{N})_{(\mathcal{U})}\} .$$

It follows that the model $\left\langle L, +, \cdot, \mathcal{P}(L)_{(\mathcal{U})} \right\rangle$ is countably saturated, and so $\mathbb{N}$ is a saturated cut in this model.

It is standard that there is a countably saturated model $\mathcal{M} = (M, E)$ of the theory ZC* + CH of cardinality $\aleph_1$ such that:

(i) $\operatorname{Th} \left\langle \mathbb{N}_{(\mathcal{M})}, +, \cdot \right\rangle \neq \operatorname{Th} \langle \mathbb{N}, +, \cdot \rangle$;

(ii) every Π^0_2 sentence which is true in $\langle \mathbb{N}, +, \cdot \rangle$ is also true in $\left\langle \mathbb{N}_{(\mathfrak{M})}, +, \cdot \right\rangle$.

Take $b \in M$ such that

$$\mathcal{M} \models \text{ `} b = \mathfrak{N}_{(\mathcal{V})} \text{ for some free ultrafilter } \mathcal{V} \text{ on } \mathbb{N}\text{'} ,$$

and let

$$\mathfrak{M} = \left\langle \mathbb{N}_{(\mathfrak{M})}, +, \cdot, \mathcal{P}(\mathbb{N})_{(\mathfrak{M})} \right\rangle$$

be the corresponding structure. Thus $\mathbb{N}_{(\mathcal{M})}$ is an initial segment of $\mathbb{N}_{(\mathfrak{M})}$.

By an argument using (ii), there exists $c \in M$ such that

$$\mathfrak{M} \models \text{ `} c \text{ is an initial segment of } \mathbb{N}_{(b)} \text{ with uncountable cofinality'} ,$$

and such that

$$\operatorname{Th} \left\langle L, +, \cdot, \mathcal{P}(L)_{(\mathfrak{M})} \right\rangle = \operatorname{Th} \langle \mathbb{N}, +, \cdot, \mathcal{P}(\mathbb{N}) \rangle ,$$

where L is the initial segment of $\mathbb{N}_{(\mathfrak{M})}$ determined by c and where

$$\mathcal{P}(L)_{(\mathfrak{M})} = \{A \cap L : A \in \mathcal{P}(\mathbb{N})_{(\mathfrak{M})}\}$$

The existence of c follows from our preliminary remark applied within $\mathcal{M}$ to a suitably chosen theory which is constructed using a further overspill argument.

It follows that $\mathbb{N}_{(\mathfrak{M})} \subset L$ and that $\mathbb{N}_{(\mathcal{M})}$ is a saturated cut in the structure $\langle L, +, \cdot, \mathcal{P}(L)_{(\mathfrak{M})}\rangle$.

Since $\mathcal{M}$ is countably saturated, $\mathrm{coi}(L \setminus \mathbb{N}_{(\mathfrak{M})}) = \aleph_1$ and the structure $\langle L, +, \cdot, \mathcal{P}(L)_{(\mathfrak{M})}\rangle$ is also countably saturated, and so it is isomorphic to $\mathfrak{N}_{(\mathcal{U})}$, say the isomorphism is given by $\pi : L \to \mathbb{N}_{(\mathcal{U})}$. Set $J = \pi(\mathbb{N}_{(\mathcal{M})})$, so that $\mathrm{Th}\,\langle J, +, \cdot\rangle = \mathrm{Th}\,\langle \mathbb{N}_{(\mathcal{M})}, +, \cdot\rangle$. Then J has the required properties. □

We prove the analogue of Lemma 9.46 for $\mathbb{O}_J^{\mathbb{N}}$.

The spaces $\mathbf{Z}(\{0,1\}^{\mathbb{N}})_{(\mathcal{U})}^{\mathbb{N}}$ and $\mathbf{Z}(\{0,1\}^{\mathbb{N}})_{(\mathcal{U})}^{a}$ are defined in an analogous way to $\mathbf{Z}(Y)_{(\mathcal{U})}^{\mathbb{N}}$ and $\mathbf{Z}(Y)_{(\mathcal{U})}^{a}$, respectively; they are regarded as sequences of elements in $\mathbf{Z}(\{0,1\}^{\mathbb{N}})_{(\mathcal{U})}$.

Lemma 11.34 *Let J be a saturated cut in $\mathfrak{N}_{(\mathcal{U})}$, and let $a \in J$.*

(i) *Suppose that $W \in \mathbb{O}_J^{\mathbb{N}}$, and that*

$$(W_b : b \le a) \in \mathbf{Z}(\{0,1\}^{\mathbb{N}})_{(\mathcal{U})}^{a}\,.$$

Then either there exists $W' \in \mathbb{O}_J^{\mathbb{N}}$ with $W' \subset W$ such that $W' \cap W_b = \emptyset \quad (b \le a)$ or there exists $W' \in \mathbb{O}_J^{\mathbb{N}}$ with $W' \subset W$ such that $W' \subset W_b$ for some $b \le a$.

(ii) *Suppose that $(f, W) \in \mathbb{O}_J$, and that*

$$(W_b : b \le a) \in \mathbf{Z}(Y)_{(\mathcal{U})}^{a}\,.$$

Then either there exists $(f', W') \in \mathbb{O}_J$ with $(f', W') \le (f, W)$ and such that $W' \cap W_b = \emptyset$ $(b \le a)$ or there exists $(f', W') \in \mathbb{O}_J$ with $(f', W') \le (f, W)$ and such that $W' \subset W_b$ for some $b \le a$.

Proof (i) This lemma can be expressed in the structure $\mathfrak{N}_J^{(\mathcal{U})}$ by coding all the relevant objects by elements of $\mathcal{P}(\mathbb{N})_{(\mathcal{U})}$.

In the degenerate case in which $J = \mathbb{N}$, necessarily $a \in \mathbb{N}$, and the sequence $(W_b : b \le a)$ is finite. By induction, we may reduce this case to that in which $a = 1$. In this special case, we are given $W_1 \in \mathbf{Z}(\{0,1\}^{\mathbb{N}})_{(\mathcal{U})}$, and we may suppose that $W_1 \subset W$. For $V \in \mathbf{Z}(\{0,1\}^{\mathbb{N}})_{(\mathcal{U})}$, define F_V to be the closure of $\{f \mid \mathbb{N} : f \in V\}$ in the compact space $\{0,1\}^{\mathbb{N}}$. Since $W \in \mathbb{O}_{\mathbb{N}}^{\mathbb{N}}$, F_W has non-empty interior in $\{0,1\}^{\mathbb{N}}$, and so either there exists $s \in \{0,1\}^{<\mathbb{N}}$ with $U_s \subset F_W$ and $U_s \subset F_{W_1}$ or there exists $s \in \{0,1\}^{<\mathbb{N}}$ with $U_s \subset F_W$ and $U_s \cap F_{W_1} = \emptyset$. In the former case, by an overspill argument, there exists $d \in \mathbb{N}_{(\mathcal{U})} \setminus \mathbb{N}$ with

$$\{f \restriction d : f \in W_1\} \supset \{f \in \{0,1\}_{(\mathcal{U})}^{d} : f \restriction \ell(s) = s\}\,.$$

Set $W' = \{f \in W_1 : f \restriction \ell(s) = s\}$, so that $W' \in \mathbb{O}_{\mathbb{N}}^{\mathbb{N}}$ and $W' \subset W_1$. In the latter case, set $W' = \{f \in W : f \restriction \ell(s) = s\}$, so that, again by an overspill argument, $W' \in \mathbb{O}_{\mathbb{N}}^{\mathbb{N}}$ and $W' \cap W_1 = \emptyset$.

To cover the general case for an arbitrary saturated cut J, we repeat the argument for the degenerate case $J = \mathbb{N}$ within the structure $\mathfrak{N}_J^{(\mathcal{U})}$.

(ii) Choose $h \in \mathbb{N}_{(\mathcal{U})}^{\mathbb{N}}$ and $f' \in (\mathbb{N} \times I)_{(\mathcal{U})}^{\mathbb{N}}$ such that h and f' are injective, $h(J) \subset J$, $f \mid J = (f' \circ h) \mid J$, and $\operatorname{supp} W_b \subset f'(\mathbb{N}_{(\mathcal{U})})$ for each $b \le a$. We have $(f', W) \le (f, W)$ in $\mathbb{O}_J$.

Set $W^* = \{g \circ f' : g \in W\}$ and, for each $b \leq a$, set $W_b^* = \{g \circ f' : g \in W_b\}$. Then $W^* \in \mathbb{O}_J^{\mathbb{N}}$ and

$$(W_b^* : b \leq a) \in \mathbf{Z}(\{0,1\}^{\mathbb{N}})_{(\mathcal{U})}^a .$$

By (i), either there exists $W^{**} \in \mathbb{O}_J^{\mathbb{N}}$ with $W^{**} \subset W^*$ such that

$$W^{**} \cap W_b^* = \emptyset \quad (b \leq a)$$

or there exists $W^{**} \in \mathbb{O}_J^{\mathbb{N}}$ with $W^{**} \subset W^*$ such that $W^{**} \subset W_b^*$ for some $b \leq a$.

Set $W' = \{g \in Y_{(\mathcal{U})} : g \circ f' \in W^{**}\}$. Then $(f', W') \in \mathbb{O}_J$, and has the required properties. □

Let J be an initial segment of $\mathbb{N}_{(\mathcal{U})}$ such that J is a saturated cut, and let $\mathcal{G}$ be a filter in $\mathbb{O}_J$. For each function $f : \mathbb{N}_{(\mathcal{U})} \to \mathbb{N}_{(\mathcal{U})} \times I_{(\mathcal{U})}$ with $(f, Y_{(\mathcal{U})}) \in \mathcal{G}$, we associate to $\mathcal{G}$ a filter $\mathcal{G}_f$ in $\mathbb{O}_J^{\mathbb{N}}$ by saying that $W \in \mathcal{G}_f$ if $(f, W') \in \mathcal{G}$, where

$$W' = \{h \in Y_{(\mathcal{U})} : h \circ f \in W\} .$$

Suppose further that $c \in \mathbb{N}_{(\mathcal{U})} \setminus J$. Then we define

$$\mathcal{G}_f^c = \{\{g \restriction c : g \in W\} : W \in \mathcal{G}_f\} ,$$

so that $\mathcal{G}_f^c$ is a filter in $\mathbb{O}_J^c$.

The following observations are useful. First, suppose that $(f_1, W_1) \leq (f, W)$ in $\mathbb{O}_J$, and set

$$W_1^* = \{h \in Y_{(\mathcal{U})} : h \mid f(\mathbb{N}_{(\mathcal{U})}) \text{ for some } g \in W_1\} .$$

Then $(f, W_1^*) \in \mathbb{O}_J$, and so $(f_1, W_1) \leq (f, W_1^*)$. The key point is that

$$\operatorname{supp} Z_\sigma \subset \sigma \cap \operatorname{supp} Z$$

for each $Z \in \mathbf{Z}(Y)$ and $\sigma \subset \mathbb{N} \times I$, where

$$Z_\sigma = \{h \in Y : n \mid \sigma = g \mid \sigma \text{ for some } g \in Z\} .$$

The claim that $(f, W_1^*) \in \mathbb{O}_J$ follows from the $\mathcal{U}$-version of this fact.

Second, suppose that $W \in \mathbb{O}_J^{\mathbb{N}}$. Then there exists $W' \subset W$ and $c \in \mathbb{N}_{(\mathcal{U})} \setminus J$ such that $W' \in \mathbb{O}_J^{\mathbb{N}}$ and the map

$$h \mapsto h \restriction c, \quad W' \to \{0,1\}_{(\mathcal{U})}^c ,$$

is an injection.

Definition 11.35 *Let J be a saturated cut in $\mathbb{N}_{(\mathcal{U})}$, and let $\mathcal{G}$ be a maximal filter in $\mathbb{O}_J$. Then $\mathcal{G}$ is J-*generic *if:*

(i) *for each sequence $(W_b : b \leq a)$ in $(\mathbf{Z}(Y))_{(\mathcal{U})}^{<\mathbb{N}}$ with $a \in J$, there exists $(f, W) \in \mathcal{G}$ such that either $W \cap W_b = \emptyset \quad (b \leq a)$ or $W \subset W_b$ for some $b \leq a$.*

(ii) *for each $A \in [\mathbb{N} \times I]_{(\mathcal{U})}^{\omega}$, there exist $(f, W) \in \mathcal{G}$ with $A \subset f(\mathbb{N}_{(\mathcal{U})})$ and $c \in \mathbb{N}_{(\mathcal{U})} \setminus J$ such that the map $h \mapsto (h \circ f) \restriction c$, $W \to \{0,1\}_{(\mathcal{U})}^c$, is injective.*

Definition 11.36 *Let J be a saturated cut in $\mathfrak{N}_{(\mathcal{U})}$, and let $c \in \mathbb{N}_{(\mathcal{U})} \setminus J$. Let $\mathcal{H}$ be a filter in $\mathbb{O}_J^c$. Then $\mathcal{H}$ is* J-generic *if, for each sequence*

$$(W_b : b \leq a) \quad \text{in} \quad \left(\bigcup \{\mathcal{P}(\{0,1\}^k) : k \in \mathbb{N}\}\right)_{(\mathcal{U})}^{<\mathbb{N}}$$

with $a \in J$, either there exists $W \in \mathcal{H}$ with $W \cap W_b = \emptyset \quad (b \leq a)$ or $W_b \in \mathcal{H}$ for some $b \leq a$.

We recall from Chapter 9 that there is a bijection $\mathcal{F} \mapsto \mathcal{F}_{\mathcal{U}}$ from the set of z-ultrafilters on X containing $\mathcal{F}_0$ onto the set of maximal $\mathbf{Z}(Y)_{(\mathcal{U})}$-filters on $Y_{(\mathcal{U})}$ (see page 231).

Suppose that $\mathcal{G}$ is a J-generic filter in $\mathbb{O}_J$. We define

$$\mathcal{F}_{\mathcal{G}} = \{Z \in \mathbf{Z}(X) : (f, Z_{\mathcal{U}}) \in \mathcal{G} \quad \text{for some} \quad f \in (\mathbb{N} \times I)_{(\mathcal{U})}^{\mathbb{N}}\}\,.$$

It follows from Lemma 11.34(i) that $\mathcal{F}_{\mathcal{G}}$ is a z-filter on X containing $\mathcal{F}_0$ and that $(\mathcal{F}_{\mathcal{G}})_{\mathcal{U}}$ is J-complete.

Lemma 11.37 *Let J be a saturated cut in $\mathfrak{N}_{(\mathcal{U})}$. Suppose that $\mathcal{G}$ is a J-generic filter in $\mathbb{O}_J$ and that $(f, W) \in \mathcal{G}$ and $c \in \mathbb{N}_{(\mathcal{U})} \setminus J$. Then $\mathcal{G}_f^c$ is a J-generic filter in $\mathbb{O}_J^c$.*

Proof Let $(W_b : b \leq a)$ be a sequence in $\left(\bigcup \{\mathcal{P}(\{0,1\}^k) : k \in \mathbb{N}\}\right)_{(\mathcal{U})}^{<\mathbb{N}}$ with $a \in J$. For each $b \leq a$, define W_b' in $\mathbf{Z}(Y)_{(\mathcal{U})}$ by

$$W_b' = \{h \in Y_{(\mathcal{U})} : (h \circ f) \restriction c \in W_b\}\,.$$

Then $\{W_b' : b \leq a\} \in \mathbf{Z}(Y)_{(\mathcal{U})}^a$. Since $(\mathcal{F}_{\mathcal{G}})_{\mathcal{U}}$ is J-complete, either there exists $W' \in (\mathcal{F}_{\mathcal{G}})_{\mathcal{U}}$ with $W' \cap W_b' = \emptyset \quad (b \leq a)$ or there exists $b \leq a$ with $W_b' \in (\mathcal{F}_{\mathcal{G}})_{\mathcal{U}}$. In the former case, set

$$W^* = \{(h \circ f) \restriction c : h \in W'\}\,.$$

It follows that $(f, W^*) \in \mathcal{G}$, and so $W^* \in \mathcal{G}_f^c$ and $W^* \cap W_b = \emptyset \quad (b \leq a)$. In the latter case, set

$$W^* = \{(h \circ f) \restriction c : h \in W_b'\}\,.$$

Then again $W^* \in \mathcal{G}_f^c$ with $W^* \subset W_b$, and so $W_b \in \mathcal{G}_f^c$. This proves that $\mathcal{G}_f^c$ is J-generic in $\mathbb{O}_J^c$. □

Each function $F \in C(\beta X)$ defines in a natural fashion an associated function $F_{\mathcal{U}} : Y_{(\mathcal{U})} \to K_{\mathcal{U}}$. The set of these functions is

$$C(Y)_{(\mathcal{U})}\,.$$

Let $H \in C(Y)_{(\mathcal{U})}$. Then the *support* of H, written $\operatorname{supp} H$, is the set of elements (c, x) in $\mathbb{N}_{(\mathcal{U})} \times I_{(\mathcal{U})}$ such that there exist $f, g \in Y_{(\mathcal{U})}$ with

$$\{(c, x)\} = \{(d, y) \in \mathbb{N}_{(\mathcal{U})} \times I_{(\mathcal{U})} : f(d, y) \neq g(d, y)\}$$

and $H(f) \neq H(g)$. We see by a pointwise argument that

$$\operatorname{supp} H = \bigcup \{\operatorname{supp} H^{-1}([r, s]_{(\mathcal{U})}) : r, s \in \mathbb{Q}_{(\mathcal{U})}, r \leq s\}\,.$$

As in Chapter 9, Fin $(\{0,1\}^{<\mathbb{N}}, \{0,1\}^{<\mathbb{N}})$ is a collection of finite partial functions, so that

$$\mathrm{Fin}\,(\{0,1\}^{<\mathbb{N}}, \{0,1\}^{<\mathbb{N}})_{(\mathcal{U})}$$

is a collection of partial functions from $\{0,1\}^{<\mathbb{N}}_{(\mathcal{U})}$ into itself, the domain of each such function belonging to $[\{0,1\}^{<\mathbb{N}}]^{<\omega}_{(\mathcal{U})}$; for $c \in \mathbb{N}_{(\mathcal{U})} \setminus J$, the set of these functions from $\{0,1\}^{c}_{(\mathcal{U})}$ into itself is denoted by

$$\mathrm{Fin}^{c}(\{0,1\}^{<\mathbb{N}}, \{0,1\}^{<\mathbb{N}})_{(\mathcal{U})}\,.$$

Again let J be a saturated cut in $\mathbb{N}_{(\mathcal{U})}$, and take $c \in \mathbb{N}_{(\mathcal{U})} \setminus J$. Let $\mathcal{H}$ be a J-generic filter in $\mathbb{O}^{c}_{J}$. For each $\rho \in \mathrm{Fin}^{c}(\{0,1\}^{<\mathbb{N}}, \{0,1\}^{<\mathbb{N}})_{(\mathcal{U})}$, we define a subset $A^{\mathcal{H}}_{\rho}$ of J as follows:

$$A^{\mathcal{H}}_{\rho} = \{a \in J : \text{there exists} \quad W \in \mathcal{G} \ \text{with} \ \rho(s)(a) = 1 \quad (s \in W)\}\,.$$

The family of these sets $A^{\mathcal{H}}_{\rho}$ is $\mathcal{P}(J)_{\mathcal{H}}$.

Let $\mathcal{G}$ be a J-generic filter in $\mathbb{O}_J$. Then $\mathcal{F}_{\mathcal{G}}$ defines a prime ideal, say Q, in $C(\beta X)$, and, by Theorem 8.36(ii), there is a unique prime ideal P in $C(\beta X)$ above Q such that $\mathbb{N}_{(P)} = J$, where we are using the fact that, by the definition, $(\mathcal{F}_{\mathcal{G}})_{\mathcal{U}}$ is J-complete. We analyse the family $\mathcal{P}(\mathbb{N})_{(P)}$ for this prime ideal P.

Lemma 11.38 *Let J be a saturated cut in $\mathfrak{N}_{(\mathcal{U})}$, and let $\mathcal{G}$ be a J-generic filter in $\mathbb{O}_J$.*

(i) *Suppose that $A, B \in \mathcal{P}(\mathbb{N})_{(P)}$. Then there exist an injective function $f : \mathbb{N}_{(\mathcal{U})} \to \mathbb{N}_{(\mathcal{U})} \times I_{(\mathcal{U})}$ such that $(f, Y_{(\mathcal{U})}) \in \mathcal{G}$ and $c \in \mathbb{N}_{(\mathcal{U})} \setminus J$ such that $A, B \in \mathcal{P}(J)_{\mathcal{H}}$, where $\mathcal{H} = \mathcal{G}^{c}_{f}$.*

(ii) *Suppose that $f : \mathbb{N}_{(\mathcal{U})} \to \mathbb{N}_{(\mathcal{U})} \times I_{(\mathcal{U})}$ is an injective function for which $(f, Y_{(\mathcal{U})}) \in \mathcal{G}$, and that $c \in \mathbb{N}_{(\mathcal{U})} \setminus J$. Then $\mathcal{P}(J)_{\mathcal{H}} \subset \mathcal{P}(\mathbb{N})_{(P)}$, where $\mathcal{H} = \mathcal{G}^{c}_{f}$.*

Proof (i) We may suppose that $A, B \notin \mathcal{P}(\mathbb{N})_{(\mathfrak{N}_J)}$, and hence that there exist $x, y \in \mathbb{I}_{(P)} \setminus \mathbb{D}_{(P)}$ with

$$A = \{a \in \mathbb{N}_{(P)} : \varepsilon^{(P)}_{x}(a) = 1\}\,, \quad B = \{a \in \mathbb{N}_{(P)} : \varepsilon^{(P)}_{y}(a) = 1\}\,.$$

Take $F, G \in C(\beta X)$ with $\pi_P(F) = x$ and $\pi_P(G) = y$. Since $\mathcal{G}$ is J-generic, there exist $W \in (\mathcal{F}_{\mathcal{G}})_{\mathcal{U}}$ and $f : \mathbb{N}_{(\mathcal{U})} \to \mathbb{N}_{(\mathcal{U})} \times I_{(\mathcal{U})}$ such that $(f, W) \in \mathcal{G}$ and

$$f(\mathbb{N}_{(\mathcal{U})}) \supset (\mathrm{supp}\, F_{\mathcal{U}}) \cup (\mathrm{supp}\, G_{\mathcal{U}})\,,$$

and also $c \in \mathbb{N}_{(\mathcal{U})} \setminus J$ with $\mathrm{supp}\, W_{(f,c)} \in J$ and such that the map

$$h \mapsto h \restriction c, \quad W_f \to \{0,1\}^{c}_{(\mathcal{U})}\,,$$

is an injection.

Define a function $\rho : \{0,1\}^{c}_{(\mathcal{U})} \to \{0,1\}^{c}_{(\mathcal{U})}$ as follows. Take $h \in \{0,1\}^{c}_{(\mathcal{U})}$. Suppose that there exists $g \in W$ with $(g \circ f) \restriction c = h$ (g is uniquely specified on $\mathrm{supp}\, F_{\mathcal{U}}$). Then we set $z = F_{\mathcal{U}}(g)$, and

$$\rho(h) = \begin{cases} \varepsilon_z^{(\mathcal{U})} \restriction c & \text{if} \quad z \notin \mathbb{D}_{(\mathcal{U})}\,, \\ \varepsilon_z^{+(\mathcal{U})} \restriction c & \text{if} \quad z \in \mathbb{D}_{(\mathcal{U})}\,. \end{cases} \tag{11.17}$$

If there is no such $g \in W$, we define $\rho(h)(a) = 0 \quad (a \leq c)$. It follows that ρ is well defined and that $\rho \in \operatorname{Fin}(\{0,1\}^{<\mathbb{N}}, \{0,1\}^{<\mathbb{N}})_{(\mathcal{U})}$. We have $A = A_\rho^{\mathcal{H}}$, and so $A \in \mathcal{P}(J)_{\mathcal{H}}$, where $\mathcal{H} = \mathcal{G}_f^c$. Similarly we see that $B \in \mathcal{P}(J)_{\mathcal{H}}$.

(ii) Let $A \in \mathcal{P}(J)_{\mathcal{H}}$. We shall show that $A \in \mathcal{P}(\mathbb{N})_{(P)}$. The result is trivial if $A \in \mathcal{P}(J)_{(\mathcal{U})}$, and so we may suppose that $A \notin \mathcal{P}(J)_{(\mathcal{U})}$.

Take $\rho : \{0,1\}^c_{(\mathcal{U})} \to \{0,1\}^c_{(\mathcal{U})}$ with $A = A_\rho^{\mathcal{H}}$. It is easy to construct pointwise a function $F \in C(\beta X)$ such that, for each $g \in Y_{(\mathcal{U})}$, we have $F_{\mathcal{U}}(g) \in \mathbb{I}_{(\mathcal{U})} \setminus \mathbb{D}_{(\mathcal{U})}$ and $\varepsilon_x^{(\mathcal{U})} \restriction c = \rho((g \circ f) \restriction c)$, where $x = F_{\mathcal{U}}(g)$. The prime ideal in $C(\beta X)$ corresponding to $\mathcal{F}_{\mathcal{G}}$ is Q. We have $x = \pi_Q(F) \in \mathbb{I}_{(Q)} \setminus \mathbb{D}_{(Q)}$, and $B \cap J = A$, where

$$B = \{a \in \mathbb{N}_{(Q)} : \varepsilon_x^{(Q)}(a) = 1\}\,.$$

Since Q is a strongly convex prime in $C(\beta X)$, it follows from Lemma 9.7(ii) that $A \in \mathcal{P}(\mathbb{N})_{(P)}$, as required. □

We are now going to refer to subsets of $\{0,1\}^{\mathbb{N}}_{(P_J)}$ as 'elements of $\mathfrak{N}_{(P_J)}$'; this is an abuse of notation, to be understood in the sense that the subset is definable in the structure $\mathfrak{N}_{(P_J)}$ (which, we recall, is a model of analysis). We shall also refer to such subsets of $\{0,1\}^{\mathbb{N}}_{(P_J)}$ as *meagre* or *comeagre* if they have these properties within $\mathfrak{N}_{(P_J)}$; these notions are not the same as those that arise from the order-topology on $\{0,1\}^{\mathbb{N}}_{(P_J)}$. In a similar way we shall refer to functions $\tau : \{0,1\}^{\mathbb{N}}_{(P_J)} \to \{0,1\}^{\mathbb{N}}_{(P_J)}$ as being 'defined in $\mathfrak{N}_{(P_J)}$'. The functions τ are 'continuous on a comeagre set' if they are continuous on these sets with respect to topologies definable in the structure $\mathfrak{N}_{(P_J)}$.

Suppose that $s \in \{0,1\}^{<\mathbb{N}}_{(P_J)}$. Then

$$U_s^{(P_J)} = \{f \in \{0,1\}^{\mathbb{N}}_{(P_J)} : f \restriction \ell(s) = s\}$$

is a basic open set in the order-topology on $\{0,1\}^{\mathbb{N}}_{(P_J)}$.

Now suppose that $c \in \mathbb{N}_{(\mathcal{U})} \setminus J$, that $W \in \mathbb{O}_J^c$, and that

$$\tau : \{0,1\}^{\mathbb{N}}_{(P_J)} \to \{0,1\}^{\mathbb{N}}_{(P_J)}$$

is continuous on a comeagre subset of $\{0,1\}^{\mathbb{N}}_{(P_J)}$ in the above sense, and let $\rho \in \operatorname{Fin}^c(\{0,1\}^{<\mathbb{N}}, \{0,1\}^{<\mathbb{N}})_{(\mathcal{U})}$. Then

$$\rho \sim \tau \quad \text{with respect to } c \text{ and } W\,,$$

if, for each $W' \subset W$ with $W' \in \mathbb{O}_J^c$ and for each $q \in \{0,1\}^{<\mathbb{N}}_{(P_J)}$ such that $\rho(f) \restriction \ell(q) = q \quad (f \in W')$, there exists $p \in \{0,1\}^{<\mathbb{N}}_{(P_J)}$ such that $\tau(g) \restriction \ell(q) = q$ for each g in a comeagre subset of $U_p^{(P_J)}$ and such that $h \mid \ell(p) = p$ for some $h \in W'$. In some of the following statements we imply that ρ and τ are as above whenever we write '$\rho \sim \tau$ with respect to c and W'.

We now come to the key technical lemma.

Lemma 11.39 *Let J be a saturated cut in $\mathfrak{N}_{(\mathcal{U})}$, and let $c \in \mathbb{N}_{(\mathcal{U})} \setminus J$ and $W \in \mathbb{O}^c_J$.*

(i) *For each τ, there exists ρ such that $\rho \sim \tau$ with respect to c and W.*

(ii) *For each ρ, there exist τ and $W' \subset W$ with $W' \in \mathbb{O}^c_J$ such that $\rho \sim \tau$ with respect to c and W'.*

Proof We first prove the result in the special case where $J = \mathbb{N}$; we shall work in $\mathfrak{M}_{\mathbb{N}}$.

(i) We are given a function $\tau : \{0,1\}^{\mathbb{N}} \to \{0,1\}^{\mathbb{N}}$ which is continuous on a comeagre set (in the usual sense) in $\{0,1\}^{\mathbb{N}}$. For each $s \in \{0,1\}^{<\mathbb{N}}$, denote by U_s the open and closed set in $\{0,1\}^{\mathbb{N}}$ specified by s, and let N_τ be the set of pairs (s,t) with $s,t \in \{0,1\}^{<\mathbb{N}}$ such that

$$\{\alpha \in U_s : \tau(\alpha) \in U_t\} \text{ is comeagre in } U_s\,.$$

The set N_τ is countable, and is enumerated as $((s_k,t_k) : k \in \mathbb{N})$, say. Since $\mathbb{N}$ is a saturated cut in $\mathfrak{N}_{(\mathcal{U})}$, it follows that there is a sequence $((s_a,t_a) : a \leq c)$ in $(\{0,1\}^{<\mathbb{N}} \times \{0,1\}^{<\mathbb{N}})^{<\mathbb{N}}_{(\mathcal{U})}$ extending the given sequence; we may suppose that, for each $a \leq c$, $s_a \in \{0,1\}^{b_1}_{(\mathcal{U})}$ and $t_a \in \{0,1\}^{b_2}_{(\mathcal{U})}$ for some $b_1, b_2 < c$.

We now work in $\mathfrak{N}_{(\mathcal{U})}$ to define a function $\rho : \{0,1\}^c_{(\mathcal{U})} \to \{0,1\}^c_{(\mathcal{U})}$ as follows. Fix $f \in \{0,1\}^c_{(\mathcal{U})}$. In the case where there is no $a \leq c$ such that $f \mid \ell(s_a) = s_a$, set $\rho(f)(a) = 0 \quad (a \leq c)$. Otherwise, let $(a_d : d \leq e)$ be the sequence of maximum length such that a_1 is the least element a with $a \leq c$ such that $f \mid \ell(s_a) = s_a$ and, for each d, a_{d+1} is the least element a with $a \leq c$ such that $f \mid \ell(s_a) = s_a$ and t_a is a proper extension of t_{a_d}. We define $\rho(f)$ to be a sequence in $\{0,1\}^c_{(\mathcal{U})}$ extending t_{a_e}. Then it is easily checked that $\rho \sim \tau$ with respect to c and W.

(ii) Recall first that $\{0,1\}^{<\mathbb{N}}$ is a partially ordered set with respect to extension; this is the partial order $\mathbb{P}_1$ for adding one Cohen real.

By shrinking W, if necessary, we may suppose that there exist $p \in \{0,1\}^{<\mathbb{N}}$ and $d \in \mathbb{N}_{(\mathcal{U})} \setminus \mathbb{N}$ such that

$$\{s \restriction d : s \in W\} = \{s \in \{0,1\}^d_{(\mathcal{U})} : s \restriction \ell(p) = p\}\,.$$

We next define by induction a decreasing sequence $(W_k : k \in \mathbb{Z}^+)$ in $\mathbb{O}^c_{\mathbb{N}}$ and a decreasing sequence (d_k) in $\mathbb{N}_{(\mathcal{U})} \setminus \mathbb{N}$ such that

$$\{s \restriction d_k : s \in W_k\} = \{s \in \{0,1\}^{d_k}_{(\mathcal{U})} : s \restriction \ell(p) = p\}\,. \tag{11.18}$$

We shall also define certain sequences $(p_{i,k} : i \in \mathbb{N})$ and $(\varepsilon_{i,k} : i \in \mathbb{N})$. Indeed, set $W_0 = W$ and $d_0 = d$. Now assume that W_k and d_k have been defined. Let $(p_{i,k} : i \in \mathbb{N})$ be a maximal antichain below p in $\{0,1\}^{<\mathbb{N}}$, and let $(\varepsilon_{i,k} : i \in \mathbb{N})$ be a sequence of elements in $\{0,1\}^k$ such that, for each $i \in \mathbb{N}$, there exists $W^* \subset W_k$ with $W^* \in \mathbb{O}^c_{\mathbb{N}}$ and with

$$\rho(f) \restriction k = \varepsilon_{i,k} \quad (f \in W^*)\,,$$

and further there exists $e \in \mathbb{N}_{(\mathcal{U})} \setminus \mathbb{N}$ with

$$\{f \restriction e : f \in W^*\} = \{f \in \{0,1\}^e_{(\mathcal{U})} : f \restriction \ell(p_{i,k}) = p_{i,k}\}.$$

Let $(p_{a,k} : a \leq c) \in (\{0,1\}^{<\mathbb{N}})^{<\mathbb{N}}_{(\mathcal{U})}$ be a sequence extending $(p_{i,k} : i \in \mathbb{N})$, and let $(\varepsilon_{a,k} : a \leq c) \in (\{0,1\}^{<\mathbb{N}})^{<\mathbb{N}}_{(\mathcal{U})}$ be a sequence extending $(\varepsilon_{i,k} : i \in \mathbb{N})$. For each $a \in \mathbb{N}_{(\mathcal{U})}$ with $a \leq c$, define

$$V_a = \bigcap \{\{f \in W : \rho(f) \restriction k = \varepsilon_{b,k} \text{ whenever } f \restriction \ell(p_{b,k}) = p_{b,k}\} : b \leq a\}.$$

For each $i \in \mathbb{N}$, we see that $V_{i+1} \subset V_i \subset W$ and that there exists $e_i \in \mathbb{N}_{(\mathcal{U})} \setminus \mathbb{N}$ with

$$\{f \restriction e_i : f \in V_i\} = \{f \in \{0,1\}^{e_i}_{(\mathcal{U})} : f \restriction \ell(p) = p\}.$$

This shows that $V_i \in \mathbb{O}^c_{\mathbb{N}}$ $(i \in \mathbb{N})$. By overspill and saturation, there must exist $e \in \mathbb{N}_{(\mathcal{U})} \setminus \mathbb{N}$ such that, for each $b \leq e$, we have $V_e \subset V_b$ and

$$\{f \restriction e : f \in V_b\} = \{f \in \{0,1\}^e_{(\mathcal{U})} : f \restriction \ell(p) = p\}.$$

We set $W_{k+1} = V_e$ and $d_{k+1} = e$ to continue the inductive definition. Thus we have sequences satisfying (11.17) for all $k \in \mathbb{Z}^+$.

Now suppose that $p_{i,k}$ and $p_{j,n}$ are compatible elements in $\{0,1\}^{<\mathbb{N}}$. Then $\varepsilon_{i,k}$ and $\varepsilon_{j,n}$ are also compatible in $\{0,1\}^{<\mathbb{N}}$, and so we may define a function $\tau : \{0,1\}^{\mathbb{N}} \to \{0,1\}^{\mathbb{N}}$ such that, whenever $i, k \in \mathbb{N}$ and $r \in \{0,1\}^{\mathbb{N}}$ with $r \mid \ell(p_{i,k}) = p_{i,k}$, necessarily

$$\tau(r) \mid \ell(\varepsilon_{i,k}) = \varepsilon_{i,k}.$$

By saturation, there is a sequence $(W_b : b \leq c)$ in

$$\left(\bigcup \{\mathcal{P}(\{0,1\}^k) : k \in \mathbb{N}\}\right)^{<\mathbb{N}}_{(\mathcal{U})}$$

consisting of subsets of $\{0,1\}^c_{(\mathcal{U})}$ which extends the sequence $(W_k : k \in \mathbb{Z}^+)$. By overspill and saturation, there exists $d \in \mathbb{N}_{(\mathcal{U})} \setminus \mathbb{N}$ such that, for each $b \leq d$, we have $W_d \subset W_b$ and

$$\{f \restriction d : f \in W_b\} = \{f \in \{0,1\}^d_{(\mathcal{U})} : f \restriction \ell(p) = p\}.$$

We take $W' = W_d$. Then $\rho \sim \tau$ with respect to c and W', as required.

In summary, we have now proved the result in the special case where $J = \mathbb{N}$. The general case follows by repeating the whole argument within the structure $\mathfrak{N}^{(\mathcal{U})}_J$. □

Definition 11.40 *Let $\mathfrak{M}$ be a model of analysis. A function $f : \mathbb{N}_{(\mathfrak{M})} \to \{0,1\}$ is $\mathfrak{M}$-generic if $f \restriction a \in \{0,1\}^a_{(\mathfrak{M})}$ for each $a \in \mathbb{N}_{(\mathfrak{M})}$ and if $\{f \restriction a : a \in \mathbb{N}_{(\mathfrak{M})}\}$ is an $\mathfrak{M}$-generic filter in $\mathbb{P}_1^{(\mathfrak{M})}$.*

Definition 11.41 *Let J be a saturated cut in $\mathfrak{N}_{(\mathcal{U})}$, let $c \in \mathbb{N}_{(\mathcal{U})} \setminus J$, and let $\mathcal{H}$ be a J-generic filter in $\mathbb{O}^c_J$.*

(i) *Let* $\rho \in \mathrm{Fin}^c(\{0,1\}^{<\mathbb{N}}, \{0,1\}^{<\mathbb{N}})_{(\mathcal{U})}$. *Then the* interpretation $\mathcal{I}(\rho, \mathcal{H})$ of ρ by $\mathcal{H}$ *is a function* $\varphi : J \to \{0,1\}$ *such that*

$$\varphi(a) = 1 \quad \text{if} \quad \{g \in \{0,1\}^c_{(\mathcal{U})} : \rho(g)(a) = 1\} \in \mathcal{H},$$
$$\varphi(a) = 0 \quad \text{if} \quad \{g \in \{0,1\}^c_{(\mathcal{U})} : \rho(g)(a) = 0\} \in \mathcal{H}.$$

(ii) *The function* $f_{\mathcal{H}}$ *is the interpretation* $\mathcal{I}(\rho, \mathcal{H})$ *in the special case where* ρ *is the identity function.*

(iii) *The filter* $\mathcal{H}$ *is* weakly $\mathfrak{N}^{(\mathcal{U})}_J$-generic *if* $f_{\mathcal{H}}$ *is* $\mathfrak{N}_{(P_J)}$*-generic.*

Let ρ be as in (i), above. Then it is easy to see that

$$\mathcal{I}(\rho, \mathcal{H}) \restriction a \in \{0,1\}^a_{(\mathcal{U})}$$

for each $a \in J$.

Definition 11.42 *Let* J *be a saturated cut in* $\mathfrak{N}_{(\mathcal{U})}$, *and let*

$$\tau : \{0,1\}^{\mathbb{N}}_{(P_J)} \to \{0,1\}^{\mathbb{N}}_{(P_J)}$$

be continuous on a comeagre subset. Suppose that $f : J \to \{0,1\}$ *is* $\mathfrak{N}_{(P_J)}$*-generic. Then the* interpretation $\mathcal{I}(\tau, f)$ of τ by f *is a function* $\varphi : J \to \{0,1\}$ *such that* $\varphi(a) = 1$ *if there exists* $b \in J$ *such that* $\{g \in U^{(\mathfrak{N})}_{f \restriction b} : \tau(g)(a) = 1\}$ *is comeagre in* $U^{(\mathfrak{N})}_{f\restriction b}$, *where* $\mathfrak{N} = \mathfrak{N}_{(P_J)}$, *and* $\varphi(a) = 0$ *otherwise.*

Since f is $\mathfrak{N}_{(P_J)}$-generic, it follows that $\varphi(a) = 0$ if and only if there exists $b \in J$ such that $\{g \in U^{(\mathfrak{N})}_{f\restriction b} : \tau(g)(a) = 0\}$ is comeagre in $U^{(\mathfrak{N})}_{f\restriction b}$. The following lemma follows easily from the above definitions.

Lemma 11.43 *Let* J *be a saturated cut in* $\mathfrak{N}_{(\mathcal{U})}$, *and let* $c \in \mathbb{N}_{(\mathcal{U})} \setminus J$. *Suppose that* $\mathcal{H}$ *is a weakly* $\mathfrak{N}^{(\mathcal{U})}_J$*-generic filter, that* $W \in \mathcal{H}$, *and that* $\rho \sim \tau$ *with respect to* c *and* W. *Then* $\mathcal{I}(\tau, f_{\mathcal{H}}) = \mathcal{I}(\rho, \mathcal{H})$. □

Definition 11.44 *Let* J *be a saturated cut in* $\mathbb{N}_{(\mathcal{U})}$, *and let* $\mathcal{G}$ *be a* J*-generic filter in* $\mathbb{O}_J$. *Then* $\mathcal{G}$ *is* $\mathfrak{N}^{(\mathcal{U})}_J$-generic *if:*

(i) *for each* $(f, W) \in \mathcal{G}$ *and* $c \in \mathbb{N}_{(\mathcal{U})} \setminus J$, $\mathcal{G}^c_f$ *is a weakly* $\mathfrak{N}^{(\mathcal{U})}_J$*-generic filter in* $\mathbb{O}^c_J$;

(ii) *for each* $(f, W) \in \mathcal{G}$, $c \in \mathbb{N}_{(\mathcal{U})} \setminus J$, *and* $\rho \in \mathrm{Fin}^c(\{0,1\}^{<\mathbb{N}}, \{0,1\}^{<\mathbb{N}})_{(\mathcal{U})}$, *there exist* $(g, V) \in \mathcal{G}$, ρ', *and* τ' *such that* $(g, V) \le (f, W)$, *such that*

$$\rho'(h) \restriction c = \rho((h \circ \Phi) \restriction c),$$

and such that $\rho' \sim \tau'$ *with respect to* $\Phi(c)$ *and* $V_{(g,\Phi(c))}$, *where* $\Phi : \mathbb{N}_{(\mathcal{U})} \to \mathbb{N}_{(\mathcal{U})}$ *is the unique function with* $f = g \circ \Phi$.

We are now in a position to give an analogue of Theorem 9.36.

Theorem 11.45 (GCH) *Let* J *be a saturated cut in* $\mathbb{N}_{(\mathcal{U})}$ *such that* $\operatorname{cof} J = \aleph_1$, *let* $\mathcal{G}$ *be an* $\mathfrak{N}^{(\mathcal{U})}_J$*-generic filter in* $\mathbb{O}_J$, *and let* $\mathcal{F}$ *be the* z*-ultrafilter on* X *given* *by $\mathcal{G}$. Then the unique prime ideal P in $C(\beta X)$ above $Q_{\mathcal{F}}$ such that $\mathbb{N}_{(P)} = J$ has the following properties:*

(i) $|\mathbb{N}_{(P)}| = |\Gamma_P| = w(K_P) = \aleph_1$;

(ii) $|K_P| = \aleph_2$;

(iii) *P is a cut prime, a valuation prime, and an exponential prime;*

(iv) *K_P is an η_1-field;*

(v) *$\mathfrak{N}_{(P)}$ is a model of analysis;*

(vi) *$\mathfrak{N}_{(P)}$ is a Cohen extension of $\mathfrak{N}_{(P_J)}$.*

Suppose further that $\mathrm{Th}\langle J, +, \cdot\rangle \neq \mathrm{Th}\langle \mathbb{N}, +, \cdot\rangle$ *and that $\mathfrak{N}_{(P)}$ is solid. Then:*

(vii) *for each prime z-ideal Q in any algebra $C(\Omega)$, $(K_Q, \leq)$ is not order-isomorphic to $(K_P, \leq)$.*

Proof Since $\mathcal{F}_{\mathcal{U}}$ is J-complete, the existence of the prime ideal P is given by Proposition 9.34.

Property (i) is immediate since $\mathbb{N}_{(P)} = J$, and property (ii) follows because $\mathcal{G}$ is $\mathfrak{N}_J^{(\mathcal{U})}$-generic. For property (iii), it follows from Lemma 9.34 that P is a cut prime and a valuation prime; since J is a saturated cut, it follows from Lemma 11.32 that P is an exponential prime. Property (iv) follows from Lemma 8.33(v) and Theorem 4.28(ii).

We prove (v) and (vi). Set $\mathfrak{N} = \mathfrak{N}_{(P_J)}$, a model of analysis. We shall apply Lemma 11.29 with $\mathcal{S} = \mathcal{P}(\mathbb{N})_{(P)}$. Let $A, B \in \mathcal{S}$. By Lemma 11.38(i), there exists $(f, Y_{(\mathcal{U})}) \in \mathcal{G}$ and $c \in \mathbb{N}_{(\mathcal{U})} \setminus J$ with $A, B \in \mathcal{P}(J)_{\mathcal{H}}$, where $\mathcal{H} = \mathcal{G}_f^c$. Since $\mathcal{G}$ is $\mathfrak{N}_J^{(\mathcal{U})}$-generic, it follows that, by reducing (f, W), we may suppose that there exist τ_A, τ_B, ρ_A, and ρ_B such that $\tau_A \sim \rho_A$ and $\tau_B \sim \rho_B$ with respect to c and W and such that $A = A_{\mathcal{H}}^{\rho_A}$ and $B = A_{\mathcal{H}}^{\rho_B}$. Let χ_A and χ_B be the characteristic functions of A and B, respectively, so that $\chi_A, \chi_B : J \to \{0,1\}$. Since $\mathcal{G}$ is $\mathfrak{N}_J^{(\mathcal{U})}$-generic, $f_{\mathcal{H}}$ is $\mathfrak{N}$-generic. Let $\mathcal{G}_0$ be the filter in $\mathbb{P}_1^{(\mathfrak{N})}$ given by $f_{\mathcal{H}}$. Then $A, B \in \mathcal{P}(\mathbb{N})_{(\mathfrak{N}[\mathcal{G}_0])}$.

Now take $C \in \mathcal{P}(\mathbb{N})_{(\mathfrak{N}[\mathcal{G}_0])}$. There exists $\tau : \{0,1\}^{\mathbb{N}}_{(P_J)} \to \{0,1\}^{\mathbb{N}}_{(P_J)}$ continuous on a comeagre subset such that $\chi_C = \mathcal{I}(\tau, f_{\mathcal{H}})$. By Lemma 11.39(i), there exists ρ such that $\rho \sim \tau$ with respect to c and W. Since $\mathcal{G}_f^c$ is weakly $\mathfrak{N}_J^{(\mathcal{U})}$-generic, $\mathcal{I}(\tau, f_{\mathcal{H}}) = \mathcal{I}(\rho, \mathcal{H})$, and so $C \in \mathcal{P}(J)_{\mathcal{H}}$. By Lemma 11.38(ii), $C \in \mathcal{P}(\mathbb{N})_{(P)}$.

Thus the conditions of Lemma 11.29 are satisfied, and (v) and (vi) follow.

(vii) Now suppose that the additional conditions are satisfied. By Theorem 11.6, K_P is rigid.

Assume towards a contradiction that Q is a prime z-ideal in an algebra $C(\Omega)$ such that $(K_Q, \leq) \cong (K_P, \leq)$. Then K_Q is an η_1-field and $w(K_Q) = \aleph_1$. By Proposition 11.27, $\mathrm{Th}\langle \mathbb{N}_{(Q)}, +, \cdot\rangle = \mathrm{Th}\langle \mathbb{N}, +, \cdot\rangle$. Since K_P is rigid, it is immediate that $\mathfrak{N}_{(Q)} \cong \mathfrak{N}_{(P)}$, and so

$$\mathrm{Th}\langle J, +, \cdot\rangle = \mathrm{Th}\langle \mathbb{N}, +, \cdot\rangle,$$

a contradiction of the hypothesis. Thus (vii) follows. $\square$

To apply Theorem 11.45, we must find a filter $\mathcal{G}$ such that the corresponding structure $\mathfrak{N}_{(P)}$ is solid. Again in this case, we can only give a consistency result, established by a forcing argument: we force with the partial order $\mathbb{O}_J$. For this, we shall require in addition that $\operatorname{coi}(\mathbb{N}_{(\mathcal{U})} \setminus J) \geq \aleph_1$, so that, by Lemma 11.30(i), the partially ordered set $\mathbb{O}_J$ is ω-closed.

Definition 11.46 *Let* $\mathfrak{M} = \langle \mathbb{N}_{(\mathfrak{M})}, +, \cdot, \mathcal{P}(\mathbb{N})_{(\mathfrak{M})} \rangle$ *be a model of analysis, and let* $\mathcal{S}$ *be a subset of* $\mathcal{P}(\mathbb{N})_{(\mathfrak{M})}$.

(i) *A filter* $\mathcal{G}$ *in* $\mathbb{P}_1^{(\mathfrak{M})}$ *is* $(\mathfrak{M}, \mathcal{S})$-generic *if* $\mathcal{G} \cap D = \emptyset$ *for each order-dense set* D *in* $\mathbb{P}_1^{(\mathfrak{M})}$ *such that* D *is definable from parameters in the structure*

$$\langle \mathbb{N}_{(\mathfrak{M})}, +, \cdot, \mathcal{P}(\mathbb{N})_{(\mathfrak{M})}, \mathcal{S} \rangle .$$

(ii) *An extension* $\mathfrak{N}$ *of* $\mathfrak{M}$ *is an* $(\mathfrak{M}, \mathcal{S})$-generic Cohen *extension of* $\mathfrak{M}$ *if* $\mathfrak{N}$ *is a Cohen extension and if, for each* $A \in \mathcal{P}(\mathbb{N})_{(\mathfrak{N})}$, *there is an* $(\mathfrak{M}, \mathcal{S})$*-generic filter* $\mathcal{G}$ *in* $\mathbb{P}_1^{(\mathfrak{M})}$ *such that* $A \in \mathcal{P}(\mathbb{N})_{(\mathfrak{M}[\mathcal{G}])}$ *and* $\mathcal{P}(\mathbb{N})_{(\mathfrak{M}[\mathcal{G}])} \subset \mathcal{P}(\mathbb{N})_{(\mathfrak{N})}$.

Suppose that $\mathfrak{M}$ is a model of analysis and that $n \in \mathbb{N}$. Then elements of $\mathbb{D}^n_{(\mathfrak{M})} \times \mathbb{I}_{(\mathfrak{M})}$ are naturally coded by $\mathcal{P}(\mathbb{N})_{(\mathfrak{M})}$, and so we may regard a function $F : \mathbb{D}^n_{(\mathfrak{M})} \to \mathbb{I}_{(\mathfrak{M})}$ as a subset of $\mathcal{P}(\mathbb{N})_{(\mathfrak{M})}$. Thus there is a natural sense in which a filter in $\mathbb{P}_1^{(\mathfrak{M})}$ is $(\mathfrak{M}, F)$-generic and an extension of $\mathfrak{M}$ is an $(\mathfrak{M}, F)$-generic Cohen extension. The following is a straightforward exercise using terms.

Lemma 11.47 *Let* $\mathfrak{M}$ *be a model of analysis, let* $n \in \mathbb{N}$, *and take a function* $F : \mathbb{D}^n_{(\mathfrak{M})} \to \mathbb{I}_{(\mathfrak{M})}$. *Suppose that* $\mathfrak{M}^*$ *is an* $(\mathfrak{M}, F)$*-generic Cohen extension of* $\mathfrak{M}$, *and that* $F^* : \mathbb{I}^n_{(\mathfrak{M}^*)} \to I_{(\mathfrak{M}^*)}$ *is a continuous function with* $F = F^* \mid \mathbb{D}^n_{(\mathfrak{M}^*)}$. *Then, for each non-empty, open subset* V_1 *of* $\mathbb{D}^n_{(\mathfrak{M})}$, *there exist* $G \in {}^*C(\mathbb{I}^n, \mathbb{I})_{(\mathfrak{M})}$ *and a non-empty, open subset* V_2 *of* V_1 *with* $G \mid V_2 = F$. □

Theorem 11.48 (GCH) *Let* J *be a saturated cut in* $\mathfrak{N}_{(\mathcal{U})}$ *such that* $\operatorname{cof} J = \aleph_1$, *let* $\mathcal{G}$ *be an* $\mathfrak{N}_J^{(\mathcal{U})}$*-generic filter* $\mathbb{O}_J$, *let* $\mathcal{F}$ *be the z-ultrafilter on* X *given by* $\mathcal{G}$, *and let* P *be the unique prime ideal* P *in* $C(\beta X)$ *above* $Q_{\mathcal{F}}$ *such that* $\mathbb{N}_P = J$. *Suppose that, for each set* $\mathcal{S} \subset \mathcal{P}(\mathbb{N})_{(P)}$ *with* $|\mathcal{S}| = \aleph_1$, *there is a subset* $\mathcal{P}(\mathbb{N})_{\mathcal{S}}$ *of* $\mathcal{P}(\mathbb{N})_{(P)}$ *such that*

$$\mathfrak{M}_{\mathcal{S}} = \langle \mathbb{N}_{(P)}, +, \cdot, \mathcal{P}(\mathbb{N})_{\mathcal{S}} \rangle$$

is a model of analysis, $\mathcal{S} \subset \mathcal{P}(\mathbb{N})_{\mathcal{S}}$, *and* $\mathfrak{N}_{(P)}$ *is an* $(\mathfrak{M}_{\mathcal{S}}, \mathcal{S})$*-generic Cohen extension of* $\mathfrak{M}_{\mathcal{S}}$. *Then* $\mathfrak{N}_{(P)}$ *is solid.*

Proof Let $n \in \mathbb{N}$, let $F : \mathbb{I}^n_{(P)} \to \mathbb{I}_{(P)}$ be a continuous function, and take $\mathcal{S}$ to be the subset of $\mathcal{P}(\mathbb{N})_{(P)}$ which codes $F \mid \mathbb{D}^n_{(P)}$. Since $\mathbb{D}_{(P)} = \mathbb{D}_{(P_J)}$, we have $\mid \mathbb{D}^n_{(P)} \mid = \aleph_1$, and so we may suppose that $|\mathcal{S}| = \aleph_1$. By hypothesis, there is a subset $\mathcal{P}(\mathbb{N})_{\mathcal{S}}$ of $\mathcal{P}(\mathbb{N})_{(P)}$ such that $\mathfrak{M}_{\mathcal{S}}$ is a model of analysis, $\mathcal{S} \subset \mathcal{P}(\mathbb{N})_{\mathcal{S}}$, and $\mathfrak{N}_{(P)}$ is an $(\mathfrak{M}_{\mathcal{S}}, \mathcal{S})$-generic Cohen extension of $\mathfrak{M}_{\mathcal{S}}$. We have $\mathbb{D}_{(\mathfrak{M}_{\mathcal{S}})} = \mathbb{D}_{(P)}$. Since $\mathfrak{N}_{(P)}$ is an $(\mathfrak{M}_{\mathcal{S}}, F \mid \mathbb{D}^n_{(P)})$-generic Cohen extension of $\mathfrak{M}_{\mathcal{S}}$, it follows from Lemma 11.47 (with $\mathfrak{M}^* = \mathfrak{N}_{(P)}$) that, for each non-empty, open subset V_1 of

$\mathbb{D}^n_{(P)}$, there exists $G \in {}^*C(\mathbb{I}^n, \mathbb{I})_{(\mathfrak{M}_S)}$ and a non-empty, open subset V_2 of V_1 with $G \mid V_2 = F$.

We *claim* that, for each function $H \in {}^*C(\mathbb{I}^n, \mathbb{I})_{(\mathfrak{M}_S)}$, there is a unique function $\widetilde{H} \in {}^*C(\mathbb{I}^n, \mathbb{I})_{(P)}$ such that $\widetilde{H} \mid \mathbb{I}^n_{(\mathfrak{M}_S)} = H$. This is a standard consequence of the definitions and the fact that the structure $\mathfrak{N}_{(P)}$ is a Cohen extension of $\mathfrak{M}_S$. Thus, with V_1, V_2, and G as above, $\widetilde{G} \in {}^*C(\mathbb{I}^n, \mathbb{I})_{(P)}$ and $\widetilde{G} \mid V_2 = F$.

It follows that $\mathfrak{N}_{(P)}$ is solid. □

We now assume that GCH holds, and seek to prove that all of the hypotheses of Theorem 11.45 can be satisfied in a generic extension of **V**.

Indeed, let us start with a saturated cut J in $\mathfrak{N}_{(\mathcal{U})}$ such that

$$\operatorname{cof} J = \operatorname{coi}(\mathbb{N}_{(\mathcal{U})} \setminus J) = \aleph_1 \quad \text{and} \quad \operatorname{Th}\langle J, +, \cdot\rangle \neq \operatorname{Th}\langle \mathbb{N}_{(\mathcal{U})}, +, \cdot\rangle\,;$$

such a cut J exists by Theorem 11.33. We fix the index set I that appears in the definition of X to have cardinality $\aleph_2$.

Our method for finding a generic extension of **V** in which the hypotheses of Theorem 11.45 are satisfied is similar to that used in the proof of Theorem 9.60 to find a suitable filter in what was then called $\mathbb{P}^I_J$. Let $\mathcal{G}$ be a **V**-generic filter in $\mathbb{O}_J$. We shall, in effect, show that $\mathcal{G}$ itself is the filter that we require. Since GCH holds in **V** and $|I| = \aleph_2$ in **V**, it follows by Lemma 11.30 that cardinals are preserved in the passage from **V** to $\mathbf{V}[\mathcal{G}]$. Since $|\mathbb{O}_J| = \aleph_2$, it again follows from Lemma 11.30 that GCH holds in $\mathbf{V}[\mathcal{G}]$.

By Lemma 11.30(i), the partial order $\mathbb{O}_J$ is ω-closed, and so the ultrafilter $\mathcal{U}$ is still an ultrafilter in $\mathbf{V}[\mathcal{G}]$. Further, the structure $\mathfrak{N}^{(\mathcal{U})}_J$ is the same, whether computed in **V** or $\mathbf{V}[\mathcal{G}]$. Let $X_\mathcal{G}$ and $Y_\mathcal{G}$ be the topological spaces X and Y, respectively, as computed in $\mathbf{V}[\mathcal{G}]$. These spaces are different from their alter egos X and Y, respectively, but the spaces $C(X_\mathcal{G})$ and $C(Y_\mathcal{G})$ are naturally isomorphic to their counterparts $C(X)$ and $C(Y)$, respectively, calculated in **V**, again because $\mathbb{O}_J$ is ω-closed. To exhibit the isomorphism between $C(X)$ and $C(X_\mathcal{G})$, note that X is a dense subspace of $X_\mathcal{G}$ and that, since $\mathbb{O}_J$ is ω-closed,

$$C(X) = \{f \mid X : f \in C(X_\mathcal{G})\}\,.$$

Similarly, $C(\beta\mathbb{N})$ as computed in $\mathbf{V}[\mathcal{G}]$ is isomorphic to $C(\beta\mathbb{N})$ as computed in **V**, and so $\mathfrak{N}_{(P_J)}$ is the same, whether it be computed in **V** or $\mathbf{V}[\mathcal{G}]$.

Let $\mathbb{O}^{(\mathcal{G})}_J$ be the partial order $\mathbb{O}_J$ as computed in $\mathbf{V}[\mathcal{G}]$. Again since $\mathbb{O}_J$ is ω-closed, $\mathbb{O}_J$ and $\mathbb{O}^{(\mathcal{G})}_J$ are isomorphic in $\mathbf{V}[\mathcal{G}]$. Indeed, suppose that $(f, W) \in \mathbb{O}_J$, and take $Z \in \mathbf{Z}(X)$ to be such that $W = Z_\mathcal{U}$. The closure of Z as computed in $X_\mathcal{G}$ is denoted by $\overline{Z}$, and we set $\widetilde{W} = (\overline{Z})_\mathcal{U}$, as computed in $\mathbf{V}[\mathcal{G}]$. Then $(f, \widetilde{W}) \in \mathbb{O}^{(\mathcal{G})}_J$, and we have an order-isomorphism

$$(f, W) \mapsto (f, \widetilde{W}), \quad \mathbb{O}_J \to \mathbb{O}^{(\mathcal{G})}_J\,.$$

It follows that $\mathcal{G}$ defines a maximal filter $\widetilde{\mathcal{G}}$ in $\mathbb{O}^{(\mathcal{G})}_J$. Since $\mathcal{G}$ is **V**-generic, it follows from Lemma 11.34(ii) that (in $\mathbf{V}[\mathcal{G}]$), the filter $\widetilde{\mathcal{G}}$ is J-generic. We *claim* that $\widetilde{\mathcal{G}}$ is $\mathfrak{N}_J^{(\mathcal{U})}$-generic. Indeed, clause (i) of Definition 11.44 is satisfied because $\mathcal{G}$ is $\mathbf{V}$-generic; clause (ii) also follows from the $\mathbf{V}$-genericity of $\mathcal{G}$ because Lemma 11.39(ii) provides (in $\mathbf{V}$) the necessary dense subsets of $\mathbb{O}_J$.

We now work in $\mathbf{V}[\mathcal{G}]$. Let $\widetilde{Q}$ be the prime ideal in $C(\beta X_{\mathcal{G}})$ given by $\widetilde{\mathcal{G}}$, and let $\widetilde{P}$ be the unique prime ideal in $C(\beta X_{\mathcal{G}})$ above $\widetilde{Q}$ such that $\mathbb{N}_{(\widetilde{P})} = J$. Also, let

$$\mathfrak{M}^{\mathcal{G}} = \langle J, +, \cdot, \mathcal{P}(\mathbb{N})^{\mathcal{G}} \rangle$$

be the resulting model of analysis which is a Cohen extension of $\mathfrak{N}_{(P_J)}$ (noting that $\mathfrak{N}_{(P_J)}$ is the same in both $\mathbf{V}$ and $\mathbf{V}[\mathcal{G}]$).

Take $\{I_1, I_2\}$ to be a partition of I, where $|I_1| = \aleph_1$ in $\mathbf{V}$. Let Y_1 and Y_2 be the spaces $\{0,1\}^{\mathbb{N}\times I_1}$ and $\{0,1\}^{\mathbb{N}\times I_2}$, respectively, computed in $\mathbf{V}$, and let $\mathbb{O}_J(I_1)$ be the partial order analogous to $\mathbb{O}_J$, with I_1 for I in the definition. Finally, let

$$\mathcal{G}_1 = \{(f, W) \in \mathbb{O}_J(I_1) : (f, W') \in \mathcal{G}\},$$

where, for $W \in \mathbf{Z}(Y_1)_{(\mathcal{U})}$, we set $W' = \{h \in Y_{(\mathcal{U})} : h \mid (Y_1)_{(\mathcal{U})} \in W\}$. Since $|I_1| = \aleph_1$, it follows by a homogeneity argument related to that required for Lemma 11.30(ii) that $\mathcal{G}_1$ is a $\mathbf{V}$-generic filter in $\mathbb{O}_J(I_1)$. (We caution that, for this to be true, it is necessary that I_1 be uncountable.)

Let

$$\mathfrak{M}^{\mathcal{G}_1} = \langle J, +, \cdot, \mathcal{P}(\mathbb{N})^{\mathcal{G}_1} \rangle$$

be the model of analysis computed in $\mathbf{V}[\mathcal{G}_1]$ from $\mathcal{G}_1$ in the same way as $\mathfrak{M}^{\mathcal{G}}$ is computed in $\mathbf{V}[\mathcal{G}]$. We clearly have $\mathcal{P}(\mathbb{N})^{\mathcal{G}_1} \subset \mathcal{P}(\mathbb{N})^{\mathcal{G}}$. Suppose that $\mathcal{S}$ is a subset of $\mathcal{P}(\mathbb{N})^{\mathcal{G}_1}$ and that $\mathcal{S} \in \mathbf{V}[\mathcal{G}_1]$. We make the *key claim* that $\mathfrak{M}^{\mathcal{G}}$ is an $(\mathfrak{M}^{\mathcal{G}_1}, \mathcal{S})$-generic Cohen extension of $\mathfrak{M}^{\mathcal{G}_1}$. Indeed, suppose that $A \subset J$ and that $A \in \mathcal{P}(\mathbb{N})^{\mathcal{G}}$, and take σ to be a term for χ_A in the forcing language associated to $\mathbb{O}_J$. It follows from Lemmas 11.38(i) and 11.39(ii) by using the $\mathbf{V}$-genericity of $\mathcal{G}$ that there exist $\tau : \{0,1\}^{\mathbb{N}}_{(P_J)} \to \{0,1\}^{\mathbb{N}}_{(P_J)}$ such that τ is continuous on a comeagre set, and such that

$$(f, W) \Vdash \text{`}\sigma = \mathcal{I}(\tau, f_{\mathcal{H}}), \quad \text{where} \quad \mathcal{H} = \mathcal{G}_f^c\text{'}.$$

The claim now follows in a routine fashion.

Finally we show that, in $\mathbf{V}[\mathcal{G}]$, the $\mathfrak{M}_J$-generic filter $\widetilde{\mathcal{G}}$ in $\mathbb{O}_J^{(\mathcal{G})}$ satisfies the hypotheses of Theorem 11.48. Indeed, suppose that $\mathcal{S} \subset \mathcal{P}(\mathbb{N})^{\mathcal{G}}$ is a set with $|\mathcal{S}| = \aleph_1$ in $\mathbf{V}[\mathcal{G}]$. By Lemma 11.30(ii), there exists $I_1 \subset I$ with $I_1 \in \mathbf{V}$ such that $|I_1| = \aleph_1$ and $\mathcal{S} \in \mathbf{V}[\mathcal{G}_1]$. Thus $\mathfrak{M}^{\mathcal{G}}$ is an $(\mathfrak{M}^{\mathcal{G}_1}, \mathcal{S})$-generic Cohen extension of $\mathfrak{M}^{\mathcal{G}_1}$. Thus $\widetilde{\mathcal{G}}$ satisfies the hypotheses of Theorem 11.48, and so $\mathfrak{M}^{\mathcal{G}}$ is solid.

We have shown that $\widetilde{\mathcal{G}}$ and J satisfy all the hypotheses of Theorem 11.45 in $\mathbf{V}[\mathcal{G}]$, and so $\mathbf{V}[\mathcal{G}]$ is the desired forcing extension of $\mathbf{V}$.

We conclude this book with the relative consistency result that provides our strongest result on the separation of classes in our classification scheme.

Theorem 11.49 *It is relatively consistent with the theory* ZFC + GCH *that there is a prime ideal P in an algebra $C(\Omega)$ such that P is an exponential prime,*

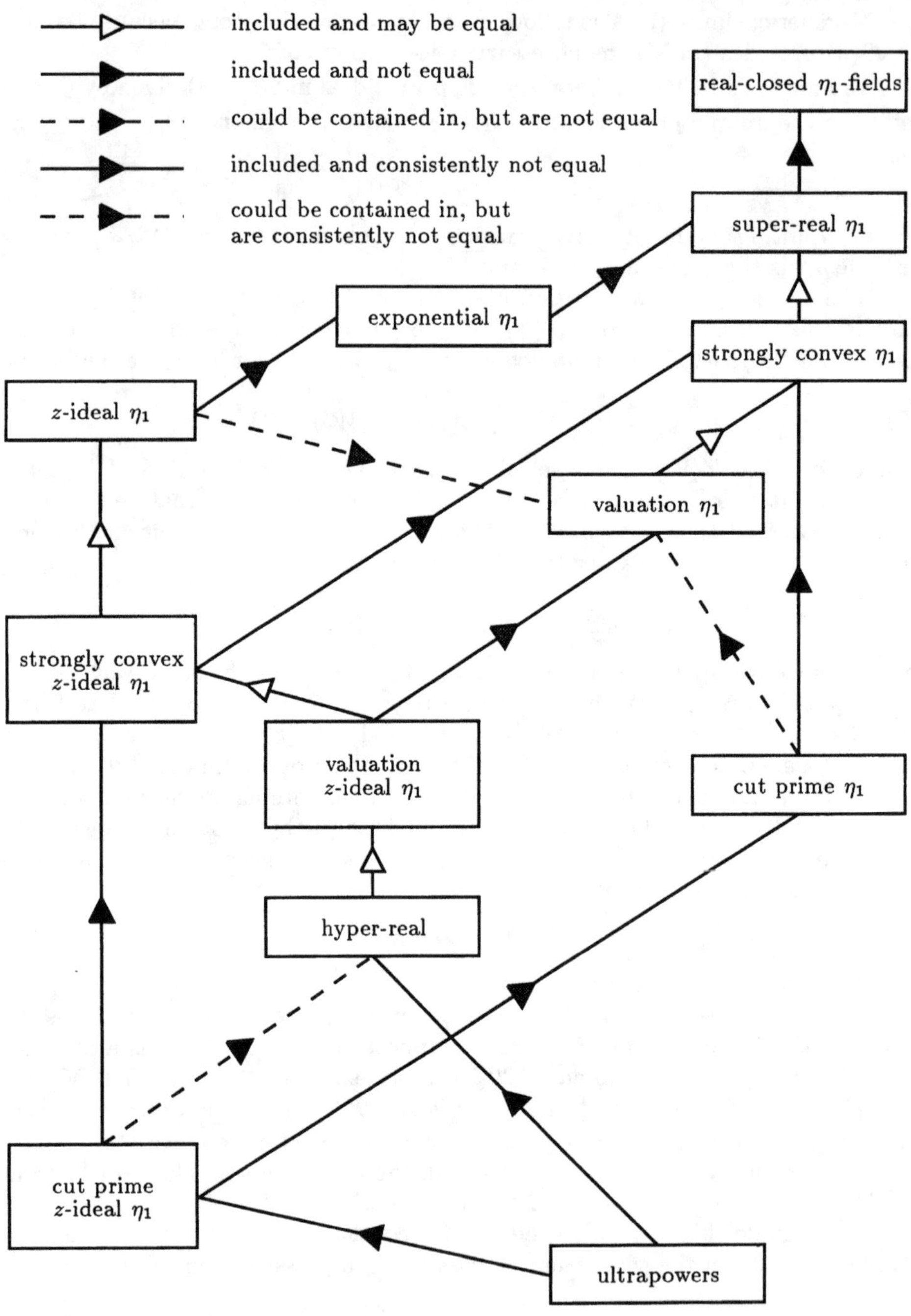

Fig. 11.5. The final relationships between the classes

a valuation prime, and a cut prime and such that the super-real η_1-field K_P has weight $\aleph_1$ and cardinality $\aleph_2$, but which is such that $(K_P, \leq)$ is not order-isomorphic to any super-real field of the form K_Q for which Q is a non-maximal, prime z-ideal. □

We summarize our results (in the theory ZFC+GCH) with the final version of our diagram, given on the previous page. The diagram contains most, but not all, of the information that we have acquired. The diagram also reflects the situation if we restrict consideration to the simplest possible non-trivial case, that in which the fields have weight $\aleph_1$, save that we cannot separate **exponential η_1** from **super-real** or **cut prime z-ideal η_1** from **ultrapowers** in this special case.

12
Open questions

In this final chapter, we collect some open questions that have arisen and have not been resolved in the course of our work.

1. Let G be a totally ordered group which is a β_1-set. Is G necessarily a β_1-group?

   Thus, we suppose that the totally ordered group G can be written as $G = \bigcup\{S_\nu : \nu < \omega_1\}$, where each S_ν is an α_1-subset of G, and we ask if $G = \bigcup\{G_\nu : \nu < \omega_1\}$, where each G_ν is a subgroup which is an α_1-set. It is not sufficient just to take G_ν to be the subgroup generated by S_ν because G_ν need not be an α_1-set.

   In the theory ZFC + CH, a β_1-set has cardinality at most $\aleph_1$, and so the result is trivially positive in this case.

2. Let K be a field (with ground field $\mathbb{R}$). We say that K is a β_1-*rational field* if $K = \bigcup\{R_\nu : \nu < \rho\}$, where $\{R_\nu : \nu < \rho\}$ is a chain of rational subfields, each of which is an α_1-set. We have the following immediate implications:

$$\begin{aligned} K \text{ is a } \beta_1\text{-field} &\Rightarrow K \text{ is a } \beta_1\text{-rational field} \\ &\Rightarrow K \text{ is a } \beta_1\text{-group} \\ &\Rightarrow K \text{ is a } \beta_1\text{-set}. \end{aligned}$$

   Can any of these implications be reversed?

   Again, we know that the four statements are trivially equivalent in the theory ZFC + CH. A related question is to ask whether a real-closed field or a super-real field which is order-isomorphic to $\mathbf{R}$ is necessarily isomorphic to $\mathbf{R}$.

3. Let K be an ordered field which is isomorphic as a ring to a super-real field. Is K already a super-real field?

   We raised two conjectures about the normability of $K^\#$ for certain ordered fields K in Chapter 5. In Theorems 5.13 and 5.15, we showed that these general conjectures would be resolved if they could be resolved for two specific examples. We have confessed that we have found the following question to be intractable.

4. (i) Is the algebra $\widehat{\mathbf{R}}^\#$ normable (with CH) ?

(ii) Is the algebra $\widetilde{\mathbf{R}}^{\#}$ normable (with CH) ?

If it should transpire that either the algebra $\widehat{\mathbf{R}}^{\#}$ or the algebra $\widetilde{\mathbf{R}}^{\#}$ is not normable, then we would have to return to our original questions of Chapter 5; we restate these questions here.

5. (i) For which ordered fields K is the algebra $K^{\#}$ normable ?

   (ii) For which non-maximal, prime ideals P in algebras $C(\Omega)$ are the algebras A_P and/or $K_P^{\#}$ normable ?

The remaining remarks summarize the relation between the various classes of real-closed η_1-fields that we introduced in Chapter 4.

By Theorem 4.27, every super-real field is real-closed, and, by Theorem 6.23, $\widehat{\mathbf{R}}$ is a real-closed η_1-field which is not isomorphic as a field to any super-real field. Thus we have

$$\textbf{super-real } \eta_1 \quad \subsetneq \quad \textbf{real-closed } \eta_1\textbf{-fields}.$$

Indeed, by Theorem 8.63, there is a real-closed η_1-field which is not order-isomorphic to any super-real field, and so the above two classes are not equal in the category of ordered sets.

By Theorem 6.30, $(\widehat{\mathbf{R}}, +)$ is not isomorphic to $(K, +)$ for any field K in the class **exponential** η_1.

6. Is $\widehat{\mathbf{R}}$ isomorphic as an ordered group or order-isomorphic to any super-real field ?

   Note that, by Corollary 6.26, in the theory ZFC + MA + $\neg$CH, $\widehat{\mathbf{R}}$ is not order-isomorphic to any super-real field, and so, effectively, the above question is asked in the theory ZFC + CH.

   It was proved in Proposition 9.10 that (with CH) there is a Cauchy complete field K in the class **cut prime** η_1 with $\operatorname{cof} K = \aleph_1$.

7. Is there a Cauchy complete field K in the class **cut prime** z**-ideal** η_1 with $\operatorname{cof} K = \aleph_1$?

   The constructions involving $\widetilde{\mathbf{R}}$ in Chapter 9 enabled us to separate several classes of our scheme, but a number of related questions remain open. The example $\widetilde{\mathbf{R}}$ shows that (with GCH):

$$\begin{aligned} \textbf{cut prime } \eta_1 \quad &\subsetneq \quad \textbf{strongly convex } \eta_1\,; \\ \textbf{cut-prime } z\textbf{-ideal } \eta_1 \quad &\subsetneq \quad \textbf{strongly convex } z\textbf{-ideal } \eta_1\,; \\ \textbf{hyper-real} \quad &\not\subset \quad \textbf{cut-prime } z\textbf{-ideal } \eta_1\,; \\ \textbf{valuation } \eta_1 \quad &\not\subset \quad \textbf{cut prime } \eta_1\,. \end{aligned}$$

In each case, the inequalities in the above relations are proved in the category of totally ordered groups.

8. Is it the case that every member of the right-hand class of these four relations is order-isomorphic to a member of the left-hand class? In particular, is $\widetilde{\mathbf{R}}$ order-isomorphic to any member of **cut prime** $\boldsymbol{\eta_1}$? Is $\widetilde{\mathbf{R}}$ order-isomorphic to an ultrapower?

9. (i) Is it true that the class **cut prime** $\boldsymbol{z}$**-ideal** $\boldsymbol{\eta_1}$ is contained in **hyper-real** ?

   (ii) Is it true that the class **cut prime** $\boldsymbol{\eta_1}$ is contained in **valuation** $\boldsymbol{\eta_1}$?

   The example $\widetilde{\mathbf{R}}$ shows that each of the two possible containments in the above question are not equalities.

   We proved in Theorem 9.61 that it is relatively consistent with the theory ZFC + GCH that there is super-real η_1-field K_P which is not in the class **exponential** $\boldsymbol{\eta_1}$, and indeed in this result P is a cut prime and a valuation prime.

10. (i) Can it be proved in ZFC (or ZFC + CH) that

$$\textbf{exponential } \boldsymbol{\eta_1} \neq \textbf{super-real } \boldsymbol{\eta_1}?$$

   (ii) Can it be proved in ZFC + GCH that every super-real η_1-field of weight $\aleph_1$ belongs to **exponential** $\boldsymbol{\eta_1}$?

11. Is $\widetilde{\mathbf{R}}$ rigid?

   We finally come to what we regard as the most interesting open question and the one that we feel has the best chance of a positive answer. First recall from a subset of our main classification diagram, Fig. 11.5. on page 338, that we have:

$$\begin{aligned} \textbf{hyper-real} &\subset \textbf{valuation } \boldsymbol{z}\textbf{-ideal } \boldsymbol{\eta_1}\,; \\ \textbf{valuation } \boldsymbol{z}\textbf{-ideal } \boldsymbol{\eta_1} &\subset \textbf{strongly convex } \boldsymbol{z}\textbf{-ideal } \boldsymbol{\eta_1}\,; \\ \textbf{strongly convex } \boldsymbol{z}\textbf{-ideal } \boldsymbol{\eta_1} &\subset \textbf{z-ideal } \boldsymbol{\eta_1}\,. \end{aligned}$$

12. Which, if any, of the above inclusions are equalities (in the theory ZFC + GCH)? We conjecture that some, perhaps all, of the inclusions are equalities in the category of ordered fields, at least if attention be restricted to fields of weight $\aleph_1$.

References

Alling, N.L. (1962*a*). On the existence of real-closed fields that are η_α-sets of power $\aleph_\alpha$. *Trans. Amer. Math. Soc.*, **103**, 341–52.

Alling, N.L. (1962*b*). On exponentially closed fields. *Proc. Amer. Math. Soc.*, **13**, 706–11.

Alling, N.L. (1987). *Foundations of analysis over surreal number fields.* North-Holland, Amsterdam.

Antonovskij, M.Y., Chudnovsky, D.V., Chudnovsky, G.V., and Hewitt, E. (1981). Rings of real-valued continuous functions II. *Math. Zeit.*, **182**, 151–86.

Bade, W.G. and Curtis, P.C., Jr. (1960). Homomorphisms of commutative Banach algebras. *Amer. J. Math.*, **82**, 589–608.

Banaschewski, B. (1957). Über die Vervollständigung geordneter Gruppen. *Math. Nachrichten*, **16**, 51–71.

Benedetti, R. and Risler, J.-J. (1990). *Real algebraic and semi-algebraic sets.* Actualités Mathématiques, Hermann, Paris.

Bonsall, F.F. and Duncan, J. (1973). *Complete normed algebras.* Springer-Verlag, Berlin.

Bourbaki, N. (1985). *Algèbre commutative.* Masson, Paris.

Brumfiel, G.W. (1979). *Partially ordered rings and semi-algebraic geometry.* London Mathematical Society Lecture Notes, **37**, Cambridge University Press.

Chang, C.C. and Keisler, H.J. (1990). *Model theory*, 3rd edn. (1st edition, 1973). North-Holland, Amsterdam.

Cherlin, G.L. and Dickmann, M.A. (1986). Real closed rings I. Residue rings of rings of continuous functions. *Fund. Math.*, **126**, 147–83.

Cohen, L.W. and Goffman, C. (1949). The topology of ordered abelian groups. *Trans. Amer. Math. Soc.*, **67**, 310–19.

Comfort, W.W. and Negrepontis, S. (1974). *The theory of ultrafilters.* Springer-Verlag, Berlin.

Conrad, P. and Dauns, J. (1969). An embedding theorem for lattice-ordered groups. *Pacific J. Maths.*, **30**, 385–98.

Conway, J.H. (1976). *On numbers and games.* London Mathematical Society Monographs, **6**, Academic Press, London.

Cutland, N. (Ed.) (1988). *Nonstandard analysis and its applications.* London Mathematical Society Student Texts, **10**, Cambridge University Press.

Dales, H.G. (1978). Automatic continuity : a survey. *Bull. London Math. Soc.*, **110**, 129–83.

Dales, H.G. (1979). A discontinuous homomorphism from $C(X)$. *Amer. J. Math.*, **101**, 647–734.

Dales, H.G. (1989). On norms on algebras. *Proceedings of the Centre for Mathematical Analysis, Australian National University, Canberra*, **21**, 61–96.

Dales, H.G. (1997) *Banach algebras and automatic continuity.* London Mathematical Society Monographs, Oxford University Press, to appear.

Dales, H.G. and Woodin, W.H. (1987). *An introduction to independence for analysts.* London Mathematical Society Lecture Note Series, **115**, Cambridge University Press.

Delfosse, J.-P. (1975). Caractérisations d'anneaux de fonctions continues. *Ann. Soc. Sci. Bruxelles, Sér. I*, **89**, 364–68.

Dow, A. (1984). On ultrapowers of Boolean algebras. *Topology Proc.*, **9**, 269-91.

Dries, L. van den, Macintyre, A., and Marker, D. (1994). The elementary theory of restricted analytic fields with exponentiation, *Annals of Maths.*, **140**, 183–205.

Erdős, P., Gillman, L. and Henriksen, M. (1955). An isomorphism theorem for real-closed fields. *Annals of Maths.*, **61**, 542–54.

Esterle, J. (1977). Solution d'un problème d'Erdős, Gillman et Henriksen et application à l'étude des homomorphismes de $C(K)$. *Acta Math. (Hungarica)*, **30**, 113–127.

Esterle, J. (1978*a*). Semi-normes sur $C(K)$. *Proc. London Math. Soc.* (3), **36**, 27–45.

Esterle, J. (1978*b*). Injection de semi-groupes divisibles dans des algèbres de convolution et construction d'homomorphismes discontinus de $C(K)$. *Proc. London Math. Soc.* (3), **36**, 59–85.

Esterle, J. (1979). Homomorphismes discontinus des algèbres de Banach commutatives séparables. *Studia Math.*, **56**, 119–41.

Esterle, J. (1984). Unpublished lectures, California Institute of Technology.

Foreman, M., Magidor, M., and Shelah, S. (1988). Martin's Maximum, saturated ideals and non-regular ultrafilters. Part II. *Annals of Maths.*, **127**, 521–45.

Frankiewicz, R. and Zbierski, P. (1994). *Hausdorff gaps and limits.* Studies in Logic and the Foundations of Mathematics, **132**, North-Holland, Amsterdam.

Fremlin, D.H. (1984). *Consequences of Martin's Axiom.* Cambridge Tracts in Mathematics, **84**, Cambridge University Press.

Fuchs, L. (1963). *Partially ordered algebraic systems.* Pergamon Press, Oxford.

Gamelin, T.W. (1969). *Uniform algebras.* Prentice Hall, Englewood Cliffs, New Jersey.

Gillman, L. and Jerison, M. (1960). *Rings of continuous functions.* Van Nostrand, New York.

Hahn, H. (1907). Über die nichtarchimedischen Grössensysteme. *S.B. Akad. Wiss. Wien IIa*, **116**, 601–55.

Hausdorff, F. (1908). Grundzüge einer Theorie der geordneten Mengen. *Math. Ann.*, **158**, 435–505.

Hausdorff, F. (1914). *Grundzüge der Mengenlehre.* Verlag von Veit, Leipzig.

Holland, C. (1963). Extensions of ordered groups and sequence completion. *Trans. Amer. Math. Soc.*, **107**, 71–82.

Hungerford, T.W. (1980). *Algebra.* Springer-Verlag, New York.

Hurd, A.E. and Loeb, P.A. (1985). *An introduction to non-standard real analysis.* Academic Press, Orlando, Florida.

Jacobson, N. (1980). *Basic algebra II.* W.H. Freeman, San Francisco.

Jech, T. (1978). *Set Theory.* Academic Press, New York.

Jech, T. (1987). *Multiple forcing.* Cambridge Tracts in Mathematics, **88**, Cambridge University Press.

Johnson, B.E. (1976). Norming $C(\Omega)$ and related algebras. *Trans. Amer. Math. Soc.*, **220**, 37–58.

Kaplansky, I. (1942). Maximal fields with valuations. *Duke Math. J.*, **9**, 303–21.

Kaplansky, I. (1949). Normed algebras. *Duke Math. J.*, **16**, 399–418.

Keisler, H.J. and Schmerl, J.H. (1991). Making the hyperreal line both saturated and complete. *J. Symbolic Logic*, **56**, 1016–25.

Kendig, K. (1977). *Elementary algebraic geometry.* Graduate Texts in Mathematics, Springer-Verlag, New York.

Kohls, C.W. (1958). Prime ideals in rings of continuous functions. *Illinois J. Math.*, **2**, 505–36.

Kreisel, G. and Krivine, J.L. (1967). *Elements of mathematical logic.* North-Holland, London.

Krull, W. (1932). Allgemeine Bewertungstheorie, *J. Reine Angew. Math.*, **167**, 160–92.

Kuhlmann, F.-V. (1995). *Valuation theory of fields, abelian groups and modules.* Gordon and Breach, New York.

Kuhlmann, F.-V., and Kuhlmann, S. (1994). On the structure of nonarchimedean exponential fields II. *Communications in Algebra*, **22**, 5079–103.

Kuhlmann, F.-V., Pank, M., and Roquette, P. (1986). Immediate and purely wild extensions of valued fields. *Manuscripta Math.*, **55**, 39–67.

Kuhlmann, S. (1996). On the structure of nonarchimedean exponential fields I. Preprint.

Kunen, K. (1980). *Set theory. An introduction to independence proofs.* North-Holland, Amsterdam.

Lam, T.Y. (1980). The theory of ordered fields, in *Ring theory and algebra III*, Lecture Notes in Pure and Applied Mathematics, **55**, (ed. B.R. McDonald), M. Dekker, New York.

Lam, T.Y. (1991). *A first course in noncommutative rings.* Graduate Texts in Mathematics, Springer-Verlag, New York.

Lang, S. (1958). *Algebraic geometry.* Interscience, New York.

Laver, R. (1982). Saturated ideals and non-regular ultrafilters. *Patras Logic Symposium*, North-Holland, Amsterdam, 297–305.

Mac Lane, S. (1939). The universality of formal power series fields. *Bull. Amer. Math. Soc.*, **45**, 888–90.

Massaza, C. (1969–70). Sul completamento dei campi ordinati. *Rend. Sem. Math. Univ. e Politec. Torino*, **29**, 329–48.

Moloney, J.J. (1990). Residue class domains of the ring of convergent sequences and of $C^\infty([0,1],\mathbf{R})$. *Pacific J. Maths*, **143**, 79–153.

Priess-Crampe, S. (1983). *Angeordnete Strukturen: Gruppen, Körper, projektive Ebenen.* Springer-Verlag, Berlin.

Ribenboim, P. (1965). On the existence of totally ordered abelian groups which are η_α-sets. *Bull. Acad. Polon. Sci. Sér. Sci. Math. Astro. Phys.*, **13**, 545–48.

Rickart, C.E. (1960). *General theory of Banach algebras.* Van Nostrand, Princeton, New Jersey.

Robinson, A. (1966). *Non-standard analysis.* North-Holland, Amsterdam.

Rosenstein, J.G. (1982). *Linear orderings.* Academic Press, New York.

Rudin, W. (1991). *Functional analysis.* 2nd edn. McGraw-Hill, New York.

Schilling, O.F.G. (1950). *The theory of valuations.* Mathematical Surveys, American Mathematical Society, Providence, Rhode Island.

Scott, D. (1969). On completing ordered fields, in *Applications of model theory to algebra, analysis and probability*, ed. W.A.J. Luxemburg, 274–78, Holt, Rinehart and Winston, New York.

Shelah, S. (1982). *Proper forcing.* Lecture Notes in Mathematics, **940**, Springer-Verlag, Berlin.

Sinclair, A.M. (1975). Homomorphisms from $C_0(\mathbb{R})$. *J. London Math. Soc.* (2), **11**, 165–74.

Stout, E.L. (1971). *Uniform algebras.* Bogden and Quigley, Tarrytown-on-Hudson, New York.

Stroyan, K.D. and Luxemburg, W.A.J. (1976). *Introduction to the theory of infinitesimals.* Academic Press, New York.

Walker, R.C. (1974). *The Stone-Čech compactification.* Springer-Verlag, Berlin.

Whaples, G. (1957). Galois cohomology of additive polynomials and n-th power mappings of fields. *Duke Math. J.*, **24**, 143–50.

Wilkie, A. (1996). Model completeness results for expansions of the ordered field of real numbers by restricted Pfaffian functions and the exponential function. *Journal Amer. Math. Soc.*, to appear.

Woodin, W.H. (1993). A discontinuous homomorphism from $C(X)$ without CH. *J. London Math. Soc.* (2), **48**, 299–315.

Zaharov, V.K. (1988). On functions connected with sequential absolute, Cantor completion and classical rings of quotients. *Periodica Mathematica Hungarica*, **19**, 113–33.

Zariski, O. and Samuel, P. (1960). *Commutative algebra, II.* Van Nostrand, Princeton, New Jersey.

Index of notation

Index

The manufacturer's authorised representative in the EU for product safety is Oxford University Press España S.A. of el Parque Empresarial San Fernando de Henares, Avenida de Castilla, 2 – 28830 Madrid (www.oup.es/en or product.safety@oup.com). OUP España S.A. also acts as importer into Spain of products made by the manufacturer.

www.ingramcontent.com/pod-product-compliance
Lightning Source LLC
LaVergne TN
LVHW011908200926
848484LV00002B/562

9780198539919